Biotechnological Applications of Polyhydroxyalkanoates

Vipin Chandra Kalia
Editor

Biotechnological Applications of Polyhydroxyalkanoates

Springer

Editor
Vipin Chandra Kalia
Department of Chemical Engineering
Konkuk University
Seoul, Republic of Korea

ISBN 978-981-13-3758-1 ISBN 978-981-13-3759-8 (eBook)
https://doi.org/10.1007/978-981-13-3759-8

Library of Congress Control Number: 2019930049

This Springer imprint is published by the registered company Springer Nature Singapore Pte Ltd.
The registered company address is: 152 Beach Road, #21-01/04 Gateway East, Singapore 189721, Singapore

Dedicated to my lovely wife, Amita

Preface

Discoveries and inventions made through the principles of physics and chemistry have always been fascinating to human beings. Most of them have revolutionized the evolution and development of the society. Plastics are one such contribution, which have been proving their worth by the wide range of applications for which these can be exploited. It is nearly impossible to imagine life without plastics. Despite being such a wonderful product, it has some associated disadvantages, the foremost being the use of petrochemicals for their production and, secondly, their nonbiodegradable nature. The net impact is their accumulation in large quantities in the environment, which is difficult to dispose in an eco-friendly manner. Thus, they are a major cause of concern among the environmentalists. Biological products are better as they can undergo degradation at a much faster pace than synthetic materials. In fact, the biological discovery which brought hope to mankind was the microbial ability to produce biodegradable plastics also known as bioplastics. In the beginning of the twentieth century, the bacterium – *Bacillus megaterium* – was found to store food under adverse environmental conditions, in the form of special molecules. These biomolecules on extraction proved to have properties like plastics. Taking advantage of the unique bacterial ability, intensive research led to the discovery of a wide range of bacteria with similar property but with varying potentials in terms of the feed usage and the yield of bioplastic. These bioplastics have been fascinating to human society as they offer eco-friendly solutions to our daily problems. However, before embarking upon establishing full-scale production facility, one needs to look up on potential economic benefits and returns on investments. The major component of the high cost of production is due to the feed itself. Replacing pure substrates such as glucose with cheaper organic matter seems to be an obvious choice. Biowastes are organic-rich materials, which can be metabolized by a diversity of microbes. Thus, microbial production of bioplastics is a wonderful solution to the issues related to plastics. There is a growing interest in the field of biodegradable polymers, especially their roles in diverse fields, such as the environment, agriculture, and human health. The question being raised is: Shall we continue to carry out curiosity-driven research, which is proving to be too uneconomical or device mechanisms to design materials with high values? There seems to be a great dilemma as to which application should be the focus of research in the future. So, at this juncture, we should evaluate the potential applications of these systems.

This book is thus a proposal to all those who are actively involved in bioplastics especially polyhydroxyalkanoates to revisit these phenomena and present strategies – opinions to exploit them for high-end biotechnological applications. This is a request to scientists to join the venture and present the various potentials of this wonderful biopolymer in environment, agriculture, and health sectors in the book entitled *Biotechnological Applications of Polyhydroxyalkanoates*. This book is based on the contributions of those research scientists who have devoted their time, money, and scientific expertise in pursuit of getting greater insights into mechanisms which can lead to the production of biopolymers with unique properties. They have presented their thoughts and research experience for the young curious minds and to provide economically fruitful products for human consumption. This book reflects the clarity of mind, sincerity of efforts, and dedication of the persons for the welfare of mankind. These contributions are expected to enable researchers with latest update in the areas of bioplastics. I must thank all the contributors and their team members for accepting my request to write for the benefit of society and trusting my abilities. I will remain indebted to all of them forever. This acknowledgment is not sufficient to express my gratitude. The appreciation shown by the readers will perhaps satisfy them that their efforts have not gone waste. Beyond the scientific participation, which enabled me to take up this assignment, I was inspired by the faith and support of my family members and associates – Mr. R. B. Kalia and late Mrs. Kanta Kalia (my parents); late Mrs. Santosh Jaichand and late Mr. Ram Swarup Jaichand (parents-in-law); Daksh and Bhrigu (my sons); Nivedita; Sunita Bhardwaj; Ravi Bhardwaj; Devender ji; Kusum and Kumud Bhua ji; V. B. Jaichand; Anuradha Dube and Nikhil Dube; Dr. Amit Ghosh (my mentor); Rup Lal, Hemant J. Purohit, Yogendra Singh, Prince Sharma, Amulya K. Panda, Jyoti, Neeru, Ritusree, Mamta Kapila, Madhurima Kahali, Raman Shukla, Sr. Pastor Hyoungmin Kim, Pastor Sangam Elliot Lee, Prof. Doo Hwan Kim, Ms. Hyesoon Cho, and Ms. Kyong Sook Lee (my friends); and Young Choi, Jasmine Park, Juyeon Oh, Jinny Jeon, Dr. Seulji Lee, Isabella Jeon, Judy Gopal, Manikandan Muthu, Iyyakkannu Sivanesan, and Diby Paul (my young friends). I must also acknowledge the support of my student friends – Sadhana, Mamtesh, Chinoo, Bhartendu, Sanjay Patel, Ashish Bhushan, Sanjay Yadav, Sanjay Gulati, Neeraj Pandey, Rahul, Virender, Shiva, Gopi, Anurag, Devashish, Subhasree, Shikha, Jyotsana, and Priyanka.

Seoul, Republic of Korea Vipin Chandra Kalia

Contents

About the Editor

Vipin Chandra Kalia is presently working as Professor, National Research Foundation (Korea), Department of Chemical Engineering, Konkuk University, Seoul, Republic of Korea. Previously, he was working as CSIR Emeritus Scientist. He has been the Chief Scientist and the Deputy Director at Microbial Biotechnology and Genomics, CSIR Institute of Genomics and Integrative Biology, Delhi. He is a Professor, AcSIR, who obtained his MSc and PhD in genetics, from Indian Agricultural Research Institute, New Delhi. He has been elected as (1) Fellow of the National Academy of Sciences (FNASc), (2) Fellow of the National Academy of Agricultural Sciences (FNAAS), and (3) Fellow of the Association of Microbiologists of India (FAMSc). His main areas of research are microbial biodiversity, bioenergy, biopolymers, genomics, microbial evolution, quorum sensing, quorum quenching, drug discovery, and antimicrobials. He has published more than 100 papers in scientific journals, such as (1) *Nature Biotechnology*, (2) *Biotechnology Advances*, (3) *Trends in Biotechnology*, (4) *Annual Review of Microbiology*, (5) *Critical Reviews in Microbiology*, (6) *Bioresource Technology*, (7) *PLoS ONE*, (8) *BMC Genomics*, (9) *International Journal of Hydrogen Energy*, and (10) *Gene*. He has authored 28 book chapters. His works have been cited 5796 times with an h index of 42 and an i10 index of 92 (http://scholar.google.co.in/citations?hl=en&user=XaUw-VIAAAAJ). He has edited 13 books: (i) *Quorum Sensing vs Quorum Quenching: A Battle with No End in Sight* (2015); (ii) *Microbial Factories – Biofuels, Waste Treatment: Vol. 1* (2015); (iii) *Microbial Factories – Biodiversity, Biopolymers, Bioactive Molecules: Vol. 2* (2015); (iv) *Waste Biomass Management – A Holistic*

Approach (2017); (v) *Drug Resistance in Bacteria, Fungi, Malaria, and Cancer* [Arora, G., Sajid, A., and Kalia, V. C. (Eds.)]; (vi) *Microbial Applications Vol. 1* [Kalia, V.C. (Ed) and Kumar, P. (Ed)] (2017); (vii) *Microbial Applications Vol. 2* [Kalia, V. (Ed)] (2017); (viii) *Metabolic Engineering for Bioactive compounds: Strategies and Processes* [Saini, A. K. and Kalia, V. C. (Eds.)] (2017) (Springer Nature); (ix) *Mining of Microbial Wealth and MetaGenomics* [Kalia, V. C., Shouche, Y., Purohit, H. J., and Rahi, P. (Eds.)] (2017) (Springer Nature); (x) *Optimization of Applicability of Bioprocesses* [Purohit, H. J., Kalia, V. C., Vaidya, A., and Khardenavis A.A. P. (Eds.)] (2018) (Springer Nature); (xi) *Soft Computing for Biological Systems* [Purohit, H. J., Kalia, V. C., More R (Eds.)] (2017) (Springer Nature); (xii) *Biotechnological Applications of Quorum Sensing Inhibitors* (2018) (Springer Nature); and (xiii) Quorum Sensing and Its Biotechnological Applications (2018) (Springer Nature). He is presently the Editor-in-Chief of the *Indian Journal of Microbiology* (2013 through 2021) and editor of (1) *Journal of Microbiology and Biotechnology* (Korea), (2) *International Scholarly Research Network ISRN Renewable Energy*, (3) *Dataset Papers in Microbiology*, and (4) *PLoS ONE*. He is a life member of the following scientific societies: (1) Society of Biological Chemists of India; (2) Society for Plant Biochemistry and Biotechnology, India; (3) Association of Microbiologists of India; (4) Indian Science Congress Association; (5) BioEnergy Society of India; and (6) the Biotech Research Society of India (BRSI). He has been a member of the American Society for Microbiology (2010–2015). He has been conferred the following awards: (i) Prof. S.R. Vyas Memorial Award, Association of Microbiologists of India (2016); (ii) ASM-IUSSF Indo-US Professorship Program, American Society of Microbiologists, USA (2014); (iii) INSA Bilateral Exchange Programme, Indian National Science Academy, India (2006); (iv) DBT Overseas Associateship, Department of Biotechnology, Government of India (2005–2006); (v) Dr. J.V Bhat Award, Association of Microbiologists of India (2012, 2015, 2016, 2017); and Faculty Research Award in Microbiology (2018). He can be contacted at vckaliaku@gmail.com.

The Dawn of Novel Biotechnological Applications of Polyhydroxyalkanoates

1

Vipin Chandra Kalia, Subhasree Ray, Sanjay K. S. Patel, Mamtesh Singh, and Gajendra Pratap Singh

Abstract

The synthetic polymers – plastics have been applied in a wide range of activities of our daily routine. However, extensive usage and their non-biodegradable nature have led to their accumulation in quantities, which are difficult to manage and a major cause of environmental pollution. Bacteria have the ability to accumulate Carbon (C) as biopolymers especially under stress conditions. The biopolymers – polyhydroxyalkanotes (PHAs) are biodegradable and have properties quite close to those possessed by plastics. PHAs have been explored in diverse fields including agriculture and medical. In the field of medicine, PHAs hold greater promise because of their usage in producing high value products, in addition to their biodegradable, biocompatible, and non-toxic nature. PHAs have been explored for their role as implants, drug carriers, tissue engineering, biocontrol agents, inhibitors of cancerous growth, and memory enhancing molecules.

Keywords

Biopolymers · Polyhdroxyalkanoates · Tissue engineering · Drug carriers · Biocontrol agents

V. C. Kalia (✉) · S. K. S. Patel
Department of Chemical Engineering, Konkuk University, Seoul, Republic of Korea

S. Ray
Microbial Biotechnology and Genomics, CSIR – Institute of Genomics and Integrative Biology (IGIB), Delhi University Campus, Delhi, India

Academy of Scientific & Innovative Research (AcSIR), New Delhi, India

M. Singh
Department of Zoology, Gargi College, University of Delhi, Delhi, India

G. P. Singh
Mathematical Sciences and Interdisciplinary Research Lab (MathSciIntR-Lab), School of Computational and Integrative Sciences, Jawaharlal Nehru University, New Delhi, India

V. C. Kalia (ed.), *Biotechnological Applications of Polyhydroxyalkanoates*,
https://doi.org/10.1007/978-981-13-3759-8_1

1.1 Introduction

Microbes have unique abilities to divert or even curtail their metabolic pathways, as soon as they encounter major modifications in their immediate vicinity. A unique case which clearly exemplifies this phenomenon is the diversion in their highly energy efficient metabolic pathway during stress conditions. Bacteria sense the nutritional imbalance, such as excess of Carbon (C) compounds and limited quantities of nitrogen, phosphorus, potassium, oxygen, and magnesium. Instead of metabolizing C compounds to generate energy through the Tricarboxylic acid cycle (TCA), bacteria divert the acetyl CoA towards polyhydroxyalkanoates (PHA) synthetic pathway (Porwal et al. 2008; Singh et al. 2009; Kumar et al. 2013; Ray and Kalia 2016, 2017). As far as microbes are concerned, they seem to store C as PHA granules and use them as energy reservoirs (Patel et al. 2011, 2012, 2015a, b, 2016; Singh et al. 2013; Kumar et al. 2014, 2015a, b, c; Bhatia et al. 2015a,b, 2016, 2017, 2018, Kalia et al. 2016; Koller et al. 2017). However, because of the unique physicochemical properties, PHAs have been explored for their potential to replace synthetic plastics. The direct usage of PHAs has been proving uneconomical. This has forced researchers to look for high value derivatives of PHAs. Hence, the emphasis has shifted towards use of PHA catabolic pathway products and their chemical modifications, which confer unique properties for biomedical applications (Hazer et al. 2012; Martinez et al. 2014; Ke et al. 2017).

1.2 Antimicrobials, Biocontrol and Anticancer Agents

Catabolic activity of PHAs results in intermediate like 3-Hydoxy acids (3HAs). It primarily involves depolymerase enzyme resulting in monomers. These intermediates can be modified to synthesize antimicrobials (Gallo et al. 2014; Kalia et al. 2019). Hydroxycarboxylic acids: 2-alkylated 3HB and β-lactones produced by transforming 3HAs, can be used as oral drugs. Antibiotics carbapenem or macrolide can be used to treat *Staphylococcus aureus* infections (Dinjaski et al. 2014). Medium chain length 3HAs prepared by *Streptomyces* strains from *Jatropha curcas* as antimicrobial agent against pathogens such as *Salmonella typhimurium, Listeria monocytogenes* and *E. coli* (Allen et al. 2012). PHA co-polymer poly (3-hydroxybutyrate-co-70%4-hydroxybutyrate) produced by *Cupriavidus* sp. also proved to have antimicrobial properties against pathogens such as *S. aureus* (Hema et al. 2013). Chlorhexidine (CHX), an efficient antifungal agent was carried through PHB/PEO fibres, showed 99–100% reduction in *E. coli* and *S. aureus* population at 1 wt% concentration (Fernandes et al. 2014). Tetracycline encapsulated in polymeric microspheres showed 85% reduction in periodontitis-causing bacteria *Actinobacillus actinomycetemcomitans* and *Porphyromonas gingivalis* (Panith et al. 2016). P3HB and P4HB can be exploited for treating skin infections and healing wounds (Shishatskaya et al. 2016). Combining 3HAs with D-peptide is effective against cancers (O'Connor et al. 2013; Sangsanoh et al. 2017). Aquaculture and livestock industry employ antibiotics at quite low doses along with the feed. This regular supplementation has turned out to be harmful to gastrointestinal microflora. This selection pressure is likely to cause evolution of drug resistant bacteria. PHAs as food supplement have been shown to act as anti-pathogenic in the intestine of giant tiger prawn (Defoirdt et al. 2007, 2009; Halet et al. 2007; Dang et al. 2009; Liu et al. 2010; Ludevese-Pascual et al. 2016).

1.3 Drug Carriers

Efficiency of the drug for treating diseases is dependent up on their delivery to the target (Nigmatullin et al. 2015). Since, the Drugs can be delivered through intravenous, subcutaneous and oral routes. The delivery system to be opted depends upon the nature and dose of the drug to be administered. The drug release is also influenced by the composition of the polymer (Kamaly et al. 2016). Nano-particles and scaffolds can prove effective for eluting drugs from PHA derived monomers (Mokhtarzadeh 2016). Nanoparticles of poly (4-hydroxybutyrate)-monomethoxypoly(ethylene glycol) were used for delivering anticancer drug cisplatin in to hippocampal HT22 cells of mouse (Shah et al. 2014). Monomers of PHA such as 3-hydroxybutyrate (3HB) can prove helpful for synthesizing novel biodegradable polymers. Dendrimers – tamsulosin, ketoprofen and clonidine, have high monodispersity and surface-functional moieties, which help these molecules to play the role of drug carriers (Parlane et al. 2016a, b). Microspheres made up of PHAs, in combination with rifampicin behave as drug carriers and hemoembolizing agents Implants such as rods made up of PHA co-polymers have efficient ability to deliver antibiotics. Nanoparticles based on docetaxel loaded with PHA copolymer – poly (3-hydroxybutyrate-co-3-hydroxyvalerate) was used for its pharmacokinetic evaluation. These nanoparticles were reported to have stability with reference to drug content and physical characteristics. Nanoparticles helped to increase the efficacy in inhibiting human breast cancer cell line (Vardhan et al. 2017).

1.4 Engineering Tissues

Chemically modified PHAs can be helpful in tissue engineering (Goonoo et al. 2017). These can be used as therapeutics and for other medical applications such as: (i) grafts, (ii) cardio-vascular valves, and (iii) nerve tissues (Chen 2011). They also find use as films, pins, sutures, screws, and scaffolds for repairing skin, cartilage and liver tissue engineering (Levine et al. 2015; Ching et al. 2016; Insomphun et al. 2016; Shishatskaya et al. 2016; Raşoga et al. 2017).

1.5 Medical Implants and Devices

The use of PHAs for medical devices is improved by developing co-polymers. These specific PHAs are relatively quite strong, and highly biocompatible (Qu et al. 2006a). The biotechnological application range gets broader since their ability to resist bacterial infections is high, they lack immunogenicity and have been found to be non-toxic. Potential medical devices developed have been implants: rivets and tacks, orthopaedic pins, stents, cardiovascular grafting, meniscus repair, cartilage repair, staples, mesh, sutured fastener, repair patches (Lobler et al. 2002; Qu et al. 2006b; Rodríguez-Contreras et al. 2017). PHA films embedded with lysozyme inhibit bacterial biofilm formation and are useful in wound dressing (Kehail and Brigham 2017).

1.6 Anti-osteoporosis Agent

3HB improves growth of osteoblasts and proves useful as an anti-osteoporosis agent. The serum alkaline phosphatise activity and ability to improve calcium deposition process are the properties by which 3HB helps in prevention of lowering of bone density and serum osteocalcin (Tokiwa and Calabia 2007; Zhao et al. 2007; Chen 2011).

1.7 Memory Enhancer

PHAs can rapidly diffuse to improve cardiac efficiency and prevent brain damage, by acting as source of energy. Parkinson and Alzheimer diseases can be cured through PHA monomers such as 3HB. It acts by preventing neuronal cell death (Camberos-Luna et al. 2016). Modified PHA monomers such as methyl esters of 3-hydroxybutyrate can be employed as drugs to protect mitochondrial damage (Zhang et al. 2013). HA can stimulate Ca^{2+} channels, which acts as an aid in enhancing memory especially patient with dementia – Alzheimer's disease (Cheng et al. 2006; Xiao et al. 2007; Zou et al. 2009; Magdouli et al. 2015).

1.8 Packaging

Use of plastics as packaging material is quite prevalent. Their use for packing food material need special attention. The specific requirements include: (i) protection from dust, contaminants, dehydration, etc., (ii) food grade quality, (iii) food stability, and (iv) degradation during (Prasad and Kochhar 2014). PHAs have the potential to meet the requirements of food grade packing material, especially material properties and permeability (Chen 2010; Chanprateep 2010; Rai et al. 2011; Wang and Chen 2017). Copolymers of PHA having high hydroxyvalerate and mcl-PHA content helps to reduce brittleness and Young's modulus, allowing it to achieve higher flexibility (Fu et al. 2014; Albuquerque and Malafaia 2018).

Packaged food needs to retain its aroma for a long storage period. PHB made films show higher barrier to aromatic compounds. Limonene, which is commonly used for testing the loss of aroma during storage was found to be retained for longer period in PHBV copolymers (Sanchez-Garcia et al. 2007). Nanocomposites of PHB or PHBV with organo-modified montmorillonite Cloisite® 30B or halloysite (HNT), bacterial cellulose nano-whiskers allowed variation in morphology, thermal and mechanical properties (Wang et al. 2005; Carli et al. 2011; Martínez-Sanz et al. 2014; Arrieta et al. 2015). More recently, polymer films with desirable characteristics such as odorless, high flexibility, nontoxicity, antimicrobial and antioxidant activities have been developed by using: (i) PHB: nanomelanin: glycerol polymer film (Kiran et al. 2017), (ii) PHBV along with natural vermiculite and organoclay (Reis et al. 2016, 2017).

1.9 Agriculture

The usage of PHA in agriculture has been exploited only to a limited extent in comparison to that in the medical field. The obvious reason for his biased attitude is the high cost associated with medical applications. Among the few fields where PHAs have found some application are: (i) mulching, (ii) nets, and (iii) bags. Mulching helps to improve and maintain good soil structure, control contamination, and regulate weeds. PHA copolymer (PHBHHx) based NodaxTM has been used to prepare agricultural mulch (Hassan et al. 2006). Another mulch being produced at commercial level is made from PHA based Mirel™ resin, Metabolix Inc. (Andrews 2014). Nets are used in greenhouse and for protecting crops from insects, birds, hails, and for creating special environmental conditions (Castellano et al. 2008; Niaounakis 2015; Guerrini et al. 2017; Ojanji 2017). PHA based bags are used for seedlings, retaining water and regulating temperature (Lu et al. 2014; Schrader et al. 2016). PHA nanomaterials specifically microspheres have found its application as nanoherbicide which have lower genotoxicity and high biodegradability increasing the herbicide efficacy (Grillo et al. 2010; Lobo et al. 2011).

1.10 Challenges in Customizing PHAs

Despite the wonderful and unique characteristics of PHAs, their real-life applications are limited (Singh et al. 2015). The major challenges include: (i) selecting a host organism to express genes involved in PHA synthesis (Singh et al. 2009), (ii) regulating co-polymer composition and production (Kumar et al. 2015c; Ray and Kalia 2016), (iii) manipulate feed composition, (iv) improving physicochemical properties, and (v) develop techniques to modify the products generated from metabolism of PHAs (Singh et al. 2015).

1.11 The Future

PHAs have the necessary potential for being applied in diverse fields. The major limitation has been the economic – feasibility of this product. Application of PHAs and their metabolic products in the field of medicine can circumvent the economic issue. The synthetic biology approach to produce these biochemical in a cell-free system has been envisaged as a viable alternative to limit costs (Opgenorth et al. 2016).

Acknowledgements This work was supported by Brain Pool grant (NRF-2018H1D3A2001746) by National Research Foundation of Korea (NRF) to work at Konkuk University.

References

Albuquerque PBS, Malafaia CB (2018) Perspectives on the production, structural characteristics and potential applications of bioplastics derived from polyhydroxyalkanoates. Int J Biol Macromol 107:615–625. https://doi.org/10.1016/j.ijbiomac.2017.09.026

Allen AD, Daley P, Ayorinde FO, Gugssa A, Anderson WA, Eribo BE (2012) Characterization of medium chain length (R)-3-hydroxycarboxylic acids produced by *Streptomyces* sp. JM3 and the evaluation of their antimicrobial properties. World J Microbiol Biotechnol 28:2791–2800. https://doi.org/10.1007/s11274-012-1089-z

Andrews M (2014) Mirel™ PHA polymeric modifiers and additives. https://www.slideshare.net/MetabolixInc/metabolix-mirel-pha-polymeric-modifiers-and-additives

Arrieta MP, Fortunati E, Dominici F, López J, Kenn JM (2015) Bionanocomposite films based on plasticized PLA-PHB/cellulose nanocrystal blends. Carbohydr Polym 121:265–275. https://doi.org/10.1016/j.carbpol.2014.12.056

Bhatia S, Yi DH, Kim HJ, Jeon JM, Kim YH, Sathiyanarayanan G, Seo H, Lee J, Kim JH, Park K (2015a) Overexpression of succinyl-CoA synthase for poly (3-hydroxybutyrate-co-3-hydroxyvalerate) production in engineered *Escherichia coli* BL21 (DE3). J Appl Microbiol 119:724–735. https://doi.org/10.1111/jam.12880

Bhatia SK, Shim Y-H, Jeon J-M, Brigham CJ, Kim Y-H, Kim H-J, Seo H-M, Lee J-H, Kim J-H, Yi D-H, Lee YK, Yang Y-H (2015b) Starch based polyhydroxybutyrate production in engineered *Escherichia coli*. Bioprocess Biosyst Eng 38:1479–1484. https://doi.org/10.1007/s00449-015-1390-y

Bhatia SK, Bhatia RK, Yang Y-H (2016) Biosynthesis of polyesters and polyamide building blocks using microbial fermentation and biotransformation. Rev Environ Sci Biotechnol 15:639–663. https://doi.org/10.1007/s11157-016-9415-9

Bhatia SK, Kim J-H, Kim M-S, Kim J, Hong JW, Hong YG, Kim H-J, Jeon J-M, Kim S-H, Ahn J, Lee H, Yang Y-H (2017) Production of (3-hydroxybutyrate-co-3-hydroxyhexanoate) copolymer from coffee waste oil using engineered *Ralstonia eutropha*. Bioprocess Biosyst Eng. https://doi.org/10.1007/s00449-017-1861-4

Bhatia SK, Yoon JJ, Kim HJ, Hong JW, Gi Hong Y, Song HS, Moon YM, Jeon JM, Kim YG, Yang YH (2018) Engineering of artificial microbial consortia of *Ralstonia eutropha* and *Bacillus subtilis* for poly(3-hydroxybutyrate-co-3-hydroxyvalerate) copolymer production from sugarcane sugar without precursor feeding. Bioresour Technol 257:92–10. https://doi.org/10.1016/j.biortech.2018.02.056

Camberos-Luna L, Gerónimo-Olvera C, Montiel T, Rincon-Heredia R, Massieu L (2016) The ketone body, β-Hydroxybutyrate stimulates the autophagic flux and prevents neuronal death induced by glucose deprivation in cortical cultured neurons. Neurochem Res 41:600–609. https://doi.org/10.1007/s11064-015-1700-4

Carli LN, Crespo JS, Mauler RS (2011) PHBV nanocomposites based on organomodified montmorillonite and halloysite: the effect of clay type on the morphology and thermal and mechanical properties. Composites Part A 42:1601–1608. https://doi.org/10.1016/j.compositesa.2011.07.007

Castellano S, Mugnozza GS, Russo G, Briassoulis D, Mistriotis A, Hemming S, Waaijenberg D (2008) Plastic nets in agriculture: a general review of types and applications. Appl Eng Agric 24:799–808. https://doi.org/10.13031/2013.25368

Chanprateep S (2010) Current trends in biodegradable polyhydroxyalkanoates. J Biosci Bioeng 110:621–632. https://doi.org/10.1016/j.jbiosc.2010.07.014

Chen GQ (2010) Plastics completely synthesized by bacteria: polyhydroxyalkanoates. In: Chen GQ (ed) Plastics from bacteria: Natural functions and applications, microbiology monographs. Springer, Berlin, pp 17–37. https://doi.org/10.1007/978-3-642-03287_5_2.

Chen GQ (2011) Biofunctionalization of polymers and their applications. In: Nyanhongo GS, Steiner W, Gubitz G (eds) Biofunctionalization of polymers and their applications. Springer, Berlin, pp 29–45. https://doi.org/10.1007/10_2010_89

Cheng S, Chen GQ, Leski M, Zou B, Wang Y, Wu Q (2006) The effect of D, L-β-hydroxybutyric acid on cell death and proliferation in L929 cells. Biomaterials 27:3758–3765. https://doi.org/10.1016/j.biomaterials.2006.02.046

Ching KY, Andriotis OG, Li S, Basnett P, Su B, Roy I, Stolz M (2016) Nanofibrous poly (3-hydroxybutyrate)/poly (3-hydroxyoctanoate) scaffolds provide a functional microenvironment for cartilage repair. J Biomater Appl 31:77–91. https://doi.org/10.1177/0885328216639749

Dang TVC, Nguyen VH, Dierckens K, Defoirdt T, Boon N, Sorgeloos P, Bossier P (2009) Novel approach of using homoserine lactone-degrading and poly-betahydroxybutyrate-accumulating bacteria to protect *Artemia* from the pathogenic effects of *Vibrio harveyi*. Aquaculture 291:23–30. https://doi.org/10.1016/j.aquaculture.2009.03.009

Defoirdt T, Halet D, Vervaeren H, Boon N, Van de Wiele T, Sorgeloos P, Bossier P, Verstraete W (2007) The bacterial storage compound poly-beta-hydroxybutyrate protects *Artemia franciscana* from pathogenic *Vibrio campbellii*. Environ Microbiol 9:445–452. https://doi.org/10.1111/j.1462-2920.2006.01161.x

Defoirdt T, Boon N, Sorgeloos P, Verstraete W, Bossier P (2009) Short-chain fatty acids and poly-β-hydroxyalkanoates: (New) biocontrol agents for a sustainable animal production. Biotechnol Adv 27:680–685. https://doi.org/10.1016/j.biotechadv.2009.04.026

Dinjaski N, Fernandez-Gutierrez M, Selvam S, Parra-Ruiz FJ, Lehman SM, San Roman J, Garcia E, Garcia JL, Garcia AJ Prieto MA (2014) PHACOS, a functionalized bacterial polyester with bactericidal activity against methicillin-resistant *Staphylococcus aureus*. Biomaterials 35:14–24. https://doi.org/10.1016/j.biomaterials.2013.09.059

Fernandes JG, Correia DM, Botelho G, Padrão J, Dourado F, Ribeiro C, Lanceros-Méndez S, Sencadas V (2014) PHB-PEO electrospun fiber membranes containing chlorhexidine for drug delivery applications. Polym Test 34:64–71. https://doi.org/10.1016/j.polymertesting.2013.12.007

Fu XZ, Tan D, Aibaidula G, Wu Q, Chen JC, Chen GQ (2014) Development of *Halomonas* TD01 as a host for open production of chemicals. Metab Eng 23:78–91. https://doi.org/10.1016/j.ymben.2014.02.006

Gallo J, Holinka M, Moucha CS (2014) Antibacterial surface treatment for orthopaedic implants. Int J Mol Sci 15:13849–13880. https://doi.org/10.3390/ijms150813849

Goonoo N, Bhaw-Luximon A, Passanha P, Esteves SR, Jhurry D (2017) Third generation poly(hydroxyacid) composite scaffolds for tissue engineering. J Biomed Mater Res Part B 105:1667–1684. https://doi.org/10.1002/jbm.b.33674

Grillo R, Melo NFS, de Lima R, Lourenço RW, Rosa AH, Fraceto LF (2010) Characterization of atrazine-loaded biodegradable poly(hydroxybutyrate-cohydroxyvalerate) microspheres. J Polym Environ 18:26–32

Guerrini S, Borreani G, Voojis H (2017) Biodegradable materials in agriculture: case histories and perspectives. In: Malinconico M (ed) Soil degradable bioplastics for a sustainable

modern agriculture. Green chemistry and sustainable technology. Springer, Berlin. https://doi.org/10.1007/978-3-662-54130-2_3
Halet D, Defoirdt T, Van Damme P, Vervaeren H, Forrez I, Van de Wiele T, Boon N, Sorgeloos P, Bossier P, Verstraete W (2007) Poly-β-hydroxybutyrate-accumulating bacteria protect gnotobiotic *Artemia franciscana* from pathogenic *Vibrio campbellii*. FEMS Microbiol Ecol 60:363–369. https://doi.org/10.1111/j.1574-6941.2007.00305.x
Hassan MK, Abou-Hussein R, Zhang X, Mark JE, Noda I (2006) Biodegradable copolymers of 3-hydroxybutyrate-co-3-hydroxyhexanoate (NodaxTM), including recent improvements in their mechanical properties. Mol Cryst Liq Cryst 447:23–341. https://doi.org/10.1080/15421400500380028
Hazer DB, Kılıçay E, Hazer B (2012) Poly(3-hydroxyalkanoate)s: diversification and biomedical applications: a state of the art review. Mater Sci Eng C32:637–647. https://doi.org/10.1016/j.msec.2012.01.021
Hema R, Ng PN, Amirul AA (2013) Green nanobiocomposite: reinforcement effect of montmorillonite clays on physical and biological advancement of various polyhydroxyalkanoates. Polym Bull 70:755–771. https://doi.org/10.1007/s00289-012-0822-y
Insomphun C, Chuah JA, Kobayashi S, Fujiki T, Numata K (2016) Influence of hydroxyl groups on the cell viability of polyhydroxyalkanoate (PHA) scaffolds for tissue engineering. ACS Biomater Sci Eng. https://doi.org/10.1021/acsbiomaterials.6b00279
Kalia VC, Prakash J, Koul S (2016) Biorefinery for glycerol rich biodiesel industry waste. Indian J Microbiol 56:113–125. https://doi.org/10.1007/s12088-016-0583-7
Kalia VC, Patel SKS, Kang YC, Lee JK (2019) Quorum sensing inhibitors as antipathogens: biotechnological applications. Biotechnol. Adv. 37:68–90. https://doi.org/10.1016/j.biotechad.2018.11.006
Kamaly N, Yameen B, Wu J, Farokhzad OC (2016) Degradable controlled-release polymers and polymeric nanoparticles: mechanisms of controlling drug release. Chem Rev 116:2602–2663. https://doi.org/10.1021/acs.chemrev.5b00346
Ke Y, Zhang XY, Ramakrishna S, He LM, Wu G (2017) Reactive blends based on polyhydroxyalkanoates: preparation and biomedical application. Mater Sci Eng C Mater Biol Appl 70:1107–1119. https://doi.org/10.1016/j.msec.2016.03.114
Kehail AA, Brigham CJ (2017) Anti-biofilm activity of solvent-cast and electrospun polyhydroxyalkanoate membranes treated with lysozyme. J Polym Environ:1–7. https://doi.org/10.1007/s10924-016-0921-1
Kiran GS, Jackson SA, Priyadharsini S, Dobson ADW, Selvin J (2017) Synthesis of Nm-PHB (nanomelanin-polyhydroxy butyrate) nanocomposite film and its protective effect against biofilmforming multi drug resistant *Staphylococcus aureus*. Sci Rep 7:9167. https://doi.org/10.1038/s41598-017-08816-y
Koller M, Marsalek L, de Sousa Dias MM, Braunegg G (2017) Producing microbial polyhydroxyalkanoate (PHA) biopolyesters in a sustainable manner. New Biotechnol 37:24–38. https://doi.org/10.1016/j.nbt.2016.05.001
Kumar P, Patel SKS, Lee JK, Kalia VC (2013) Extending the limits of *Bacillus* for novel biotechnological applications. Biotechnol Adv 31:1543–1561. https://doi.org/10.1016/j.biotechadv.2013.08.007
Kumar P, Singh M, Mehariya S, Patel SKS, Lee JK, Kalia VC (2014) Ecobiotechnological approach for exploiting the abilities of *Bacillus* to produce co-polymer of polyhydroxyalkanoate. Indian J Microbiol 54:151–157. https://doi.org/10.1007/s12088-014-0457-9
Kumar P, Mehariya S, Ray S, Mishra A, Kalia VC (2015a) Biodiesel industry waste: a potential source of bioenergy and biopolymers. Indian J Microbiol 55:1–7. https://doi.org/10.1007/s12088-014-0509-1
Kumar P, Mehariya S, Ray S, Mishra A, Kalia VC (2015b) Biotechnology in aid of biodiesel industry effluent (glycerol): biofuels and bioplastics. In: Kalia VC (ed) Microbial factories. Springer, New Delhi, pp 105–119. https://doi.org/10.1007/978-81-322-2598-0
Kumar P, Ray S, Patel SKS, Lee JK, Kalia VC (2015c) Bioconversion of crude glycerol to polyhydroxyalkanoate by *Bacillus thuringiensis* under non-limiting nitrogen conditions. Int J Biol Macromol 78:9–16. https://doi.org/10.1016/j.ijbiomac.2015.03.046

Levine AC, Sparano A, Twigg FF, Numata K, Nomura CT (2015) Influence of cross-linking on the physical properties and cytotoxicity of polyhydroxyalkanoate (PHA) sccaffolds for tissue engineering. ACS Biomater Sci Eng 1:567–576. https://doi.org/10.1021/acsbiomaterials.6b00279

Liu Y, De Schryver P, Van Delsen B, Maignien L, Boon N, Sorgeloos P, Verstraete W, Bossier P, Defoirdt T (2010) PHB-degrading bacteria isolated from the gastrointestinal tract of aquatic animals as protective actors against luminescent vibriosis. FEMS Microbiol Ecol 74:196–204. https://doi.org/10.1111/j.1574-6941.2010.00926.x

Lobler M, Sab M, Kunze C, Schmitz KP, Hopt UT (2002) Biomaterial implants induce the inflammation marker CRP at the site of implantation. J Biomed Mater Resear Part A 61:165–167. https://doi.org/10.1002/jbm.10155

Lobo FA, de Aguirre CL, Silva MS, Grillo R, de Melo NFS, de Oliveira LK, de Morais LC, Campos V, Rosa AH, Fraceto LF (2011) Poly (hydroxybutyrate-co-hydroxyvalerate) microspheres loaded with atrazine herbicide: screening of conditions for preparation, physicochemical characterization, and in vitro release studies. Polym Bull 67:479–495

Lu H, Madbouly SA, Schrader JA, Kessler MR, Grewell D, Graves WR (2014) Novel bio-based composites of polyhydroxyalkanoate (PHA)/distillers dried grains with solubles (DDGS). RSC Adv 4:39802–39808. https://doi.org/10.1039/C4RA04455J

Ludevese-Pascual G, Laranja JLQ, Amar EC, Sorgeloos P, Bossier P, De Schryver P (2016) Poly-beta-hydroxybutyrate-enriched *Artemia* sp. for giant tiger prawn *Penaeus monodon* larviculture. Aquaculture 23:422–429. https://doi.org/10.1111/anu.12410

Magdouli S, Brar SK, Blais JF, Tyagi RD (2015) How to direct the fatty acid biosynthesis towards polyhydroxyalkanoates production? Biomass Bioenergy 74:268–279. https://doi.org/10.1016/j.biombioe.2014.12.017

Martinez V, Dinjaski N, De Eugenio LI, De la Pena F, Prieto MA (2014) Cell system engineering to produce extracellular polyhydroxyalkanoate depolymerase with targeted applications. Int J Biol Macromol 71:28–33. https://doi.org/10.1016/j.ijbiomac.2014.04.013

Martínez-Sanz M, Villano M, Oliveira C, Albuquerque MG, Majone M, Reis M, Lopez-Rubio A, Lagaron JM (2014) Characterization of polyhydroxyalkanoates synthesized from microbial mixed cultures and of their nanobiocomposites with bacterial cellulose nanowhiskers. New Biotechnol 31:364–376. https://doi.org/10.1016/j.nbt.2013.06.003

Mokhtarzadeh A (2016) Recent advances on biocompatible and biodegradable nanoparticles as gene carriers. Expert Opin Biol Ther 16:771–785. https://doi.org/10.1517/14712598.2016.1169269

Niaounakis M (2015) Biopolymers: applications and trends. Elsevier. https://doi.org/10.1016/B978-0-323-35399-1.00004-1

Nigmatullin R, Thomas P, Lukasiewicz B, Puthussery H, Roy I (2015) Polyhydroxyalkanoates, a family of natural polymers, and their applications in drug delivery. J Chem Technol Biotechnol 90:1209–1221. https://doi.org/10.1002/jctb.4685

O'Connor S, Szwej E, Nikodinovic-Runic J, O'Connor A, Byrne AT, Devocelle M, O'Donovan N, Gallagher WM, Babu R, Kenny ST, Zinn M (2013) The anti-cancer activity of a cationic antimicrobial peptide derived from monomers of polyhydroxyalkanoate. Biomaterials 34:2710–2718. https://doi.org/10.1016/j.biomaterials.2012.12.032

Ojanji W (2017) Plastic ban to hit seedlings sector as firms adopt new technology. The Daily Nation. https://www.nation.co.ke/business/seedsofgold/Plastic-ban-to-hit-seedlings-sector/2301238-3973850-qvm0ipz/index.html

Opgenorth PH, Korman TP, Bowie JU (2016) A synthetic biochemistry module for production of bio-based chemicals from glucose. Nat Chem Biol 12:393–395. https://doi.org/10.1038/nchembio.2062

Panith N, Assavanig A, Lertsiri S, Bergkvist M, Surarit R, Niamsiri N (2016) Development of tunable biodegradable polyhydroxyalkanoates microspheres for controlled delivery of tetracycline for treating periodontal disease. J Appl Polym Sci 133:44128–44141. https://doi.org/10.1002/app.44128

Parlane NA, Chen S, Jones GJ, Vordermeier HM, Wedlock DN, Rehm BH, Buddle BM (2016a) Display of antigens on polyester inclusions lowers the antigen concentration required for a bovine tuberculosis skin test. Clin Vaccine Immunol 23:19–26. https://doi.org/10.1128/CVI.00462-15

Parlane NA, Gupta SK, Rubio-Reyes P, Chen S, Gonzalez-Miro M, Wedlock DN, Rehm BH (2016b) Self-assembled protein-coated polyhydroxyalkanoate beads: properties and biomedical applications. ACS Biomater Sci Eng 3:3043–3057. https://doi.org/10.1021/acsbiomaterials.6b00355

Patel SKS, Singh M, Kalia VC (2011) Hydrogen and polyhydroxybutyrate producing abilities of *Bacillus* spp. from glucose in two stage system. Indian J Microbiol 51:418–423. https://doi.org/10.1007/s12088-011-0236-9

Patel SKS, Singh M, Kumar P, Purohit HJ, Kalia VC (2012) Exploitation of defined bacterial cultures for production of hydrogen and polyhydroxybutyrate from pea-shells. Biomass Bioenergy 36:218–225. https://doi.org/10.1016/j.biombioe.2011.10.027

Patel SKS, Kumar P, Singh S, Lee JK, Kalia VC (2015a) Integrative approach for hydrogen and polyhydroxybutyrate production. In: Kalia VC (ed) Microbial factories: waste treatment. Springer, New Delhi, pp 73–85. https://doi.org/10.1007/978-81-322-2598-0_5

Patel SKS, Kumar P, Singh S, Lee JK, Kalia VC (2015b) Integrative approach to produce hydrogen and polyhydroxybutyrate from biowaste using defined bacterial cultures. Bioresour Technol 176:136–141. https://doi.org/10.1016/j.biortech.2014.11.029

Patel SKS, Lee JK, Kalia VC (2016) Integrative approach for producing hydrogen and polyhydroxyalkanoate from mixed wastes of biological origin. Indian J Microbiol 56:293–300. https://doi.org/10.1007/s12088-016-0595-3

Porwal S, Kumar T, Lal S, Rani A, Kumar S, Cheema S, Purohit HJ, Sharma R, Patel SKS, Kalia VC (2008) Hydrogen and polyhydroxybutyrate producing abilities of microbes from diverse habitats by dark fermentative process. Bioresour Technol 99:5444–5451. https://doi.org/10.1016/j.biortech.2007.11.011

Prasad P, Kochhar A (2014) Active packaging in food industry: a review. IOSR J Environ Sci, Toxicol Food Technol 8:01–07

Qu XH, Wu Q, Chen GQ (2006a) In vitro study on hemocompatibility and cytocompatibility of poly (3-hydroxybutyrate-co-3-hydroxyhexanoate). J Biomater Sci Polym Edition 17:1107–1121. https://doi.org/10.1163/156856206778530704

Qu XH, Wu Q, Liang J, Zou B, Chen GQ (2006b) Effect of 3-hydroxyhexanoate content in poly (3-hydroxybutyrate-co-3-hydroxyhexanoate) on *in vitro* growth and differentiation of smooth muscle cells. Biomaterials 27:2944–2950. https://doi.org/10.1016/j.biomaterials.2006.01.013

Rai R, Keshavarz T, Roether J, Boccaccini AR, Roy I (2011) Medium chain length polyhydroxyalkanoates, promising new biomedical materials for the future. Mater Sci Eng R 72:29–47. https://doi.org/10.1016/j.mser.2010.11.002

Rașoga O, Sima L, Chirițoiu M, Popescu-Pelin G, Fufă O, Grumezescu O, Socol M, Stănculescu A, Zgură I, Socol G (2017) Biocomposite coatings based on Poly (3-hydroxybutyrate-co-3-hydroxyvalerate)/calcium phosphates obtained by MAPLE for bone tissue engineering. Appl Surf Sci 417:204–212. https://doi.org/10.1016/j.apsusc.2017.01.205

Ray S, Kalia VC (2016) Microbial co-metabolism and polyhydroxyalkanoate co-polymers. Indian J Microbiol 57:39–47. https://doi.org/10.1007/s12088-016-0622-4

Ray S, Kalia VC (2017) Co-metabolism of substrates by *Bacillus thuringiensis* regulates polyhydroxyalkanoate co-polymer composition. Bioresour Technol 224:743–747. https://doi.org/10.1016/j.biortech.2016.11.089

Reis DCC, Lemos-Morais AC, Carvalho LH, Alves TS, Barbosa R (2016) Assessment of the morphology and interaction of PHBV/clay Bionanocomposites: uses as food packaging. Macromol Symp 367:113–118. https://doi.org/10.1002/masy.201500143

Reis DCC, Oliveira TA, Carvalho LH, Alves TS, Barbosa (2017) The influence of natural clay and organoclay vermiculite on the formation process of bionanocomposites with poly (3-hydroxybutyrate-co-3-hydroxyvalerate). Matéria (Rio J) 22:1186. https://doi.org/10.1590/S1517-707620170004.0220

Rodríguez-Contreras A, García Y, Manero JM, Rupérez E (2017) Antibacterial PHAs coating for titanium implants. Eur Polym J 91:470–480. https://doi.org/10.1016/j.eurpolymj.2017.03.004

Sanchez-Garcia MD, Gimenez E, Lagaron JM (2007) Novel PET nanocomposites of interest in food packaging applications and comparative barrier performance with biopolyester nanocomposites. J Plas Film Sheet 23:133–148. https://doi.org/10.1177/8756087907083590

Sangsanoh P, Israsena N, Suwantong O, Supaphol P (2017) Effect of the surface topography and chemistry of poly(3-hydroxybutyrate) substrates on cellular behavior of the murine neuroblastoma Neuro2a cell line. Polym Bull 10:4101–4118. https://doi.org/10.1007/s00289-017-1947-9

Schrader JA, Behrens JJ, Michel M, Grewell D (2016) Bioplastics and biocomposites for horticulture containers: processing, properties, and manufacturing potential. In: Schrader JA, Kratsch HA, Graves WR (eds) Bioplastic container cropping systems: green technology for the green industry. Sustainable Hort Res Consortium, Ames. pp 67–95. https://www.researchgate.net/profile/James_Schrader/publication/311983321_Degradability_of_Bioplastic_Containers_in_Soil_and_Compost/links/5866a3b508ae8fce490f1ed6/Degradability-of-Bioplastic-Containers-in-Soil-and-Compost.pdf

Shah M, Ullah N, Choi MH, Yoon SC (2014) Nanoscale poly (4-hydroxybutyrate)-mPEG carriers for anticancer drugs delivery. J Nanosci Nanotechnol 14(11):8416–8421. https://www.ncbi.nlm.nih.gov/pubmed/25958538

Shishatskaya EI, Nikolaeva ED, Vinogradova ON, Volova TG (2016) Experimental wound dressings of degradable PHA for skin defect repair. J Mater Sci Mater Med 27:165. https://doi.org/10.1007/s10856-016-5776-4

Singh M, Patel SKS, Kalia VC (2009) *Bacillus subtilis* as potential producer for polyhydroxyalkanoates. Microb Cell Factories 8:38. https://doi.org/10.1186/1475-2859-8-38

Singh M, Kumar P, Patel SKS, Kalia VC (2013) Production of polyhydroxyalkanoate co-polymer by *Bacillus thuringiensis*. Indian J Microbiol 53:77–83. https://doi.org/10.1007/s12088-012-0294-7

Singh M, Kumar P, Ray S, Kalia VC (2015) Challenges and opportunities for the customizing polyhydroxyalkanoates. Indian J Microbiol 55:235–249. https://doi.org/10.1007/s12088-015-0528-6

Tokiwa Y, Calabia BP (2007) Biodegradability and biodegradation of polyesters. J Polym Environ 15:259–267. https://doi.org/10.1007/s10924-007-0066-3

Vardhan H, Mittal P, Adena SKR, Upadhyay M, Mishra B (2017) Development of long-circulating docetaxel loaded poly (3-hydroxybutyrate-co-3-hydroxyvalerate) nanoparticles: optimization, pharmacokinetic, cytotoxicity and in vivo assessments. Int J Biol Macromol 103:791–801. https://doi.org/10.1016/j.ijbiomac.2017.05.125

Wang Y, Chen GQ (2017) Polyhydroxyalkanoates: sustainability, production, and industrialization. In: Tang C, Ryu CY (eds) Sustainable polymers from biomass. Wiley-VCH Verlag GmbH & Co. KGaA, Weinheim, pp 11–33

Wang S, Song C, Chen G, Guo T, Liu J, Zhang B, Takeuchi S (2005) Characteristics and biodegradation properties of poly(3-hydroxybutyrate-co-3-hydroxyvalerate)/organophilic montmorillonite (PHBV/OMMT) nanocomposites. Polym Degrad Stab 87:69–76. https://doi.org/10.1016/j.polymdegradstab.2004.07.008

Xiao XQ, Zhao Y, Chen GQ (2007) The effect of 3-hydroxybutyrate and its derivatives on the growth of glial cells. Biomaterials 28:3608–3616. https://doi.org/10.1016/j.biomaterials.2007.04.046

Zhang J, Qian C, Shaowu L, Xiaoyun L, Yongxi Z, Ji-Song G, Jin-Chun C, Qiong W, Guo-Qiang C (2013) 3-hydroxybutyrate methyl ester as a potential drug against Alzheimer's disease via mitochondria protection mechanism. Biomaterials 34:7552–7562. https://doi.org/10.1016/j.biomaterials.2013.06.043

Zhao YH, Li HM, Qin LF, Wang HH, Chen GQ (2007) Disruption of the polyhydroxyalkanoate synthase gene in *Aeromonas hydrophila* reduces its survival ability under stress conditions. FEMS Microbiol Lett 276:34–41. https://doi.org/10.1111/j.1574-6968.2007.00904.x

Zou XH, Li HM, Wang S, Leski M, Yao YC, Yang XD, Huang QJ, Chen GQ (2009) The effect of 3-hydroxybutyrate methyl ester on learning and memory in mice. Biomaterials 30:1532–1541. https://doi.org/10.1016/j.biomaterials.2008.12.012

Strategy for Biosynthesis of Polyhydroxyalkonates Polymers/Copolymers and Their Application in Drug Delivery

2

Shashi Kant Bhatia, Puneet Wadhwa, Ravi Kant Bhatia, Sanjay Kumar Singh Patel, and Yung-Hun Yang

Abstract

Drug delivery technology is an emerging field to improve health, and there is considerable interest to use biodegradable biomaterial in drug delivery. Polyhydroxyalkonate (PHA) is a natural origin polymer and getting attention for drug delivery due to its biodegradability and biocompatible properties. PHA with different monomer compositions has different physical and chemical properties which further affect drug loading and release. PHA polymers of different monomer composition can be synthesized using various approaches such as metabolic engineering, precursor feedings, etc. PHA polymers in their intrinsic form or block copolymers forms can be used to prepare different structure such as microsphere, naonparticles, nanofiber, and scaffold, etc. Drug and pharmaceutically active molecule loaded PHA material carrier can be delivered through oral, intravenous or subcutaneous route. Selection of drug delivery route depends on the nature and required dose of drug. Drug release from the polymeric material occurs through osmosis, diffusion or biodegradation mechanism. Composition of polymeric material affects its drug release behavior. This chapter discusses the various trends in PHA polymer production and their application in drug delivery.

S. K. Bhatia (✉) · Y.-H. Yang
Department of Biological Engineering, College of Engineering, Konkuk University, Seoul, Republic of Korea

P. Wadhwa
Department of Oral and Maxillofacial Surgery, College of Dental Science, Guro Hospital, Korea University, Seoul, Republic of Korea

R. K. Bhatia
Department of Biotechnology, Himachal Pradesh University, Shimla, India

S. K. S. Patel
Department of Chemical Engineering, Konkuk University, Seoul, Republic of Korea

V. C. Kalia (ed.), *Biotechnological Applications of Polyhydroxyalkanoates*,
https://doi.org/10.1007/978-981-13-3759-8_2

Keywords
Polyhydroxyalkonates · Polymers · Copolymers · Drug delivery

2.1 Introduction

Drug delivery has been getting immense attention since last few decades as it is facing many challenges. For effective results drug should be delivered at certain concentrations, below this drug is ineffective and beyond this range drugs can lead to side effects. Controlled and targeted drug delivery improves the efficacy of drug as it helps in drug release at predetermined rate (Tiwari et al. 2012). Drug delivery using biopolymeric material has gained great interest due to their biodegradable and biocompatible nature. Different polymeric materials have been reported for drug delivery such as poly lactic acid (PLA), polyhydroxyalkonates (PHA) and poly glycolic acid (PGA) etc. (Bhatia et al. 2016; Danafar et al. 2017; Khalil et al. 2017; Michalak et al. 2017; Ray and Kalia 2017a, b, c, d). Among all these biodegradable polymers, use of PHA is more advantageous due to many superior properties such as tunable mechanical properties, biocompatibility and biodegradability (Kalia et al. 2003, 2016; Reddy et al. 2003; Porwal et al. 2008; Patel et al. 2011, 2016; Singh et al. 2009, 2013, 2015; Panith et al. 2016; Ray et al. 2018). PHA is produced and stored as granules by microorganisms under various stress conditions as reserved food which can be used as a carbon source during starvation (Kumar et al. 2009, 2014, 2016; Bhatia et al. 2015b; Patel et al. 2015). Phasin is the main protein associated with PHA granule, it promote growth and control size and distribution of granule (Mezzina and Pettinari 2016; Seo et al. 2016). PHA has been studied for many health related applications such as heart valve engineering, cartilage engineering, vascular engineering, and drug delivery. Poly(3-hydroxybutyrate) P(3HB) is the PHA accumulated by most of microbes normally (Bhatia et al. 2017; Bhatia et al. 2018). P(3HB) is brittle in nature and has poor properties which make it unsuitable for various applications (Kumar et al. 2013, 2015). Thermal and physical properties of PHA polymer depends on the side chain of monomer units (Jeon et al. 2017; Sathiyanarayanan et al. 2017). Polymers with short side chain length monomer units are brittle in nature while flexibility increases with the increase of side chain length (Brigham and Sinskey 2012). Only few microbes related to *Pseudomonas* have been reported for medium chain length PHA accumulation (Sathiyanarayanan et al. 2017). Different type of copolymers can be produced by varying carbon sources, adding various copolymer precursors and by using metabolic engineering approach (Fig. 2.1) (Patel et al. 2012; Yang et al. 2012; Bhatia et al. 2015a; Bhatia et al. 2019a). PHA polymers have enormous potential to be used as a drug carrier in targeted drug delivery system (TDS). It can be used as it is, mixed with other polymers to prepare graft and copolymer, or functionalized by using various chemical and metabolic

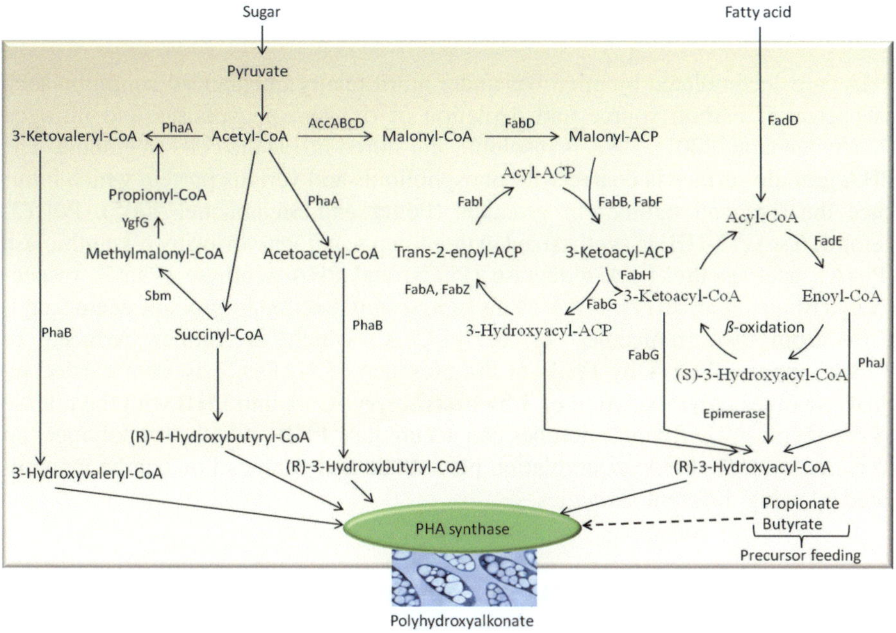

Fig. 2.1 Sketchmatic diagram representing different pathways involved in PHA production. Various enzymes have role in PHA synthesis are: PhaA (ß-ketothiolase), PhaB (acetoacetyl-CoA reductase), PhaC (PHA synthase), Sbm (sleeping beauty mutase), YgfG (methylmalonyl-CoA decarboxylase), AccABCD (acetyl-CoA carboxylase), FabD (malonyl transferase), FabB, F, H (3-oxoacyl-ACP synthase I, II, III respectively), FabG (3-oxoacyl-ACP reductase), FabA (ß-hydroxydecanoyl thioester dehydrase), FabZ (3-hydroxymyristol-ACP dehydratase), FabI (enoyl-ACP reductase), FadD (acyl-CoA synthase), FadE (acyl-CoA dehydrogenase), PhaJ (enoyl-CoA hydratase) (Yang et al. 2013; Bhatia et al. 2017; Bhatia et al. 2015a)

engineering approaches (Bhatia et al. 2019b). Various statures such as nanoparticles, nonofibre, scaffolds, microspheres can be prepared by combining PHA with other pharmaceutically active molecule for TDS (Lu et al. 2011; Masaeli et al. 2014). These drug imbedded particles can be administrated using various routes such as oral, intravenous or subcutaneous, and it depends on the nature of drug molecule. Drug carrier able to release drugs when they come in contact with physiological solution inside the body. Drug release occurs mainly by osmosis, diffusion or degradation of polymeric material (Kamaly et al. 2016).

The main purpose of this chapter is to review the new advancement in drug delivery using PHA. This chapter provides information about various strategies used to produce polymers and copolymers, PHA suture prepared for drug delivery, method of drug delivery, drug release mechanism and properties of biomaterial needed for effective drug release.

2.2 Strategy for Biosynthesis of Polyhydroxyalkonates

PHAs are accumulated by microbes under nutritionally imbalanced condition such as excess of carbon source and depletion of oxygen, phosphorus and nitrogen (Verlinden et al. 2007). PHA is insoluble and stored inside the cells as granule. The PHA granule surface is coated with phospholipids and various protein which influence the size and number of granules (Pötter and Steinbüchel 2005). Poly(3-ketothiolase) P(3-HB) is synthesized in three step reactions carried by β-ketothiolase (PhaA), acetoacethyl-CoA reductase (PhaB) and PHA synthase (PhaC), respectively (Bhatia et al. 2017). PhaA catalyze first step reaction to produce acetoacetyl-CoA from two molecules of acetyl-CoA, which is further reduced to 3-hydroxybutyryl-CoA by PhaB in the presence of NADH. Last step carried by PhaC involves polymerization of 3-hydroxybutyryl-CoA into PHB with the release of coenzyme-A. Different microbes can accumulate PHB and other copolymers up to various w/w%. PHA accumulation percentage and composition can be manipulated by using different strategies.

2.2.1 Enhancement of PHA Accumulation and Production

Different approaches have been reported to enhance PHA accumulation and production such as improvement of carbon source utilization range, and morphology change. Many well-known PHA accumulators are able to utilize only limited carbon source. *Ralstonia eutropha* is a chemolithotrophic bacteria and can easily shift in between autotrophic and heterotrophic mode of nutrition (Volodina et al. 2016). It can utilize various carbohydrates (gluconate, n-acetylglucosamine and fructose), lipids and organic acids as carbon source and can accumulate PHA in all mode of nutrition. Most of the carbon source are costly and have impact on production cost. To make the PHA production process more economic there is need to extend its capacity to utilize inexpensive carbon source such as lignocelluloses, cellulose, hemicelluloses and sugars derived from these polymers (glucose, xylose, mannose, and arabinose etc.). Wild type *R. eutropha* is unable to utilize glucose as it fails to enter in the cell. Franz et al. prepared mutant strain of *R. eutropha* able to utilize glucose by incubating it with higher concentration of glucose. Mutant *R. eutropha* showed elevated level of glucose-6-phosphate dehydrogenase (Franz et al. 2012). Sichwart et al. extended *R. eutropha* capability towards mannose and glucose by using metabolic engineering. Heterologous expression of glucose transporter (*glf*) from *Zygomonas mobilis* and glucose kinase (*glk*) from *E. coli* results in increased glucose utilization, while overexpression of *glf* and phosphomannose isomerase (*pmi*) leads to increased growth on mannose (Sichwart et al. 2011). *R. eutropha* is able to utilize various fatty acids and organic acids so there is no need to extend its capability, by simply optimizing culture condition PHA productivity can be increased. Other non PHA accumulator can also be engineered to biosynthesize PHA. Engineered *E. coli* strain heterologously expressing PHA synthesizing gene operon from *R. eutropha* further engineered to utilize starch as carbon source by using α-amylase of *Bacillus* (Bhatia et al. 2015b).

To increase PHA accumulation Wu et al. suggested a morphology change approach (Wu et al. 2016). Most of bacteria divide by binary fission (equally divide into two), and morphology is directly linked to growth and PHA storage. By using metabolic engineering approach *E. coli* binary fission mechanism changed to multiple-fission by deletion of various genes (*minC* and *mind*) related to fission (Wu et al. 2016). Resulted engineered strain allows development of fission rings (Z-rings) and enabled cells to divide in many daughter cells. Change in morphology resulted in increase in biomass production and PHA accumulation. Microbes can accumulate more inclusion bodies if its cellular stiffness is lowered and cell expansion is allowed. MreB an actin homolog is required to maintain the rod shape morphology in bacterial cells. Deletion of *mreB* in *E. coli* results in change in morphology from rod to circular shape. Overexpression of PHA synthesis operon *phaABC* in these engineered *E. coli* cells leads to enhanced production of PHA (Jiang and Chen 2016).

2.2.2 Approaches to Produce Copolymers

PHB has limited applications due to its poor physical properties. There are different types of PHA copolymers having superior properties and may have application in different field. PHA is classified as short and medium chain length according to carbon number, and homopolymer and copolymer according to monomer units. Microbial capability to biosynthesize different types of polymers can be extended by using various approaches such as precursor feeding and metabolic engineering (Fig. 2.1) (Bhatia et al. 2015a).

2.2.2.1 Precursor Feeding

Most of microbes are able to produce P(3HB) using various carbon source such as fructose, glucose, and xylose etc. When substrate is changed it can also produce various copolymers. Yang *et al.* used engineered *E. coli* strain expressing propionyl-CoA transferase gene (*pct*) which helps cell to utilize propionic acid for P(3-hydroxybutyrate-*co*-3-hydroxyvalerate) P(3HB-*co*-3HV) production (Yang et al. 2012). The engineered strain is able to produce copolymer P(3HB-*co*-3HV) when it is fed with glucose, and propionate a precursor for hydroxyvalerate (HV) monomer unit. Similarly to produce P(3HB-*co*-3HV) copolymer other researchers expressed genes succinyl-coA synthase (*sucD*) and acetyl-coA acetyltransferase (*atoAD*) in PHA accumulating *E. coli* strain and fed this strain with glucose and propionate and process results in increased fraction of 3HV monomer in copolymer (Bhatia et al. 2015a; Jeon et al. 2017). To produce copolymers poly(3-hydroxybutyrate-*co*-3-hydroxyhexanoate) (P(3HB-*co*-3HHx)) having favorable physical and chemical properties, Jeon et al. engineered *R. eutropha* strains by deleting native PHA synthase and overexpressing phaC2 from *Rhodococcus aetherivorans* I24. Resulting strain was cultured on butyrate as carbon source, and process resulted into accumulation of P(3HB-*co*-3HHx) copolymer (Jeon et al. 2014).

2.2.2.2 Metabolic Engineering to Synthesize Copolymer Precursors

Addition of different precursors in growth media to produce various copolymers is a cost effective process. These precursors are toxic to cells and inhibit growth, which further effect the productivity. Metabolic engineering of microbes to biosynthesize various precursors on their own can be a possible solution to improve productivity. Srirangan et al. reported an approach to produce a copolymer P(3HB-*co*-3HV) from unrelated carbon source (glucose and glycerol) without adding any additional precursor. A sleeping beauty mutase (*sbm*) was expressed in *E. coli* to produce propionyl-CoA a precursor for 3HV (Srirangan et al. 2016). Propionyl-CoA condensed with acetyl-CoA to produce 3-ketovaleryl-CoA by PhaA and the resulting thioester is further channeled into PHA biosynthetisis pathway through PhaB and PhaC, respectively (Fig. 2.1). Yang et al. reported P(3HB-*co*-3HV) production directly from glucose by direct conversion of 2-ketobutyrate oxidase into propionate. *E. coli* strain was engineered with pyruvate oxidase (*poxB*) and propionyl-CoA synthetase (*prpE*) which are involved in 2-ketobutyrate conversion into propionate and propionate to propionyl-CoA, respectively. Engineered strain able to produce P(3HB-*co*-3HV) copolymer containing 5.95 mol% of 3HV (Yang et al. 2014).

2.3 Biodegradability and Biocompatibility of PHA

To use PHA for biomedical application it should be biocompatible and biodegradable within clinically reasonable time period. PHA is known to degradable under in vitro and in vivo conditions. In in-vitro most of aliphatic polyesters are easily mineralized by aerobic and anaerobic microbes which are widely distributed in nature. Synthetic polymers are resistant to microbial degradation due to large molecular mass and presence of aromatic and halogen substitution. Degradation of PHA depends on its composition and occurs through random scission on amorphous and crystalline region of polymers (Freier et al. 2002). Varying fraction of 3HV in P(3HB-*co*-3HV) affect its degradation property as 3HV effects crystallinity, porosity and molecular weight of the polymer. To increase polymers degradation rate these can be blended with other material like gelatin, as it may help to decrease crystallinity and increase porosity of material which provide more surface for hydrolytic attack (Wang et al. 2005). In in-vivo biological active molecules such as enzymes and lipids effects the degradation, and the host reaction depends on the various product released on its degradation. Qu et al. performed in-vivo tissue reaction and biodegradation study of PHB, and PHBHHx, and reported polymer degradation occur in order PHB<PHBHHx. PHBHHx elicited very weak immunological response while acute response was observed against PHB (Qu et al. 2006).

Along with the biodegradability of PHA, biocompatibility, toxicity and its ability to support the growth and proliferation of cells are also important properties (Ali and Jamil 2016). Host release various pro- and anti-inflammatory mediators which can cause acute inflammation or complete rejection of biomaterial. Kose et al. studied PHB with many cells like chondrocytes, osteoblasts and fibroblasts and showed tremendous biocompatibility without any side effect (Kose et al. 2003). Unmodified mcl-PHA also reported to show biocompatible conduct with mesenchymal stromal cells (Naveen et al. 2015). Cellular response of PHA degradation products has an important role in deciding biocompatibility. Sun et al. studied cellular response of various oligo-hydroxyalkonates (OHAs) and found that concentration lower than 20 mg/L have no effect on cell viability while concentration more than 40 mg/L reduce cell viability and proliferation (Sun et al. 2007). Cytotoxicity of OHAs decrease with the length of OHAs side chain, and mcl PHA are preferred material for biomedical applications. The degradation of PHA also varies from tissue to tissue, and its structure such as nanofibers, film and scaffolds (Brigham and Sinskey 2012). Owing to favorable properties of PHA it can be used for various applications such as drug carrier, tissue and nerve regeneration etc.

2.4 PHA in Drug Delivery

Drug delivery technology play major role in human health. The conventional therapeutic systems have drawbacks such as toxicity, inconsistency, and delivery of drug to targeted and non-targeted tissue (Masood et al. 2015). Targeted and controlled drug delivery of pharmacologically active agent is the main goal of research in developing new drug delivery technology. Targeted drug delivery ensures the proper dosage of drug for required time at specific site without affecting other tissue and organ (Bae and Park 2011). Polymeric materials are identified as important candidate for controlled drug release at specific site. These polymer based drug delivery system can be used to deliver various pharmacologically active molecules such as antibiotics, antioxidants, hormones, contraceptive, neurotrophic factors and other gene therapy agent (Table. 2.1). PHA natural polymer is getting more attention due to its biodegradable nature, biocompatibility and can be turned into nanoparticles, microsphere and nanofibers like structure.

2.4.1 Intrinsic PHA in Drug Delivery

PHA is a hydrophobic material and can be molded into various structures such as nanoparticles, microsphere, nanofibers and scaffolds which can be used for drug encapsulation (Table. 2.1). Nanoparticles based drug delivery system has recently got attention in various medical fields. In cancer therapy drug delivery to solid tumors is a challenging job. Recently nonotechnology become a promising technology in development of new techniques for drug delivery system.

2.4.1.1 Nanoparticles

Nanoparticles are sub-micronsized particles (1–1000 nm), formed by self-aggregation of amphiphilic particles comprised of inner hydrophilic and outer hydrophobic domains (Tao and Uhrich 2006). Nanoparticles can be used for the delivery of hydrophilic and hydrophobic drug, proteins, vaccines and other

Table 2.1 Various intrinsic form of PHA used for delivery of various drugs

Intrinsic PHA	Structure	Drug	Application	References
P(3HB)	Microsphere	Aclarubicim	Carrier for controlled drug release	Kazuhiko et al. (1986)
	Microsphere	Gentamycin	Encapsulation material	Francis et al. (2011)
	Microsphere	Progestron	Hormone delivery	Gangrade and Price (1991)
	Microsphere	Rifampicin	Chemoimobilization agent	Kassab et al. (1997) and Shishatskaya et al. (2008)
	Microparticle	Sulphamethizole	Carrier for controlled drug release	Brophy and Deasy (1986)
	Nanoparticle	TGX221, and PI3K inhibitor	Used as carrier to increase stability of drug	Lu et al. (2011)
	Tube	Paclitaxel and triamcinolone acetonide	Inhibit fibrosis process by reducing fibroblast activity in a glaucoma drainage system	Lobler et al. (2011)
	Microsphere	Tetracyclin	Provide non-uniform surface and increase the cells viability	Francis et al. (2010b)
P(3HB3HV)	Nanoparticle	Ellipticine	Increase drug cytotoxic effect	Masood et al. (2013)
	Disc	Gentamycin	Antibiotic delivery to prevent orthopedic infection	Rossi et al. (2004)
	Nanoparticle	Paclitaxel	Drug carrier and increase cytotoxicity to cell line	Vilos et al. (2013)
	Microsphere	Progestron	Hormone delivery	Gangrade and Price (1991)
	Microsphere	Tetracyclin	Controlled drug release system for periodontal disease	Sendil et al. (1999)
	–	Sulbactam-cefoperazone	Biocompatible, and osteoconductive due to its piezoelectric property	Yagmurlu et al. (1999)
P(3HB4HB)	Nanofibres	Paclitaxel	Act as biocompatible drug eluting stents	Lee et al. (2017)
P(3HB3HHx)	Nanoparticle	5-Fluorouracil	Controlled drug release	Lu et al. (2010)
	Nanoparticle	Rhodamine B isothiocyanate	Suitable for hydrophobic drug delivery	Yao et al. (2008)

biological molecules. Shah et al. prepared nanoparticles of poly(4-hydroxybutyrate)-mPEG (P(4HB)-mPEG) for delivery of anticancer drug cisplatin to mouse hippocampal HT22 cells (Shah et al. 2014). Different type of PHA nanoparticles prepared by using PHB and PHBHHx were screened for drug release behavior of rhodamine B isothiocyanate (RBITC). Drug loading efficiency for each particle was over 75%

with sustained release time of 20 days in macrophage (Xiong et al. 2010). Phosphoinositide-3-kinases (PI3Ks) is responsible for different kind of human cancer, although many type of inhibitors has been reported to inhibit its activity but poor solubility and stability of these inhibitors is the main problem. Lu et al. used PHA nanoparticles for sustained release of PI3Ks to stop the proliferation of cancer cell (Lu et al. 2011).

2.4.1.2 Microsphere

Microspheres are small spherical particles (1–1000 μm), sometimes also referred as microparticles. To prolong drug release of ibuprofen Bazzo et al. prepared microsphere of poly(hydroxybutyrate-*co*-hydroxyvalerate) (PHBV) and poly(l-lactide) (PLA) using oil-in-water (O/W) emulsion solvent evaporation technique (Bazzo et al. 2012). Microsphere particles were spherical in shape carrying drug particles on its surface. Pure PHBV microsphere surface was porous and rough while its blend with PLA results in smooth surfaced microsphere. Microspheres prolonged drug release of ibuprofen in first 2 h and maximum level reached after 6 h of administration (Bazzo et al. 2012). Chaturvedi et al. prepared blend microsphere of PHB/ cellulose acetate phthalate (CAP) using solvent evaporation technique for colon delivery of anticancer drug 5- fluorouracil (Chaturvedi et al. 2011). Tetracycline is an effective antibiotic for periodontal disease and its controlled release can make it more effective. Sendila et al. used PHBV microsphere to load tetracycline in its acidic and neutral form to make it more effective (Sendil et al. 1999).

2.4.1.3 Scaffolds

Scaffold is a three dimensional structure act as a temporary extracellular matrix to provide support to cells and new tissue being formed (Lee and Shin 2007). Repair of injured bone tissue requires several growth factors to provide signals to local tissue to trigger the healing process (Mouriño and Boccaccini 2009). Many researchers have tried to include growth factors in scaffolds to stimulate cellular adhesion, differentiation and proliferation. Cheng et al. used scaffolds of poly(3-hydroxybutyrate-*co*-4-hydroxybutyrate) (P3HB4HB) containing 0–40 mol% 4HB to study its potential as a material for tissue and blood vessel regenerations. This material has potential to induce elastin formation and can be used as scaffold for artificial blood vessel engineering (Cheng et al. 2008). Schwaan cells interaction with scaffold is important for nerve regeneration. Masaeli et al. prepared electrospun scaffold by combining PHB and PHBV with collagen I and neuromimetic peptides. Cells grown on this scaffold shows increase in proliferation and metabolic activity (Masaeli et al. 2014).

2.4.1.4 Nanofibers

Biometric nanofibers emerged as a potential candidate for drug delivery. Different techniques have been developed including fabrication, phase separation and self-assembly for nanofibres production (Sharma et al. 2015). During last few decades various nanofibers based drug delivery system has been reported for different bioactive molecules such as growth factors, protein and therapeutic drugs. To prepare drug loaded nanofibres different methods can be used such as post treatment of nanofibers, coaxial electrospinning, electrospininning of drug-polymer blend and immobilization of drug loaded nanocarriers on nanofibers (Sharma et al. 2015). Kuppan et al. prepared nonofibers of R-spondin-1-PHBV using electrospun technique and used these for skin regeneration (Kuppan et al. 2011). The wound contracture using R-spondin-1-PHBV was significantly higher as compared no normal PHBV nanofiber. Paclitaxel is an anti-tumor agent and used for treatment of various cancers such as ovarian, breast, head and neck, leukemia, and melanoma. Current formulations methods used for paclitaxel in the human body may results in various side effects such as neurotoxicity hypersensitivity reaction, cardiotoxicity and nephrotoxicity etc. Lee et al. developed an alternate method of paclitaxel delivery using nanofibers of P3HB4HB (Lee et al. 2017).

2.4.2 Functionalized PHA in Drug Delivery

2.4.2.1 Functionalization of PHA Granule Associated Proteins

PHA is stored inside the microbial cells in the form of granules of various size. These granules are comprised of hydrophobic core surrounded by a boundary layer and various proteins. Protein engineering technique can be used to attach a functional protein to PHA granules to improve its function in drug delivery system. Phasin protein (PhaP), a PHA granule associated protein able to bind with granules by hydrophobic interaction. Human α1-acid glycoprotein (hAGP) a plasma protein produced and secreted by the liver. hAGP can be used as legend for target delivery to liver as it can be recognized through its five carbohydrate chain. EGFRs a member of tyrosine kinase is expressed by many cancer cells such as hepatocellular carcinoma and human astrocytoma cells (Oberoi et al. 2013; Masood 2016). Yao et al. developed a targeted drug delivery system (TDDS) by fusing PhaA protein genes with hAGP and hEGF legend and expressed them in *Pichia pastoris* GS115 and *Escherichia coli* BL21 (DE3), respectively (Yao et al. 2008). These fusion proteins rPhaP-hAGP and rPhaP-hEGF were purified from their respective host and combined with anticancer drug (rhodamin B isothiocynate) containing PHA nanoparticles. When these functionalized PHA nanoparticles were used in TDDS, a considerable amount of these drug loaded particle was found in cancerous cells (Yao et al. 2008).

PHA synthase (PhaC) is enzyme involved in polymerization of hydroxyacyl-CoA into PHA. The main property of PhaC is that after polymerization it remains covalently attached to polymer through specific amino acids residue. This property of PHA synthase can be used to fuse it other bioactive molecules (Paik et al. 2005). Various cancer cells express $\alpha_v\beta_3$ integrins which helps in cell adhesion to extracellular matrix. Arg-Gly-Asp (RGD) motif have been reported to these integrins. Using genetic engineering PhaC can be fused with RGD, expressed in PHA producing *E. coli* (Fig. 2.2). These functionalized PHA granules purified and further used to target cancer cells that express integrin (Kim et al. 2009).

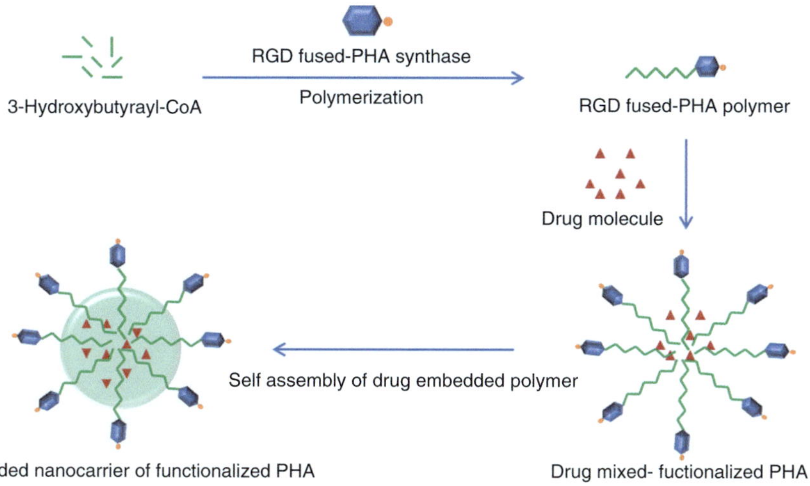

Fig. 2.2 Schematic drawing representing preparation of tumor targeted drug loaded PHA carrier, and their self-assembly into nano-sized micelles

2.4.2.2 Functionalized PHA Graft Polymer

PHA modification through graft and copolymerization is another technique that can be used to prepare functionalized PHA for drug delivery (Li and Loh 2017). Various targeting moieties such as proteins, legend, and other bioactive molecule can also be attached to functionalized PHA nanoparticles. Domenek et al. prepared functionalized polymer of poly(3-hydroxyoctanoate-*co*-3-hydroxyundecenoate) (PHOU) by performing oxidation of double bound of PHOU, and carboxyl group of resulting poly(3-hydroxyoctanoate-*co*-3-hydroxy-9-carboxydecanoate) (PHOD) was further esterified with PEG using a single step reaction. Grafting results in increased hydrophilic nature of polymer and made it more soluble in polar solvents. These change in property made this polymer a suitable polymer for drug delivery (Domenek et al. 2007). Shah et al. prepared amphiphilic block copolymer P(3 HV-*co*-4HB)-b-mPEG by performing poly(3-hydroxyvalerate-*co*-4-hydroxybutyrate) transesterification with monomethoxypoly(ethylene glycol) (mPEG) in the presence of catalyst bis(2-ethylhexanoate) tin and used these functionalized polymers for delivery of anti-cancerous ciplastin (Table 2.2) (Shah et al. 2012).

Zhou et al. developed a method for siRNA delivery using functional PHA (Zhou et al. 2012). Functionalized PHA graft polymer was constructed by using Michael addition reaction between acrylated monomethoxy-poly(hydroxyalkanoates) (mPHA-acrylated) and poly(ethyleneimine) (bPEI). Grafted polymers able to bind siRNA, protect its degradation from nuclease and release it when there is a lower concentration of polyanionic heparin. Zhou et al. studied targeted delivery capacity of grafted polymer for siRNA against luciferase in cultured A549-Luc and MCF-7-Luc cells (Zhou et al. 2012).

Table 2.2 Various modified form of PHA used for delivery of various drugs

Modified PHA	Structure	Drug	Application	References
P(3HB)/45S5 Bioglass	microsphere	Gentamycin	Material help in controlled drug release	Francis et al. (2010a)
P(4HB)-b-mPEG	Nonoparticles	Cisplatin	Carrier helps in sustained release of drug	Shah et al. (2014)
P(3HV-*co*-4HB)-b-PEG	Nonoparticles	Cisplatin	Carrier increase drug loading efficiency	Shah et al. (2012)
P(3HB-*co*-3HHx)-b-PEG	Nonoparticles	Rapamycin	Nonocarriers helps in sustained release of drug	Lu et al. (2014)
PHBV/PLA	microsphere	Ibuprofen	Porolong the drug release	Bazzo et al. (2012)
PHB and cellulose acetate phthalate	microspheres	5-fluorouracil	Porolong the drug release	Chaturvedi et al. (2011)
P(3HB-*co*-3HV)/wollastonite composite	microspheres	Gentamycin	Controlled drug relesae	Li and Chang (2005)

2.4.2.3 Functionalized PHA Block Copolymer

Li et al. synthesized triblock copolymer of PHB and poly(ethylene oxide) (PEO). First, PHB-diol was prepared by its transesterification with diethylene glycol after that coupling reaction was performed with two chain of methoxy-PEO-monocarboxylic acid. Synthesized triblock copolymer PEO-PHB-PEO is extracted using repeated precipitation and fractionation (Li et al. 2003a). Triblock PEO-PHB-PEO can assemble into spherical micelle having dense core surrounded by the soluble PEG outer shells, which make them suitable for drug delivery (Li et al. 2005). Li et al. prepared inclusion complex of cyclodextrin with PEO-PHB-PEO and reported increase in thermal stability of cyclodextrin (Li et al. 2003b). Loh et al. synthesized a multiblock thermo-sensitive block copolymer of poly-[(R)-3-hydroxybutyrate] (PHB), poly(ethylene glycol) (PEG), and poly(propylene glycol) (PPG) which have properties to undergo reverse sol-gel transition upon temperature change at low copolymer concentration (Loh et al. 2007). This material can act as a suitable material for injectable drug delivery as it can be formulated at low temperature and can form gel on subcutaneous injection (Loh et al. 2007).

2.5 Route for Drug Delivery

Various drug delivery routes are in trend to target specific site and make the drug more effective such as oral, subcutaneous, and intravenous administration. Oral drug administration is the most common method as it provides maximum active surface area for drug absorption.

2.5.1 Oral Route

In oral drug release there is no control on drug release and its fluctuation in concentration may lead to some side effects. Rate of drug absorption depend on physiochemical nature of drug, gastro intestinal pH, presence and absence of food etc. Main limitation of the oral drug delivery system is restricted gastric residence time. Rapid gastro intestinal transit reduces the efficacy of drug as it prevents complete release of drug in absorption zone i.e. stomach and upper part of the small intestine (Surana and Kotecha 2010). To overcome rapid transit limitation various new techniques has been introduced such as mucoadhesive systems, swelling, and floating drug dosage system etc. (Chen et al. 2015; Huanbutta et al. 2016; Mansuri et al. 2016). Yu et al. reported use of various PHA nonocapusles as a carrier for oral delivery of various proteins (insulin, thyroid stimulating hormone, gastin and somatostatin etc.) (Yu et al. 2014).

2.5.2 Intravenous Route

Intravenous (IV) route involves the direct delivery of drugs into veins using a needle. IV method provides control on dosage in case of situation where patient need medication quickly such as heart attack, stroke and poisoning. Some drugs require IV route, because they can be affected or degraded by enzymes in mouth and stomach. To make them more effective they are directly sent to blood stream by the IV route.

2.5.3 Subcutaneous Route

This route involves delivery below the dermis and epidermis layer of the skin. Subcutaneous layer has few blood vessels and drugs injected here shows slow rate of absorption. This route used to deliver drugs when other delivery methods are not suitable and drug required in small volume (1–2 ml). Some vaccine and allergy shots are administrated through this route. Wang et al. developed a transdermal drug delivery system for tamsulosin (used against prostatic hyperplasia), using PHA with polyamidoamine dendrimer (Wang et al. 2003).

2.6 Drug Release Mechanism

Drug release from polymeric material depends on various mechanisms like osmosis, diffusion and degradation etc. In osmotic drug release mechanism osmotic pressure acts as a main driving force. To prepare osmotic based drug release system, active pharmaceutical agents are positioned at the core of the tablet and then tablet coated with a semi-permeable membrane. Most of oral drug delivery tablets work on osmotic release mechanism. Cellulose acetate is the most widely used material and its permeability is controlled by its degree of acetylation (Francis 2011). After oral intake the hydrostatic pressure inside the tablet which led to expulsion of drug solution out. Diffusion based drug delivery system mainly use to type of devices, i.e. monolithic and reservoir device. In monolithic device, pharmaceutically active agent is mixed with polymeric material and drug is released from it through dissolution and diffusion in physiological fluid. In reservoir device, drug is placed in the core of film and its release totally depends upon thickness and porosity of the film. Biopolymers used in drug delivery are biodegradable and drug release mainly occurs through surface or bulk erosion. Surface erosion occurs when erosion increase more than water penetration rate into polymer, while bulk erosion happen when the water penetration rate exceed more than erosion.

2.7 Factors Affect Drug Release

The drug release behavior of biopolymer is depends on properties such as crystallinity, molecular weight, thermal stability and methods used in their preparation. Polymers have two types of structure region, i.e. crystalline and amorphous. Crystalline region is tougher than amorphous region, as this reason has increased intra-molecular interaction which makes it compact and impermeable. Reduced permeability affects polymer hydrolysis and drug dissolution which further affect drug release. Drug release from copolymer P(3HB-*co*-3HV) is generally higher than homopolymer. Molecular weight of various PHA polymers depends on the microorganism used, carbon source, method of fermentation and procedure used for polymer extraction. Varying chain length of copolymers affects the hydrophobicity of the polymers, which further influences the drug release kinetics. High molecular weight polymers are compact and irregular in shape, which entangled with each other's and results in reduced surface area and face diffusion problem (Pouton and Akhtar 1996). Khang et al. prepared P(3HB) microsphere of different size (65 kDa, 135 kDa, 450 kDA) with anti-cancer drug 5-fluoro-2′-deoxyuridinerelease and studied drug release behavior of these microsphere. Microsphere prepared using low molecular weight PHA (65 kDa) showed better drug release capacity as compared to others one (Khang et al. 2001). Incorporation of drug in polymer matrics affects it thermal properties which further affects drug release from polymers. Processing conditions also affects drug release as porosity, viscosity and drug distribution very with processing condition. Type of solvent also have important role, as use of chloroform results in microsphere having porous and irregular surface while use of methylene chloride led to uniform surface microsphere.

2.8 Conclusion

With the advancement in medical health sector, there is an increase in interest for drug delivery using biodegradable and biocompatible biopolymer. PHA is a suitable biomaterial for drug delivery due to its wide range of properties and release of nontoxic product on degradation. PHA based carrier has been successfully investigated for the delivery of antibiotics and many other pharmaceutically active molecules. PHA material are crystalline in nature which results in the rapid drug release in an uncontrolled manner. There is a need to work on to improve its properties by surface modification and functionalization without affecting its biodegradability and biocompatibility properties to ensure controlled drug release. With the advancement of technology and production of PHA polymers having a different monomer composition and properties, it may emerge as a next generation biomaterial with a wide range of applications.

2.9 Opinion

Comparatively a little research work has been done on biomedical application of PHA irrespective of its biodegradability and biocompatible behavior. Most of the PHA produced through bacterial fermentation, and only commercial sugars were utilized as carbon source which makes the production process costly. There is need to work on PHA production using abundantly available carbon sources such as lignocelluloses, cellulose and other waste material. Microbial production of PHA is also advantageous as PHA having unique monomer composition can be produced by economic way. Isolation of new microbes which can accumulate novel PHA using wide range of carbon source will be helpful in this regard. Functionalization of PHA may further improve their properties to meet the standard required for a drug carrier. Human immune response against biomaterial is also a matter of concern and need to be addressed more precisely using animal models. With day by day progress in biomaterial science research, we can hope that PHA based drug delivery system will play a vital role in future medical care.

Acknowledgement The authors acknowledge the KU Research Professor Program of Konkuk University, Seoul, South Korea for providing financial support to Dr. Shashi Kant Bhatia.

References

Ali I, Jamil N (2016) Polyhydroxyalkanoates: current applications in the medical field. Front Biol 11:19–27. https://doi.org/10.1007/s11515-016-1389-z

Bae YH, Park K (2011) Targeted drug delivery to tumors: myths, reality and possibility. J Control Release 153:198–205. https://doi.org/10.1016/j.jconrel.2011.06.001

Bazzo GC, Macedo AT, Crenca JP, Silva VE, Pereira EM, Zétola M, Pezzini BR (2012) Microspheres prepared with biodegradable PHBV and PLA polymers as prolonged-release system for ibuprofen: in vitro drug release and in vivo evaluation. Braz J Pharma Sci 48(4):773–780. https://doi.org/10.1590/S1984-82502012000400021

Bhatia SK, Yi DH, Kim HJ, Jeon JM, Kim YH, Sathiyanarayanan G, Seo H, Lee J, Kim JH, Park K (2015) Overexpression of succinyl-CoA synthase for poly (3-hydroxybutyrate-co-3-hydroxyvalerate) production in engineered *Escherichia coli* BL21 (DE3). J Appl Microbiol 119:724–735. https://doi.org/10.1111/jam.12880

Bhatia SK, Shim Y-H, Jeon J-M, Brigham CJ, Kim Y-H, Kim H-J, Seo H-M, Lee J-H, Kim J-H, Yi D-H, Lee YK, Yang Y-H (2015b) Starch based polyhydroxybutyrate production in engineered *Escherichia coli*. Bioprocess Biosyst Eng 38:1479–1484. https://doi.org/10.1007/s00449-015-1390-y

Bhatia SK, Bhatia RK, Yang Y-H (2016) Biosynthesis of polyesters and polyamide building blocks using microbial fermentation and biotransformation. Rev Environ Sci Biotechnol 15:639–663. https://doi.org/10.1007/s11157-016-9415-9

Bhatia SK, Kim J-H, Kim M-S, Kim J, Hong JW, Hong YG, Kim H-J, Jeon J-M, Kim S-H, Ahn J, Lee H, Yang Y-H (2017) Production of (3-hydroxybutyrate-co-3-hydroxyhexanoate) copolymer from coffee waste oil using engineered *Ralstonia eutropha*. Bioprocess Biosyst Eng. https://doi.org/10.1007/s00449-017-1861-4

Bhatia SK, Yoon JJ, Kim HJ, Hong JW, Gi Hong Y, Song HS, Moon YM, Jeon JM, Kim YG, Yang YH (2018) Engineering of artificial microbial consortia of *Ralstonia eutropha* and *Bacillus subtilis* for poly(3-hydroxybutyrate-co-3-hydroxyvalerate) copolymer production from sugarcane sugar without precursor feeding. Bioresour Technol 257:92–10. https://doi.org/10.1016/j.biortech.2018.02.056

Bhatia SK, Gurav R, Choi TR, Jung HR, Yang SY, Moon YM, Song HS, Jeon JM, Choi KY, Yang YH (2019a) Bioconversion of plant biomass hydrolysate into bioplastic (polyhydroxyalkanoates) using Ralstonia eutropha 5119. Bioresour Technol 271:306–315. https://doi.org/10.1016/j.biortech.2018.09.122

Bhatia SK, Wadhwa P, Hong JW, Hong YG, Jeon JM, Lee ES, Yang YH (2019b) Lipase mediated functionalization of poly(3-hydroxybutyrate-co-3-hydroxyvalerate) with ascorbic acid into an antioxidant active biomaterial. Int J Biol Macromol 123:117–123. https://doi.org/10.1016/j.ijbiomac.2018.11.052

Brigham CJ, Sinskey AJ (2012) Applications of polyhydroxyalkanoates in the medical industry. Int J Biotechnol Wellness Ind 1:52–60. http://www.lifescienceglobal.com/pms/index.php/ijbwi/article/view/85

Brophy MR, Deasy PB (1986) In vitro and in vivo studies on biodegradable polyester microparticles containing sulphamethizole. Int J Pharm 29:223–231. https://doi.org/10.1016/0378-5173(86)90119-5

Chaturvedi K, Kulkarni AR, Aminabhavi TM (2011) Blend microspheres of poly (3-hydroxybutyrate) and cellulose acetate phthalate for colon delivery of 5-fluorouracil. Ind Eng Chem Res 50:10414–10423. https://doi.org/10.1021/ie2011005

Chen Y-C, Ho H-O, Liu D-Z, Siow W-S, Sheu M-T (2015) Swelling/floating capability and drug release characterizations of gastroretentive drug delivery system based on a combination of hydroxyethyl cellulose and sodium carboxymethyl cellulose. PLoS One 10:e0116914. https://doi.org/10.1371/journal.pone.0116914

Cheng ST, Chen ZF, Chen GQ (2008) The expression of cross-linked elastin by rabbit blood vessel smooth muscle cells cultured in polyhydroxyalkanoate scaffolds. Biomaterials 29(31):4187–4194. https://doi.org/10.1016/j.biomaterials.2008.07.022

Danafar H, Rostamizadeh K, Davaran S, Hamidi M (2017) Drug-conjugated PLA–PEG–PLA copolymers: a novel approach for controlled delivery of hydrophilic drugs by micelle formation. Pharm Dev Technol 22:947–957. https://doi.org/10.3109/10837450.2015.1125920

Domenek S, Langlois V, Renard E (2007) Bacterial polyesters grafted with poly (ethylene glycol): behaviour in aqueous media. Polym Degrad Stab 92:1384–1392. https://doi.org/10.1016/j.polymdegradstab.2007.03.001

Francis L (2011) Biosynthesis of polyhydroxyalkanoates and their medical applications. University of Westminster, London. http://westminsterresearch.wmin.ac.uk/id/eprint/9506

Francis L, Meng D, Knowles JC, Roy I, Boccaccini AR (2010a) Multi-functional P(3HB) microsphere/45S5 Bioglass-based composite scaffolds for bone tissue engineering. Acta Biomater 6:2773–2786. https://doi.org/10.1016/j.actbio.2009.12.054

Francis L, Meng D, Locke IC, Mordan N, Salih V, Knowles JC, Boccaccini AR, Roy I (2010b) The influence of tetracycline loading on the surface morphology and biocompatibility of films made from P(3HB) microspheres. Adv Eng Mater 12:B260–B268. https://doi.org/10.1002/adem.201080020

Francis L, Meng D, Knowles J, Keshavarz T, Boccaccini AR, Roy I (2011) Controlled delivery of gentamicin using poly(3-hydroxybutyrate) microspheres. Int J Mol Sci 12:4294–4314. https://doi.org/10.3390/ijms12074294

Franz A, Rehner R, Kienle A, Grammel H (2012) Rapid selection of glucose-utilizing variants of the polyhydroxyalkanoate producer *Ralstonia eutropha* H16 by incubation with high substrate levels. Lett Appl Microbiol 54:45–51. https://doi.org/10.1111/j.1472-765X.2011.03171.x

Freier T, Kunze C, Nischan C, Kramer S, Sternberg K, Saß M, Hopt UT, Schmitz K-P (2002) In vitro and in vivo degradation studies for development of a biodegradable patch based on poly(3-hydroxybutyrate). Biomaterials 23:2649–2657. https://doi.org/10.1016/S0142-9612(01)00405-7

Gangrade N, Price JC (1991) Poly(hydroxybutyrate-hydroxyvalerate) microspheres containing progesterone: preparation, morphology and release properties. J Microencapsul 8:185–202. https://doi.org/10.3109/02652049109071487

Huanbutta K, Limmatvapirat S, Sungthongjeen S, Sriamornsak P (2016) Novel strategy to fabricate floating drug delivery system based on sublimation technique. AAPS PharmSciTech 17:693–699. https://doi.org/10.1208/s12249-015-0398-6

Jeon J-M, Brigham CJ, Kim Y-H, Kim H-J, Yi D-H, Kim H, Rha C, Sinskey AJ, Yang Y-H (2014) Biosynthesis of poly (3-hydroxybutyrate-co-3-hydroxyhexanoate)(P (HB-co-HHx)) from butyrate using engineered *Ralstonia eutropha*. Appl Microbiol Biotechnol 98(12):5461–5469. https://doi.org/10.1007/s00253-014-5617-7

Jeon J-M, Kim H-J, Bhatia SK, Sung C, Seo H-M, Kim J-H, Park H-Y, Lee D, Brigham CJ, Yang Y-H (2017) Application of acetyl-CoA acetyltransferase (AtoAD) in Escherichia coli to increase 3-hydroxyvalerate fraction in poly(3-hydroxybutyrate-co-3-hydroxyvalerate). Bioprocess Biosyst Eng 1:9. https://doi.org/10.1007/s00449-017-1743-9

Jiang X-R, Chen G-Q (2016) Morphology engineering of bacteria for bio-production. Biotechnol Adv 34:435–440. https://doi.org/10.1016/j.biotechadv.2015.12.007

Kalia VC, Chauhan A, Bhattacharyya G, Rashmi (2003) Genomic databases yield novel bioplastic producers. Nat Biotechnol 21:845–846. https://doi.org/10.1038/nbt0803-845

Kalia VC, Prakash J, Koul S (2016) Biorefinery for glycerol rich biodiesel industry waste. Indian J Microbiol 56:113–125. https://doi.org/10.1007/s12088-016-0583-7

Kamaly N, Yameen B, Wu J, Farokhzad OC (2016) Degradable controlled-release polymers and polymeric nanoparticles: mechanisms of controlling drug release. Chem Rev 116:2602–2663. https://doi.org/10.1021/acs.chemrev.5b00346

Kassab AC, Xu K, Denkbas EB, Dou Y, Zhao S, Piskin E (1997) Rifampicin carrying polyhydroxybutyrate microspheres as a potential chemoembolization agent. J Biomater Sci Polym Ed 8:947–961. https://www.ncbi.nlm.nih.gov/pubmed/9399144

Kazuhiko J, Masahiro N, Miho K (1986) Controlled release of aclarubicim, an anticancer antibiotic, from poly-β-hydroxybutyric acid microspheres. J Control Release 4:25–32. https://doi.org/10.1016/0168-3659(86)90030-1

Khalil IR, Burns ATH, Radecka I, Kowalczuk M, Khalaf T, Adamus G, Johnston B, Khechara MP (2017) Bacterial-derived polymer poly-γ-glutamic acid (γ-PGA)-based micro/nanoparticles as a delivery system for antimicrobials and other biomedical applications. Int J Mol Sci 18:313. https://doi.org/10.3390/ijms18020313

Khang G, Kim SW, Cho JC, Rhee JM, Yoon SC, Lee HB (2001) Preparation and characterization of poly(3-hydroxybutyrate-co-3-hydroxyvalerate) microspheres for the sustained release of 5-fluorouracil. Biomed Mater Eng 1:89–103. https://www.ncbi.nlm.nih.gov/pubmed/11352116

Kim H-N, Lee J, Kim H-Y, Kim Y-R (2009) Enzymatic synthesis of a drug delivery system based on polyhydroxyalkanoate-proteinblock copolymers. Chem Commun (46):7104–7106. https://doi.org/10.1039/b912871a

Kose GT, Korkusuz F, Korkusuz P, Purali N, Ozkul A, Hasirci V (2003) Bone generation on PHBV matrices: an in vitro study. Biomaterials 24:4999–5007. https://doi.org/10.1016/S0142-9612(03)00417-4

Kumar T, Singh M, Purohit HJ, Kalia VC (2009) Potential of *Bacillus* sp. to produce polyhydroxybutyrate from biowaste. J Appl Microbiol 106:2017–2023. https://doi.org/10.1111/j.1365-2672.2009.04160.x

Kumar P, Patel SK, Lee JK, Kalia VC (2013) Extending the limits of *Bacillus* for novel biotechnological applications. Biotechnol Adv 31:1543–1561. https://doi.org/10.1016/j.biotechadv.2013.08.007

Kumar P, Singh M, Mehariya S, Patel SK, Lee JK, Kalia VC (2014) Ecobiotechnological approach for exploiting the abilities of *Bacillus* to produce co-polymer of polyhydroxyalkanoate. Indian J Microbiol 54:151–157. https://doi.org/10.1007/s12088-014-0457-9

Kumar P, Ray S, Patel SK, Lee JK, Kalia VC (2015) Bioconversion of crude glycerol to polyhydroxyalkanoate by *Bacillus thuringiensis* under non-limiting nitrogen conditions. Int J Biol Macromol 78:9–16. https://doi.org/10.1016/j.ijbiomac.2015.03.046

Kumar P, Ray S, Kalia VC (2016) Production of co-polymers of polyhydroxyalkanoates by regulating the hydrolysis of biowastes. Bioresour Technol 200:413–419. https://doi.org/10.1016/j.biortech.2015.10.045

Kuppan P, Vasanthan KS, Sundaramurthi D, Krishnan UM, Sethuraman S (2011) Development of poly(3-hydroxybutyrate-co-3-hydroxyvalerate) fibers for skin tissue engineering: effects of topography, mechanical, and chemical stimuli. Biomacromolecules 12:3156–3165. https://doi.org/10.1021/bm200618w

Lee S-H, Shin H (2007) Matrices and scaffolds for delivery of bioactive molecules in bone and cartilage tissue engineering. Adv Drug Deliv Rev 59:339–359. https://doi.org/10.1016/j.addr.2007.03.016

Lee Y-F, Sridewi N, Ramanathan S, Sudesh K (2017) The Influence of electrospinning parameters and drug loading on polyhydroxyalkanoate (PHA) nanofibers for drug delivery. Intl J Biotechnol Wellness Ind 4:103–113. http://www.lifescienceglobal.com/pms/index.php/ijbwi/article/view/3560

Li H, Chang J (2005) Preparation, characterization and in vitro release of gentamicin from PHBV/wollastonite composite microspheres. J Control Release 107:463–473. https://doi.org/10.1016/j.jconrel.2005.05.019

Li Z, Loh XJ (2017) Recent advances of using polyhydroxyalkanoate-based nanovehicles as therapeutic delivery carriers. Wiley Interdiscip Rev Nanomed Nanobiotechnol 9:e1429. https://doi.org/10.1002/wnan.1429

Li J, Li X, Ni X, Leong KW (2003a) Synthesis and characterization of new biodegradable amphiphilic poly(ethylene oxide)-b-poly[(R)-3-hydroxy butyrate]-b-poly(ethylene oxide) triblock copolymers. Macromolecules 36:2661–2667. https://doi.org/10.1021/ma025725x

Li X, Li J, Leong KW (2003b) Preparation and characterization of inclusion complexes of biodegradable amphiphilic Poly(ethylene oxide)–Poly[(R)-3-hydroxybutyrate]–Poly(ethylene oxide) triblock copolymers with cyclodextrins. Macromolecules 36:1209–1214. https://doi.org/10.1021/ma0213347

Li J, Ni X, Li X, Tan NK, Lim CT, Ramakrishna S, Leong KW (2005) Micellization phenomena of biodegradable amphiphilic triblock copolymers consisting of poly(beta-hydroxyalkanoic acid) and poly(ethylene oxide). Langmuir 2:8681–8685. https://doi.org/10.1021/la0515266

Lobler M, Sternberg K, Stachs O, Allemann R, Grabow N, Roock A, Kreiner CF, Streufert D, Neffe AT, Hanh BD, Lendlein A, Schmitz KP, Guthoff R (2011) Polymers and drugs suitable for the development of a drug delivery drainage system in glaucoma surgery. J Biomed Mater Res B Appl Biomater 97:388–395. https://doi.org/10.1002/jbm.b.31826

Loh XJ, Goh SH, Li J (2007) New biodegradable thermogelling copolymers having very low gelation concentrations. Biomacromolecules 8(2):585–593. https://doi.org/10.1021/bm0607933

Lu X-Y, Zhang Y, Wang L (2010) Preparation and in vitro drug-release behavior of 5-fluorouracil-loaded poly(hydroxybutyrate-co-hydroxyhexanoate) nanoparticles and microparticles. J Appl Polym Sci 116:2944–2950. https://doi.org/10.1002/app.31806

Lu XY, Ciraolo E, Stefenia R, Chen GQ, Zhang Y, Hirsch E (2011) Sustained release of PI3K inhibitor from PHA nanoparticles and in vitro growth inhibition of cancer cell lines. Appl Microbiol Biotechnol 89:1423–1433. https://doi.org/10.1007/s00253-011-3101-1

Lu X-Y, Li M-C, Zhu X-L, Fan F, Wang L-L, Ma J-G (2014) Microbial synthesized biodegradable PHBHHxPEG hybrid copolymer as an efficient intracellular delivery nanocarrier for kinase inhibitor. BMC Biotechnol 14:4–4. https://doi.org/10.1186/1472-6750-14-4

Mansuri S, Kesharwani P, Jain K, Tekade RK, Jain NK (2016) Mucoadhesion: a promising approach in drug delivery system. React Funct Polym 100:151–172. https://doi.org/10.1016/j.reactfunctpolym.2016.01.011

Masaeli E, Wieringa PA, Morshed M, Nasr-Esfahani MH, Sadri S, van Blitterswijk CA, Moroni L (2014) Peptide functionalized polyhydroxyalkanoate nanofibrous scaffolds enhance Schwann cells activity. Nanomed Nanotechnol Biol Med 10:1559–1569. https://doi.org/10.1016/j.nano.2014.04.008

Masood F (2016) Polymeric nanoparticles for targeted drug delivery system for cancer therapy. Mater Sci Eng C 60:569–578. https://doi.org/10.1016/j.msec.2015.11.067

Masood F, Chen P, Yasin T, Fatima N, Hasan F, Hameed A (2013) Encapsulation of Ellipticine in poly-(3-hydroxybutyrate-co-3-hydroxyvalerate) based nanoparticles and its in vitro application. Mater Sci Eng C Mater Biol Appl 33:1054–1060. https://doi.org/10.1016/j.msec.2012.11.025

Masood F, Yasin T, Hameed A (2015) Polyhydroxyalkanoates – what are the uses? Current challenges and perspectives. Crit Rev Biotechnol 35:514–521. https://doi.org/10.3109/07388551.2014.913548

Mezzina MP, Pettinari MJ (2016) Phasins, multifaceted polyhydroxyalkanoate granule-associated proteins. Appl Environ Microbiol 82:5060–5067. https://doi.org/10.1128/AEM.01161-16

Michalak M, Kurcok P, Hakkarainen M (2017) Polyhydroxyalkanoate-based drug delivery systems. Polym Int 66:617–622. https://doi.org/10.1002/pi.5282

Mouriño V, Boccaccini AR (2009) Bone tissue engineering therapeutics: controlled drug delivery in three-dimensional scaffolds. J R Soc Interface 7(43):209–227. https://doi.org/10.1098/rsif.2009.0379

Naveen SV, Tan IKP, Goh YS, Balaji Raghavendran HR, Murali MR, Kamarul T (2015) Unmodified medium chain length polyhydroxyalkanoate (uMCL-PHA) as a thin film for tissue engineering application – characterization and in vitro biocompatibility. Mater Lett 141:55–58. https://doi.org/10.1016/j.matlet.2014.10.144

Oberoi P, Jabulowsky RA, Bähr-Mahmud H, Wels WS (2013) EGFR-targeted granzyme B expressed in NK cells enhances natural cytotoxicity and mediates specific killing of tumor cells. PLoS One 8:e61267. https://doi.org/10.1371/journal.pone.0061267

Paik H-j, Kim Y-R, Orth RN, Ober CK, Coates GW, Batt CA (2005) End-functionalization of poly(3-hydroxybutyrate) via genetic engineering for solid surface modification. Chem Commun 5(15):1956–1958. https://doi.org/10.1039/b415809a

Panith N, Assavanig A, Lertsiri S, Bergkvist M, Surarit R, Niamsiri N (2016) Development of tunable biodegradable polyhydroxyalkanoates microspheres for controlled delivery of tetracycline for treating periodontal disease. J Appl Polym Sci 133:44128. https://doi.org/10.1002/app.44128

Patel SK, Singh M, Kalia VC (2011) Hydrogen and polyhydroxybutyrate producing abilities of *Bacillus* spp. from glucose in two stage system. Indian J Microbiol 51:418–423. https://doi.org/10.1007/s12088-011-0236-9

Patel SKS, Singh M, Kumar P, Purohit HJ, Kalia VC (2012) Exploitation of defined bacterial cultures for production of hydrogen and polyhydroxybutyrate from pea-shells. Biomass Bioenergy 36:218–225. https://doi.org/10.1016/j.biombioe.2011.10.027

Patel SK, Kumar P, Singh M, Lee JK, Kalia VC (2015) Integrative approach to produce hydrogen and polyhydroxybutyrate from biowaste using defined bacterial cultures. Bioresour Technol 176:136–141. https://doi.org/10.1016/j.biortech.2014.11.029

Patel SK, Lee JK, Kalia VC (2016) Integrative approach for producing hydrogen and polyhydroxyalkanoate from mixed wastes of biological origin. Indian J Microbiol 56:293–300. https://doi.org/10.1007/s12088-016-0595-3

Porwal S, Kumar T, Lal S, Rani A, Kumar S, Cheema S, Purohit HJ, Sharma R, Singh Patel SK, Kalia VC (2008) Hydrogen and polyhydroxybutyrate producing abilities of microbes from diverse habitats by dark fermentative process. Bioresour Technol 99:5444–5451. https://doi.org/10.1016/j.biortech.2007.11.011

Pötter M, Steinbüchel A (2005) Poly(3-hydroxybutyrate) granule-associated proteins: impacts on poly(3-hydroxybutyrate) synthesis and degradation. Biomacromolecules 6:552–560. https://doi.org/10.1021/bm049401n

Pouton CW, Akhtar S (1996) Biosynthetic polyhydroxyalkanoates and their potential in drug delivery. Adv Drug Deliv Rev 18:133–162. https://doi.org/10.1016/0169-409X(95)00092-L

Qu X-H, Wu Q, Zhang K-Y, Chen GQ (2006) In vivo studies of poly(3-hydroxybutyrate-co-3-hydroxyhexanoate) based polymers: biodegradation and tissue reactions. Biomaterials 27:3540–3548. https://doi.org/10.1016/j.biomaterials.2006.02.015

Ray S, Kalia VC (2017a) Biomedical applications of polyhydroxyalkanoates. Indian J Microbiol 57:261–269. https://doi.org/10.1007/s12088-017-0651-7

Ray S, Kalia VC (2017b) Co-metabolism of substrates by *Bacillus thuringiensis* regulates polyhydroxyalkanoate co-polymer composition. Bioresour Technol 224:743–747. https://doi.org/10.1016/j.biortech.2016.11.089

Ray S, Kalia VC (2017c) Microbial cometabolism and polyhydroxyalkanoate co-polymers. Indian J Microbiol 57:39–47. https://doi.org/10.1007/s12088-016-0622-4

Ray S, Kalia VC (2017d) Polyhydroxyalkanoate production and degradation patterns in *Bacillus* Species. Indian J Microbiol 57:387–392. https://doi.org/10.1007/s12088-017-0676-y

Ray S, Sharma R, Kalia VC (2018) Co-utilization of crude glycerol and biowastes for producing polyhydroxyalkanoates. Indian J Microbiol 58:33–38. https://doi.org/10.1007/s12088-017-0702-0

Reddy CSK, Ghai R, Rashmi KVC (2003) Polyhydroxyalkanoates: an overview. Bioresour Technol 87:137–146. https://doi.org/10.1016/S0960-8524(02)00212-2

Rossi S, Azghani AO, Omri A (2004) Antimicrobial efficacy of a new antibiotic-loaded poly(hydroxybutyric-co-hydroxyvaleric acid) controlled release system. J Antimicrob Chemother 54:1013–1018. https://www.ncbi.nlm.nih.gov/pubmed/15537698

Sathiyanarayanan G, Bhatia SK, Song H-S, Jeon J-M, Kim J, Lee YK, Kim Y-G, Yang Y-H (2017) Production and characterization of medium-chain-length polyhydroxyalkanoate copolymer from Arctic psychrotrophic bacterium *Pseudomonas* sp. PAMC 28620. Int J Biol Macromol 97:710–720. https://doi.org/10.1016/j.ijbiomac.2017.01.053

Sendil D, Gürsel I, Wise DL, Hasırcı V (1999) Antibiotic release from biodegradable PHBV microparticles. J Control Release 59:207–217. https://doi.org/10.1016/S0168-3659(98)00195-3

Seo H-M, Kim J-H, Jeon J-M, Song H-S, Bhatia SK, Sathiyanarayanan G, Park K, Kim KJ, Lee SH, Kim HJ, Yang Y-H (2016) In situ immobilization of lysine decarboxylase on a biopolymer by fusion with phasin. Process Biochem 51:1413–1419. https://doi.org/10.1016/j.procbio.2016.07.019

Shah M, Ullah N, Choi MH, Kim MO, Yoon SC (2012) Amorphous amphiphilic P(3HV-co-4HB)-b-mPEG block copolymer synthesized from bacterial copolyester via melt transesterification: Nanoparticle preparation, cisplatin-loading for cancer therapy and in vitro evaluation. Eur J Pharm Biopharm 80:518–527. https://doi.org/10.1016/j.ejpb.2011.11.014

Shah M, Ullah N, Choi MH, Yoon SC (2014) Nanoscale poly (4-hydroxybutyrate)-mPEG carriers for anticancer drugs delivery. J Nanosci Nanotechnol 14(11):8416–8421. https://www.ncbi.nlm.nih.gov/pubmed/25958538

Sharma J, Lizu M, Stewart M, Zygula K, Lu Y, Chauhan R, Yan X, Guo Z, Wujcik EK, Wei S (2015) Multifunctional nanofibers towards active biomedical therapeutics. Polymers 7:186–219. https://doi.org/10.3390/polym7020186

Shishatskaya EI, Goreva AV, Voinova ON, Inzhevatkin EV, Khlebopros RG, Volova TG (2008) Evaluation of antitumor activity of rubomycin deposited in absorbable polymeric microparticles. Bull Exp Biol Med 145:358–361. https://www.ncbi.nlm.nih.gov/pubmed/19039944

Sichwart S, Hetzler S, Broker D, Steinbuchel A (2011) Extension of the substrate utilization range of *Ralstonia eutropha* strain H16 by metabolic engineering to include mannose and glucose. Appl Environ Microbiol 77:1325–1334. https://doi.org/10.1128/AEM.01977-10

Singh M, Patel SK, Kalia VC (2009) *Bacillus subtilis* as potential producer for polyhydroxyalkanoates. Microb Cell Factories 8:38. https://doi.org/10.1186/1475-2859-8-38

Singh M, Kumar P, Patel SK, Kalia VC (2013) Production of polyhydroxyalkanoate co-polymer by *Bacillus thuringiensis*. Indian J Microbiol 53:77–83. https://doi.org/10.1007/s12088-012-0294-7

Singh M, Kumar P, Ray S, Kalia VC (2015) Challenges and opportunities for customizing polyhydroxyalkanoates. Indian J Microbiol 55:235–249. https://doi.org/10.1007/s12088-015-0528-6

Srirangan K, Liu X, Tran TT, Charles TC, Moo-Young M, Chou CP (2016) Engineering of Escherichia coli for direct and modulated biosynthesis of poly (3-hydroxybutyrate-co-3-hydroxyvalerate) copolymer using unrelated carbon sources. Sci Rep 6:36470. https://doi.org/10.1038/srep36470

Sun J, Dai Z, Zhao Y, Chen GQ (2007) In vitro effect of oligo-hydroxyalkanoates on the growth of mouse fibroblast cell line L929. Biomaterials 28:3896–3903. https://doi.org/10.1016/j.biomaterials.2007.05.011

Surana AS, Kotecha RK (2010) An overview on various approaches to oral controlled drug delivery system via gastroretention. Int J Pharm Sci Rev Res 2:68–72. http://www.globalresearchonline.net/journalcontents/volume2issue2/Article%20014.pdf

Tao L, Uhrich KE (2006) Novel amphiphilic macromolecules and their in vitro characterization as stabilized micellar drug delivery systems. J Colloid Interface Sci 298:102–110. https://doi.org/10.1016/j.jcis.2005.12.018

Tiwari G, Tiwari R, Sriwastawa B, Bhati L, Pandey S, Pandey P, Bannerjee SK (2012) Drug delivery systems: an updated review. Int J Pharm Investig 2:2–11. https://doi.org/10.4103/2230-973x.96920

Verlinden RA, Hill DJ, Kenward M, Williams CD, Radecka I (2007) Bacterial synthesis of biodegradable polyhydroxyalkanoates. J Appl Microbiol 102(6):1437–1449. https://doi.org/10.1111/j.1365-2672.2007.03335.x

Vilos C, Morales FA, Solar PA, Herrera NS, Gonzalez-Nilo FD, Aguayo DA, Mendoza HL, Comer J, Bravo ML, Gonzalez PA, Kato S, Cuello MA, Alonso C, Bravo EJ, Bustamante EI, Owen GI, Velasquez LA (2013) Paclitaxel-PHBV nanoparticles and their toxicity to endometrial and primary ovarian cancer cells. Biomaterials 34:4098–4108. https://doi.org/10.1016/j.biomaterials.2013.02.034

Volodina E, Raberg M, Steinbüchel A (2016) Engineering the heterotrophic carbon sources utilization range of *Ralstonia eutropha* H16 for applications in biotechnology. Crit Rev Biotechnol 36:978–991. https://doi.org/10.3109/07388551.2015.1079698

Wang Z, Itoh Y, Hosaka Y, Kobayashi I, Nakano Y, Maeda I, Umeda F, Yamakawa J, Kawase M, Yag K (2003) Novel transdermal drug delivery system with polyhydroxyalkanoate and starburst polyamidoamine dendrimer. J Biosci Bioeng 95:541–543. https://doi.org/10.1016/S1389-1723(03)80059-2

Wang Y-W, Wu Q, Chen J, Chen G-Q (2005) Evaluation of three-dimensional scaffolds made of blends of hydroxyapatite and poly(3-hydroxybutyrate-co-3-hydroxyhexanoate) for bone reconstruction. Biomaterials 26:899–904. https://doi.org/10.1016/j.biomaterials.2004.03.035

Wu H, Fan Z, Jiang X, Chen J, Chen GQ (2016) Enhanced production of polyhydroxybutyrate by multiple dividing *E. coli*. Microb Cell Factories 15:016–0531. https://doi.org/10.1186/s12934-016-0531-6

Xiong YC, Yao YC, Zhan XY, Chen GQ (2010) Application of polyhydroxyalkanoates nanoparticles as intracellular sustained drug-release vectors. J Biomater Sci Polym Ed 21:127–140. https://doi.org/10.1163/156856209X410283

Yagmurlu MF, Korkusuz F, Gursel I, Korkusuz P, Ors U, Hasirci V (1999) Sulbactam-cefoperazone polyhydroxybutyrate-co-hydroxyvalerate (PHBV) local antibiotic delivery system: in vivo effectiveness and biocompatibility in the treatment of implant-related experimental osteomyelitis. J Biomed Mater Res 46:494–503. https://www.ncbi.nlm.nih.gov/pubmed/10398010

Yang YH, Brigham CJ, Song E, Jeon JM, Rha CK, Sinskey AJ (2012) Biosynthesis of poly(3-hydroxybutyrate-co-3-hydroxyvalerate) containing a predominant amount of 3-hydroxyvalerate by engineered *Escherichia coli* expressing propionate-CoA transferase. J Appl Microbiol 113:815–823. https://doi.org/10.1111/j.1365-2672.2012.05391.x

Yang JE, Choi SY, Shin JH, Park SJ, Lee SY (2013) Microbial production of lactate-containing polyesters. Microb Biotechnol 6:621–636. https://doi.org/10.1111/1751-7915.12066

Yang JE, Choi YJ, Lee SJ, Kang K-H, Lee H, Oh YH, Lee SH, Park SJ, Lee SY (2014) Metabolic engineering of *Escherichia coli* for biosynthesis of poly (3-hydroxybutyrate-co-3-hydroxyvalerate) from glucose. Appl Microbiol Biotechnol 98:95–104. https://doi.org/10.1007/s00253-013-5285-z

Yao YC, Zhan XY, Zhang J, Zou XH, Wang ZH, Xiong YC, Chen J, Chen GQ (2008) A specific drug targeting system based on polyhydroxyalkanoate granule binding protein PhaP fused with targeted cell ligands. Biomaterials 29:4823–4830. https://doi.org/10.1016/j.biomaterials.2008.09.008

Yu PHF, Chan AHY, Ho KP (2014) Synthesis and use of polyhydroxyalkanoate (pha) nanocapsules as a protein carrier. Google Patents, https://www.google.com/patents/US20140205671

Zhou L, Chen Z, Chi W, Yang X, Wang W, Zhang B (2012) Mono-methoxy-poly (3-hydroxybu tyrate-co-4-hydroxybutyrate)-graft-hyper-branched polyethylenimine copolymers for siRNA delivery. Biomaterials 33:2334–2344. https://doi.org/10.1016/j.biomaterials.2011.11.060

Applications of Polyhydroxyalkanoates and Their Metabolites as Drug Carriers

3

Vipin Chandra Kalia, Subhasree Ray, Sanjay K. S. Patel, Mamtesh Singh, and Gajendra Pratap Singh

Abstract

Drug delivery systems based on biodegradable biomaterials have proved attractive to improve their efficacy. Biopolymers such as Polyhydroxyalkanoates (PHAs) have the potential to be used as materials for preparing nanoparticles as drug carriers. This review describes the recent developments in drug delivery systems based on PHAs derived from microbial fermentations. Various strategies for the synthesis of microspheres, encapsulations, nano-constructs, have been presented, with their potential applications in medical domain.

Keywords

Polyhydroxyalkanoates · Nanoparticles · Drug carriers · Bacteria

3.1 Introduction

Antipathogenic compounds, nanomaterial hybrids and drug delivery are rapidly developing areas in biotechnology and pharmacology field, which have made an outstanding progress in recent years (Lionzo et al. 2007; Alvarez-Lorenzo and

V. C. Kalia (✉) · S. K. S. Patel
Department of Chemical Engineering, Konkuk University, Seoul, Republic of Korea

S. Ray
Microbial Biotechnology and Genomics, CSIR – Institute of Genomics and Integrative Biology (IGIB), Delhi, India

Academy of Scientific & Innovative Research (AcSIR), New Delhi, India

M. Singh
Department of Zoology, Gargi College, University of Delhi, Delhi, India

G. P. Singh
Mathematical Sciences and Interdisciplinary Research Lab (MathSciIntR-Lab), School of Computational and Integrative Sciences, Jawaharlal Nehru University, New Delhi, India

V. C. Kalia (ed.), *Biotechnological Applications of Polyhydroxyalkanoates*,
https://doi.org/10.1007/978-981-13-3759-8_3

Concheiro 2013; Kamaly et al. 2016; Pramual et al. 2016; Tibbitt et al. 2016; Patel et al. 2016b, 2017a, b, 2018a, b, c, d; Kalia et al. 2019; Kumar et al. 2018a, b; Otari et al. 2018). To enhance the efficiency of the drugs, it has been suggested that they should be delivered in a well regulated and targeted manner (Xiong et al. 2010). This has obliged the search of a novel drug delivery system to ensure greater bioavailability of hydrophobic drugs. This can be achieved by encapsulating them within the biodegradable material-based nanoparticles (Masood et al. 2012). This gains further importance since the use of silicone polymers have been found to be dangerous to the extent of causing cancers (Shrivastav et al. 2013). Being biocompatible and biodegradable in nature, Polyhydroxyalkanoates (PHAs) have attracted various medical agencies for their potential clinical applications (Valappil et al. 2006). Several prokaryotic micro-organisms produce PHA as reserve material from pure sugars and biowaste materials (Porwal et al. 2008; Singh et al. 2009, 2013, 2015; Patel et al. 2011, 2012, 2015a, b, 2016a; Kumar et al. 2013, 2014, 2015; Ray and Kalia 2017). Various homo-polymers and co-polymers of PHAs produced by microbial fermentation have been recommended to be utilized as drug carriers and tissue engineering materials (Khang et al. 2001; Kose et al. 2003; Timbart et al. 2004; Yao et al. 2008; Vilos et al. 2012; Saranya and Radha 2014). Hence, they should be used as raw material for delivering hydrophobic drugs, which will be more effective (Nigmatullin et al. 2015) (Table 3.1).

3.2 Applications of PHA as Drug Carriers

The suitability of PHA as drug carrier stems from the observations that these biopolymers exist not only in microbes but in plants and animals as well. Another important feature which is imperatively required is the biocompatibility of the drug carriers, which is evident from the observations such as – the metabolic products of PHB and R-3-hydroxybutanoic acid are produced in liver by fatty acid oxidation. This ketone body produced by liver mitochondria acts as fuel for brain. 2-hydroxybutyric acid is a component of human blood. PHB complexed to calcium phosphate acts as an ion channel through cellular membranes (Reusch 1989; Zinn et al. 2001). PHAs have been successfully implemented in numerous biomedical fields including tissue engineering, drug delivery, scaffolds preparations. These PHAs based self assembled systems have been shown to be fully biocompatible.

3.3 Anticancer Carriers

P(3HB) produced by *Bacillus cereus* SPV have proved helpful in the synthesis of spherical polymeric nano-constructs (SPNs) for the release of anticancer drug molecule. SPNs are made up of hydrophobic polymeric core where the drug is entrapped. Docetaxel (DTXL) is anti-cancer drug molecule used for the treatment of various neoplastic diseases. P(3HB)-DTXL was prepared by emulsion evaporation method. The drug loading efficiency was found to be up to 17.4% and the encapsulation efficiency was 43.3%. These SPNs as characterized through SEM, exhibited a diameter of 185.8 nm and the *polydispersity index* (PDI) was 0.169. However, the

Table 3.1 Polyhydroxyalkanoates (PHAs) and their metabolites as drug carriers

Bio-products	Source	Particles	Method	Characteristics of nanoparticles (NPs) Size (nm)	PDI	DLE (%)	Zeta potential (mV)	Applications	References
P(3HB)	*Bacillus cereus* SPV	DTXL-loaded P(3HB)	Sonication -EDM	201	0.25	17.4, 43.3	7	Drug carrier for DTXL (Docetaxel) (anticancer drug)	Mascolo et al. (2016)
P(3HB)	*Hyphomicrobium zavarzinii*	MPs	SEM	7.5	–	18	–	Encapsulation of tramadol into PHB microspeheres (epidural analgesic treatment alternative in patients)	Salman et al. (2003)
P(3HB-3HV)	*Bacillus cereus* FB11	EPT loaded with PHBV-12 NPs	EDM-SEM	208	0.52	42.13	82	Drug carrier for elipticine (anticancer drug)	Masood et al. (2013a)
P(3HB-3HV):: 95:5; 89:11; 85:15	*Bacillus cereus* FA11, FC11	NPs	EDM-SEM	209, 184, 283	0.12, 0.15, 0.28	45.55 39.32–45.65	13.17, 13.41,11.53	Drug carrier for elipticine (anticancer drug)	Masood et al. (2013b)
								Encapsulation of Ellipticine (EPT) for cancer therapy	
								Increases the cytotoxic effect of EPT by increasing its bioavailability	
								A promising vehicle for the admistration of hydrophophobic drugs for treatment of cancer	

(continued)

Table 3.1 (continued)

Bio-products	Source	Particles	Method	Characteristics of nanoparticles (NPs)				Applications	References
				Size (nm)	PDI	DLE (%)	Zeta potential (mV)		
P(3HB-3HV)	Synthetic	NPs	Double-SEM	–	–	–	–	NPs loaded with quercetin (Polyphenolic flavonoid)	Vidal et al. (2016)
P(3HB-3HV)	Synthetic	NPs	Double EDM-SEM	261	–	39	6.0	Paclitaxel-loaded PHBV shows toxicity to endometrial and primary ovarian cancer cells	Vilos et al. (2013)
P(3HB-3HV)::90.2:9.8	Synthetic	NPs	SEM	116	2.0	–	–	Drug carrier for Dexamethasone acetate	Lionzo et al. (2007)
PHB	Synthetic	NPs	EDM	178	–	25.16	1.4	Delivery of a hydrophobic photosensitizer drug in photodynamic therapy (cancer treatment)	Pramual et al. (2016)
P(3HB-3HV):: 88:12; 80:20				191		46.64	1.2		
				196		44.33	1.5		
P(3HB-3HV)-mPEG;	Synthetic	NPs	SEM	112–162	0.22–0.38	29–57	18–27	Drug carrier for thymoquinone	Shah et al. (2010)
P(3HB-4HB)-mPEG									
P(3HB-3HV)::63:37	Hydrogen oxidizing microorgnisms	NPs	SEM	700–750	–	–	32	Drug carrier for Doxorubicin (DOX) [encapsulation of cytostatic drug] Effective against HeLa tumor cells	Murueva et al. (2013)
P(3HB-4HV)::84:16							29		

P(3HB); P(3HB-3HV); P(3HB-3HHx)	Synthetic	NPs	EDM	–	–	97	–	PI3K inhibitor (TGX221) loaded with PHA NPs to block proliferation of cancer cell lines	Lu et al. (2011)
P(3HB-3HHx)	Synthetic	NPs	SEM	180	0.13–0.16	35–53	20.2–28.8	Base matrix for the production of nanotherapeutic including antineoolastic agent Etoposide	Kilicay et al. (2011)
P(3HB-3HO)	*Sinorhizobium fredii*	NPs	SEM	200–300	0.24–0.33	27.4–29.6	–	Folic acid (FA) as targeting ligand, doxorubicin (DOX) (anticancer drug)	Zhang et al. (2010)
PHA-PhaP	*Pichia pastoris*; *E. coli*	NPs	–	–	–	–	–	Drug carrier for Rhodamine B isothiocyanate (RBITC)	Yao et al. (2008)
PhaP-RGD peptide	*Ralstonia eutropha* H16	NPs	–	70–600	–	–	–	Nanocarrier with cancer targeting capability	Kim et al. (2009)
3HB	Synthetic	Conjugate	Anionic oligomerization	–	–	–	–	Ibuprofen-oligo(3-hydroxybutyrate) conjugates (anticancer potential of the drug) to inhibit proliferation of HT-29 and HCT-116 colon cancer cells	Zawidlak-Węgrzyńska et al. (2010)

(continued)

Table 3.1 (continued)

Bio-products	Source	Particles	Method	Characteristics of nanoparticles (NPs)				Applications	References
				Size (nm)	PDI	DLE (%)	Zeta potential (mV)		
P(3HB-3HV)	Synthetic	Nano-capsules	Double-SEM	–	–	–	–	Entrapment of therapeutically active proteins for cancer therapy	Baran et al. (2002)
								L-asparaginase, catalase, glucose oxidase, BSA	
P(3HB-3HV) +P(3HB-4HB)	*Alacaligenes latus,*	PHA rods carrying 50% (w/w) sulperazone or Duocid	–	–	–	–	–	Implant-related osteomyelitis (IRO)	Gürsel et al. (2001)
	Ralstonia eutropha							Antibacterial property against *Staphylococus aureus*	

DLE Drug loading efficiency, *SEM* Solvent extraction/evaporation method, *EDM* Emulsification diffusion method, *SDM* Solvent displacement method, *mPEG* Polyethylene glycol, *PHA* Polyhydroxyalkanoate, *PHB* Polyhydroxybutyrate, *3HA* 3-hydroxyalkanoic acid, *3HB* 3-hydroxybutyric acid, *3HV* 3-hydroxyvaleric acid, *3HO* 3-hydroxyoctanoate, *3HD* 3-hydroxydecanoic acid, *4HB* 4-hydroxybutyric acid, *P(3HB-3HV)* Poly(3-hydroxybutyrate-co-3-hydroxyvalerate), *P(3HB-4HB-3HV)* Poly(3-hydroxybutyrate-co-4-hydroxybutyrate-co-3-hydroxyvalerate), *P(3HB-4HB)* Poly(3-hydroxybutyrate-co-4-hydroxybutyrate), *P(3HB-3HV-3HHx)* Poly(3-hydroxybutyrate-co-3-hydroxyvalerate-co-3-hydroxyhexanoate), *P(3HB-3HO)* Poly(3-hydroxybutyrate-co-3-hydroxyoctanoate), *P(3HB-3HV-DHB)* Poly(3-hydroxybutyrate-co-3-hydroxyvalerate-co-2,3-dihydroxybutyrate), *P(3HB-3HHx)* Poly(3-hydroxybutyrate-co-3-hydroxyhexanoate), *OA* Octanoic acid, *UA* Undecanoic acid, *NPs* Nanoparticles zeta potential of DTXL-P(3HB) was −54.7 mV. The hydrophobic nature of P(3HB) enables it to support the entrapment of larger drug molecules. The drug release profile of DTXL-P(3HB) was less than 60% of their cargo molecules at 96 h (Mascolo et al. 2016). Similarly, PHA co-polymers produced by *B. cereus* FA11, FC11 and FB11 have been exploited for novel therapeutic tool for cancer therapy. Ellipticine (EPT) is a natural plant alkaloid (5,11-dimethyl-6H-pyrido[4,3-b] carbazole), which was extracted from the leaves of the plant – *Ochrosia elliptica* and encapsulated in three nano-carriers made up of copolymers of PHAs: P(3HB-3HV)- 5 mol%, P(3HB-3HV)-11 mol%, and P(3HB-3HV)-15 mol% for drug delivery. These three PHBV based Nano-particles (NPs) can be produced by method involving emulsification and solvent evaporation. The NPs which were characterized for their physiochemical properties showed a Zeta potential value of −13.17 mV, −9.60 mV and−12.49 mV. The diameters of the EPT loaded NP were 184–198 nm and 208–283 nm (Masood et al. 2013a). It showed anticancer activity against brain and non-small-cell-lung tumors (Hagg et al. 2004). Due to its small size and hydrophobic nature, EPT interacts and accumulates on the cell DNA and its membrane surface (Masood et al. 2013a). The encapsulation efficiency of PHBV-S, −11 and−15 were in the range of 39–45%, which regulated the valerate content in PHBV NPs. It was responsible for controlling the drug encapsulation efficiency. Here PHBV-S showed higher drug encapsulation efficiency as compared to other two PHBV NPs. These three nano-carriers showed high biocompatibility without influencing the survival of cancerous cells (Masood et al. 2013a, b). Another factor, which proves helpful in high drug encapsulation efficiency is its partition coefficient in hydrophobic and aqueous (external) phases (Mittal et al. 2007; Kamaly et al. 2016). PHB microspheres prepared by solvent evaporation method proved effective in drug delivery. These microspheres could release antitumor drug rubomycin, which acted to inhibit proliferation of Ehrlich's carcinoma in mice (Shishatskaya et al. 2008). Novel conjugates formed by ibuprofen with oligo(3HB) synthesized via anionic polymerization inhibit proliferation of colorectal cancer cells such as HT-29 and HCT 116 (Zawidlak-Wegrzynska et al. 2010).

3.4 Antibacterial Carriers

Tramadol encapsulated in a PHB microspheres was used in epidural administration for pain relief (Salman et al. 2003). Chemical coupling of P(3HB-3HV) and P(3HB-4HB) to monomethoxy poly(ethylene glycol) (mPEG) via trans-esterification reactions were used as nanocarriers for drug delivery. The hydrophobic drug used for encapsulation was Thymoquinone (TQ), which is the necessary component of black seed essential oil, and show anti-inflammatory and anti-cancer activities. The average NP size was found to 112–162 nm with a Zeta potential in the range of −18 to −27 mV. The biocompatibility and cytotoxicity effect of these NPs were evaluated in prenatal rat NIH/3T3 fibroblast cell line and neuronal hippocampal cells. The NPs showed reduced cytotoxicity. So, this novel system is likely to provide a safe and nontoxic system for encapsulating hydrophobic drugs, which is regarded as promising vehicle for drug delivery (Shah et al. 2010, 2012; Murueva et al. 2013; Schneider-stock et al. 2014). Similarly, homo-polymers like P(4HB) complex with m-PEG was used for cancer therapy by encapsulating cisplastin. The encapsulation for drug cisplastin showed effect of suppression on cell growth and also enhanced apoptotic process (Shah et al. 2014). Cisplastin intercalates between two genomic DNA strands. P(3HV-4HB)-b-mPEG-diblock copolymer encapsulated with cisplastin proves to be an effective way for site-specific delivery of drug. Within the cytoplasm, the NP loaded with the drug released slowly, which resulted in a sustained therapeutic effect (Liu et al. 2007; Panyam and Labhasetwar 2004). This encapsulating drug results in activation and cleavage of apoptogenic proteins (Shah et al. 2012). PHB based nano-drug carriers, release drugs in a uniform manner in comparison to those administered ensuring reduced local irritation (Des Rieux et al. 2006).

The efficiency of Tetracycline deposited in PHA co-polymers matrix in the form of microspheres was effective against periodontal disease-related microorganisms. Encapsulation efficiency was in the range of 51–65% (Sendil et al. 1999). Composite microspheres prepared by mixing copolymers of PHA i.e., PHBV and bioactive wollastonite, were used to encapsulate gentamicin. The release of gentamicin was slower than that recorded with microspheres prepared from PHBV alone. The release was sustained for a period of 22 days compared to 8 days (Li and Chang 2005). This antibiotic based PHA implants were effectively applied in treating implant related osteomyclitis (rods) (Yagmurlu et al. 1999; Gürsel et al. 2001; Türesin et al. 2001). PHA rods prepared from PHBV and poly(3hydroxybutyrate-co-4-hydroxybutyrate) (P3/4HB) carrying Sulperazone or Duocid were implanted in tibia of rabbits carrying *Staphylococcus aureus* as infecting agent. No infection could be detected within 6 weeks of treatment (Gürsel et al. 2001). Similarly, controlled release of antibiotics was reported in other cases (Yagmurlu et al. 1999; Gürsel et al. 2001; Danhier et al. 2012; Pacheco et al. 2015; Li and Loh 2017).

Paclitaxel-loaded P(3HB-3HV) proved toxic to endometrial and primary ovarian cancer cells (Vilos et al. 2013). PHA co-polymers contacting 3HD conjugated with D-peptide (DP18) improved the anticancer activities of the peptide and anticancer hydrophobic drugs (O'Connor et al. 2013). Several epidemiological studies reported that the use of non-steroid anti-inflammatory drugs causes reduction in breast cancer and colon cancer (Brasky et al. 2016). Ibuprofen exhibits anti-inflammatory, analgesic and antipyretic activity (Lou and Zhu 2016). 3HB monomers have proved effective in enhancing calcium levels within the cells by protecting the mitochondria. This could be a suitable drug for nervous system and bone diseases. The mechanism revealed that arrangements of hydroxyl monomers are more important for the antiproliferation effect of a PHA (Shrivastav et al. 2013).

3.5 Wound Dressers

Generally, dressings are used in the treatment of various wounds. But their non-resorbant properties make them unfavourable for dressings. However, resorbable wound dressings have proved effective in wound healing (Sombatmankhong et al.

2006; Li et al. 2008; Francesko and Tzanov 2010). 3HB/4HB monomers have proved helpful in skin repair. These monomers have been developed as polymer films and electrospun membranes to treat inflammation. It also enhances angiogenic properties of skin and accelerates wound healing process (Shishatskaya et al. 2016). Similarly, anti-inflammatory drugs such as diclofenac and dexamethasone encapsulated within PHA fasten the wound healing process and other skin defects (Murueva et al. 2014). P(3HB-3HHx) produced by microbial fermentation was used as scaffolds for tarsal repair in eyelid reconstruction in rat (Zhou et al. 2010).

3.6 Conclusion

PHAs are one of the most promising carriers in drug delivery. With favourable size and properties, they can cross easily several physiological barriers to reach disease affected area with the effective drug. The tailored biodegradability biopolymers have improved bioavailability, encapsulation efficiency and compatibility of a wide range of drugs. PHA and its co-polymers produced by microbial fermentation are suitable candidate for drug carrier. In this study, we summarized the most recent progress of using PHA based drug carrier. The significance of using these drug carriers in the cancer therapy, antibacterial properties and wound dresser is also discussed. Different monomeric compositions of PHA have made it possible to develop a wide range of drug carrier system.

Acknowledgement This work was supported by Brain Pool grant (NRF-2018H1D3A2001746) by National Research Foundation of Korea (NRF) to work at Konkuk University.

References

Alvarez-Lorenzo C, Concheiro A (2013) From drug dosage forms to intelligent drug-delivery systems: a change of paradigm. In: Alvarez-Lorenzo C, Concheiro A (eds) Smart materials for drug delivery. RSC Smart Mater Drug Deliv 1:1–32. https://doi.org/10.1039/9781849736800-00001

Baran ET, Ozer N, Hasirci V (2002) Poly (hydroxybutyrate-co-hydroxyvalerate) nanocapsules as enzyme carriers for cancer therapy: an in vitro study. J Microencapsul 19:363–376. https://doi.org/10.1080/02652040110105355

Brasky TM, Cohn DE, Bernardo BM (2016) Aspirin and endometrial cancer risk. Gynecol Oncol Rep 17:1–2. https://doi.org/10.1016/j.gore.2016.04.007

Danhier F, Ansorena E, Silva JM, Coco R, Le Breton A, Préat V (2012) PLGA-based nanoparticles: an overview of biomedical applications. J Control Release 161:505–522. https://doi.org/10.1016/j.jconrel.2012.01.043

Des Rieux A, Fievez V, Garinot M, Schneider YJ, Preat V (2006) Nanoparticles as potential oral delivery systems of proteins and vaccines: a mechanistic approach. J Control Release 116:1–27. https://doi.org/10.1016/j.jconrel.2006.08.013

Francesko A, Tzanov T (2010) Chitin, chitosan and derivatives for wound healing and tissue engineering. In: Nyanhongo G, Steiner W, Gubitz GM (eds) Biofunctionalization of polymers and their applications. Adv Biochem Eng Biotechnol. Springer, Berlin. 125:1–27. https://doi.org/10.10007/10_2010_93

Gürsel I, Korkusuz F, Türesin F, Alaeddinoğlu NG, Hasırcı V (2001) In vivo application of biodegradable controlled antibiotic release systems for the treatment of implant-related osteomyelitis. Biomaterials 22:73–80. https://doi.org/10.1016/S0142-9612(00)00170-8

Hägg M, Berndtsson M, Mandic A, Zhou R, Shoshan MC, Linder S (2004) Induction of endoplasmic reticulum stress by ellipticine plant alkaloids. Mol Cancer Ther 3:489–497

Kalia VC, Patel SKS, Kang YC, Lee JK (2019) Quorum sensing inhibitors as antipathogens: biotechnological applications. Biotechnol. Adv. 37:68–90. https://doi.org/10.1016/j.biotechad.2018.11.006

Kamaly N, Yameen B, Wu J, Farokhzad OC (2016) Degradable controlled-release polymers and polymeric nanoparticles: mechanisms of controlling drug release. Chem Rev 116:2602–2663. https://doi.org/10.1021/acs.chemrev.5b00346

Khang G, Kim SW, Cho JC, Rhee JM, Yoon SC, Lee HB (2001) Preparation and characterization of poly (3-hydroxybutyrate-co-3-hydroxyvalerate) microspheres for the sustained release of 5-fluorouracil. Bio-med Mater Eng 11:89–103

Kılıçay E, Demirbilek M, Türk M, Güven E, Hazer B, Denkbas EB (2011) Preparation and characterization of poly (3-hydroxybutyrate-co-3-hydroxyhexanoate) (PHBHHX) based nanoparticles for targeted cancer therapy. Eur J Pharm Sci 44:310–320. https://doi.org/10.1016/j.ejps.2011.08.013

Kim HN, Lee J, Kim HY, Kim YR (2009) Enzymatic synthesis of a drug delivery system based on polyhydroxyalkanoate-protein block copolymers. Chem Commun (Camb) 14:7104–7106. https://doi.org/10.1039/b912871a

Köse GT, Ber S, Korkusuz F, Hasirci V (2003) Poly (3-hydroxybutyric acid-co-3-hydroxyvaleric acid) based tissue engineering matrices. J Mater Sci Mater Med 14:121–126. https://doi.org/10.1023/A:1022063628099

Kumar P, Patel SKS, Lee JK, Kalia VC (2013) Extending the limits of *Bacillus* for novel biotechnological applications. Biotechnol Adv 31:1543–1561. https://doi.org/10.1016/j.biotechadv.2013.08.007

Kumar P, Singh M, Mehariya S, Patel SKS, Lee JK, Kalia VC (2014) Ecobiotechnological approach for exploiting the abilities of *Bacillus* to produce co-polymer of polyhydroxyalkanoate. Indian J Microbiol 54:151–157. https://doi.org/10.1007/s12088-014-0457-9

Kumar P, Ray S, Patel SKS, Lee JK, Kalia VC (2015) Bioconversion of crude glycerol to polyhydroxyalkanoate by *Bacillus thuringiensis* under non-limiting nitrogen conditions. Int J Biol Macromol 78:9–16. https://doi.org/10.1016/j.ijbiomac.2015.03.046

Kumar A, Kim I-W, Patel SKS, Lee J-K (2018a) Synthesis of protein-inorganic nanohybrids with improved catalytic properties using co (PO). Indian J Microbiol 58:100–104. https://doi.org/10.1007/s12088-017-0700-2

Kumar A, Park GD, Patel SKS, Kondaveeti S, Otari S, Anwar MZ, Kalia VC, Singh Y, Kim SC, Cho BK, Sohn JH, Kim DR, Kang YC, Lee JK (2018b) SiO microparticles with carbon nanotube-derived mesopores as an efficient support for enzyme immobilization. Chem Eng J. https://doi.org/10.1016/j.cej.2018.11.052

Li H, Chang J (2005) Preparation, characterization and in vitro release of gentamicin from PHBV/wollastonite composite microspheres. J Control Release 107:463–473. https://doi.org/10.1016/j.jconrel.2005.05.019

Li Z, Loh XJ (2017) Recent advances of using polyhydroxyalkanoate-based nanovehicles as therapeutic delivery carriers. WIREs Nanomed Nanobiotechnol 9(3). https://doi.org/10.1002/wnan.1429

Li XT, Zhang Y, Chen GQ (2008) Nano fibrous polyhydroxyalkanoate matrices as cell growth supporting materials. Biomaterials 29:3720–3728. https://doi.org/10.1016/j.biomaterials.2008.06.004

Liu Y, Shipton MK, Ryan J, Kaufman ED, Franzen S, Feldheim DL (2007) Synthesis, stability, and cellular internalization of gold nanoparticles containing mixed peptide– poly (ethylene glycol) monolayers. Anal Chem 79:2221–2229. https://doi.org/10.1021/ac061578f

Lionzo MIZ, Re MI, Guterres SS, Pohlmann AR (2007) Microparticles prepared with poly(hydroxybutyrate-co-hydroxyvalerate) and poly(epsiloncaprolactone) blends to control

the release of a drug model. J Microencapsul 24:175–186. https://doi.org/10.1080/02652040701233556
Lou Y, Zhu J (2016) Carboxylic acid nonsteroidal anti-inflammatory drugs (NSAIDs). In: Lamberth C, Dinges J (eds) Bioactive carboxylic compound classes: pharmaceuticals and agrochemicals. Wiley-VCH Verlag GmbH & Co. KGaA, Weinhein. https://doi.org/10.1002/9783527693931.ch16
Lu XY, Ciraolo E, Stefenia R, Chen GQ, Zhang Y, Hirsch E (2011) Sustained release of PI3K inhibitor from PHA nanoparticles and in vitro growth inhibition of cancer cell lines. App Microbiol Biotechnol 89:1423–1433. https://doi.org/10.1007/s00253-011-3101-1
Mascolo DD, Basnett P, Palange AL, Francardi M, Roy I, Decuzzi P (2016) Tuning core hydrophobicity of spherical polymeric nanoconstructs for docetaxel delivery. Polym Int 65:741–746. https://doi.org/10.1002/pi.5072
Masood F, Chen P, Yasin T, Fatima N, Hasan F, Hameed A (2012) Encapsulation of Ellipticine in poly-(3-hydroxybutyrate-co-3-hydroxyvalerate) based nanoparticles and its in vitro application. Mater Sci Eng C 33:1054–1060. https://doi.org/10.1016/j.msec.2012.11.025
Masood F, Chen P, Yasin T, Fatima N, Hasan F, Hameed A (2013a) Encapsulation of ellipticine in poly-(3-hydroxybutyrate-co-3-hydroxyvalerate) based nanoparticles and its in vitro application. Mater Sci Eng 33:1054–1060. https://doi.org/10.1016/j.msec.2012.11.025
Masood F, Chen P, Yasin T, Hasan F, Ahmad B, Hameed A (2013b) Synthesis of poly-(3-hydroxybutyrate-co-12 mol% 3-hydroxyvalerate) by *Bacillus cereus* FB11: its characterization and application as a drug carrier. J Mater Sci Mater Med 24:1927–1937. https://doi.org/10.1007/s10856-013-4946-x
Mittal G, Sahana DK, Bhardwaj V, Kumar MNVR (2007) Estradiol loaded PLGA nanoparticles for oral administration: effect of polymer molecular weight and copolymer composition on release behavior in vitro and in vivo. J Control Release 119:77–85. https://doi.org/10.1016/j.jconrel.2007.01.016
Murueva AV, Shishatskaya EI, Kuzmina AM, Volova TG, Sinskey AJ (2013) Microparticles prepared from biodegradable polyhydroxyalkanoates as matrix for encapsulation of cytostatic drug. J Mater Sci Mater Med 24:1905–1915. https://doi.org/10.1007/s10856-013-4941-2
Murueva AV, Shershneva AM, Shishatskaya EI, Volova TG (2014) The use of polymeric microcarriers loaded with anti-inflammatory substances in the therapy of experimental skin wounds. Bull Exp Biol Med 157:597–602. https://doi.org/10.1007/s10517-014-2624-8
Nigmatullin R, Thomas P, Lukasiewicz B, Puthussery H, Roy I (2015) Polyhydroxyalkanoates, a family of natural polymers, and their applications in drug delivery. J Chem Technol Biotechnol 90:1209–1221. https://doi.org/10.1002/jctb.4685
Otari SV, Patel SKS, Kim S-Y, Haw JR, Kalia VC, Kim I-W, Lee J-K (2018) Copper ferrite magnetic nanoparticles for the immobilization of enzyme. Indian J Microbiol. https://doi.org/10.1007/s12088-018-0768-3
O'Connor S, Szwej E, Nikodinovic-Runic J, O'Connor A, Byrne AT, Devocelle M, O'Donovan N, Gallagher WM, Babu R, Kenny STM, Zinn M, Zulian QR, O'Conner KE (2013) The anticancer activity of a cationic anti-microbial peptide derived from monomers of polyhydroxyalkanoate. Biomaterials 34:2710–2718. https://doi.org/10.1016/j.biomaterials.2012.12.032
Pacheco DP, Amaral MH, Reis RL, Marques AP, Correlo VM (2015) Development of an injectable PHBV microparticles-GG hydrogel hybrid system for regenerative medicine. Int J Pharm 478:398–408. https://doi.org/10.1016/j.ijpharm.2014.11.036
Panyam J, Labhasetwar V (2004) Sustained cytoplasmic delivery of drugs with intracellular receptors using biodegradable nanoparticles. MolPharm 1:77–84. https://doi.org/10.1021/mp034002c
Patel SKS, Singh M, Kalia VC (2011) Hydrogen and polyhydroxybutyrate producing abilities of *Bacillus* spp. from glucose in two stage system. Indian J Microbiol 51:418–423. https://doi.org/10.1007/s12088-011-0236-9
Patel SKS, Singh M, Kumar P, Purohit HJ, Kalia VC (2012) Exploitation of defined bacterial cultures for production of hydrogen and polyhydroxybutyrate from pea-shells. Biomass Bioenergy 36:218–225. https://doi.org/10.1016/j.biombioe.2011.10.027

Patel SKS, Kumar P, Singh S, Lee JK, Kalia VC (2015a) Integrative approach for hydrogen and polyhydroxybutyrate production. In: Kalia VC (ed) Microbial factories: waste treatment. Springer, New Delhi, pp 73–85. https://doi.org/10.1007/978-81-322-2598-0_5

Patel SKS, Kumar P, Singh S, Lee JK, Kalia VC (2015b) Integrative approach to produce hydrogen and polyhydroxybutyrate from biowaste using defined bacterial cultures. Bioresour Technol 176:136–141. https://doi.org/10.1016/j.biortech.2014.11.029

Patel SKS, Lee JK, Kalia VC (2016a) Integrative approach for producing hydrogen and polyhydroxyalkanoate from mixed wastes of biological origin. Indian J Microbiol 56:293–300. https://doi.org/10.1007/s12088-016-0595-3

Patel SKS, Choi SH, Kang YC, Lee J-K (2016b) Large-scale aerosol-assisted synthesis of biofriendly Fe O yolk-shell particles: a promising support for enzyme immobilization. Nanoscale 8:6728–6738. https://doi.org/10.1039/C6NR00346J

Patel SKS, Otari SV, Kang YC, Lee JK (2017a) Protein-inorganic hybrid system for efficient histagged enzymes immobilization and its application in L-xylulose production. RSC Adv 7:3488–3494. https://doi.org/10.1039/c6ra24404a

Patel SKS, Choi SH, Kang YC, Lee J-K (2017b) Eco-friendly composite of Fe O -reduced graphene oxide particles for efficient enzyme immobilization. ACS Appl Mater Interfaces 9:2213–2222. https://doi.org/10.1021/acsami.6b05165

Patel SKS, Anwar MZ, Kumar A, Otari SV, Pagolu RT, Kim S-Y, Kim I-W, Lee J-K (2018a) Fe O yolk-shel particle-based laccase biosensor for efficient detection of 2,6-dimethoxyphenol. Biochem Eng J 132:1–8. https://doi.org/10.1016/j.bej.2017.12.013

Patel SKS, Kim JH, Kalia VC, Lee JK (2018b) Antimicrobial ativity of amino-derivatized cationic polysaccharides. Indian J Microbiol. https://doi.org/10.1007/s12088-018-0764-7

Patel SKS, Lee JK, Kalia VC (2018c) Nanoparticles in biological hydrogen production: an overview. Indian J Microbiol 58:8–18. https://doi.org/10.1007/s12088-017-0678-9

Patel SKS, Otari SV, Li J, Kim DR, Kim SC, Cho B-K, Kalia VC, Kang YC, Lee J-K (2018d) Synthesis of cross-linked protein-metal hybrid nanoflowers and its application in repeated batch decolorization of synthetic dyes. J Harad Mater 347:442–450. https://doi.org/10.1016/j.jhazmat.2018.01.003

Pramual S, Assavanig A, Bergkvist M, Batt CA, Sunintaboon P, Lirdprapamongkol K, Svasti J, Niamsiri N (2016) Development and characterization of bio-derived polyhydroxyalkanoate nanoparticles as a delivery system for hydrophobic photodynamic therapy agents. J Mater Sci Mater Med 27:40. https://doi.org/10.1007/s10856-015-5655-4

Porwal S, Kumar T, Lal S, Rani A, Kumar S, Cheema S, Purohit HJ, Sharma R, Patel SKS, Kalia VC (2008) Hydrogen and polyhydroxybutyrate producing abilities of microbes from diverse habitats by dark fermentative process. Bioresour Technol 99:5444–5451. https://doi.org/10.1016/j.biortech.2007.11.011

Ray S, Kalia VC (2017) Co-metabolism of substrates by *Bacillus thuringiensis* regulates polyhydroxyalkanoate co-polymer composition. Bioresour Technol 224:743–747. https://doi.org/10.1016/j.biortech.2016.11.089

Reusch RN (1989) Poly-β-hydroxybutyrate/calcium polyphosphate complexes in eukaryotic membranes. Proc Soc Exptl Biol Med 191:377–381 doi:NA

Salman MA, Sahin A, Onur MA, Oge K, Kassab A, Aypar U (2003) Tramadol encapsulated into polyhydroxybutyrate microspheres: in vitro release and epidural analgetic effect in rats. Acta Anaesthesiol Scand 47:1006–1012. https://doi.org/10.1034/j.1399-6576.2003.00180.x

Saranya S, Radha KV (2014) Review of nanobiopolymers for controlled drug delivery. Polym Plast Technol Eng 53:1636–1646. https://doi.org/10.1080/03602559.2014.915035

Schneider-Stock R, Fakhoury IH, Zaki AM, El-Baba CO, Gali-Muhtasib HU (2014) Thymoquinone: fifty years of success in the battle against cancer models. Drug Discov Today 19:18–30. https://doi.org/10.1016/j.drudis.2013.08.021

Sendil D, Gürsel I, Wise DL, Hasirci V (1999) Antibiotic release from biodegradable PHBV microparticles. J Control Release 59:207–217. https://doi.org/10.1016/S0168-3659(98)00195-3

Shah M, Naseer MI, Choi MH, Kim MO, Yoon SC (2010) Amphiphilic PHA–mPEG copolymeric nanocontainers for drug delivery: preparation, characterization and in vitro evaluation. Int J Pharm 400:165–175. https://doi.org/10.1016/j.ijpharm.2010.08.008

Shah M, Ullah N, Choi MH, Kim MO, Yoon SC (2012) Amorphous amphiphilic P (3HV-co-4HB)-b-mPEG block copolymer synthesized from bacterial copolyester via melt transesterification: nanoparticle preparation, cisplatin-loading for cancer therapy and in vitro evaluation. Eur J Pharm Biopharm 80:518–527. https://doi.org/10.1016/j.ejpb.2011.11.014

Shah M, Ullah N, Choi MH, Yoon SC (2014) Nanoscale poly (4-hydroxybutyrate)-mPEG carriers for anticancer drugs delivery. J NanosciNanotechnol 14:8416–8421. https://doi.org/10.1166/jnn.2014.9924

Shishatskaya EI, Goreva AV, Voinova ON, Inzhevatkin EV, Khlebopros RG, Volova TG (2008) Evaluation of antitumor activity of rubomycin deposited in absorbable polymeric microparticles. Bull Exp Biol Med 145:358–361. https://doi.org/10.1007/s10517-008-0091-9

Shishatskaya EI, Nikolaeva ED, Vinogradova ON, Volova TG (2016) Experimental wound dressings of degradable PHA for skin defect repair. J Mater Sci Mater Med 27:165. https://doi.org/10.1007/s10856-016-5776-4

Shrivastav A, Kim HY, Kim YR (2013) Advances in the applications of polyhydroxyalkanoate nanoparticles for novel drug delivery system. Biomed Res Int 2013:1–12, Article ID 581684. https://doi.org/10.1155/2013/581684

Singh M, Patel SKS, Kalia VC (2009) *Bacillus subtilis* as potential producer for polyhydroxyalkanoates. Microb Cell Factories 8:38. https://doi.org/10.1186/1475-2859-8-38

Singh M, Kumar P, Patel SKS, Kalia VC (2013) Production of polyhydroxyalkanoate co-polymer by *Bacillus thuringiensis*. Indian J Microbiol 53:77–83. https://doi.org/10.1007/s12088-012-0294-7

Singh M, Kumar P, Ray S, Kalia VC (2015) Challenges and opportunities for the customizing polyhydroxyalkanoates. Indian J Microbiol 55:235–249. https://doi.org/10.1007/s12088-015-0528-6

Sombatmankhong K, Suwantong O, Waleetorncheepsawat S, Supaphol P (2006) Electrospunfiber mats of poly (3-hydroxybutyrate), poly (3-hydroxybutyrate-co-3-hydroxyvalerate), and their blends. J Polym Sci B Polym Phys 44:2923–2933. https://doi.org/10.1002/polb.20915

Tibbitt MW, Dahlman JE, Langer R (2016) Emerging frontiers in drug delivery. J Am Chem Soc 138:704–717. https://doi.org/10.1021/jacs.5b09974

Timbart L, Renard E, Langlois V, Guerin P (2004) Novel biodegradable copolyesters containing blocks of poly (3-hydroxyoctanoate) and poly (ε-caprolactone): synthesis and characterization. Macromol Biosci 4:1014–1020. https://doi.org/10.1002/mabi.200400104

Türesin F, Gürsel I, Hasirci V (2001) Biodegradable polyhydroxyalkanoate implants for osteomyelitis therapy: in vitro antibiotic release. J Biomater Sci Polym Ed 12:195–207

Valappil SP, Misra SK, Boccaccini AR, Roy I (2006) Biomedical applications of polyhydroxyalkanoates, an overview of animal testing and in vivo responses. Exp Rev Med Dev 3:853–868. https://doi.org/10.1586/17434440.3.6.853

Vidal LS, Rojas C, Bouza Padín R, Pérez Rivera M, Haensgen A, González M, Rodríguez-Llamazares S (2016) Synthesis and characterization of polyhydroxybutyrate-co-hydroxyvalerate nanoparticles for encapsulation of quercetin. J Bioact Compat Polym 31:439–452. https://doi.org/10.1177/0883911516635839

Vilos C, Constandil L, Herrera N, Solar P, Escobar-Fica J, Velásquez LA (2012) Ceftiofur-loaded PHBV microparticles: a potential formulation for a long-acting antibiotic to treat animal infections. Electron J Biotechnol 15:1. https://doi.org/10.2225/vol15-issue4-fulltext-2

Vilos C, Morales FA, Solar PA, Herrera NS, Gonzalez-Nilo FD, Aguayo DA, Mendza HL, Comer J, Bravo ML, Gonzalez PA, Kato SP (2013) Paclitaxel-PHBV nanoparticles and their toxicity to endometrial and primary ovarian cancer cells. Biomaterials 34:4098–4108. https://doi.org/10.1016/j.biomaterials.2013.02.034

Xiong YC, Yao YC, Zhan XY, Chen GQ (2010) Application of polyhydroxyalkanoates nanoparticles as intracellular sustained drug-release vectors. J Biomater Sci Polym Ed 21:127–140. https://doi.org/10.1163/156856209X410283

Yagmurlu MF, Korkusuz F, Gürsel I, Korkusuz P, Örs Ü, Hasirci V (1999) Sulbactam-cefoperazone polyhydroxybutyrate-co-hydroxyvalerate (PHBV) local antibiotic delivery system: in vivo effectiveness and biocompatibility in the treatment of implant related experimental osteomyelitis. J Biomed Mater Res 46:494–503

Yao YC, Zhan XY, Zhang J, Zou XH, Wang ZH, Xiong YC, Chen J, Chen GQ (2008) A specific drug targeting system based on polyhydroxyalkanoate granule binding protein PhaP fused with targeted cell ligands. Biomaterials 29:4823–4830. https://doi.org/10.1016/j.biomaterials.2008.09.008

Zawidlak-Węgrzyńska B, Kawalec M, Bosek I, Łuczyk-Juzwa M, Adamus G, Rusin A, Filipczak P, Glowala-Kosinska M, Wolanska K, Krawczyk Z, Kurcok P (2010) Synthesis and antiproliferative properties of ibuprofen–oligo (3-hydroxybutyrate) conjugates. Eur J Med Chem 45:1833–1842. https://doi.org/10.1016/j.ejmech.2010.01.020

Zhang C, Zhao L, Dong Y, Zhang X, Lin J, Chen Z (2010) Folate-mediated poly (3-hydroxybuty rate-co-3-hydroxyoctanoate) nanoparticles for targeting drug delivery. Eur J Pharm Biopharm 76:10–16. https://doi.org/10.1016/j.ejpb.2010.05.005

Zhou J, Peng SW, Wang YY, Zheng SB, Wang Y, Chen GQ (2010) The use of poly (3-hydroxyb utyrate-co-3-hydroxyhexanoate) scaffolds for tarsal repair in eyelid reconstruction in the rat. Biomaterials 31:7512–7518. https://doi.org/10.1016/j.biomaterials.2010.06.044

Zinn M, Witholt B, Egli T (2001) Occurrence, synthesis and medical application of bacterial polyhydroxyalkanoate. Adv Drug Deliv Rev 53:5–21. https://doi.org/10.1016/S0169-409X(01)00218-6

Polyhydroxyalkanoates Applications in Antimicrobial Agents Delivery and Wound Healing

4

Veronica S. Giourieva, Rigini M. Papi, and Anastasia A. Pantazaki

Abstract

Bacterial infections constitute a significant and increasing problem in healthcare originated from both bacterial colonization and biofilm development in biomaterials, a prelude to the onset of systemic infection and consequently the dysfunction of implanted devices. In addition, a serious issue is the growing incidence of multi-drug resistant bacteria, which cannot be killed using multiple antibiotics, coined by the term "superbugs". Enormous efforts have focused on developing drugs which, at one hand, exhibit as characteristics great effectiveness and at the other hand, make the evolutionary resistant bacteria susceptible to them. The development of novel, more potent antibiotics able to overcome this acquired resistance remains until now, a provocation for the scientific world because novel modes of antibacterial activity are necessary. Biopolymeric antimicrobials have appeared as a hopeful nominee for investigation and evolution in the field of antibacterial therapy. Poly([R]-3-hydroxyalkanoates) (PHAs), the wide family of the naturally-synthesizing bacterial bio-polyesters, constitute one of the principal nominees, which dominates in prospective for utilization in the field of biomedical applications. They constitute a large class of biodegradable biopolymers that exhibit biocompatibility obtaining minimal tissue toxicity. Exploiting these properties, PHAs has been used as a matrix to construct slow-release formulations of antibiotic delivery, providing them antimicrobial, antifungal, antibiofilm, anti-inflammatory and virucidal properties dependent on the conjugated/enclosed therapeutic agent. Moreover, the antimicrobial activity that PHAs themselves, or their derivatives have been reported to exhibit should not be ignored. Wound healing is a novel medical area where PHAs are also extended

V. S. Giourieva · R. M. Papi · A. A. Pantazaki (✉)
Laboratory of Biochemistry, Department of Chemistry, Aristotle University of Thessaloniki, Thessaloniki, Greece
e-mail: natasa@chem.auth.gr

V. C. Kalia (ed.), *Biotechnological Applications of Polyhydroxyalkanoates*,
https://doi.org/10.1007/978-981-13-3759-8_4

based on the same principle of the drug delivery for rapid tissue regeneration. This review covers all this collective information.

Keywords
Polyhydroxyalkanoates · Antibacterial agent · Biofilm · PHA derivatives · Wound healing · Drug delivery

4.1 Introduction

A variety of bacteria naturally produce polyhydroxyalkanoates (PHAs), which are hydrophobic polyesters stored as intracellular granules. PHAs biosynthesis mainly occurs under conditions of a nutrient limitation excluding carbon source, while recently renewable feedstocks or alternative carbon sources are used as growth media, such as molasses and used palm oil (Mozejko and Ciesielski 2013). Following the trend, genetically modified strains or mixed cultures are also applied for PHAs production (Johnson et al. 2009; Mifune et al. 2010; Castro-Mayorga et al. 2014, 2016a; Jeon et al. 2014; Tappel et al. 2014). Their physicochemical properties can be modified by cross-linking or by incorporating selected functional groups, like hydroxyl groups, carboxyl groups, polyethylene glycol, or other copolymers. More specifically, depending on their conjugates they present modified crystallinity, thermodynamics and flexibility (Chen and Wang 2002; Luef et al. 2015).

PHAs have diverse useful properties like biocompatibility, biodegradability, and ability to support cell proliferation and adhesion, profitable for a wide range of medical applications. Moreover, their ability to compose copolymers with organic or inorganic additives allows their exploitation in various domains like wound healing, prevention of agglomeration, drug delivery, tissue engineering and nerve injury repair (Luef et al. 2015).

Other applications of PHAs include food packaging, dye removing from wastewater, material coating, creation of waste bags, agricultural plastics (Luef et al. 2015; Li et al. 2016), and as antibacterial agents (Dinjaski et al. 2014).

It is noteworthy that bactericidal properties have been reported and attributed to some polymers since 1965 (Cornell and Donaruma 1965). The use of antimicrobial polymers attracts of a great interest, constituting a family of an alternative to the conventional bactericidal drugs. In the last 10 years, a remarkable increase occurred in the number of the polymers used as disinfectants approved by the Food and Drug Administration (FDA), demonstrating the compelling need for discovery of environmentally friendly disinfectants to be used as alternatives to antibiotics (Siedenbiedel and Tiller 2012). Although a number of medicinal applications currently use polymeric and biodegradable systems based on polylactic acid, polyglycolic acid as well as on their copolymers (Gomes and Reis 2004), PHAs. Thanks to their biodegradability and minimal tissue toxicity, they constitute dominant nominees for utilization in various prospective applications in the biomedical domain including this of antibacterial one (Chen and Wu 2005a). Monomer constitution of

PHAs and their chemical structure remarkably affect their mechanical characteristics and physical properties (Chen 2009). In combination with the current FDA approval given for P4HB [poly-(4-hydroxylbutyrate)] to be used in clinical trials, it is very hopeful the expectation for an excellent prospect of PHAs applications in the medical fields as antimicrobial agents (Wu et al. 2009).

Browsing in the literature showed that the plethora of PHAs applications as antimicrobial agents is based on their property to be capable of bio-functionalization thanks to which they are used rather for antimicrobial drugs delivery. Thus, there is scarce information for antibacterial activity of PHAs themselves. In this review PHAs' applications involved them as antimicrobial delivery systems or antimicrobial factors themselves are included.

4.2 Bacterial Infections

Many bacteria have the ability to establish protection against antimicrobial agents leading to difficulties in the remedy of bacterial infections. Therefore, a plethora of antimicrobial drugs of diverse families or groups demonstrated to present clinical interest. The mechanisms of action of these antibacterial agents are variable and it depends on the affected targets ending to inhibit bacterial growth or to provoke their death. Bacterial infections, which are provoked by pathogenic bacteria, are considered as one of the principal cause of universal morbidity and then mortality, mainly by *Staphylococcus aureus* (Kiedrowski and Horswill 2011). The strong pathogenic properties of *S. aureus* (DeLeo and Chambers 2009), in combination with its ability to provoke long-standing infections, principally persistent on medical implants or on host tissues colonization, are accepted as a serious healthcare threat (Darouiche 2004).

Bacterial infections, which are provoked by pathogenic bacteria, are considered as one of the principal cause of universal morbidity, which sometimes is ended in mortality, mainly by *S. aureus* (Tenover 2006). It is impressive the fact that year 2012 have been recorded worldwide 450,000 cases of multidrug-resistant tuberculosis, provoked by Gram-negative bacteria or methicillin-resistant *S. aureus* that prevented the effectiveness of the treatment of the disease (Fischbach and Walsh 2009; Donadio et al. 2010; Gwynn et al. 2010). Infections attributed to some bacteria such as a methicillin-resistant strain of *S. aureus* (MRSA), a vancomycin-resistant strain of *Enterococcus faecium* (VRE) and fluoroquinolone-resistant *Pseudomonas aeruginosa* are quickly growing in the United States health centers and hospitals. *S. aureus* is the dominant causative of both surgical site and periprosthetic joint infections, whereas the prevalence of methicillin-resistant *S. aureus* is increasing (Aggarwal et al. 2014; Gallo et al. 2014). Additionally even more terrifying are the circumstances of pan-antibiotic-resistant (pan-drug-resistant) infections, involving *Acinetobacter* species, multidrug-resistant (MDR) *P. aeruginosa* and carbapenem-resistant *Klebsiella* species (Souli et al. 2008; Arias and Murray 2009). Due to the rapid appearance of infectious diseases and simultaneously to the ability of bacteria to develop resistance against to conventional antimicrobial drugs, there is an exigent need to develop innovative antimicrobial therapies under the umbrella of biomedical applications (Donadio et al. 2010).

However, the investigation to discover innovative antibiotic classes has not followed the increasing threat originated from the bacterial resistance. Strategies for the synthesis of antibiotic candidates capable to act through nonconventional mechanisms on novel targets might possess a more considerable potential to confront and overcome bacterial resistance (Gwynn et al. 2010). Although antibiotic resistance comprising Gram-(+) and Gram-(−) pathogens, keep going to augment in hospitals and in the society worldwide, the total of lately approved agents has been reduced, since only a few new antibiotics have been approved. It is therefore understood why antibiotics should be utilized more cautiously and why there is the need to develop and use new antimicrobial agents.

Another factor that influences the bacterial resistance is their ability to form biofilms, which has been associated with their persistence to establish chronic infections/inflammations since antibiotics cannot easily penetrate into biofilms. The need to confront bacterial biofilms has prompted to the investigation of effective strategies for the development of surfaces possessing antimicrobial properties suitable to impede biofilm expansion. These surfaces might repel the microbial population preventing their attachment to them or inhibiting their growth completely until killing them (Siedenbiedel and Tiller 2012).

Inhibition of biofilm formation and simultaneously reducing local immune response are the two main eliminating factors for the usage of biomaterials in medicine. The constructed anti-infective surface must be sufficient to inhibit even low content of bacterial populations. Bacteria have a high survival strategy in general. They are able to adhere and populate all natural and synthetic surfaces (Costerton et al. 2003; Busscher and van der Mei 2012; Gallo et al. 2014), acquiring various virulent components of the cell-surface of bacteria that assist adhesion to other cells or to bio-surfaces, usually of the host they are infecting or living in. These adhesins allow free-floating bacteria to accumulate biomaterials. Consequently, the reversibly attached bacteria may express biofilm-specific genes (Laverty et al. 2013; Gallo et al. 2014). During the phase of adhesion, the expression and secretion of a protective slime occur, making bacteria extremely resistant to the antibiotic diffusion affecting host immune system (Costerton et al. 1999; Fux et al. 2005; Gallo et al. 2014). This transition phase is the final occasion for taking of clinically preventative measures.

4.3 PHAs as Antimicrobial Agents

The monomeric units of PHAs ([R]-3-hydroxylkanoates, R-HAs) are hydroxyl-substituted fatty acids and for more than 80 years, and have been considered as antibacterial agents due to their function as anionic surfactants. Upon the fatty acid adsorption by cells, this is followed by alterations in cell permeability that cause an inhibitory action. It is noteworthy that not all the R-HAs but only some of them show antimicrobial/or antiviral activities, a fact that seems curious. (Burke et al. 1989; Sandoval et al. 2005; Ruth et al. 2007). This has been attributed to the length of the carbon chain, which seems to possess a meaningful role in the antimicrobial potency (Desbois and Smith 2010; Huang et al. 2011; Fischer et al. 2012; Desbois and Lawlor 2013).

One of the most powerful applications of the PHAs monomers (R-HAs), due to the presence of a chiral center and of two simply modified functional groups (–OH and –COOH), is their utilization as precious synthons for the synthesis of antibiotics such as beta-lactams (Ohashi and Hasegawa 1992a) except drugs, peptides, vitamins, biofuels etc. (Chiba and Nakai 1985; Seebach et al. 2001). When starvation conditions occur for bacteria, PHAs can be catabolized to generate monomers, the R-HAs, which can be used more efficiently by the bacteria (Jendrossek and Handrick 2002; Ruth et al. 2007). The synthesis of antibiotics, flavors, vitamins, and pheromones can be based on R-HAs as starting materials (Seuring and Seebach 1978; Seidel and Seebach 1982; Sutter and Seebach 1983; Ohashi and Hasegawa 1992b; Lee et al. 2002; Ruth et al. 2007). R-HAs can be produced both enzymatically and chemically by depolymerizing biodegradable PHAs and tailored with functional groups. Degradation of PHAs by PHA depolymerases results in intermediates such as 3-hydroxy alkanoic acids (3-HAs), which can be modified to synthesize antimicrobial agents (Gallo et al. 2014).

R-HAs have been shown to exhibit some antimicrobial, and maybe in the future insecticidal and/or compelling antiviral activities due to their relevance to short-chain fatty acids (SCFAs) (Chen and Wu 2005b; Shiraki et al. 2006; Ren et al. 2010). For example, (R)-3-hydroxy-*n*-phenylalkanoic acids can be used against *Listeria monocytogenes* (Sandoval et al. 2005). Antimicrobial activity of R-HAs produced from PHAs was tested on the growth of different bacteria species including *L. innocua*, *L. monocytogenes*, *L. ivanovii*, *E. coli* MG1655, *Salmonella enterica* and *S. aureus* RN4220. The PHAs were produced in a *P. putida* strain, then they were degraded to R-HAs, which were secreted to the culture supernatant from where subsequently purified, and separated (Sandoval et al. 2005). Some of the produced R-HAs were assessed in respect of their antibacterial activity by carrying out spectrophotometrical measurements. Medium chain length (MCL) R-HAs demonstrated that R-3 hydroxy octanoic acid, (*R*)-3-hydroxy-8-nonenoic acid, and (*R*)-3-hydroxy-10-undecenoic acid seemed to exhibit greater antimicrobial activities against the growth of both microorganisms, *Listeria* species and *S. aureus* in comparison to their racemic mixtures or their non-hydroxylated free fatty acid analogs (Ruth et al. 2007). Treatment with (R)-3-HO, (R)-3-hydroxy-8-nonenoic acid and (R)-3-hydroxy-10-undecenoic acid showed growth inhibition of the *Listeria* species and *S. aureus* with a MIC (Minimum Inhibitory Concentration) 1–5 mM while they did not affect the growth rate of *E. coli* and *S. enterica*. These works open new routes for the production of diverse enantiomerically pure R-HAs suitable to be employed for antimicrobial implementations (Ruth et al. 2007).

Douglas and colleagues have managed to produce R-HAs synthons indicating antibacterial activity against strains of *E. coli*, as well as bacteriostatic action against *Listeria monocytogenes* and *Salmonella typhimurium* (Allen et al. 2012). Recently in the study of Radivojevic et al. (2016), eighteen HAs compounds were synthesized from PHAs and were checked for antimicrobial activity. PHA mainly consisting of 3-HO monomers was used to synthesize halogen derivatives with antimicrobial activity against *E. coli*, *S. typhimurium*, *S. aureus*, *L. monocytogenes* and *P.*

aeruginosa, as well as against fungi *Candida albicans* and *Micrococcus gypsum.* The MIC values of the alkanoic acids were 2.7–7 mM for bacteria. Antifungal activity was higher with values 0.15–6.32 mM. Generally, the compounds synthesized were more active against fungi rather than bacteria. In addition, 3-HO reduced *C. albicans* hyphae formation which is a structure vital for biofilm creation. Therefore, these compounds could be used for antifungal activity due to low MIC and no cytotoxicity (Radivojevic et al. 2016).

A second generation novel biopolymer was produced by incorporating thioester groups on the side chain, poly-3-hydroxy-acetyl thioalkanoate-*co*-3-hydroxy alkanoate (PHACOS), (Escapa et al. 2011). The antibacterial properties of the biopolymer PHACOS were analyzed in in vitro and in vivo models, compared to the non-reactive poly(3-hydroxy octanoate-*co*-hydroxy hexanoate) (PHO-HHx) and poly(ethylene terephthalate) (PET) (Dinjaski et al. 2014). To examine the bactericidal activity of PHACOS, *P. aeruginosa* and *S. aureus* were used. This functionalized with natural manner-bacterial PHA (PHACOS) was reported to exhibit antibacterial properties, since it was demonstrated that the polymer is capable to inhibit bacterial growth of methicillin-resistant *S. aureus* (MRSA), selectively and efficiently, in both groups of experiments, in in vitro and in vivo models. This potentiality has been attributed to the functionalization of the side chains by introducing thioester groups, which constitute the as high as 77% of the biopolymer. They observed that 6-acetylthiohexanoic acid, which is a monomer of PHACOS, had a MIC of 40 μM, while hexanoic acid had a MIC of 0.7 mM. However, it is not well-established if the antibacterial activity of PHACOS is due to 6-acetyl thio-hexanoic acid or due to hydroxyl-derivatives (monomers). ESEM (Environmental Scanning Electron Microscopy) showed that biofilm formation of *S. aureus* on PHACOS was three times less than biofilm formation of *P. aeruginosa* while on PHO-HHx both strains formed equal biofilms. Also, the CFUs (Colony Forming Units) measurements of *S. aureus* grown on PHACOS were significantly lower compared to *P. aeruginosa* after 16 h of incubation. Moreover, this functionalized PHA PHACOS was proposed to display a contact active surface mode of antimicrobial action (Dinjaski et al. 2014).

4.4 PHAs as Delivery System of Antimicrobial Agents

4.4.1 PHB

Thanks to the exploitation of excellent properties of PHAs, their biodegradability and biocompatibility several applications have emerged involving PHAs as carriers of antibiotics.

There is a lot of interest in constructing biomaterials that resemble the extracellular matrix of the cells for tissue engineering and drug delivery purposes. PHB [poly(3-hydroxybutyric acid)] nanofibers were constructed with electrospinning and were seeded with antibiotic kanamycin (Naveen et al. 2010). In this method, a liquid drop was subjected to a high electric field from a tip that acts as an electrode.

Consequently, the antibiotic kanamycin was positioned between PHB nanofibers. The antibacterial effect of this formulation was tested against the bacterium *S. aureus*. Fibroblasts cells grown on the nanofibers showed 87% viability and appropriate spreading, which indicate the PHB's biocompatibility. An inhibition zone of 2.1 cm confirms the antibacterial activity of antibiotic-seeded PHB nanofibers against *S. aureus*. Within the first 2 h, a 70% drug release was measured and within 8 h, it was completely discharged. These properties designate that PHB nanofibers loaded with kanamycin have the potential of short-term drug release and delivery as well as the potential for tissue engineering processes (Naveen et al. 2010).

Cupriavidus necator (known as *Ralstonia eutropha*) is considered to be one of the greatest PHB producers. It has been reported that silver nanoparticles (AgNPs) affect positively *C. necator* growth dynamics (Schacht et al. 2013; Castro-Mayorga et al. 2018). AgNPs were simultaneously produced in *C. necator* during PHB biosynthesis while antimicrobial properties were evaluated. It was found that PHB films loaded with low amounts of silver showed antimicrobial effect on *Salmonella enterica* and *L. monocytogenes* (bacterial population of *S. enterica* was reduced below detection limit (33 CFU/mL), while the viable cells of *L. monocytogenes* showed a reduction of 2.5 log units (99.8%) compared with the control). In addition, incorporation of AgNPs did not cause any effect on the thermal degradation, the optical properties and the biodegradability of the polymer (Castro-Mayorga et al. 2018).

Carvacrol is considered to be an antioxidant and antimicrobial agent. Conjugation of carvacrol with poly(lactic acid)/PHB blends resulted in higher antibacterial activity against *S. aureus* rather than *E. coli* (Burgos et al. 2017; Ke et al. 2017).

Sophorolipids/PHB conjugates were reported to have bactericidal activity against *Propionibacterium acnes*. The antibacterial activity was sophorolipids (surface active glycolipids compounds synthesized by yeasts) content dependent (Solaiman et al. 2015; Ke et al. 2017). In addition, PHB/PEO (poly(ethylene oxide)) fiber surface with 1% chlorhexidine, an antimicrobial agent against Gram-(-) and Gram-(+) bacteria for oral protection, exhibited a 99.7% *S. aureus* reduction within 24 h in comparison to 59.3% of reduction when PHB/PEO was used. An abrupt reduction of *E. coli* was observed with PHB/PEO conjugated with 1% of chlorhexidine (Fernandes et al. 2014; Ke et al. 2017). Fibrous surfaces of poly(5,5-dimethyl-3-(3′-triethoxysilylpropyl)hydantoin)/PHB after bleach treatment resulted in inactivation of 92% *S. aureus* and 85% *E. coli* within 30 min, respectively (Fan et al. 2015; Ke et al. 2017).

Another substance which was reported to have antimicrobial activity is caffeic acid. 1800 μg/mL resulted in bactericidal activity of *S. aureus* within 24 h treatment. PHB fibrous materials conjugated with caffeic acid obtained same bactericidal effect against *S. aureus* to caffeic acid, thus, inhibiting *E. coli* growth rate slower than that of *S. aureus* (Ignatova et al. 2016; Ke et al. 2017).

The most common nanoparticles with antibacterial effect once released from the matrix are metal oxide nanomaterials. PHB/ZnO matrices were reported to inhibit the *E. coli* and *S. aureus* growth (Díez-Pascual and Díez-Vicente 2014; Ke et al. 2017).

Biopolymers PHAs in combination with microbial bio-preservatives (like bacteriocins) play a substantial role in the field of food technology and constitute auspicious biomaterials for applications such as food preservation and environmental safeguard. In this context, films were fabricated based on PHB possessing antimicrobial activity by incorporation of eugenol, from 10 to 200 μg/g of PHB and their antimicrobial activity was evaluated against a spectrum of spoilage bacteria, foodborne pathogens, and fungi, including *S. aureus, E. coli, S. typhimurium, Bacillus cereus, Aspergillus flavus, Aspergillus niger, Penicillium sp.,* and *Rhizopus sp.*

Furthermore the synergistic antimicrobial activity of the films was also investigated after the addition of crude pediocin. Despite the absence of antifungal activity of crude pediocin, the presence of antimicrobial PHB film containing also eugenol (80 μg/g) caused fungal growth inhibition, and sporulation, for the fungi *A. niger, A. flavus, Penicillium* sp., and *Rhizopus* sp. up to 7 days in the case of molds in potato dextrose broth. Utilizing crude pediocin in combination with eugenol-incorporated PHB film could operate in synergistic mode of action leading to an effective barrier toward food contaminating microorganisms (Narayanan and Ramana 2013). Table 4.1 summarizes PHB polymers as delivery system of various therapeutic agents, their functionality and the microorganism affected.

4.4.2 PHAs, Their Derivatives and Copolymers

The impact of PHBV (poly-(hydroxybutyrate-co-hydroxyvalerate))-gentamicin complexes was evaluated for their implication in biofilm-forming *S.* species. The increment of HV content from 8% to 12% was correlated with the faster antibiotic release in vitro, while increment of gentamicin content decreased the absorbance from the bacteria and release rate of the antibiotic from the biomaterial. An increased reduction of the adherent *S. aureus* and gentamicin-resistant *Staphylococcus haemolyticus* was observed within the 48 h exposure to the PHBV-gentamicin polymers. For 6 weeks the polymers PHBV 8% and PHBV 12% exhibited sustained antibiotic release above the MIC of *S. aureus* and *S. haemolyticus*. No adverse effects on the blood cell integrity was reported (Rossi et al. 2004).

Microorganisms obtain the ability to be accumulated in implant surfaces and form biofilms, which lead to biomaterial infections (Gallo et al. 2014; Rodríguez-Contreras et al. 2017). A promising approach to deal with this problem is the incorporation of antibiotics to the implant surface and antibiotic controlled release. PHAs are promising polymer candidates for antibiotic adhesion. They are synthesized in bacteria without production of toxic byproducts. In addition, their hydrolytic degradation by-products are found naturally in the blood constituents (ketone bodies) (Williamson et al. 1961; Rodríguez-Contreras et al. 2017).

In a recent study, gentamicin loaded PHAs coatings and the antibiotic release phase were investigated. The PHAs were synthesized in *Cupriavidus necator*. Prepared Titanium (Ti) surfaces were coated with gentamicin-loaded PHAs to create surfaces exhibiting antibacterial functionality. Bacterial growth inhibition was performed both for Gram-negative and -positive bacteria (*E. coli* and *S. aureus*

Table 4.1 PHB polymers as delivery system of various antimicrobial agents

Antimicrobial agent	Functionality	Microorganism	References
Kanamycin	Antibacterial	*S. aureus*	Naveen et al. (2010)
AgNPS	Antibacterial	*Salmonella enterica*, *Listeria monocytogenes*	Castro-Mayorga et al. (2018)
Carvacrol (PLA/PHB blend)	Antibacterial	*S. aureus*	Burgos et al. (2017) and Ke et al. (2017)
Sophorolipids	Antibacterial	*Propionibacterium acnes*	Solaiman et al. (2015) and Ke et al. (2017)
Chlorhexidine (PHB/PEO blend)	Antibacterial	*S. aureus*, *E. coli*	Fernandes et al. (2014) and Ke et al. (2017)
poly(5,5-dimethyl-3-(3′-triethoxysilyl propyl) hydantoin) (chlorine treatment)	Antibacterial	*S. aureus*, *E. coli*	Fan et al. (2015) and Ke et al. (2017)
Caffeic acid	Antibacterial	*S. aureus*, *E. coli*	Ignatova et al. (2016) and Ke et al. (2017)
ZnO	Antibacterial	*S. aureus*, *E. coli*	Díez-Pascual and Díez-Vicente (2014) and Ke et al. (2017)
Eugenol, pediocin	Antibacterial	*S. aureus*, *E. coli*, *Salmonella typhimurium*, *B. cereus*	Narayanan and Ramana (2013)
	Antifungal	*Aspergillus flavus*, *A. niger*, *Penicillium sp.*, *Rhizopus sp*	
Chitosan or chito-oligosaccharides	Antibacterial	*E. coli*, *P. aeruginosa*, methicillin-resistant *S. aureus*, *S. aureus*	Hu et al. (2003)
Carboxyl groups (alkaline treatment)	Antibacterial (tissue engineering)	*S. aureus*	Karahaliloglu et al. (2015) and Ke et al. (2017)
Ostholamide, collagen	Antibacterial	*P. aeruginosa*, *S. aureus*	Kandhasamy et al. (2017) and Ke et al. (2017)
Tetracyclin (PHB microspheres)	Antibacterial (oral drug)	*Porphyromonas gingivalis*, *Actinobacillus actinomycetemcomitans*	Panith et al. (2016)

respectively). The copolymer PHB and PHBV (poly(3-hydroxybutyric acid-*co*-3-hydroxyvaleric acid)) was used in this study. Biopolymer embedding with antibiotic was able to be degraded in a controlled manner. Usage of copolymer PHBV for covering of Ti surfaces reflects in a more homogeneous and constant drug release (Rodríguez-Contreras et al. 2017).

A PHA formulation, prepared by incorporation of Nisin, into the highly biodegradable copolymer poly(3-hydroxy butyrate-*co*-3-hydroxy hexanoate) (P(HB-*co*-HHx)), was reported to exhibit efficacious antibacterial activity. This Nisin is a polycyclic antibacterial peptide composed of 34 amino acid residues. This peptide that is produced by the bacterium *Lactococcus lactis* and is employed as a food preservative has the rare amino acids lanthionine (Lan), methyllanthionine (MeLan), didehydroalanine (Dha), and didehydroalanine (Dha), and didehydroaminobutyric acid (Dhb). Subsequently, P(HB-*co*-HHx) films containing embedded Nisin were tested for quantification of their antimicrobial activity on agar plate against *Micrococcus luteus*, by recording the zone inhibition size of bacterial growth. Moreover, Nisin release from the copolymer films in liquid cultures and their long-time antibacterial activity was confirmed. The threshold of Nisin concentration against bacteria growing, for prolonged and powerful inhibition was 25 μg/g, suggesting the establishment of a technological foundation for the PHA application in the areas of food industry and biomedicine (Lou et al. 2016).

AgNPs have strong antimicrobial activity (Rai et al. 2009; Moritz and Geszke-Moritz 2013; Castro-Mayorga et al. 2017), and the suitability of PHBV18/AgNP coating as antiviral surfaces was shown. Castro-Mayorga and collaborators (Castro-Mayorga et al. 2016b) had showed that PHBV loaded with small amounts of AgNP obtained prolonged antibacterial activity against *L. monocytogenes* and *S. enterica*. Silver nanoparticles (AgNPs) were investigated for their effect on norovirus surrogates (family *Caliviridae*). In addition, in order to develop virucidal biopolymers AgNP coating on PHBV3 films was examined. Clear reduction of virus titers was observed when AgNP were used in a dose dependent manner. AgNP stabilized with a polymer matrix such as PHBV18 might prevent the AgNP aggregation resulting in more sufficient virucidal activity (Castro-Mayorga et al. 2014, 2017). AgNP/PHBV3/PHBV18 films were constructed and antiviral activity was checked. It was found that low addition of AgNP into the film had virucidal activity against norovirus surrogates without any effect on the optical properties of the films (Castro-Mayorga et al. 2017).

MCL-PHAs lack functional groups of hydrophilic nature like carboxyl, imines and hydroxyl groups making them ideal candidates for various modifications. Graft copolymerization is a method to add functional groups (Kim et al. 2007; Chung et al. 2012) which often result in changes of the physicochemical properties of the biopolymer (Kim et al. 2008; Chung et al. 2012). In this study, vinylimidazole (imidazole rings or its derivatives have shown antibacterial, and antifungal activities) was grafted with PHO and tested for antimicrobial activity against the bacteria *E. coli* ATCC 25922, *S. aureus* 6538P and yeast *C. albicans* ATCC 11651. PHO was produced by *Pseudomonas oleovorans* ATCC 29347. Treatment with 2% (w/v) vinylimidazole-PHO copolymer resulted in 90% reduction of cell survival after 12 h treatment. However, PHO did not exhibit any considerable antibacterial functionality against the microorganisms tested (Chung et al. 2012). Collective information is summarized in Table 4.2, concerning the PHAs (or their derivatives and copolymers)-conjugated/or enclosed therapeutic agents for obtaining antimicrobial or other functionalities, against specific microbial strains.

Table 4.2 PHAs (or their derivatives and copolymers)-conjugated/enclosed agents for obtaining antimicrobial or other activities are presented. Functionality against specific microbial strains was included

PHA polymer	Conjugated group	Functionality	Microorganism	References
(R)-3HAs (PHA derivative)	–	Antifungal, antibacterial	*C. albicans*, *Microsporum gypsum*, *Escherichia coli* O157:H7, *Listeria monocytogenes* ATCC 7644, *Salmonella typhimurium* ATCC 14028, *L. innocua*, *L. ivanovii*, *S. aureus* RN4220	Ruth et al. (2007), Allen et al. (2012), and Radivojevic et al. (2016)
PHACOS	Thioester groups	Antibacterial, antibiofilm	Methicillin-resistant *S. aureus*	Dinjaski et al. (2014)
PHBV	Gentamicin Chitosan/chitooligosaccharides Curcumin	Anti-inflammatory, antibacterial, anticancer	*E. coli*, *P. aeruginosa*, methicillin-resistant *S. aureus*, *S. aureus*, *B. subtilis*, *S. haemolyticus*	Hu et al. (2003), Rossi et al. (2004), Veleirinho et al. (2012), Pramanik et al. (2015), and Rodríguez-Contreras et al. (2017)
PHBV3/PHBV18	AgNPs	Antibacterial, virucidal	*Salmonella enterica*, *Listeria monocytogenes*, Norovirus (Family Caliviridae)	Castro-Mayorga et al. (2014, 2016b, 2017)
P(HB-*co*-HHx)	Nisin Lysozyme	Antibacterial, antibiofilm	*Micrococcus luteus*, *Rhodococcus opacus* PD630	Lou et al. (2016) and Kehail and Brigham (2017)
PHO	Vinylimidazole	Antibacterial, antifungal	*E. coli* ATCC 25922, *S. aureus* 6538P, *C. albicans* ATCC 11651	Chung et al. (2012)
P(3HB-*co*-4HB)	Dispersin B Yellow pigment (*Cupriavidus sp.* USMAHM13)	Antibiofilm	*S. epidermidis*, *B. subtilis*	Marcano et al. (2015) and Ismail et al. (2017)

4.5 Antibacterial Activity Against Biofilms

Bacteria are present either as free cells or form biofilms and affect human health. The term "biofilm" was firstly used for microorganisms adhering to wet surfaces (Costerton et al. 1978). Nowadays, biofilms are designated as organized bacteria communities that interact with each other and are connected through a self-produced polymeric matrix. This matrix contains polysaccharides synthesized by the bacteria communities of the biofilm which was mentioned to be relevant to antibiotic resistance (Costerton et al. 1987; Donlan 2002). Biofilms may consist of one or more species (Pamp et al. 2007; Paraje 2011). It is a beneficial form to bacteria as its formation provides nutrient availability, protection from environment, metabolic cooperation and gene exchange (Costerton et al. 1987; Davey and O'Toole 2000; Paraje 2011). Biofilms come in a variety of shapes and sizes such as mushroom-like and flat multicellular structures (Costerton et al. 1987; Pamp et al. 2007; Paraje 2011).

Biofilm formation occurs in several steps, shown in Fig. 4.1. Firstly, pioneer bacteria adhere to free-floating bacteria directed to a quiescent surface with zero velocity. Some of these bacteria separate as no chemical bonds occur, while others start forming chemical attachments, they reproduce and their colonies produce a polysaccharide matrix. Subsequently, biofilm parts are disturbed and the bacterial population is capable to be disseminated and form novel biofilms. The polysaccharide matrix acts as a protective coating and traps nutrient molecules (Webb et al. 2003; Paraje 2011). This matrix is loosely organized in the mature biofilm. Few hours to several weeks may take for biofilm formation (Paraje 2011).

Microbial communities populate the human body in balance and are beneficial to it. However, when this balance is diminished malfunction of the human body occurs. It is estimated that more than 60% of the microbial infections are caused by biofilms, as well as chronic infections are related to biofilm (Costerton et al. 2003). In addition, it is considered that 99% of all bacteria endure in biofilm state (Costerton et al. 1987; Donlan 2002). Some of the signs indicative of biofilm infection are bacteria embodied in self-generated polysaccharide matrix, tolerance to innate and adaptive immune responses, tolerance to antibiotic dosage despite sensitivity of free-floating bacteria and presence of chronic infections (Paraje 2011).

The structural form of biofilm provides resistance towards antimicrobial agents and immune responses. In contrast to free-living bacteria, the MIC of biofilm growing bacteria are 100–1000 fold higher (Mah and O'Toole 2001; Paraje 2011). However, the molecular mechanisms of antibiotic resistance of biofilms are not well established. How do biofilms provide protection against antibiotics?

Some theories mention that biofilms provide a physical barrier against antibiotics. In biofilms rapidly growing cells are more susceptible to antimicrobials in contrast to cells located in the biofilm center which are divided slowly (Costerton et al. 1999), are differentiated into persistent cells, and mutations and increased horizontal gene transmission occur (Paraje 2011).

Biofilm formation starts with the attachment of planktonic bacteria. The most critical period for long-term implants is the first 6 h of blood exposure where biomaterials are usually colonized by pathogenic bacteria (Hetrick and Schoenfisch

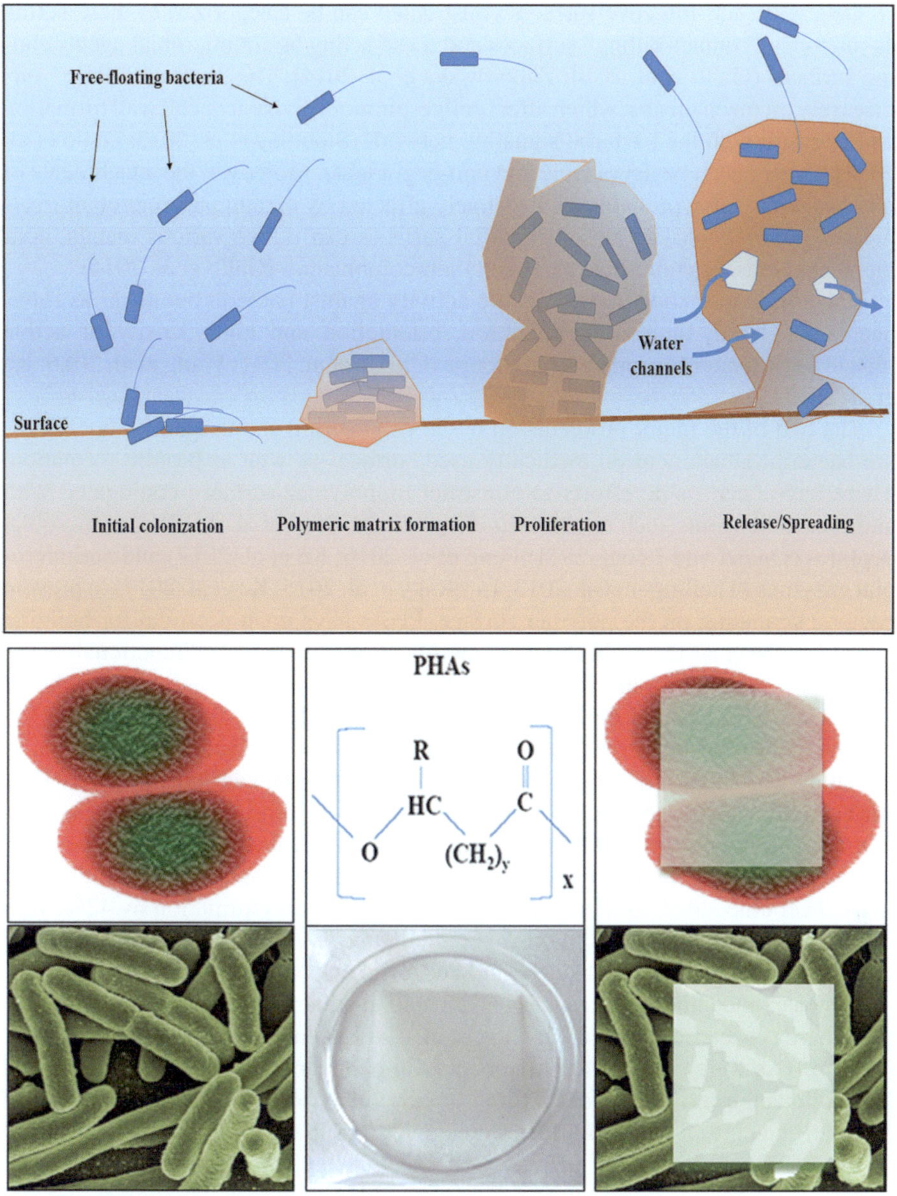

Fig. 4.1 Representative images showing biofilm formation, where PHAs were reported to be involved in drug delivery for antibiofilm activity

2006; Ke et al. 2017; Loh 2017). Thus, anti-adhesive surfaces could be a solution to the problem (Shah et al. 2013; Zhu and Jun Loh 2015; Ke et al. 2017). Antibiotic-loaded surfaces obtain significant antimicrobial activity to a broad range of bacteria (Dhand et al. 2017; Ke et al. 2017).

Generally, anti-infective surfaces constructed can be categorized as these acting as microbes "contact-killing" surfaces and these acting as antimicrobial agent- eluting surfaces (Gallo et al. 2014; Zaborowska et al. 2014). The "contact-killing" surfaces rely on mechanisms, which affect cell respiration, division or cell wall formation and interfere with the bacterial signaling network (Stoodley et al. 2013; Gallo et al. 2014). Coating can be degradable and non-degradable. Moreover, the attachments of host cells may also probably be negatively affected by certain anti-microbial treatments. The technologies of antibacterial surfaces can utilize various metals, non-metal elements, organic substances and their combination (Gallo et al. 2014).

Bacteriophages show more specific activity against bacteria but as far as it has been considered, their immobilization, orientation and high density of active phages in biomaterials remain a challenge (Clokie et al. 2011; Wang et al. 2016; Ke et al. 2017).

The two of the major problems on tissue engineering and usage of biopolymers are bacterial attachment on medically used surfaces as well as biofilm formation. There have been made efforts to construct biopolymer surfaces conjugated with antimicrobial agents such as chitosan (Ke et al. 2017; Li et al. 2017), antimicrobial peptides (Onaizi and Leong 2011; Peng et al. 2016; Ke et al. 2017) and antimicrobial enzymes (Thallinger et al. 2013; Lipovsky et al. 2015; Ke et al. 2017) to prevent bacteria spreading on the polymer surface. PHAs have great potential for building precise polymer surfaces via chemical modification with expected functionalities as they lack functional groups for coupling reactions and they are highly hydrophobic (Kai et al. 2015; Li and Loh 2015; Ke et al. 2016, 2017, Li et al. 2016, 2017).

Due to biofilms innate resistance it is very difficult to prevent their formation. In the study of Kehail and Brigham, lysozyme was immobilized on P(HB-*co*-HHx) surface and was tested for its anti-biofilm activity against *Rhodococcus opacus* PD630. The polymer was produced from *C. necator*. A linear correlation between increment of HHx monomer and protein absorption was observed. In addition, the PHA sheet embedded with lysozyme inhibited the biofilm formation by 42%. This indicate that P(HB-*co*-HHx) could potentially be employed as wound dressing thanks to its high protein intake (Kehail and Brigham 2017).

Chitosan and chitooligosaccharides are able to adhere the bacteria cell wall suppressing the bacteria metabolism (Iida et al. 1987; Koneman et al. 1997; Hu et al. 2003). PHAs were subjected to treatment with ozone to conjugate acrylic acid and then followed esterification of chitosan or chitooligosaccharides. Amino groups of chitosan and chitoologosaccharides were conjugated to carboxyl groups of the PHB or the PHBV. Antibacterial activity was checked against *E. coli*, *P. aeruginosa*, methicillin-resistant *S. aureus* and *S. aureus*. PHA polymer membranes did not affect the growth of the four mentioned bacteria, while the membranes grafted with chitosan or chitooligosaccharides resulted in bacteria death after 18 h incubation at 37 °C. The order of the antibacterial activity of the grafted membranes was higher against *E. coli*, followed by *P. aeruginosa*, *S. aureus* and methicillin-resistant *S. aureus* (Hu et al. 2003).

The methodology of the construction of biomaterials with immobilized antibacterial agents that affect biofilm formation was applied (Marcano et al. 2015).

Dispersin B, an enzyme formed by *Actinobacillus actinomycetemcomitans* pathogen with strong anti-biofilm formation activity, was immobilized on a PHA-based asymmetric membrane. The PHA (poly (3- hydroxybutyrate-*co*-4-hydroxybutyrate) named P(3HB-*co*-4HB) with 13% 4HB content is biodegradable and its removal was not a necessity. The anti-biofilm activity was checked on *S. epidermidis* strain. Disk diffusion test was applied to check antiseptic and disinfectant properties of the PHA-based membrane. There was no significant inhibition of *S. epidermidis* biofilm formation even though deconstruction of the biofilm occurred in the sample where PHA-membrane with Dispersin B was used in comparison to non-loaded PHA membranes. This indicates that maybe the amount of the enzyme was not efficient to constrain biofilm formation. Also, a 12% deconstruction of already formed *S. epidermidis* biofilms was observed. An explanation for this occurrence might be that Dispersin B loaded membranes are more efficient to destruct preformed biofilms rather than inhibit their formation. These observations do not agree with Kaplan et al. (2004) who showed biofilm inhibition with the usage of Dispersin B. The loading capacity of Dispersin B according to Marcano et al. was 2 μg cm^{-2} (6 μg/disc), while Kaplan et al. used 7 μg/cm^{-2}. They also differ in the membrane construction method (Kaplan et al. 2004; Marcano et al. 2015).

Biofilm formation which is known to occur ubiquitously in the environment and pose risks of infection has been well known as a determinant of bacterial virulence. Many relevant pathogens with increasing medical importance were evaluated for their biofilm-forming potential. Therefore, there is need for antibiofilm polymers that have the ability to inhibit bacterial biofilm formation (Ismail et al. 2017). Recently towards this direction optimization studies lead to the simultaneous production 53 wt% of a PHA copolymer P(3HB-*co*-4HB) and of an antimicrobial yellow pigment 0.35 g/L concentration from *Cupriavidus* sp. USMAHM13. Antibiofilm assays demonstrated that the yellow pigment enrobed by P(3HB-*co*-4HB) copolymer had considerable inhibitory action on bacteria proliferation and colonization from 6 h forward attaining 100% inhibition within 12 h, therefore effectively impeding the biofilm formation (Ismail et al. 2017).

4.6 PHAs Applications

4.6.1 PHAs as Delivery Systems for Oral Use

A promising tool in therapeutics of bacterial or other infections is the use of antimicrobial peptides with the only limiting step being their protection against proteolysis. During their oral administration, antimicrobial peptides are subjected to proteolysis thanks to the enzymes localized in the gastrointestinal (GI) tract. Therefore, polymeric carriers such as PHAs are needed to encapsulate antibacterial peptides until their uptake by the enterocytes (Delie and Blanco-Príeto 2005).

PHA-based drug delivery systems, for oral administration have started in the early 1980s by Korsatko's group and the production of 7-hydroxyethyltheophylline pills incorporated within the PHB polymeric network. The pastilles were easily prepared from the polymer and the drug by compounding them and compressing (Korsatko et al. 1983). Although there are reports on PHA tablet formulations, their number is significantly lower in comparison with these of biodegradable implants and microencapsulation ones. This may be attributed to the relatively great PHAs production cost, which outweighs the accepted benefits (Williams and Martin 2005).

PHB and the co-polymer P(3HB-*co*-3HV) were used in the study of Brophy and Deasy (1986) to suspend sulphamethizole by a solvent evaporation method and produce oral capsules. L(-) polylactic acid was also used to overcoat one of the products and reduce the burst effect. The products were tested in vivo using six dogs and showed a similar release in comparison to a conventional product of a analogue dose, expect for the over coated product where the in vitro release rate was smaller. The bioavailability of the formulations under investigation was similar (Brophy and Deasy 1986).

Chlorhexidine (CHX) is one of the most efficient antimicrobial agents against Gram(+) and Gram(−) bacteria and fungi, and are non-toxic toward mammalian cells (Chen et al. 2008; Verraedt et al. 2010). It is used in several products for oral protection and in dentistry. In the work of Fernandes et al. (2014) PHB/PEO fibre membranes containing different amounts of CHX were produced by electrospinning and their antibacterial activity was studied against *E. coli* and *S. aureus*. The results were very promising since samples with 1 wt% CHX presented a reduction of 100% and 99.69% against *E. coli* and *S. aureus*, respectively. The results are very promising and electrospinning will also enable the incorporation of a wide range of pharmaceutical drugs into the fibers (Fernandes et al. 2014).

Periodontitis is a bacterial infection of gum and other tissues supporting teeth (Pihlstrom et al. 2005; Panith et al. 2016). Tetracycline is an antibiotic commonly used for periodontitis treatment (Seymour and Heasman 2005; Panith et al. 2016). Here, tetracycline was encapsulated into polymeric microspheres via double emulsion-solvent evaporation method. The antibacterial activity was tested against two commonly periodontitis-causing bacteria *Porphyromonas gingivalis* and *Actinobacillus actinomycetemcomitans*. The results suggested the potential use of PHB microspheres loaded with tetracycline for slow-release delivery system. PHB was obtained from Sigma-Aldrich. The tetracycline loaded-PHB microspheres possessed powerful antibacterial activity causing 85% growth reduction of *P. gingivalis* even within the first 2 h, while within the first 12 h the bacterium population had been erratically diminished. However, bacterial viability of *A. actinomycetemcomitans* reduced to less than 10^5 CFU/mL (from an initial 10^8 CFU/mL) within the first 12 h indicating stronger antibacterial activity of tetracycline-charged microspheres against *P. gingivalis*. This type of microspheres allow prolonged tetracycline release to kill bacteria inside the defected periodontal pocket, during mechanical cleaning for effective confrontation of infections, indicating potential future application in periodontal treatment (Panith et al. 2016).

For example, alkaline treatment of PHB, which generated carboxylic groups on the surface reduced *S. aureus* growth rate greater than 60% within the first 48 h, indicating a possible application as an antibacterial surface for dental tissue engineering (Karahaliloglu et al. 2015; Ke et al. 2017).

4.6.2 PHAs in Wound Healing

Over the past years, PHAs and their conjugates have been used in various medical applications due to their biodegradability and low cytotoxicity. One of their applications is wound dressing containing medical constituents. Sometimes, however, these structures are infected by bacteria that form biofilms resulting in chronic wounds. A new approach to this problem is the construction of biomaterials with immobilized antibacterial agents that affect biofilm formation. The conjugation of growth factors important for various stages of wound healing with biomaterials such as liposomes, poly-ethylene-glycol (PEG), hydrogels, ethyl cellulose etc. was also reported and summarized in the review (Stejskalová and Almquist 2017).

A major problem in biomaterials usage clinically is the tendency of the latter to be infected by microorganisms. These bacterial infections and their persistence are correlated with biofilm formations. In addition, more and more bacterial strains evolve antibiotic resistance making it difficult to elucidate them (Dinjaski et al. 2014).

One of the major concerns of modern reconstructive medicine is the selection of proper material for more efficient skin regeneration and to be more specific wound healing. In recent years, studies revealed that cells (fibroblasts, keratinocytes) cultured on biodegradable scaffolds provide more efficient skin regeneration. In addition, there is much scientific evidence that the biodegradable polymers such as PHAs could be used for skin engineering (Shishatskaya et al. 2016).

PHB-gelatin nanomesh precoated with ostholamide (which is synthesized from osthole a natural coumarin) and collagen obtain mechanical stability in wound healing. There was observed that the coated nanomesh exhibited sustained release rate of ostholamide and subsequently, showed antibacterial activity against the above mentioned bacteria indicating a potential use in wound healing (Kandhasamy et al. 2017; Ke et al. 2017).

Cell adhesion and proliferation occurs during wound healing, which requires fibroblasts and formation of extracellular matrix (ECM). In some cases, these processes are very slow due to a medical condition or extensive tissue damage and the usage of biodegradable scaffolds may assist the regeneration process. PHBV is a biocompatible and biodegradable polymer but is also hydrophobic and with a low degradation rate (5% during incubation in the medium after 28 days), while the regeneration process requires higher degradation rate. A hybrid PHBV/chitosan nano fibrous polymer pad was constructed for skin regeneration and specifically for treatment of cutaneous wounds (Veleirinho et al. 2012). For this purpose, PHBV was conjugated with chitosan which is highly hydrophilic, enhances degradation rate of the biopolymer, and has anti-inflammatory, anti-bacterial and wound healing effect. Two-hybrid scaffolds of different PHBV/chitosan ratios were used the 4:1 and 2:3 (w/w). Scanning electron microscopy showed that the hybrid with more chitosan had higher porosity and higher fiber diameter. Fibroblast viability increased in both hybrids. After 7 days the fibroblast viability was higher with PHBV/chitosan 4:1 (w/w) usage while in PHBV/chitosan 2:3 was decreased probably due to the higher degradation rate of the latter. SEM and CLSM images of fibroblast showed proliferative activity in both hybrid scaffolds after 12 h, 2 days and 7 days. In vivo studies in

Wistar rats with skin-thickness wounds indicated PHBV/chitosan (4:1) decreased inflammation leading to proliferative phase. In addition, after 14 days no traces of the polymer scaffolds were found. These findings suggest that the PHBV/chitosan polymer scaffolds could be used for wound healing processes (Veleirinho et al. 2012).

PHAs are one of the most promising candidates for treating skin infections and healing wounds (Shishatskaya et al. 2016). Using the degradable copolymer composed by hydroxyl derivatives of carboxylic acids (PHAs) and the copolymer 3-hydroxybutyrate/4-hydroxybutyrate [P(3HB/4HB)], a wound dressing material was produced. This microbial copolymer [P(3HB/4HB)] was synthesized in a culture of *Cupriavidus eutrophus* B10646 strain (Shishatskaya et al. 2016).

Scaffolds in the form of membranes were constructed with the electro-spunning method using the P(3HB/4HB) copolymer and evaluated in vivo for their wounds healing ability. Then, the polymerized membranes were seeded with 10^5 fibroblasts cells and tested on model skin-defect Wistar rats. The copolymers showed high elasticity, high rate of biodegradability in vivo and low rate of crystallinity. Histological examination of rats showed that fibroblasts-seeded membranes promoted the processes of vascularization, tissue regeneration and in turn leading in wound healing. Moreover, secretion of extracellular matrix proteins by the cells onto the polymeric membrane surface, formed a layer and encouraged the movement of epidermal cells from the neighboring tissues towards the wound. The healing process underneath the P(3HB/4HB) membrane inoculated with cells progress faster compared to the wounds without cells-inoculated membrane (1.4 times) and 3.5 times faster compared to the wounds healing the control (no membrane or cells were used, natural process occurred). Complete wound healing was observed after 14 days when membrane seed with cells was used. It should be highlighted the significance of nonwoven membranes constructed from degradable copolymers P(3HB/4HB) in diminishing inflammation, augmenting skin angiogenic properties contributing to the facilitation of wound healing process (Shishatskaya et al. 2016).

An interesting hybrid hydrophilic-hydrophobic conjugate has been reported to promote faster and more efficient wound healing. This conjugate was a guar gum/PHAs-curcumin blend (GPCC) film which was constructed and checked for bactericidal activity and as candidate wound dressing (Pramanik et al. 2015). Guar gum is a water-soluble polysaccharide and is used as a wound healing material (Prabaharan 2011). Curcumin is a polyphenolic phyto component with anticancer, antimicrobial and wound healing activities (Teow et al. 2016). The PHBV copolymer, isolated from *Alkaliphilus oremlandii* ohILAs strain, was used. The bactericidal activity was investigated by inhibition zone measurements. Pure guar and guar gum/PHBV films showed no significant antimicrobial activity while curcumin-guar gum/PHBV conjugates prevented bacterial growth viability. The effect of the blended film was more clear in Gram-(+) (*B. substilis, S. aureus*) rather than Gram-negative bacteria (*E. coli* XL1B, *Enterobacter aeregenes, Vibrio vulnificus, P. aeruginosa*) (Pramanik et al. 2015). Curcumin and curcumin blended PHBV films exhibited 17.5 ± 0.57 mm zone of inhibition in Gram-(+) bacteria and

15–14 ± 0.77 mm zone of inhibition in Gram-negative strains. In addition, it was observed that both curcumin and curcumin charged films showed antimicrobial activity (Pramanik et al. 2015). A possible explanation for the bactericidal activity of curcumin is that it decreases bundling of FtsZ protofilaments leading to diffusion of cell proliferation and consequently to bactericidal activity (Rai et al. 2008).

No cytotoxicity was observed for curcumin-guar gum/PHBV blends. In addition to wound healing, a 90% wound recession was observed after 7 days in mice treated with curcumin-guar gum/PHBV films, while in control mice a 55% reduction and scar formation occurred. To conclude, a curcumin-guar gum/PHBV blend was successfully constructed and showed antibacterial activity as well as the potential of wound healing (Pramanik et al. 2015).

Wound healing is a complicated procedure involving biochemical, molecular and cellular responses. A PHA-PCL hydrogel (poly(3-hydroxy alkanoates)-*co*-(6-hydroxy hexanoate) (PHA-PCL) composite hydrogel) composite was constructed to evaluate its wound healing potential. The PHA part was synthesized in *Pseudomonas putida* Bet001 strain. Cell viability did not exhibit any important difference between the groups treated with PHA-PCL hydrogel compared to the negative control group, even after 72 h. A representative image for the PHA use in wound healing is depicted in Fig. 4.2.

A significant wound contraction was observed within 7 days–10 ($p < 0.005$) and the reduction has further occurred after 14 days when wounds were dressed with PHA-PCL hydrogel. The faster healing process in PHA-PCL hydrogel appeared to be due to the widespread degradation in tissue granulation and reduction in

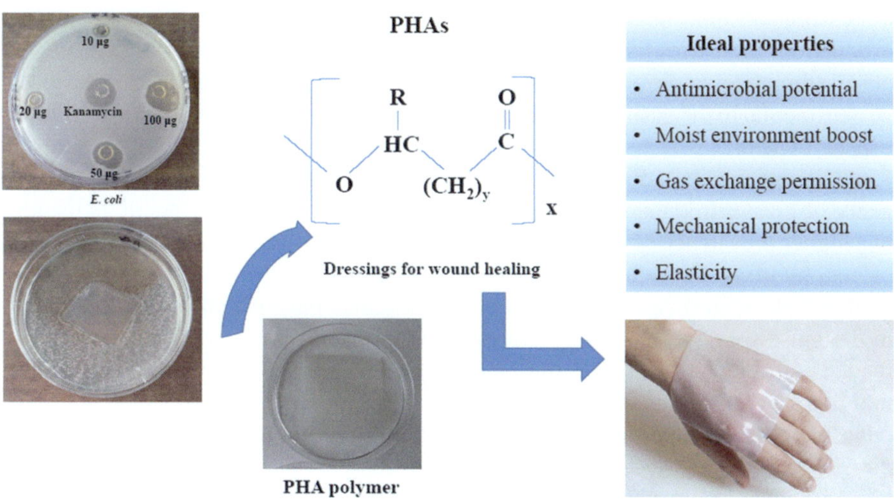

Fig. 4.2 Representative image of a PHA film containing antibacterial agent (center), of which the bactericidal activity was investigated against E.coli by zone inhibition measurements (left). PHA dressing for would healing with suitable properties (right)

Table 4.3 PHAs polymers as wound dressings

PHA polymer type	PHA origin	Activity	References
Hybrid PHBV/chitosan nanofibrous	Commercial product	Treatment of cutaneous wounds	Veleirinho et al. (2012)
P3HB/P3HO	*Bacillus cereus* SPV/*Pseudomonas mendocina*	Tissue engineering/drug delivery system (possibility)	Basnett et al. (2013)
P(3HB/4HB)	*Cupriavidus eutrophus* B10646	Anti-inflammatory, wound healing	Shishatskaya et al. (2016)
Guar gum/PHAs-curcumin blend (GPCC) film	*Alkaliphilus oremlandii* ohILAs	Bactericidal activity wound dressing	Pramanik et al. (2015)
PHA-PCL hydrogel	*Pseudomonas putida* Bet001	Cutaneous wound healing	Gumel et al. (2015)

mononuclear inflammatory cells. Angiogenesis and collagen deposition promotes faster-wound closure letting fewer scars. To conclude, PHA-PCL hydrogel is efficient in cutaneous wound healing and could potentially be used in humans and animals (Gumel et al. 2015). Table 4.3 summarizes PHAs polymers used as wound dressings.

4.6.3 PHAs as Bacteriostatic Agents in Animal Production

Additionally, due to the hazard of antibiotic resistance emergence, there is an increasing consciousness that antibiotics must be utilized more cautiously in animal production. Industries like livestock and aquaculture utilize antibiotics at quite low doses administered along with the feed. This tactical administration has proved to be detrimental to the gastrointestinal microflora leading to the evolution of antibiotic-resistant bacteria.

At the other side, a diminished utilization of antibiotics could reflect in an emergence of greater frequency of pathogenic bacteria, a fact that could result, in its turn, in a higher prevalence of infections (Defoirdt et al. 2009). Since PHAs are polymers of beta-hydroxy short-chain fatty acids (SCFAs), which have long been known to demonstrate bacteriostatic activity, PHAs could also have the ability to decompose in the gut displaying biocontrol effects similar to the SCFAs when delivered with feed. These biocontrol effects may include controlling pathogenic bacteria in animal hosts by inhibiting growth, interfering with nutrient transport, interruption of outer membrane properties and affecting macromolecule formation (Defoirdt et al. 2009) Moreover, these compounds also down-regulate the expression of some specific virulence factors and thus affect positively the gastrointestinal prosperity of the host organism. Previous studies have shown that butyrate down-regulates a set of virulence-related pathogenicity regions of *Salmonella enterica* and *Vibrio campbellii* which includes a number of genes (Defoirdt et al. 2006; Gantois et al. 2006). For this reason, these substances attract currently an increasing interest to be used in animal production as biocontrol agents (Defoirdt et al. 2009).

4.7 Conclusions: Opinion

Last decade a wide range of research activities focus on investigating novel functional biomaterials. The naturally-occurring polymers derived from diverse technologies and sources of production as well may lead to a variety of physicochemical characteristics and properties. Exploiting these properties with selected therapeutics leads to smart formulations. Several formulations of protracted-therapeutic action (antibacterial, anti-biofilm, anti-inflammatory etc.) were obtained, based commonly on polymeric materials, degradable in the corps and biocompatible. Research-based implementations turned towards the construction of contemporary wound dressings, which possess diverse features necessary in the stage of wound healing, in which they must be utilized. All modern biomedical requirements turn towards resorbable wound dressings, which can be advantageous in the entire spectrum of wound healing in both early and late phases. The design and use of wound dressings constructed by resorbable polymer and with various biodegradation rates, a feature assured by the use of PHAs copolymers, constitutes one of the principal objectives of the biomedical research. This advantage in combination with a selected antibacterial activity-possessing agent, enables them precious tools in the bio-medicinal area applications, including oral drugs and skin and wound healing, and motivate research to produce naturally functionalized bacterial PHA with antibacterial properties to confront even antibiotic-resistant bacteria.

References

Aggarwal V, Bakhshi H, Ecker N, Parvizi J, Gehrke T, Kendoff D (2014) Organism profile in periprosthetic joint infection: pathogens differ at two arthroplasty infection referral centers in Europe and in the United States. J Knee Surg 27:399–406. https://doi.org/10.1055/s-0033-1364102

Allen AD, Daley P, Ayorinde FO, Gugssa A, Anderson WA, Eribo BE (2012) Characterization of medium chain length (R)-3-hydroxycarboxylic acids produced by *Streptomyces* sp. JM3 and the evaluation of their antimicrobial properties. World J Microbiol Biotechnol 28:2791–2800. https://doi.org/10.1007/s11274-012-1089-z

Arias CA, Murray BE (2009) Antibiotic-resistant bugs in the 21st century – a clinical super-challenge. N Engl J Med 360:439–443. https://doi.org/10.1056/NEJMp0804651

Basnett P, Ching KY, Stolz M, Knowles JC, Boccaccini AR, Smith C, Locke IC, Keshavarz T, Roy I (2013) Novel poly(3-hydroxyoctanoate)/poly(3-hydroxybutyrate) blends for medical applications. React Funct Polym 73:1340–1348. https://doi.org/10.1016/j.reactfunctpolym.2013.03.019

Brophy MR, Deasy PB (1986) In vitro and in vivo studies on biodegradable polyester microparticles containing sulphamethizole. Int J Pharm 29:223–231. https://doi.org/10.1016/0378-5173(86)90119-5

Burgos N, Armentano I, Fortunati E, Dominici F, Luzi F, Fiori S, Cristofaro F, Visai L, Jiménez A, Kenny JM (2017) Functional properties of plasticized bio-based poly(lactic acid)-poly(hydroxybutyrate) (PLA-PHB) films for active food packaging. Food Bioprocess Technol 10:770–780. https://doi.org/10.1007/s11947-016-1846-3

Burke TR, Knight M, Chandrasekhar B (1989) Solid-phase synthesis of viscosin, a cyclic depsipeptide with antibacterial and antiviral properties. Tetrahedron Lett 30:519–522

Busscher HJ, van der Mei HC (2012) How do bacteria know they are on a surface and regulate their response to an adhering state? PLoS Pathog 8:e1002440. https://doi.org/10.1371/journal.ppat.1002440

Castro-Mayorga JL, Martínez-Abad A, Fabra MJ, Olivera C, Reis M, Lagarón JM (2014) Stabilization of antimicrobial silver nanoparticles by a polyhydroxyalkanoate obtained from mixed bacterial culture. Int J Biol Macromol 71:103–110. https://doi.org/10.1016/j.ijbiomac.2014.06.059

Castro-Mayorga J, Fabra M, Cabedo L, Lagaron JM (2016a) On the use of the electrospinning coating technique to produce antimicrobial polyhydroxyalkanoate materials containing in situ-stabilized silver nanoparticles. Nanomaterials 7:4. https://doi.org/10.3390/nano7010004

Castro-Mayorga JL, Fabra MJ, Lagaron JM (2016b) Stabilized nanosilver based antimicrobial poly(3-hydroxybutyrate-*co*-3-hydroxyvalerate) nanocomposites of interest in active food packaging. Innov Food Sci Emerg Technol 33:524–533. https://doi.org/10.1016/j.ifset.2015.10.019

Castro-Mayorga JL, Randazzo W, Fabra MJ, Lagaron JM, Aznar R (2017) Antiviral properties of silver nanoparticles against norovirus surrogates and their efficacy in coated polyhydroxyalkanoates systems. LWT-Food Sci Technol 79:503–510. https://doi.org/10.1016/j.lwt.2017.01.065

Castro-Mayorga JL, Freitas F, Reis M, Prieto A, Lagaron JM (2018) Biosynthesis of silver nanoparticles and polyhydroxybutyrate nanocomposites of interest in antimicrobial applications. Int J Biol Macromol 108:426–435. https://doi.org/10.1016/j.ijbiomac.2017.12.007

Chen G-Q (2009) A microbial polyhydroxyalkanoates (PHA) based bio- and materials industry. Chem Soc Rev 38:2434–2446. https://doi.org/10.1039/b812677c

Chen LJ, Wang M (2002) Production and evaluation of biodegradable composites based on PHB–PHV copolymer. Biomaterials 23:2631–2639. https://doi.org/10.1016/S0142-9612(01)00394-5

Chen G-Q, Wu Q (2005a) The application of polyhydroxyalkanoates as tissue engineering materials. Biomaterials 26:6565–6578. https://doi.org/10.1016/j.biomaterials.2005.04.036

Chen G-Q, Wu Q (2005b) Microbial production and applications of chiral hydroxyalkanoates. Appl Microbiol Biotechnol 67:592–599. https://doi.org/10.1007/s00253-005-1917-2

Chen L, Bromberg L, Hatton TA, Rutledge GC (2008) Electrospun cellulose acetate fibers containing chlorhexidine as a bactericide. Polymer (Guildf) 49:1266–1275. https://doi.org/10.1016/j.polymer.2008.01.003

Chiba T, Nakai T (1985) A synthetic approach to (+)-thienamycin from methyl (R)-3-hydroxybutanoate. A new entry to (3 R, 4 R)-3-[(R)-1-hydroxyethyl]-4-acetoxy-2-azetidinone. Chem Lett 14:651–654. https://doi.org/10.1246/cl.1985.651

Chung GM, Kim WH, Kim RB, Kim BY, Rhee HY (2012) Biocompatibility and antimicrobial activity of poly (3-hydroxyoctanoate) grafted with vinylimidazole. Int J Biol Macromol 50:310–316. https://doi.org/10.1016/j.ijbiomac.2011.12.007

Clokie MR, Millard AD, Letarov AV, Heaphy S (2011) Phages in nature. Bacteriophage 1:31–45. https://doi.org/10.4161/bact.1.1.14942

Cornell RJ, Donaruma LG (1965) 2-Methacryloxytropones. Intermediates for the synthesis of biologically active polymers. J Med Chem 8:388–390

Costerton JW, Geesey GG, Cheng K-J (1978) How bacteria stick. Sci Am 238:86–95. https://doi.org/10.1038/scientificamerican0178-86

Costerton JW, Cheng KJ, Geesey GG, Ladd TI, Nickel JC, Dasgupta M, Marrie TJ (1987) Bacterial biofilms in nature and disease. Annu Rev Microbiol 41:435–464. https://doi.org/10.1146/annurev.mi.41.100187.002251

Costerton JW, Stewart PS, Greenberg EP (1999) Bacterial biofilms: a common cause of persistent infections. Science 284:1318–1322

Costerton W, Veeh R, Shirtliff M, Pasmore M, Post C, Ehrlich G (2003) The application of biofilm science to the study and control of chronic bacterial infections. J Clin Invest 112:1466–1477. https://doi.org/10.1172/JCI200320365

Darouiche RO (2004) Treatment of infections associated with surgical implants. N Engl J Med 350:1422–1429. https://doi.org/10.1056/NEJMra035415

Davey ME, O'toole GA (2000) Microbial biofilms: from ecology to molecular genetics. Microbiol Mol Biol Rev 64:847–867. https://doi.org/10.1128/MMBR.64.4.847-867.2000

Defoirdt T, Halet D, Sorgeloos P, Bossier P, Verstraete W (2006) Short-chain fatty acids protect gnotobiotic *Artemia franciscana* from pathogenic *Vibrio campbellii*. Aquaculture 261:804–808. https://doi.org/10.1016/j.aquaculture.2006.06.038

Defoirdt T, Boon N, Sorgeloos P, Verstraete W, Bossier P (2009) Short-chain fatty acids and poly-β-hydroxyalkanoates: (New) biocontrol agents for a sustainable animal production. Biotechnol Adv 27:680–685. https://doi.org/10.1016/j.biotechadv.2009.04.026

DeLeo FR, Chambers HF (2009) Reemergence of antibiotic-resistant *Staphylococcus aureus* in the genomics era. J Clin Invest 119:2464–2474. https://doi.org/10.1172/JCI38226

Delie F, Blanco-Príeto MJ (2005) Polymeric particulates to improve oral bioavailability of peptide drugs. Molecules 10:65–80. https://doi.org/10.3390/10010065

Desbois A, Lawlor K (2013) Antibacterial activity of long-chain polyunsaturated fatty acids against *Propionibacterium acnes* and *Staphylococcus aureus*. Mar Drugs 11:4544–4557. https://doi.org/10.3390/md11114544

Desbois AP, Smith VJ (2010) Antibacterial free fatty acids: activities, mechanisms of action and biotechnological potential. Appl Microbiol Biotechnol 85:1629–1642. https://doi.org/10.1007/s00253-009-2355-3

Dhand C, Venkatesh M, Barathi VA, Harini S, Bairagi S, Goh Tze Leng E, Muruganandham N, Low KZW, Fazil MHUT, Loh XJ, Shirivasa DK, Liu SP, Beuerman RW, Verma NK, Ramakrishna S, Lakshminarayanan R (2017) Bio-inspired crosslinking and matrix-drug interactions for advanced wound dressings with long-term antimicrobial activity. Biomaterials 138:153–168. https://doi.org/10.1016/j.biomaterials.2017.05.043

Díez-Pascual A, Díez-Vicente A (2014) Poly(3-hydroxybutyrate)/ZnO bionanocomposites with improved mechanical, barrier and antibacterial properties. Int J Mol Sci 15:10950–10973. https://doi.org/10.3390/ijms150610950

Dinjaski N, Fernández-Gutiérrez M, Selvam S, Parra-Ruiz FJ, Lehman SM, San Román J, García E, García JL, García AJ, Prieto MA (2014) PHACOS, a functionalized bacterial polyester with bactericidal activity against methicillin-resistant *Staphylococcus aureus*. Biomaterials 35:14–24. https://doi.org/10.1016/j.biomaterials.2013.09.059

Donadio S, Maffioli S, Monciardini P, Sosio M, Jabes D (2010) Antibiotic discovery in the twenty-first century: current trends and future perspectives. J Antibiot (Tokyo) 63:423–430. https://doi.org/10.1038/ja.2010.62

Donlan RM (2002) Biofilms: microbial life on surfaces. Emerg Infect Dis 8:881–890. https://doi.org/10.3201/eid0809.020063

Escapa IF, Morales V, Martino VP, Pollet E, Avérous L, García JL, Prieto MA (2011) Disruption of β-oxidation pathway in *Pseudomonas putida* KT2442 to produce new functionalized PHAs with thioester groups. Appl Microbiol Biotechnol 89:1583–1598. https://doi.org/10.1007/s00253-011-3099-4

Fan X, Jiang Q, Sun Z, Li G, Ren X, Liang J, Huang TS (2015) Preparation and characterization of electrospun antimicrobial fibrous membranes based on polyhydroxybutyrate (PHB). Fibers Polym 16:1751–1758. https://doi.org/10.1007/s12221-015-5108-1

Fernandes JG, Correia DM, Botelho G, Padrão J, Dourado F, Ribeiro C, Lanceros-Méndez S, Sencadas V (2014) PHB-PEO electrospun fiber membranes containing chlorhexidine for drug delivery applications. Polym Test 34:64–71. https://doi.org/10.1016/j.polymertesting.2013.12.007

Fischbach MA, Walsh CT (2009) Antibiotics for emerging pathogens. Science 325:1089–1093. https://doi.org/10.1126/science.1176667

Fischer CL, Drake DR, Dawson DV, Blanchette DR, Brogden KA, Wertz PW (2012) Antibacterial activity of sphingoid bases and fatty acids against gram-positive and gram-negative bacteria. Antimicrob Agents Chemother 56:1157–1161. https://doi.org/10.1128/AAC.05151-11

Fux CA, Costerton JW, Stewart PS, Stoodley P (2005) Survival strategies of infectious biofilms. Trends Microbiol 13:34–40. https://doi.org/10.1016/j.tim.2004.11.010

Gallo J, Holinka M, Moucha CS (2014) Antibacterial surface treatment for orthopaedic implants. Int J Mol Sci 15:13849–13880. https://doi.org/10.3390/ijms150813849

Gantois I, Ducatelle R, Pasmans F, Haesebrouck F, Hautefort I, Thompson A, Hinton JC, Van Immerseel F (2006) Butyrate specifically down-regulates *Salmonella* pathogenicity island 1 gene expression. Appl Environ Microbiol 72:946–949. https://doi.org/10.1128/AEM.72.1.946-949.2006

Gomes ME, Reis RL (2004) Biodegradable polymers and composites in biomedical applications: from catgut to tissue engineering. Part 1 available systems and their properties. Int Mater Rev 49:261–273. https://doi.org/10.1179/095066004225021918

Gumel AM, Razaif-Mazinah MRM, Anis SNS, Annuar MSM (2015) Poly (3-hydroxyalkanoates)-*co*-(6-hydroxyhexanoate) hydrogel promotes angiogenesis and collagen deposition during cutaneous wound healing in rats. Biomed Mater 10:045001. https://doi.org/10.1088/1748-6041/10/4/045001

Gwynn MN, Portnoy A, Rittenhouse SF, Payne DJ (2010) Challenges of antibacterial discovery revisited. Ann N Y Acad Sci 1213:5–19. https://doi.org/10.1111/j.1749-6632.2010.05828.x

Hetrick EM, Schoenfisch MH (2006) Reducing implant-related infections: active release strategies. Chem Soc Rev 35:780–789. https://doi.org/10.1039/b515219b

Hu S-G, Jou C-H, Yang M-C (2003) Antibacterial and biodegradable properties of polyhydroxyalkanoates grafted with chitosan and chitooligosaccharides via ozone treatment. J Appl Polym Sci 88:2797–2803. https://doi.org/10.1002/app.12055

Huang CB, Altimova Y, Myers TM, Ebersole JL (2011) Short- and medium-chain fatty acids exhibit antimicrobial activity for oral microorganisms. Arch Oral Biol 56:650–654. https://doi.org/10.1016/j.archoralbio.2011.01.011

Ignatova M, Manolova N, Rashkov I, Markova N (2016) Quaternized chitosan/κ-carrageenan/caffeic acid-coated poly(3-hydroxybutyrate) fibrous materials: preparation, antibacterial and antioxidant activity. Int J Pharm 513:528–537. https://doi.org/10.1016/j.ijpharm.2016.09.062

Iida J, Une T, Ishihara C, Nishimura K, Tokura S, Mizukoshi N, Azuma I (1987) Stimulation of non-specific host resistance against Sendai virus and *Escherichia coli* infections by chitin derivatives in mice. Vaccine 5:270–274

Ismail I, Gurusamy TP, Ramachandran H, Al-Ashraf Amirul A (2017) Enhanced production of poly(3-hydroxybutyrate-*co*-4-hydroxybutyrate) copolymer and antimicrobial yellow pigmentation from *Cupriavidus* sp. USMAHM13 with antibiofilm capability. Prep Biochem Biotechnol 47:388–396. https://doi.org/10.1080/10826068.2016.1252925

Jendrossek D, Handrick R (2002) Microbial degradation of polyhydroxyalkanoates. Annu Rev Microbiol 56:403–432

Jeon JM, Brigham CJ, Kim YH, Kim H-J, Yi D-H, Kim H, Rha C, Sinskey A, Yang Y-H (2014) Biosynthesis of poly(3-hydroxybutyrate-*co*-3-hydroxyhexanoate) (P(HB-*co*-HXX)) from butyrate using engineered *Ralstonia eutropha*. Appl Microbiol Biotechnol 98:5461. https://doi.org/10.1007/s00253-014-5617-7

Johnson K, Jiang Y, Kleerebezem R, Muyzer G, van Loosdrecht M (2009) Enrichment of a mixed bacterial culture with a high polyhydroxyalkanoate storage capacity. Biomacromolecules 10:670. https://doi.org/10.1021/bm8013796

Kai D, Low ZW, Liow SS, Abdul Karim A, Ye H, Jin G, Li K, Loh XJ (2015) Development of lignin supramolecular hydrogels with mechanically responsive and self-healing properties. ACS Sustain Chem Eng 3:2160–2169. https://doi.org/10.1021/acssuschemeng.5b00405

Kandhasamy S, Perumal S, Madhan B, Umamaheswari N, Banday JA, Perumal PT, Santhanakrishnan VP (2017) Synthesis and fabrication of collagen-coated ostholamide electrospun nanofiber scaffold for wound healing. ACS Appl Mater Interfaces 9:8556–8568. https://doi.org/10.1021/acsami.6b16488

Kaplan JB, Ragunath C, Velliyagounder K, Fine DH, Ramasubbu N (2004) Enzymatic detachment of *Staphylococcus epidermidis* biofilms. Antimicrob Agents Chemother 48:2633–2636. https://doi.org/10.1128/AAC.48.7.2633-2636.2004

Karahaliloglu Z, Ercan B, Taylor EN, Chung S, Denkbaş EB, Webster TJ (2015) Antibacterial nanostructured polyhydroxybutyrate membranes for guided bone regeneration. J Biomed Nanotechnol 11:2253–2263. https://doi.org/10.1166/jbn.2015.2106

Ke Y, Zhang XY, Ramakrishna S, He LM, Wu G (2016) Synthetic routes to degradable copolymers deriving from the biosynthesized polyhydroxyalkanoates: a mini review. Express Polym Lett 10:36–53. https://doi.org/10.3144/expresspolymlett.2016.5

Ke Y, Liu C, Zhang X, Xiao M, Wu G (2017) Surface modification of polyhydroxyalkanoates toward enhancing cell compatibility and antibacterial activity. Macromol Mater Eng 302:1700258. https://doi.org/10.1002/mame.201700258

Kehail AA, Brigham CJ (2017) Anti-biofilm activity of solvent-cast and electrospun polyhydroxyalkanoate membranes treated with lysozyme. J Polym Environ. https://doi.org/10.1007/s10924-016-0921-1

Kiedrowski MR, Horswill AR (2011) New approaches for treating staphylococcal biofilm infections. Ann N Y Acad Sci 1241:104–121. https://doi.org/10.1111/j.1749-6632.2011.06281.x

Kim DY, Kim HW, Chung MG, Rhee YH (2007) Biosynthesis, modification, and biodegradation of bacterial medium-chain-length polyhydroxyalkanoates. J Microbiol 45:87–97

Kim HW, Chung MG, Kim YB, Rhee YH (2008) Graft copolymerization of glycerol 1,3-diglycerolate diacrylate onto poly(3-hydroxyoctanoate) to improve physical properties and biocompatibility. Int J Biol Macromol 43:307–313. https://doi.org/10.1016/j.ijbiomac.2008.07.002

Koneman EW, Allen SD, Janda WM, Sohreckenberger PC, Winn WC (1997) Color atlas and textbook of diagnostic microbiology. Lippincott, New York, 1736 p

Korsatko W, Wabnegg B, Braunegg G, Lafferty MR, Strempfl F (1983) Poly-D-(-)-3-hydroxybutyric acid (PHBA) a biodegradable carrier for long term medication dosage 1. Pharm Ind 45:525–527

Laverty G, Gorman SP, Gilmore BF (2013) Biomolecular mechanisms of staphylococcal biofilm formation. Future Microbiol 8:509–524. https://doi.org/10.2217/fmb.13.7

Lee SY, Park SH, Lee Y, Lee S (2002) Production of chiral and other valuable compounds from microbial polyester. In: Doi Y, Steinbüchel A (eds) Biopolymers. Wiley-VCH, Germany, pp 375–387. https://doi.org/10.1002/3527600035.bpol4014

Li Z, Loh XJ (2015) Water soluble polyhydroxyalkanoates: future materials for therapeutic applications. Chem Soc Rev 44:2865–2879. https://doi.org/10.1039/c5cs00089k

Li Z, Yang J, Loh XJ (2016) Polyhydroxyalkanoates: opening doors for a sustainable future. NPG Asia Mater 8:e265. https://doi.org/10.1038/am.2016.48

Li Y, Na R, Wang X, Liu H, Zhao L, Sun X, Ma G, Cui F (2017) Fabrication of antimicrobial peptide-loaded PLGA/Chitosan composite microspheres for long-acting bacterial resistance. Molecules 22:E1637. https://doi.org/10.3390/molecules22101637

Lipovsky A, Thallinger B, Perelshtein I, Ludwig R, Sygmund C, Nyanhongo GS, Guebitz GM, Gedanken A (2015) Ultrasound coating of polydimethylsiloxanes with antimicrobial enzymes. J Mater Chem B 3:7014–7019. https://doi.org/10.1039/C5TB00671F

Loh XJ (2017) Latest advances in antibacterial materials. J Mol Eng Mater 5:1740001. https://doi.org/10.1142/S2251237317400019

Lou Q, Ma Y, Che X, Zhong J, Sun X, Zhang H (2016) Preparation and characterization of polyhydroxyalkanoate bioplastics with antibacterial activity. Sheng Wu Gong Cheng Xue Bao 32:1052–1059. https://doi.org/10.13345/j.cjb.150488

Luef KP, Stelzer F, Wiesbrock F (2015) Poly (hydroxy alkanoate)s in medical applications. Chem Biochem Eng Q 29:287–297. https://doi.org/10.15255/CABEQ.2014.2261

Mah T-FC, O'Toole GA (2001) Mechanisms of biofilm resistance to antimicrobial agents. Trends Microbiol 9:34–39. https://doi.org/10.1016/S0966-842X(00)01913-2

Marcano A, Ba O, Thebault P, Crétois R, Marais S, Duncan AC (2015) Elucidation of innovative antibiofilm materials. Colloids Surf B Biointerfaces 136:56–63. https://doi.org/10.1016/j.colsurfb.2015.08.007

Mifune J, Nakamura S, Fukui T (2010) Engineering of *pha* operon on *Cupriavidus necator* chromosome for efficient biosynthesis of poly(3-hydroxybutyrate-*co*-3-hydroxyhexanoate) from vegetable oil. Polym Degrad Stab 95:1305–1312. https://doi.org/10.1016/j.polymdegradstab.2010.02.026

Moritz M, Geszke-Moritz M (2013) The newest achievements in synthesis, immobilization and practical applications of antibacterial nanoparticles. Chem Eng J 228:596–613. https://doi.org/10.1016/j.cej.2013.05.046

Mozejko J, Ciesielski S (2013) Saponified waste palm oil as an attractive renewable resource for mcl-polyhydroxyalkanoate synthesis. J Biosci Bioeng 116:485–492. https://doi.org/10.1016/j.jbiosc.2013.04.014

Narayanan A, Ramana KV (2013) Synergized antimicrobial activity of eugenol incorporated polyhydroxybutyrate films against food spoilage microorganisms in conjunction with pediocin. Appl Biochem Biotechnol 170:1379–1388. https://doi.org/10.1007/s12010-013-0267-2

Naveen N, Kumar R, Balaji S, Uma TS, Natrajan TS, Sehgal PK (2010) Synthesis of nonwoven nanofibers by electrospinning – a promising biomaterial for tissue engineering and drug delivery. Adv Eng Mater 12:380–387. https://doi.org/10.1002/adem.200980067

Ohashi T, Hasegawa J (1992a) D-(−)-β-hydroxycarboxylic acids as raw materials for captopril and beta lactams. In: Collins A, Sheldrake G, Crosby J (eds) Chirality in industry. Wiley, New York, pp 269–278

Ohashi T, Hasegawa J (1992b) New preparative methods for optically active β- hydroxycarboxylic acids. In: Collins AN, Sheldrake GN, Crosby J (eds) Chirality in industry. Wiley, New York, pp 249–268

Onaizi SA, Leong SSJ (2011) Tethering antimicrobial peptides: current status and potential challenges. Biotechnol Adv 29:67–74. https://doi.org/10.1016/j.biotechadv.2010.08.012

Pamp SJ, Gjermansen M, Tolker-Nielsen T (2007) The biofilm matrix: a sticky framework. In: Kjelleberg S, Givskov M (eds) The biofilm mode of life: mechanisms and adaptations. HorizonBioscience, Norfolk, pp 37–69

Panith N, Assavanig A, Lertsiri S, Bergkvist M, Surarit R, Niamsiri N (2016) Development of tunable biodegradable polyhydroxyalkanoates microspheres for controlled delivery of tetracycline for treating periodontal disease. J Appl Polym Sci 133:44128–44141. https://doi.org/10.1002/app.44128

Paraje MG (2011) Antimicrobial resistance in biofilms. In: Mendez-Vilas A (ed) Science against microbial pathogens: communicating current research and technological advances. Formatex Research Center, Badajoz, pp 736–744 ISBN 978-84-939843-1-1

Peng L-H, Huang Y-F, Zhang C-Z, Niu J, Chen Y, Chu Y, Jiang Z-H, Gao J-Q, Mao Z-W (2016) Integration of antimicrobial peptides with gold nanoparticles as unique non-viral vectors for gene delivery to mesenchymal stem cells with antibacterial activity. Biomaterials 103:137–149. https://doi.org/10.1016/j.biomaterials.2016.06.057

Pihlstrom BL, Michalowicz BS, Johnson NW (2005) Periodontal diseases. Lancet 366:1809–1820. https://doi.org/10.1016/S0140-6736(05)67728-8

Prabaharan M (2011) Prospective of guar gum and its derivatives as controlled drug delivery systems. Int J Biol Macromol 49:117–124. https://doi.org/10.1016/j.ijbiomac.2011.04.022

Pramanik N, Mitra T, Khamrai M, Bhattacharyya A, Mukhopadhyay P, Gnanamani A, Basu RK, Kundu PP (2015) Characterization and evaluation of curcumin loaded guar gum/polyhydroxyalkanoates blend films for wound healing applications. RSC Adv 5:63489–63501. https://doi.org/10.1039/C5RA10114J

Radivojevic J, Skaro S, Senerovic L, Vasiljevic B, Guzik M, Kenny ST, Maslak V, Nikodinovic-Runic J, O'Connor KE (2016) Polyhydroxyalkanoate-based 3-hydroxyoctanoic acid and its derivatives as a platform of bioactive compounds. Appl Microbiol Biotechnol 100:161–172. https://doi.org/10.1007/s00253-015-6984-4

Rai D, Singh JK, Roy N, Panda D (2008) Curcumin inhibits FtsZ assembly: an attractive mechanism for its antibacterial activity. Biochem J 410:147–155. https://doi.org/10.1042/BJ20070891

Rai M, Yadav A, Gade A (2009) Silver nanoparticles as a new generation of antimicrobials. Biotechnol Adv 27:76–83. https://doi.org/10.1016/j.biotechadv.2008.09.002

Ren Q, Ruth K, Thöny-Meyer L, Zinn M (2010) Enatiomerically pure hydroxycarboxylic acids: current approaches and future perspectives. Appl Microbiol Biotechnol 87:41–52. https://doi.org/10.1007/s00253-010-2530-6

Rodríguez-Contreras A, García Y, Manero JM, Rupérez E (2017) Antibacterial PHAs coating for titanium implants. Eur Polym J 90:66–78. https://doi.org/10.1016/j.eurpolymj.2017.03.004

Rossi S, Azghani AO, Omri A (2004) Antimicrobial efficacy of a new antibiotic-loaded poly(hydroxybutyric-*co*-hydroxyvaleric acid) controlled release system. J Antimicrob Chemother 54:1013–1018. https://doi.org/10.1093/jac/dkh477

Ruth K, Grubelnik A, Hartmann R, Egli T, Zinn M, Ren Q (2007) Efficient production of (R)-3-hydroxycarboxylic acids by biotechnological conversion of polyhydroxyalkanoates and their purification. Biomacromolecules 8:279–286. https://doi.org/10.1021/bm060585a

Sandoval A, Arias-Barrau E, Bermejo F, Caedo L, Naharro G, Olivera ER, Luengo JM (2005) Production of 3-hydroxy-n-phenylalkanoic acids by a genetically engineered strain of *Pseudomonas putida*. Appl Microbiol Biotechnol 67:97–105. https://doi.org/10.1007/s00253-004-1752-x

Schacht VJ, Neumann LV, Sandhi SK, Chen L, Henning T, Klar PJ, Theophel K, Schnell S, Bunge M (2013) Effects of silver nanoparticles on microbial growth dynamics. J Appl Microbiol 114:25–35

Seebach D, Albert M, Arvidsson P, Rueping M, Schreiber J (2001) From the biopolymer PHB to biological investigations of unnatural β- and γ-peptides. Chimia (Aarau) 55:345–353

Seidel W, Seebach D (1982) Grahamimycin A1 synthesis and determination of configuration and chirality. Tetrahedron Lett 23:159–162

Seuring B, Seebach D (1978) Syntheses and determinations of the absolute-configurations of nor-pyrenophorin, pyrenophorin, and vermiculine. Liebigs Ann Chem 12:2044–2073

Seymour RA, Heasman PA (2005) Tetracyclines in the management of periodontal diseases. J Clin Periodontol 22:22–35. https://doi.org/10.1111/j.1600-051X.1995.tb01767.x

Shah SR, Tatara AM, D'Souza RN, Mikos AG, Kasper FK (2013) Evolving strategies for preventing biofilm on implantable materials. Mater Today 16:177–182. https://doi.org/10.1016/j.mattod.2013.05.003

Shiraki M, Endo T, Saito T (2006) Fermentative production of (R)-(−)-3-hydroxybutyrate using 3-hydroxybutyrate dehydrogenase null mutant of *Ralstonia eutropha* and recombinant *Escherichia coli*. J Biosci Bioeng 102:529–534. https://doi.org/10.1263/jbb.102.529

Shishatskaya EI, Nikolaeva ED, Vinogradova ON, Volova TG (2016) Experimental wound dressings of degradable PHA for skin defect repair. J Mater Sci Mater Med 27:165. https://doi.org/10.1007/s10856-016-5776-4

Siedenbiedel F, Tiller JC (2012) Antimicrobial polymers in solution and on surfaces: overview and functional principles. Polymers (Basel) 4:46–71. https://doi.org/10.3390/polym4010046

Solaiman DKY, Ashby RD, Zerkowski JA, Krishnama A, Vasanthan N (2015) Control-release of antimicrobial sophorolipid employing different biopolymer matrices. Biocatal Agric Biotechnol 4:342–348. https://doi.org/10.1016/j.bcab.2015.06.006

Souli M, Galani I, Giamarellou H (2008) Emergence of extensively drug-resistant and pandrug-resistant gram-negative bacilli in Europe. Euro Surveill 13:19045. https://doi.org/10.2807/ese.13.47.19045-en

Stejskalová A, Almquist BD (2017) Using biomaterials to rewire the process of wound repair. Biomater Sci 5:1421–1434. https://doi.org/10.1039/C7BM00295E

Stoodley P, Hall-Stoodley L, Costerton B, DeMeo P, Shirtliff M, Gawalt E, Kathju S (2013) Biofilms, biomaterials, and device-related infections. In: Ratner B, Hoffman AS, Schoen FJ, Lemons JE (eds) Biomaterials science: an introduction to materials in medicine. Academic, Waltham, pp 565–583 ISBN 0-12-582460-2

Sutter MA, Seebach D (1983) Synthesis of (2E, 4E, 6S, 7R,10E, 12E, 14S, 15R)-6,7,14,15-tetramethyl-8,16-dioxa-2,4,10,12-cyclohexadecatet- raene-1,9-diones-A model system for elaiophylin. Liebigs Ann Chem 14:939–949. https://doi.org/10.1002/chin.198339350

Tappel R, Pan W, Bergey N, Wang Q, Patterson I, Ozumba O, Matsumoto K, Taguchi S, Nomura C (2014) Engineering *Escherichia coli* for improved production of short-chain-length-*co*-medium-chain-length poly[(R)-3-hydroxyalkanoate] (scl- R co -mcl PHA) Copolymers from renewable nonfatty acid feedstocks. Sustain Chem Eng 2:1879. https://doi.org/10.1021/sc500217p

Tenover FC (2006) Mechanisms of antimicrobial resistance in bacteria. Am J Med 119:S3–S10. https://doi.org/10.1016/j.amjmed.2006.03.011

Teow S-Y, Liew K, Ali SA, Khoo AS-B, Peh S-C (2016) Antibacterial action of curcumin against *Staphylococcus aureus*: a brief review. J Trop Med 2016:1–10. https://doi.org/10.1155/2016/2853045

Thallinger B, Prasetyo EN, Nyanhongo GS, Guebitz GM (2013) Antimicrobial enzymes: an emerging strategy to fight microbes and microbial biofilms. Biotechnol J 8:97–109. https://doi.org/10.1002/biot.201200313

Veleirinho B, Coelho DS, Dias PF, Maraschin M, Ribeiro-do-Valle RM, Lopes-da-Silva JA (2012) Nanofibrous poly(3-hydroxybutyrate-*co*-3-hydroxyvalerate)/chitosan scaffolds for skin regeneration. Int J Biol Macromol 51:343–350. https://doi.org/10.1016/j.ijbiomac.2012.05.023

Verraedt E, Pendela M, Adams E, Hoogmartens J, Martens JA (2010) Controlled release of chlorhexidine from amorphous microporous silica. J Control Release 142:47–52. https://doi.org/10.1016/j.jconrel.2009.09.022

Wang C, Sauvageau D, Elias A (2016) Immobilization of active bacteriophages on polyhydroxyalkanoate surfaces. ACS Appl Mater Interfaces 8:1128–1138. https://doi.org/10.1021/acsami.5b08664

Webb JS, Thompson LS, James S, Charlton T, Tolker-nielsen T, Koch B, Givskov M, Kjelleberg S, Al WET, Acteriol JB (2003) Cell death in *Pseudomonas aeruginosa* biofilm development. J Bacteriol 185:4585–4592. https://doi.org/10.1128/JB.185.15.4585

Williams SF, Martin DP (2005) Applications of Polyhydroxyalkanoates (PHA) in medicine and pharmacy. In: Doi Y, Steinbüchel A (eds) Biopolymers online. Wiley-VCH Verlag GmbH & Co. KGaA, Weinheim. https://doi.org/10.1002/3527600035.bpol4004

Williamson DH, Mellanby J, Krebs HA (1961) Enzymic determination of D(-)-β-hydroxybutyric acid acetoacetic acid in blood. Biochem J 82:90–96

Wu Q, Wang Y, Chen G-Q (2009) Medical application of microbial biopolyesters polyhydroxyalkanoates. Artif Cells Blood Substit Biotechnol 37:1–12. https://doi.org/10.1080/10731190802664429

Zaborowska M, Welch K, Brånemark R, Khalilpour P, Engqvist H, Thomsen P, Trobos M (2014) Bacteria-material surface interactions: methodological development for the assessment of implant surface induced antibacterial effects. J Biomed Mater Res Part B Appl Biomater 103:179–187. https://doi.org/10.1002/jbm.b.33179

Zhu X, Jun Loh X (2015) Layer-by-layer assemblies for antibacterial applications. Biomater Sci 3:1505–1518. https://doi.org/10.1039/C5BM00307E

Polyhydroxyalkanoates Applications in Drug Carriers

5

Christos Papaneophytou, George Katsipis, Eleftherios Halevas, and Anastasia A. Pantazaki

Abstract

Drug delivery carriers (DDCs) have been successfully used in the previous years for the controlled release of therapeutic compounds for the treatment and cure as well as the diagnosis of many diseases including cardiovascular disorders, HIV infection, Alzheimer's disease, rheumatoid arthritis, and cancer. Several biodegradable/biocompatible polymers including poly- (lactic acid-co-glycolic acid), poly-L-lactic acid, and poly-L-lysine have been used as drug carriers for the transportation and controlled release of therapeutic compounds to the tissue or cells of action. Amongst the various biodegradable polymers tested as DDCs, polyhydroxyalkanoates (PHAs) were extensively investigated due to their unique biochemical properties. The practical applications of PHAs in biomedical/pharmaceutical areas have been approved by FDA in 2007. PHAs are suitable for in vivo therapeutic applications including tissue coating, tissue regeneration devices, stents, tubing, bone or tissue cement wound dressings, and drug delivery. PHAs are biodegradable, biocompatible, and hydrophobic while they are used to produce microspheres, microcapsules, and nanoparticles as well as films and porous matrices. Most importantly, PHAs have significant advantages over other biodegradable polymers, since their degradation products, including oligomers and monomers, are not toxic to the surrounding tissues. The variety of PHAs, along with their biocompatibility and their excellent biochemical and physical properties, make them excellent biopolymers for use in drug delivery. The activity of some anti-cancer agents was enhanced when they combined with PHAs or HA

C. Papaneophytou
Department of Life and Health Sciences, School of Sciences and Engineering, University of Nicosia, Nicosia, Cyprus

G. Katsipis · E. Halevas · A. A. Pantazaki (✉)
Laboratory of Biochemistry, Department of Chemistry, Aristotle University, Thessaloniki, Greece
e-mail: natasa@chem.auth.gr

V. C. Kalia (ed.), *Biotechnological Applications of Polyhydroxyalkanoates*, https://doi.org/10.1007/978-981-13-3759-8_5

monomers. PHA-based microspheres, nanoparticles, non-woven membranes, and hydrogels have been successfully used as carriers of therapeutic agents for several diseases including cancer and inflammatory diseases.

Keywords

Drug delivery carriers · Microspheres · Nanoparticles · Hydrogels · Micelles · Polyhydroxyalkanoates

5.1 Introduction

Administration of free drugs, either orally or through intravenous injection, is associated with limitations such as high toxicity, low solubility, and low membrane permeability that restrict their use. In addition, the administration of free drugs in several cases results in the development of different side effects. In general, an ideal drug should exhibit (i) high therapeutic efficiency at low concentrations, (ii) limited side effects and (iii) target specific action (Dutta 2007). However, the process of identification and development of new drugs is not only expensive but also time-consuming (Tan et al. 2016). Thus, there were some efforts to improve the efficiency of "old" drugs using a variety of techniques such as dose titration (Hsu and Laddu 1994), personalized drug therapy (Vitezić et al. 2016), and therapeutic drug monitoring (Sanathanan and Peck 1991; Kang and Lee 2009) but the success rate was low. Modern research on drug development focuses not only on the identification of new molecules to cure or prevent disease but also on the elimination of side effects caused by drugs in various tissues and organs. The most promising approach to develop an ideal drug is the encapsulation of therapeutic compounds in biocompatible delivery carriers (Dutta 2007). Drug delivery carriers (DDCs), aiming to deliver drugs to the desired site of the body at a desirable time frame and at a predetermined rate and in order to cure a disease, have evolved since the early 1960s in the form of macroscopic controlled drug delivery devices and implants of macro scale dimensions (Hoffman 2008; Park 2014; Abdul Karim and Loh 2015; Kai et al. 2015; Ye et al. 2015; Chee et al. 2016; Karim et al. 2016; Pushpamalar et al. 2016; Li and Loh 2017). In general, a DDC is a device or formulation that facilitates the transportation of a therapeutic agent into the body and increases both its safety and efficacy by controlling the release of the drug to the target tissue in a specific time and rate (Jain 2008).

Encapsulation of therapeutic agents into drug carries in order to be delivered to a specific part of the body and subsequently released from the carrier in a slow and controlled rate, has been successfully used for the cure and control of diseases in both animals and humans (Tiwari et al. 2012). Most importantly, this alternative strategy allows the administration of highly toxic, hydrophobic or low membrane-permeable drugs to the tissue of interest. In addition, the drug concentrations in healthy tissues are reduced, whilst their concentrations in target tissues or cells (e.g., tumors) is increased (Sur et al. 2014). In general, therapeutic agents are dispensed through a no immunogenic, biodegradable carrier that must be remained intact until it reaches the target tissue and subsequently releases the drug at a sustained rate. Therefore, the drug carriers must be elaborately designed to deliver the loaded drug to the target tissue only, without dispersing their load randomly in the body, especially when toxic drugs, such as antitumor agents, are being delivered (Dutta 2007). Currently, there is an increased interest in the formulation of DDCs, especially to administrate antitumor agents directly to cancer cells without affecting the normal cells (Peer et al. 2007; Fang et al. 2011). However, the development of DDCs is challenging because these carriers should be able to cross several cellular barriers to deliver the drug to the target side. In general, most cellular barriers prohibit the transportation of drug molecules or particles with undesired surface properties and/or large sizes (Li and Loh 2017). For this purpose, a variety of materials, including natural biopolymers (e.g., hyaluronic acid, chitosan, and collagen), bio-ceramics (e.g., bioactive glass and hydroxyapatite), and synthetic biodegradable polymers [e.g., poly(N-vinyl pyrrolidone) (PVP), poly(ethylene glycol) (PEG), poly(lactic acid) (PLA), poly(glycolic acid) (PGA), and PLA/PGA copolymers] has been used for the development of DDCs (Liechty et al. 2010; Zhao 2011). In addition, different techniques including spray and dip coating, compression, and encapsulation have been employed by pharmaceutical companies to combine therapeutic compounds with polymers (Liechty et al. 2010; Zhao 2011). Polymers are excellent candidates for the development of DDCs because they are flexible, and therefore it is possible to manipulate multiple functionalities that are essential for efficient drug delivery. The polymers that are used for the manufacturing of DDCs must exhibit the appropriate physical properties, must be stable and most importantly biocompatible. However, some of the polymers that are widely used for the synthesis of DDCs, (e.g., silicone), probably promote carcinogenesis (Cammas et al. 1999). In addition, various drug delivery systems have been developed using gels, cyclodextrins, liposomes, proliposomes, and prodrugs, among others, for the delivery of drugs through the nasal and pulmonary routes (Tiwari et al. 2012). However, many therapeutic compounds cannot be easily or effectively loaded into these carriers (Sur et al. 2014). Thus, there is a great need for non-toxic, biocompatible, and biodegradable polymers for the development of safer DDCs. Among the available biodegradable polymers, there is an increasing interest in the group of PHAs (Shrivastav et al. 2013; Ray and Kalia 2017). PHAs are natural bacterial products that can be synthesized inexpensively from waste carbon sources, like used oils, fruit peels, whey, and municipal wastewaters (Vega-Castro et al. 2016; Bengtsson et al. 2017; Sharma et al. 2017). They have also proven to be biocompatible and are exploited for various medical applications in the human body (Grage et al. 2009; Chang et al. 2014) and in vitro cell cultures (Shah et al. 2012; Li et al. 2015). More than 150 different monomers of PHAs have been discovered so far, that can be integrated into homo- or copolymeric materials, but hydroxyl butyrate (HB), hydroxyl hexanoate (HHx) and hydroxyl pentanoate [mostly known as hydroxyl valerate (HV)] are the most studied of them (Shrivastav et al. 2013; Li and Loh 2017).

PHAs are excellent candidates for the development of DDCs because they are biodegradable, biocompatible, and thermo-processable, they can be produced easily and their properties are easily modified and they degrade through surface erosion (Lenz and Marchessault 2005). PHAs can also be turned into microspheres, microcapsules, nanoparticles, films, and porous matrices (Nobes et al. 1998; Orts et al. 2008). Moreover, it has been demonstrated that PHA-based delivery systems are excellent candidates in treating highly resistant infections because they can deliver the appropriate amount of antibiotic agents at infection sites (Gould et al. 1987; Gursel et al. 2002).

In this chapter, the state-of-the-art knowledge of the usage of PHAs in the development of drug delivery carriers including micro- and nano- spheres, hydrogels, and non-woven membranes is summarized. The properties of these biopolymers can be easily manipulated by varying the nature of the monomer incorporated into PHA homopolymers and the nature and/or the ratio of different hydroxyalkanoates (HA) monomers incorporated into PHA heteropolymers. In addition, strategies for the modification of surface properties of these materials and for the sustained release of the drug to the tissue or cells of interest are highlighted.

5.2 Anticancer

Bioactive peptides exhibiting different modes of action; such as antioxidant, antiproliferative and antimicrobial, are found in many species (Suarez-Jimenez et al. 2012). Even though many peptides have been used as antibacterial agents, exceptional interest has been recently concentrated on the use of peptides as anti-cancer agents (Mader and Hoskin 2006). 3-R hyrdroxyalkanoates (3HAs) monomers have successfully been used as conjugates to enhance the efficiency of peptides exhibiting anti-cancer activity. D-Peptide 18 (DP18; WKLFKKIPKFLHLAKKF) is a well-studied peptide showing cytotoxic activity against human cancer cell lines (Shin et al. 2001). The conjugation of the HA monomer (R)-3-hydroxy-decanoic acid (3HD) to a series of DP-18 derivatives, aiming to develop novel peptides with enhanced anti-cancer activity, has been investigated (O'Connor et al. 2013). In more detail, the anti-proliferative activity of 3HD-cojugated peptides was assessed in vitro using different types of human cancer cell lines. The addition of the 3HD to the sequence of DP18 peptides enhanced the antiproliferative activity of all peptides, in comparison with their unconjugated analogues. These effects can be probably attributed to the greater uptake of the conjugated peptides into the cells. Interestingly, conjugation of decanoic acid to peptides did not increase their anticancer activity.

Additionally, proliferation studies of ibuprofen conjugates with oligo(3-hydroxybutyrate) (3HB) in HT-29 and HCT 116 colon cancer cells showed enhanced anti-proliferative activity compared to free ibuprofen, indicating the potential of similar conjugates in the improvement of the bioavailability and cytotoxicity of selected drugs with similar anti-proliferative profile (Zawidlak et al. 2010).

5.3 PHB Microspheres

The design of injection systems in the forms of micro particles (including microspheres and microcapsules) for the targeted delivery of a therapeutic compound to a tissue and the controlled drug release from microspheres' core, is an extremely challenging aim of study in the medical field (Shishatskaya et al. 2008b; Artsis et al. 2010). The main advantage of administrating therapeutic agents using drug delivery micro particles is that the concentration of the drug can be adjusted at the desired level for a long-term period, ranging from several weeks to several months, resulting in a more effective pharmaceutical action (Kost and Langer 2001; Artsis et al. 2010). The controlled release of therapeutic agents including proteins, peptides, anesthetics, antibiotics, anticancer agents, anti-inflammatory agents, hormones, steroids, and vaccines from microsphere or microcapsule DDCs composed of biodegradable materials, has been successfully used for both the prevention and cure of diseases in both animals and human with minimal side effects (Gould et al. 1987; Nobes et al. 1998; Orts et al. 2008).

A variety of methods have been described for the encapsulation of therapeutic agents in polymeric microspheres and microcapsules such as single- (Hamoudeh et al. 2007; Murugesan et al. 2007) or multiple- emulsion (Hussein et al. 2013) or phase separation (Mayya et al. 2003). The choice of encapsulation technique is based on the nature of the therapeutic molecule, the properties of the polymeric material, as well as on the duration and type of therapy. In addition, the formulation process must follow the rules of ideal controlled drug delivery systems, including high encapsulation efficiency, low burst release, stability of the loaded therapeutic agent, while syringeability and injectability are also major requirements for microspheres to be exploited as DDCs (Ravi et al. 2008; Cilurzo et al. 2011).

The water-in-oil-in-water (w/o/w) double emulsion method followed by solvent extraction/evaporation is the favorable technique for the encapsulation of hydrophilic molecules. This encapsulation method provides protection of the hydrophilic substance against denaturation by eliminating its contact with the organic solvent (Vilar et al. 2012). On the other hand, hydrophobic therapeutic agents are usually encapsulated within hydrophobic polymeric microspheres by the oil-in-water (O/W) emulsion solvent evaporation method (McGinity and O'Donnell 1997; Freitas et al. 2005). During this process, a hydrophobic drug is diluted in a solution of a volatile organic solvent and a hydrophobic polymer. The emulsification of the mixture is carried out within an aqueous phase while the organic solvent is removed, resulting in the precipitation of microspheres containing the drug. The final properties of the microspheres, and the release profile of drug, are affected by the hydrophilicity of drug, the nature of the oil/organic phase, the type and amount of used surfactant, the ratio of oil to water, the ratio of drug to polymer, the temperature and the stirring speed (McGinity and O'Donnell 1997). The main advantages of this encapsulation technique are the high reproducibility and the ability to manipulate the micro particle properties such as the size of spheres (Freitas et al. 2005; Tian et al. 2012).

To attain a sustained release of the therapeutic compounds from polymeric microspheres, an accurate control of the micro particles morphology, structure and size are required (Ravi et al. 2008; Ansary et al. 2016). Thus, there is a need for biocompatible, biodegradable, and non-toxic polymers. Currently, polymers of monocarboxylic acid derivatives (i.e. polylactides and polyglycolactides), are widely used for the development of bio-resorbable microspheres for the controlled drug release as they have been approved by the US food and drug administration (FDA) for the formulation of DDCs as well as for other medical applications (Makadia and Siegel 2011; Monnier et al. 2016). During the past years, different therapeutic molecules were successfully encapsulated into a variety of synthetic and natural biodegradable polymers, including PLA, PLGA, poly (ε-caprolactone-co-glycolic acid), and PHAs (Narayani and Panduranga Rao 1996; Sendil et al. 1999; Carino et al. 2000; Avgoustakis et al. 2002; Le Ray et al. 2003).

Among the various biodegradable polymers tested in controlled drug release, poly(3-hydroxy-butyrate)/(PHB) found to be an excellent candidate for the development of drug delivery carriers because it is biocompatible and biodegradable (Luef et al. 2015). Degradation of PHB results in the formation of the monomer, D-(-)-3-hydroxybutyric acid which is normally found in blood. Most importantly, microspheres of various sizes can be produced using PHB. To design efficient microspheres for drug delivery is essential to understand the relationship among the size of microspheres, the drug loading efficiency, and the properties of the loaded drug (Luef et al. 2015). Several drugs with various pharmacological activities were encapsulated into polymeric controlled release systems composed of PHB homopolymer. The encapsulation of 7-hydroxethyltheophylline in tablets composed of PHB homopolymer (Mw ~2000 kDa) was also studied (Korsatko et al. 1983). Following subcutaneous implantation of the PHB tablets containing 7-hydroxethyltheophylline, a quick release of the drug as well as a decrease in the weight of polymeric tablets was observed. However, PHB polymers with a molecular mass higher than 100 kDa were inappropriate for long-term medication dosage.

The release of methyl red from PHB matrices has been also studied (Pouton and Akhtar 1996). Methyl red was being released from PHB matrix for more than 7 days into two phases: a rapid one (called the "*burst effect*") and a steady-released phase. It has been proposed that drug release from such matrices can be controlled by modifying both the porosity and the rate of erosion of the polymeric matrix (Pouton and Akhtar 1996).

Kawaguchi et al. studied the release of various antitumor prodrugs of 5-fluoro-2′deoxyuridine (FUdR), with different water solubility, from PHB microspheres, with a diameter ranging from 100 to 300 μm, of different molecular mass, i.e. 65, 135, and 450 kDa. The antitumor effects of these prodrugs were tested against P388 leukemia in mice. The results of this study demonstrated that: (i) the molecular mass of PHB, (ii) the chemical properties of the encapsulated drug and (iii) the drug content, had a significant effect on the release rate of the drug from the polymeric matrix. The release rates of lipophilic prodrugs from the PHB microspheres was slower compared to those of hydrophilic drugs regardless the molecular weight of the polymeric matrix. In addition, the release rates of drugs were faster from the microspheres composed of the 65 kDa PHB compared to those composed of either 135 or 450 kDa PHB. Finally, the higher the FUdR-analogs loading in the polymeric spheres, the faster was the drug release, while the PHB microsphere itself showed excellent compatibility and low toxicity with mice. Most importantly, a single injection of the encapsulated FUdR-analogs in PHB microspheres, (composed of the 450 kDa polymer), resulted in a higher antitumor activity against P388 leukemia in mice compared to free prodrugs (Kawaguchi et al. 1992).

The in vitro release of rifampicin from PHB microspheres of different diameters ranging from 5 to 100 μm was also studied (Kassab et al. 1997). Similar to the study of Kawaguchi et al., the rate of drug release from the PHB microspheres was affected by the size of the polymer and drug loading. Interestingly, a drug loading of 407.6 mg rifampicin/g PHB was achieved, resulting in a rapid release rate of the drug, as approximately 90% of the drug was released in 24 h. The same research team reported the potential application of PHB microsphere as alternative chemoembolization agents after in vivo experiments with dogs (Kassab et al. 1999).

Salman et al. reported the encapsulation of tramadol, an analgesic, drug into PHB homopolymer microspheres. A continuous release of tramadol from PHB microspheres for 7 days was observed in vitro, while a prolonged in vivo epidural analgesic effect was noticed in rats for almost a day (Salman et al. 2003).

Bonartsev et al. synthesized PHB microspheres for the encapsulation of dipyridamole, a vasodilator/antithrombotic drug, and indomethacin, an anti-inflammatory drug. Drug release kinetic studies revealed a continuous and uniform release of both therapeutic compounds from polymeric microspheres for more than 1 month (Bonartsev et al. 2006, 2007). Shishatskaya et al. described the encapsulation of rubomycin, an effective antitumor drug, in PHB microspheres for the evaluation of its antitumor efficiency in a mouse model. In vivo studies revealed the continuous release of rubomycin from microspheres, resulting in the inhibition of proliferation rate of Ehrlich's carcinoma and improved survival of mice with tumors, without any side effects (Shishatskaya et al. 2008a).

Overall, these studies revealed that the release of the therapeutic compounds from PHB microspheres (and microcapsules) takes places via degradation of the polymeric matrix and diffusion of the drug through the surface of the spheres, while the two processes operate simultaneously (Kamaly et al. 2016). In several cases, the high initial drug release (burst effect) from the microspheres has been correlated with the high amounts of drug at the surface of the polymeric matrixes (Salman et al. 2003). In addition, the rate of drug release depends on its nature, the molecular weight and thickness of PHB and the ration initial loading of the drug in the polymeric matrix. In general, the size of microspheres affects the rate of drug release (Salman et al. 2003).

Even though several other drugs, including antitumor drugs (paclitaxel), anti-inflammatory drugs (prednisolone, dexamethasone, flurbiprofen), and antibiotics (levofloxacin, rifampicin, ciprofloxacin, metronidazole), have been successfully encapsulated in PHB microspheres for continuous administration to the target tissue (Artsis et al. 2010) in several cases, diffusion of the drug has been detected from the amorphous zones of microspheres (Monnier et al. 2016). It has been suggested that, by changing the properties of the PHB, the size of microspheres, and the initial amount of drug it is possible to control the drug release rate. Thus, the attention has been turned to copolymers of PHB with other hyrdoxyalkanoates, especially with 3-hydroxyvalerate (HV). Copolymerization of PHB with HV results in the poly-(3-hydroxybutyrate-co-3-hydroxyvalerate) [P(HB-HV)] with lower melting and lower glass temperatures compared to PHB. In addition, a more elastic polymer can be produced by increasing the amount of HV in the P(HB-HV) copolymer (Modi et al. 2011). Due to its improved physicochemical properties (biodegradability and biocompatibility) P(HB-HV) have been tested as material from the development of polymeric drug delivery microspheres (Hazer et al. 2012; Scheithauer et al. 2015). Most importantly, P(HB-HV) is suitable for the development of long-term drug carries as its degradation rate is slower than that of the commonly used synthetic biodegrade polymers (e.g., PLA and PLGA) (Amass et al. 1998; Monnier et al. 2016). P(HB-HV) is hydrophobic (Li et al. 2014), while it degrades slowly in vitro (Kose et al. 2003). The release of encapsulated drug from P(HB-HV) microparticles is based mainly on diffusion from the amorphous zones of the spheres, but not on degradation of polymeric matrix, and therefore copolymers of P(HB-HV) are more suitable for the development of drug delivery systems compared to PHB homopolymers (Li et al. 2014; Scheithauer et al. 2015; Monnier et al. 2016). Moreover, as P(HB-HV) (i) does not induce inflammation, (ii) its degradation does not produce any harmful acidic products, and (iii) the degradable products are normal components of the blood, is considered as an excellent candidate for the development of drug delivery carriers (Kose et al. 2003; Pich et al. 2006). Several therapeutic agents including anticancer drugs and antibiotics have been successfully encapsulated in PHB-HV microspheres as described in the following paragraphs.

Panith et al. reported the synthesis of PHB and P(HB-HV) microspheres for the encapsulation of tetracycline (TC) in order to treat periodontal disease. TC was loaded into PHB homopolymer and three different P(HB-HV) copolymers composed of 5, 12 or 50% (w/w) HV by the double emulsion-solvent evaporation method. The effect of molecular weight of polyvinyl alcohol (PVA) that was used as a surface stabilizer on encapsulation efficiency, drug loading, and drug release profile as well as on the size of microspheres was investigated. The results of this study revealed that the TC release profile was affected by both the HV percentage in the copolymer, as well as by the molecular weight of PVA. In addition, the highest TC loading was recorded when PHB and PVA of lower molecular weights were used. These polymeric microspheres containing TC exhibited a high killing activity against bacteria causing periodontitis (Panith et al. 2016).

Sendil et al. synthesized P(HB-HV) microcapsules and microspheres for the controlled release of TC, in both its neutral and hydrochloric-complex form (TC-HCl). The highly soluble acidic tetracycline was neutralized with NaOH in order to increase the encapsulation efficiency and decrease its releasing rate. The insoluble neutralized tetracycline was encapsulated much more easily and in higher amounts (51.9–65.3%) compared to the soluble TC-HCl (Sendil et al. 1999).

The sustained release of the anti-cancer drug 5-fluorouracil (5-FU) from P(HB-HV) microspheres for the treatment of cancer has been reported (Khang et al.

2001). Gangrade and Price reported the development of microspheres composed of either PHB homopolymeror P(HB-HV) copolymer, containing 24% or 9% (w/w) HV, for the encapsulation and sustained release of progesterone. Interestingly, progesterone was sufficiently encapsulated in microspheres regardless the composition of the polymer. The slowest release of progesterone in vitro was recorded from the P(HB-HV) microspheres containing 9% HV. It should be noted that the porosity of this copolymer was lower compared to that of PHB homopolymer and P(HB-HV) copolymer containing 24% HV (Gangrade and Price 1991).

Chen and Davis examined the release profile of diazepam, an anxiolytic drug, from microspheres (with a diameterranging from 30 to 40 μm) composed of P(HB-HV) copolymer containing 21% HV. The release of the drug from microspheres was examined for 30 days and a high initial burst effect was observed. This effect was not observed when the P(HB-HV) microspheres were coated with gelatin (Chen and Davis 2002).

The development of P(HB-HV) microspheres for sustained release of flurbiprofen, a nonsteroidal anti-inflammatory drug, has also been reported. The properties of microspheres were optimized by statistically designed experiments, using as independent variables the polymer and stabilizer concentration. The results revealed that although these two factors affected the properties of microspheres, they did not influence the release profile of flurbiprofen from microspheres (Coimbra et al. 2008).

Duran et al. reported the encapsulation of rifampicin in P(HB-HV) microspheres. The encapsulation efficiency, as well as the drug release rate from microspheres, were affected by the initial P(HB-HV), emulsifier (PVA) and rifampicin concentrations. Cytotoxicity experiments revealed that encapsulation of rifampicin in P(HB-HV) microspheres decreased the toxicity of the drug, without affecting its antimicrobial activity (Duran et al. 2008).

Li et al. have recently reported the development of multifunctional P(HB-HV) microspheres for the encapsulation of superparamagnetic iron oxide nanoparticles (SPIONs). The encapsulation efficiency of SPIONs into the polymeric matrix was increased after their surface modification with lauric acid. The biocompatibility of the synthesized magnetic P(HB-HV) microspheres was verified using HT-29 cells (adherent cells) and both Jurkat cells (suspension cells). The widely used antibiotic TC-HCl was successfully loaded into the microspheres and released at a sustained rate (Li et al. 2016).

The above studies showed that, in several cases, the surface of PHB or P(HB-HV) microspheres is porous and rough resulting in a high initial burst release (Monnier et al. 2016). It has been demonstrated that the properties of PHB or P(HB-HV) microspheres, as well as the drug release profile, may be improved by their physical (direct) blending with other polymers without any chemical reaction. The combination of the properties of different polymeric material resulted in the formation of new, more suitable materials for the sustained release of drugs from the polymeric microspheres. It should be noted that blending two different polymers usually leads to completely different behavior, compared to copolymers composed of these polymers (Imre and Pukánszky 2013).

Chaturvedi et al. used the solvent evaporation method to synthesize blend microspheres of PHB and cellulose acetate phthalate (CAP) in compositions of 2:1, 1:1, and 1:3 (w/w) for the sustained release of the anticancer drug, 5-FU from the treatment of colon cancer. The in vitro studies revealed a pH-dependent release of 5-FU from blend microspheres. In addition, blend microspheres delivered the drug more efficiently to the colon than the microspheres composed of PHB homopolymer (Chaturvedi et al. 2011).

Bazzo et al. reported the improved release of ibuprofen from P(HB-HV)/PLA blended microspheres. Microspheres composed of different ratios of P(HB-HV)-to-PLA were synthesized using the oil-in-water emulsion solvent evaporation method. The blended microspheres composed of P(HB-HV)/PLA at a 30/70 ration resulted to better ibuprofen release profile. In vivo experiments in rats revealed a sustained release of ibuprofen from the microspheres, while the maximum drug release was recorded 6 h after administration (Bazzo et al. 2012).

Zhu et al. compared the release profiles of hepatocyte growth factor (HGF) from microspheres composed of PLGA, a copolymer of P(HB-HV) containing 8% HV, and a blend of PLGA and P(HB-HV) (1:1 w/w). Microspheres were composed using the water-in-oil-in-water emulsion solvent evaporation method. The optimum HGF release profiles were observed from blended P(HB-HV)/PLGA microspheres. In addition, blended microspheres showed a better degradation rate compared to pure PLGA or P(HB-HV) copolymer. HGF was efficiently being released for more than 40 days, as was demonstrated by Hep3B cell inhibition assay (Zhu et al. 2009).

Monnier et al., developed blended microspheres of P(HB-HV) and poly(ethylene glycol)/(PEG) in order to eliminate the drug initial burst release from P(HB-HV) microspheres. The effects of PEG concentration and molecular weight on biocompatibility, degradation, encapsulation efficiency, and drug release rate were investigated (Monnier et al. 2016). PEG, a nontoxic and approved by the FDA polymer, is routinely used in polymeric blending for the encapsulation of therapeutic compounds and other bioactive molecules (Baird et al. 2010). Heparin was used as a model drug and encapsulated into P(HB-HV) or blended P(HB-HV)/PEG (of various PEG concentrations and molecular weights) microspheres using the double emulsion solvent evaporation technique. Blending of P(HB-HV) microspheres with PEG resulted in the decrease of their degradation rate and in a significant reduction of the drug initial burst effect. In addition, the drug release profiles could be adjusted by changing the molecular weight and the PEG concentration (Monnier et al. 2016).

In another study, Li et al. improved the drug release profiles of TC-HCl and daidzein (an anti-osteoporosis drug), by combining chitosan and P(HB-HV) microspheres in order to treat periodontitis and promote bone regeneration respectively (Scheithauer et al. 2015).

Overall, microspheres composed of biodegradable polymers have been widely used for the encapsulation of a variety of therapeutic agents, and to provide controlled and sustained drug release rates to the targeted tissue or cells. PHB photopolymers, PHB copolymers with other HAs, as well as their blends with other biocompatible polymers, are considered as excellent materials for the development of drug-encapsulated microspheres due to their high biodegradability and biocompatibility.

Even though PHB has been successfully used for the development of micro-scale drug carries, its main drawbacks are its stiff polymer chains and high crystallinity that, in several cases, results in a low drug encapsulation or fast drug release (Fig. 5.1).

To overcome these limitations, copolymers of PHB with HV (at various proportions) have been subsequently used as polymeric materials for the development of drug delivery microspheres. With their improved biophysical properties over PHB and other biodegradable polymers such as PLA and PLGA, and due to its unique biocompatibility P(HB-HV) became an attractive polymer the for long-term drug release systems. However, a high drug initial burst effect has been observed from P(HB-HV) in many studies. This burst effect has been eliminated by the direct blending of PHB and P(HB-HV) microspheres with other biodegradable polymers including PLA, PLGA, and chitosan (Fig. 5.1). Blended polymers have several advantages over copolymers because they combine the properties of the two polymers without while for their preparation does not involve the use of chemicals (Shah et al. 2012). In addition to the use of PHB-based polymers of the development of microspheres for the controlled drug delivery, these polymers can be used for the development of carriers for the sustained release of inhibitors or activators of enzymes (Artsis et al. 2010).

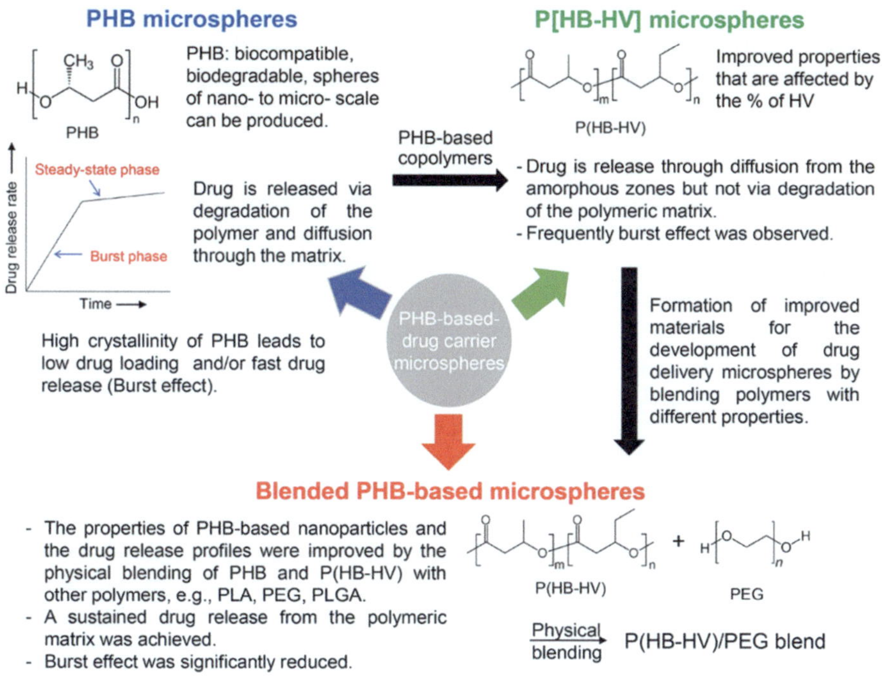

Fig. 5.1 Utilization of PHB-based polymers for the development of drug-delivery carrier microspheres

5.4 Encapsulation

5.4.1 Synthesis of Capsules

In order for highly efficient and side-effect-free therapies to be applied in vivo, there is an increasing need for specific transfer of pharmaceutical compounds. Encapsulation, as a cutting-edge field in nanotechnology, is a fine technique for the delivery of drugs, metal ions, biomolecules, and other compounds, in order to increase their bio-distribution. The main scope of these particles is to deliver their cargo specifically to the target-tissue while preserving their integrity during digestion or in the systemic circulation. Controlled release of the pharmaceutical compounds in physiological conditions is also a core challenge during the synthesis of microcapsules (MCs) or nanocapsules (NCs). It is also important that the selected encapsulating agent is biocompatible, possesses suitable physical properties, and is cost-efficient (Shrivastav et al. 2013). PHAs are most-promising compounds, covering all the aforementioned advantages demanded from an encapsulating agent (Errico et al. 2009).

NCs (size <1 μm) or MCs (size >1 μm) are particles consisting of a polymeric coat surrounding compounds in liquid, solid or molecular dispersion (de Morais et al. 2016). The method usually employed for encapsulation with PHAs is selected based on the hydrophobic or hydrophilic properties of the carrier compound. Emulsion-solvent evaporation is the most widely used method to encapsulate hydrophobic compounds in preformed polymers, which constitute the most challenging group of molecules for good and selected bio-distribution (Nobes et al. 1998; Martínez Rivas et al. 2017). In this method, a biphasic system of an organic solvent and an aqueous solution (non-solvent) is used. The organic solvent is usually chloroform (CF) (Heathman et al. 2014; Wu et al. 2014) or dichloromethane (DCM) (Gürsel and Hasirci 1995; Shah et al. 2011; Murueva et al. 2013), in which the PHAs and the hydrophobic drug are dissolved. The aqueous phase contains one or more emulsifying agents like polyvinyl alcohol (PVOH) (Shah et al. 2011; Murueva et al. 2013), polyvinyl acetate (PVAc) (Peng et al. 2016), Tween-80 (Gürsel and Hasirci 1995; Kilicay et al. 2016), and cationic cetyltrimethyl ammonium bromide (CMAT) (Lu et al. 2010), and hydrophilic drugs. The two phases are mixed (oil in water or water in oil) and magnetic stirring or sonication is employed for the formation of the capsules and for the aspiration of the organic solvent (Minost et al. 2012; Martínez Rivas et al. 2017).

If hydrophilic molecules are to be encapsulated, a double-emulsion method is usually employed. In this method, the aqueous solvent which contains the hydrophilic compound is firstly sonicated with the organic solvent and the mix is then added in an aqueous solution of an emulsifier, to form the capsules. The double emulsion solvent evaporation technique is a better alternative to the standard procedure, in order to avoid denaturation of the biological effector, but lacks the entrapment efficiency of the standard method, as compounds may diffuse in the second aqueous treatment (Embleton and Tighe 1993).

In either case, the capsules' formation is based on the rapid diffusion of the solvent into the aqueous phase which is accompanied by turbulence in the medium. It is important that the solvents employed are mutually and total miscible and that the mixture of the two solvents be a poor solvent for the used PHA. The capsules are then isolated by evaporation of the aqueous system either by rotary evaporation or freeze drying or, in some cases, by centrifugation at 22,000 g (Minost et al. 2012; Martínez Rivas et al. 2017).

After synthesis of the capsules, it is important to calculate the yield of the employed methodology, for optimization of the physicochemical parameters of the process. The production yield is calculated as the percentage (%) of the particles mass against total mass of the starting materials used in the synthesis. For compound-loaded particles, compound's mass should be included in the total starting mass calculation (El-Say 2016). In addition, it is important to evaluate the entrapment/encapsulation efficiency (*E*) and the time-related release of the pharmacological compound in physiological buffers. *E* is defined as the percentage (%) of the encapsulated mass of the compound against its initial mass employed in the reaction. Another measure that is usually calculated, is the percentage of drug loading (DL), which equals the percentage of compound's mass against capsules' mass (Pramual et al. 2016). Both of these values can be easily determined by measuring the remaining/non-loaded drug at the emulsion system by spectrophotometry (Murueva et al. 2013), immune-assays (Peng et al. 2016) or other suitable methods that in accordance with the compounds' physicochemical or biochemical properties. Alternatively, a known capsules' mass is dissolved in a strong solvent, like ethanol, and the quantity of the entrapped compound is determined per gram of capsules (Peng et al. 2010). Release study is examined in vitro at common physiological buffers like phosphate buffered saline (PBS), pH 7.4. The capsules are incubated usually at 37 °C for several minutes or hours and the compound of interest is determined at the supernatant after centrifugation of the capsules' dispersion (Murueva et al. 2013). Dialysis bags, e.g., bags of known-size pores, are sometimes involved in the process. The dispersed capsules are enclosed in a dialysis bag and incubated for various time lapses inside a buffer or plain water. The released compound is then determined at the dialysis medium (Peng et al. 2016).

5.4.2 Characterization Techniques for PHA Capsules

For the NC or MC to be exploited as pharmaceutical interventions, it is important that they are characterized by their physicochemical properties. A plethora of techniques (Fig. 5.2) has been established for their characterization in order to determine important values like size, distribution of sizes, stability, surface morphology, and the entrapment and release of the pharmacological compound of interest (Bootz et al. 2004; de Morais et al. 2016).

Size and particle size distribution (PSD) or particle distribution index (PDI) ($PDI = \langle d^2 \rangle / \langle d \rangle^2$) are usually determined by dynamic light scattering (DLS) or scanning electron microscopy (SEM). DLS is based on the diffraction of

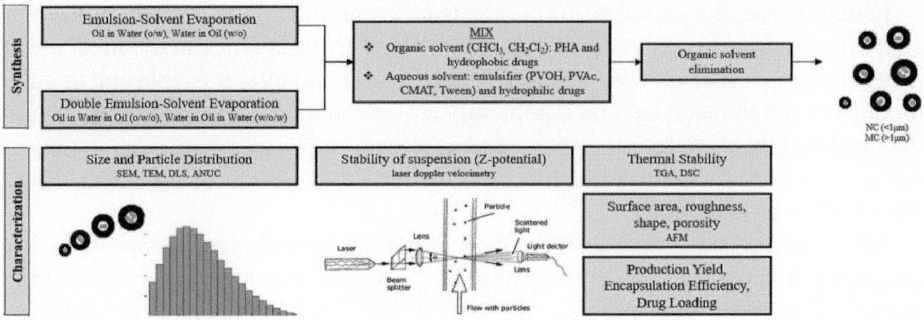

Fig. 5.2 General methodology for receiving and characterizing PHA capsules. *$CHCl_3$* chloroform, *CH_2Cl_2* dichloromethane, *PVOH* poly-(vinyl-alcohol), *PVAc* poly-(vinyl-acetate), *CTAB* cationic cetyltrimethyl ammonium bromide, *SEM* scanning electron microscope, *TEM* transmission electron microscopy, *DLS* dynamic light scattering, *ANUC* analytical ultracentrifugation, *TGA* thermogravimetric analysis, *DSC* differential scanning calorimetry, *AFM* atomic force microscopy

consecutive short shots of a monochromatic laser beam from the capsules, while they are randomly diffusion in the medium (Brownian motion). Larger particles diffuse at a slower rate and the algorithms employed can determine both the sizes and the PSD of the capsules. While SEM is also a fine technique for observing the population of obtained capsules, DLS is equipped with sophisticated software, capable of simultaneously providing the PSD. On the other hand, measurement of PSD with SEM can be achieved with simpler software and a lower number of particles than in DLS and provide rich information on the NC appearance. Analytical ultracentrifugation (ANUC) is another, less-employed sensitive technique, which can provide information on PSD, based on the rate of capsules' sedimentation during ultracentrifugation (Bootz et al. 2004; de Morais et al. 2016).

Zeta (ζ) potential in another important value for characterization, that provides information about the stability of capsules' suspension, their morphology, and surface. High ζ-potential (negative or positive) is related to higher electrostatic repulsions between particles in a dispersion. Z-potential is measured by applying an electric field across the dispersion, like in an electrophoretic apparatus. Particles are then accelerated due to coulombian forces to opposite-charge poles, and their velocity is correlated with the ζ-potential value. The velocity can be estimated with a laser Doppler velocimeter. Another important notice is that high positive values of ζ-potential are usually associated with high cytotoxicity, due to the negatively-charged nature of cell membranes (Clogston and Patri 2011; de Morais et al. 2016).

Thermal stability of the capsules can be determined with several methods, like thermo-gravimetric analysis (TGA) and differential scanning calorimetry (DSC) (Díez-Pascual and Díez-Vicente 2014). In TGA, a known mass is heated in a thermogravimetric analyzer at a constant heating rate. Mass changes due to thermal decomposition are determined simultaneously and correlated with the temperature (Coats and Redfern 1963). DSC measures heat exchanges during preset temperature programs and reveals crucial information on the structural properties of capsules, and thermal events accompanying them. The sample studied is heated or cooled at a controlled rate and the heat is monitored quantitatively (Bunjes and Unruh 2007). Important aspects of the capsules like surface morphology, roughness, shape, porosity, electrical properties and chemical composition, can be evaluated by atomic force microscopy (AFM) (de Morais et al. 2016). During AFM analysis, the surface area is scanned by a sharp tip of the instrument and the interaction forces are translated into image. AFM techniques give the opportunity of thorough, in situ analysis of the biomolecular mechanisms in their native environment. AFM can image surfaces in a buffer solution with outstanding sensitivity, to the detail of single biomolecules without fixation or staining (Müller and Dufrêne 2008).

5.4.3 Factors Affecting PHA Capsules Properties

During PHA-based synthesis of capsules with the emulsion-solvent evaporation method, a number of variables are employed, i.e. the composition of the organic phase and its proportion to the aqueous phase, molar composition of monomer(s) of the PHA, concentrations of the emulsifier and drug, and agitation conditions. Several studies have been published addressing this issue, alongside the synthesis of the desired particles (Bootz et al. 2004; de Morais et al. 2016).

Smaller ratios of organic to aqueous phase volumes and/or higher emulsifier's concentration give smaller NC with a lower PDI in standard emulsion-solvent evaporation technique (Lu et al. 2010; Kilicay et al. 2016). Diffusion of the organic phase in the aqueous phase is a core effect during NC formation. Lowering the viscosity of the organic phase and increasing its dispersion should assist this phenomenon, thus providing smaller NC (Lu et al. 2010). Another technique for lowering non-solvent proportion is by substituting part of the organic solvent with a more polar solvent like ethanol. An organic phase of 30% chloroform, 70% ethanol was employed during the synthesis of P(HB-HV)-NC, giving smaller nanoparticles (−456%) and lower PDI (−290%) than plain chloroform. This effect was attributed to the lowering of the surface tension, which was found to affect the size exponentially (Poletto et al. 2008). In addition, the molecular weight of the emulsifier used proportionally increases the sizes and the ζ-potential of the NC, but lowers DLE (Pramual et al. 2016) and increasing its concentration increases ζ-potential (Kilicay et al. 2016).

When double emulsion-solvent evaporation was used to synthesize MC of P(HV-HV), with diazepam as the encapsulated drug, increasing oil/water ration from 1/10 to 3/10, decreased the mean diameter of the received capsules. However, further increase reversed this phenomenon, and MC sizes increased. In all cases, PDI increased when oil/water increased (Chen and Davis 2002).

Lower concentrations of polymer and drug tend to produce smaller NC and MC, but this effect is less crucial when increasing emulsifier concentrations beyond an optimum value. Moreover, higher drug feeding increases the drug loaded per particle, as well as its release in vitro (Lu et al. 2010; Murueva et al. 2013). Higher polymer concentrations increase the mix viscosity and thus hinder stirring efficiency and droplets breakage into smaller particles, increasing NC size (Vardhan et al. 2017). In addition, it decreases the yield and does not increase significantly the DL in P(HB-HV)-MC with progesterone (Gangrade and Price 1991), or *E* in Docetaxel-loaded P(HB-HV)-NC (Vardhan et al. 2017). Drug loading also affects ζ-potential. Doxorubicin (DOX) entrapment significantly lowers ζ-potential in P(3HB)-NC, but this decrease is adversely correlated with DL of DOX (Murueva et al. 2013). However, curcumin entrapment or ConA conjugation increases ζ-potential in P(HB-HHx)-NC, due to the masking of the carboxylic groups on NC surface (Kilicay et al. 2016).

When using co-polymeric matrices, controlling the composition of the two PHA is important for the properties of the NC produced. Usually, different PHAs induce various effects on the final product. For instance, incorporating HV into PHB-NC increases the DL of 2,7-dichlorofluorescein and simultaneously delays the full drug release in PBS. Moreover, the product becomes more thermally unstable, in comparison with both plain PHB and unloaded NC (Gürsel and Hasirci 1995).

In addition, the capsules' diameter, shape, and smoothness of the NC is drastically affected when different polymers are employed (Table 5.1). For example, the addition of 6.5% mol of HV in P3HB-NC slightly decreased the P3HB-NC size and smoothens their surface, while ensuring spherical geometry. Increasing the molar percentage of HV produces larger and rougher NC. 7% mol of HHx or 6.1% mol of 4HB also produces smooth NC, but of larger diameters. In any occasion, inducing a second monomer to the polymeric matrix significantly decreases the ζ-potential of the NC (Murueva et al. 2013). In addition, higher agitation speed and more aggressive sonication are considered most favorable for the production of small NC and MC with lower PDI, but with a negative effect on DL/*E* values (Lu et al. 2010; Heathman et al. 2014). On the contrary, lower agitation delivers non-loaded P(HB-HHx)-NC with lower ζ-potential values (Kilicay et al. 2016). In addition, increased agitation time is positively correlated with smaller NC and with lower PDI. This effect could be attributed to the increase in energy supply, which probably increases the shear forces transmitted into the emulsion, leading to rapid breakdown of larger particles into smaller ones (Lu et al. 2010; Althuri et al. 2013). However aggressive sonication should be handled with caution and optimized, as it simultaneously leads to degradation of the PHA employed (Lu et al. 2010).

Table 5.1 General effects [positive (+) or negative (−)] of synthetic physicochemical parameters on the properties of synthesized capsules

	Polymer concentration	Drug concentration	Emulsifier concentration	$V_{Organic}/V_{aqeous}$	Agitation speed	Agitation duration
Size	+	+	−	+	−	−
PDI	+	+	−	+	−	−
|Z-potential|	−	+/−	+	N/A	−	N/A
E/DL	+	+	−	N/A	−	N/A

N/A not available information on the literature, +/−: depends on the drug used

5.4.4 Properties of PHA Capsules

PHA-NC loaded with drugs are most usually received in diameters below 250 nm (Xiong et al. 2010), down to 95 nm (Heathman et al. 2014). Sizes or PDI of PHA-MC are rarely reported, possibly due to them being an older "trend" in capsule-science when sophisticated equipment for size calculation was less available. However, diameters down to few μm have been reported (Atkins 1997; Grillo et al. 2010), as well as higher values, e.g., 100 μm (Gangrade and Price 1991; Kassab et al. 1997). This section focuses mainly on loaded-NC, due to the more thorough and analytic reports on them.

When DOX, an anti-cancer agent has been loaded in several PHA-NC, P3HB-NC and P(HB-HV) (93.5%/6.5% mol) gave the lowest diameters (700–750 nm). The ζ-potential values of all NCs studied were low; below −10 mV, which proves their good stability. Especially for PHB-NC and P(HB-HV)-NC values were −11 mV and −25 mV, respectively (Murueva et al. 2013). However, much lower diameters have been reported in other cases. P(3HB)-NC, and mixed polymers with PHV of different molecular weights: P[HB-(12 kDa)HV]-NC and P[3HB-(50 kDa)HV]-NC loaded with the photochemotherapeutic agent 5,10,15,20-Tetrakis(4-hydroxyphenyl)-21H,23H-porphine (pTHPP), were received in diameters below of 200 nm. High M_w PVOH as well as higher concentrations of drug to be loaded, increased the diameters of NC. Unfortunately, in all cases, ζ-values were very low, which proves that the NC synthesized would be prone to flocculation and unstable in dispersions (Pramual et al. 2016). In PHB-NC loaded with bone morphogenetic protein-2 (BMP-2), the particles received were in diameters of about 250 nm and with excellent ζ-potentials (−22 mV) (Peng et al. 2016). P(HB-HHx)-NC (88%/12% mol) loaded with concanavaline A (ConA), were in diameters of about 280 nm (ζ = −20 mV) and curcumin co-loading improved the properties of the NC (d = 250 nm, ζ = −28 mV) (Kilicay et al. 2016). PDI is most of the times as low as 0.2 (Kilicay et al. 2016; Peng et al. 2016).

Although generally considered thermally unstable, PHAs are in most cases thermally stable up to 150–170 °C (Gogolewski et al. 1993; Mousavioun et al. 2010) and PHA-NCs are stable up to 150–250 °C (Chen et al. 2007; Yu and Qin 2014), which are sufficiently high for biological applications. On the contrary, several loaded- and non-loaded-P(HB-HV)-MCs were studied for their Tm and it was found that MCs have a lower Tm value than the pure PHAs employed – especially the loaded ones (Gürsel and Hasirci 1995).

5.5 Pharmacokinetics and Biodegradation of PHAs

Reports of metabolism and pharmacokinetic properties of PHAs are sparse and most of them have been recorded before 2000. Although bacterial metabolic pathways of PHAs are well understood and extensively reviewed (Anderson and Dawes 1990), their degradation inside animal bodies is still a matter of speculation. It is generally acceptable though, that the degradation of PHAs both in physiological buffers and when implanted in vivo, is slow and processes in a significant rate only if chemical treatment, mix with atactic or hydrophilic polymers or irradiation precedes the implantation (Anderson and Dawes 1990; Cheng et al. 2003).

PHAs can possibly be degraded by the enzymatic activity of bacterial-secreted specific hydrolases assisting to hydrolysis of PHAs and they have been reported (Anderson and Dawes 1990). Saito et al. have studied PHB integrity when incubated with rat tissue extracts. In both alkaline and neutral Tris-HCl buffered extracts, all tissues were able to degrade PHB. They found, however, that it was degraded more significantly by an alkaline-favored hydrolase in liver extracts, and by a neutral-favored hydrolase in kidney extracts (Saito et al. 1991). Hydrolysis of P(HB-HV)/PCL (80:20) has been also reported in the presence of porcine pancreatin in PBS solution. Pancreatin is a mix of gastrointestinal enzymes, containing lipases, amylases, α-chymotrypsin, trypsin and proteases (Freier et al. 2002). PHB/PEG films degrade in the presence of lysozyme at 37 °C (half-life of 25 days), while no significant hydrolysis is noticed after 50 days in PBS (Cheng et al. 2003).

Spontaneous hydrolysis, occurred via stochastic scissions on the PHAs chain, is another mechanism for slow degradation. Generally, in physiological buffers like PBS (pH 7.4), PHB is degraded slowly, with a half-time of 56 weeks at 37 °C. A higher percentage of atactic PHB structure favors degradation, as this type of polymer is more susceptible to hydrolysis (Freier et al. 2002). In addition, a small, possibly spontaneous hydrolysis takes place in alkaline Tris-HCl buffers (Saito et al. 1991). In addition, blending PEG with P(HB-HHx) at in vivo implants accelerated hydrolysis of P(HB-HHx), attributed to the more hydrophilic nature of PEG (Qu et al. 2006b).

Encapsulation can be exploited as a way for enhancing the pharmacokinetic features of various pharmacological compounds studied. Sustained, prolonged release of the anti-cancer agent is crucial in chemotherapy. Rapamycin, a mTOR inhibitor, entrapped in polyethylene glycol (PEGylated) and non-PEGylated P(HB-HHx), when tested in prostate cancer cell line (PC3) cells, distributed better in cells than plain rapamycin. Moreover, the anti-proliferative effect insisted for more than 48 h and PEGylated-NC proved better anti-cancer agents that non-PEGylated ones, probably due to their increased hydrophobicity; a factor known to disturb the cellular interaction (Lu et al. 2014).

When DOX was loaded with folic acid (FA)-PEGylated P(HB-hydroxy-octanoate)-NC, a release of the entrapped substance was noticed in PBS, pH 7.4 for 11 days. Ovarian cancer (HeLa) cells treated with the NC, were forced to apoptosis more in a more evident way than that of plain DOX. In vivo experiments in rats injected with HeLa cells, provided a better insight into the improved pharmacological properties of P(HB-HO)-NC; 77% of the tumor formation has been inhibited in animals treated intravenously with NC, while an inhibition of 31% was noticed in animals treated with plain DOX. FA and PEGylation also enhanced the effect, as P(HB-HO)-DOX nanoparticles inhibited the growth formation by 53%. These effects may be attributed to enhanced permeability and retention effect, and high binding affinity of NC via the folate receptor of HeLa cells (Chan et al. 2016).

In another study by Vardhan et al. antineoplastic agent docetaxel (DOC) was enclosed in P(HB-HV)-NC. The drug release was slow, with only 11% of its quantity released after 48 h, suggesting prolonged action and fewer counter-effects. Overall, the drug was predicted to reach its maximum concentration at 72 h (16 μg/mL), in comparison with 6 h (41 μg/mL) of plain DOC, with a mean retention time of 129 h compared to three of plain DOC. In breast cancer cell line studied for a maximum of 4 days, uptake of DOC-loaded NC was noticed in cells treated with NC, as well as inhibition of growth up to 65%, while plain DOC could not inhibit the cell growth in the same degree (Vardhan et al. 2017).

5.6 Bio-distribution and Toxicity Studies in Animals

PHA capsules or scaffolds based on PHA are recently most studied for a number of in vivo applications, mainly anti-cancer treatment (Lu et al. 2014; Shah et al. 2014; Chan et al. 2016; Vardhan et al. 2017) and osteogenesis (Lü et al. 2013; Peng et al. 2016; Qi et al. 2017). Other applications tested through years include sutures, wound dressings, scaffolds for nerve and soft tissues repair or replacement, and stents. Most of these tests proved PHAs to be biocompatible and to possess the required physicochemical properties for the selected biomedical applications. PHAs can be exploited as suitable platforms for carrying enzymes, growth factors and other organic or inorganic materials (Valappil et al. 2006).

Maybe the most thorough research on the bio-distribution of injected PHAs has been introduced as back as 1991 (Saito et al. 1991). The scientists injected intravenously ^{14}C-labelled PHB granules and then studied possible side-effects and distribution on the tissues. The PHB-bolus did not cause any detectable side-effects on the animals up to 2 months after the injection. After 2.5 h most of the radioactivity (86%) has been detected in the liver, with 2.5% and 2.4% of it detected in the spleen and lungs. After 2 months post-treatment, most of the initial radioactivity (61%) was retained in the liver, with minimal amounts in other tissues (Saito et al. 1991). Qu et al. implanted PHB and P(HB-HHx) discs at the backs of New Zealand white rabbits. The rabbits did not exhibit any health problem during the process of the study (6 months). P(HB-HHx) elicited a mild tissue response, with a thin capsule surrounding the implant, while the capsule surrounding PHB was thicker and a number of inflammatory cells were counted. Moreover, P(HB-HHx) lost weight faster, but without losing its initial form, while PHB was degraded in several large pieces. No neutrophilic cells or macrophages have been spotted in the capsules, implying no toxicity throughout the study. The low acidity of the monomers released during hydrolysis can explain this tolerance (Qu et al. 2006a).

In vivo, Kazuhiko et al. found that PHB microspheres (100 μm) could cause acute inflammation when injected in mice thigh, but the symptoms withdrew 7 days after the injection. In another study P(HB-HV) microspheres injection caused similar effects. However, it was hypothesized by the authors that trauma of implantation or injection could be blamed for proven the inflammatory reaction, as P(HB-HV) polymers were to be biocompatible in vitro and in vivo (Kazuhiko et al. 1986). Also, when tested at several cell lines like macrophages, fibroblasts, Kupffer cells, and hepatocytes, PHB particles are generally well-tolerated (Shrivastav et al. 2013).

In C3H/J tumor model bearing female mice, animals treated with DOC-loaded NC developed tumors with lower volumes than both control and plain DOC-treated animals. Moreover, the NC did not exhibit hemolytic effects and all animals studied survived the tests (Vardhan et al. 2017).

5.7 Nanoparticles

The ability to develop and control polymeric particles to nanoscale dimensions, in combination with the potential applications of nanotechnology to medicine (nanomedicine), resulted in a further reduction of the size of controlled drug delivery carriers from macro- and micro- scale to the nano-scale (Kreuter 1978). Nanoparticles are employed for the selected and targeted drug delivery of chemotherapeutic agents directly to the tumor tissue without affecting normal organs (Yu et al. 2016). The main advantage of drug administration by nanoparticles is that due to their nanoimensions these particles can cross several cellular barriers including hematoencephalic, endothelial or epithelial barriers (Li and Loh 2017).

Drug delivery nanoparticles should be biodegradable, stable, no immunogenic, easy to fabricate, cost-effective, and able to release their payloads only at the target site (Poste and Kirsh 1983). In general, nanoparticles are divided into nanocapsules (reservoir system) and nanospheres (matrix system) (Fonseca et al. 2015). Nanocapsules are composed of a polymeric shell and a core containing the therapeutic compound (Singh and Lillard 2009). On the contrary, nanospheres are composed of a solid polymer matrix while the therapeutic compound is either heterogeneously dispersed or molecularly dissolved within the matrix. Polymeric nanoparticles have been successfully used for the targeted and continued release of anti-cancer drugs, primarily because the drug release rate can be easily adjusted by changing the type and therefore the properties of the polymeric portion of nanoparticles (Gursel et al. 2001). In vivo, the leak of loaded therapeutic compounds from the polymeric nanoparticles takes place as the polymer degrades and/or as water diffuses into the polymer leading to a sustained release of the drug at a controlled rate (Mahapatro and Singh 2011).

In addition to PHA microspheres, the use of nanoparticles composed of PHA for the encapsulation and delivery of therapeutic compounds has been widely explored for the delivery of therapeutic agents (Table 5.2). An important advantage of PHA-based nanoparticles is their ability to encapsulate hydrophobic and highly toxic chemotherapeutic (Masood et al. 2013b).

The development of nanoparticles composed of PLA and of either PHB homopolymer or its copolymer with 3-hydroxyhexanoate (HHx) [P(HB-HHx)] for the encapsulation of rhodamine B isothiocyanate (RBITC), a lipid-soluble colorant, has been reported (Xiong et al. 2010). The encapsulation efficiency of RBITC was higher (>75%) in both types of PHB-based nanoparticles compare to that of PLA-nanoparticles. Macrophage endocytosis led to an intracellular RBITC-sustained

Table 5.2 Encapsulation of a variety of therapeutic compounds in PHA-based nanoparticles

System	Loaded drug	Main findings	References
PHB[a] and P(HB-HHx)[b]	RBITC[c]	Drug release from PHA-based nanoparticles lasted longer compared to PLA nanoparticles	Xiong et al. (2010)
PHB and P(HB-HHx)	TGX221	Encapsulation of TGX221 into PHA-based nanoparticles enhanced its bioavailability	Lu et al. (2011)
PHB	DTXL[d]	Loading of DTXL into PHB nanoparticles resulted in an increased therapeutic index	Di Mascolo et al. (2016)
P(HB-HV)[e]	Paclitaxel	Paclitaxel-loaded PHA nanoparticles exhibited a high cytotoxic activity against cancer cell lines	Vilos et al. (2013)
P(HB-HV)	Ellipticine	A two-ford increased of the cytotoxic activity of ellipticine against A549 cancer cells was observed	Masood et al. (2013a)
PHB and P(HB-HV)	pTHPP[f]	A sustained release of pTHPP from PHA nanoparticles was observed	Pramual et al. (2016)
PEG-P(3HB-4HB)[g]	Cisplatin	Accumulation of cisplatin in tumor cells enhanced their apoptosis	Shah et al. (2012)
FA[h]-P(HB-HO)[i]	DOX[j]	The binding of FA-PHA nanoparticles to FA receptor in HeLa cells	Zhang et al. (2016)
FA-P(HB-HHx)	Etoposide	Conjugation of FA to PHA nanoparticles increased their selective binding to the cancer cells	Kılıçay et al. (2011)
PEI[k]-P(HB-HHx)	Rho-B[l]	PEI coating increased interaction between PHA nanoparticles and target cells	Wu et al. (2014)

[a]*PHB* Poly[(R)-3-hydroxybutyric acid]
[b]*P(HB-HHx)* poly[hydroxybutyrare-co-hydroxyhexanoate]
[c]*RBITC* rhodamine B isothiocyanate
[d]*DTXL* docetaxel
[e]*P(HB-HV)* poly[hydroxybutyrare-co-valerate]
[f]*pTHPP* 5,10,15,20-Tetrakis(4-hydroxy-phenyl)-21H, 23H-porphine
[g]PEG-P(3HB-4HB): poly (ethylene glycol)-poly(3-hydroxybutyrate-co-4-hydroxybutyrate)
[h]*FA* Folic Acid
[i]*P(HB-HO)* poly[hydroxybutyrare-co-hydroxyoctanoate]
[j]*DOX* doxorubicin
[k]*PEI* polyethylenimine
[l]*Rho-B* rhodamine-B

release from PHB-based nanoparticles for at least 20 days while the drug release from PLA nanoparticles lasted for 15 days (Xiong et al. 2010).

Lu et al. reported the encapsulation of TGX221, a phosphoinositide-3-kinase inhibitor, in PHA-based nanoparticles in order to inhibit the proliferation of cancer cell lines. A sustained release of TGX221 from PHA nanoparticles was observed resulting in a significantly slower growth rate of cancer cells lines, compared to the cells treated with free TGX221 (Lu et al. 2011).

Di Mascolo et al. synthesized nanoparticles (200 nm diameter) composed of PHB or PLGA for the loading and control release of docetaxel (DTXL), an anti-drug for the treatment of neoplastic diseases. The two DTXL-loading nanoparticles were compared in terms of encapsulation efficiency, drug release profile, and cytotoxicity. DTXL loading was higher into nanoparticles composed of the more hydrophobic PHB (approximately twofold higher than that of PLGA), while a sustained and slower drug release was obverted from PHB-nanoparticles. Encapsulation of DTXL into PHB nanoparticles resulted in an improved bioavailability by increasing the blood longevity and resulting in an enhanced therapeutic index. No significant differences were observed in terms of chemotherapeutic efficacy and cytotoxicity between PHB- and PLGA- nanoparticles, both loaded with DTXL, indicating the potential uses of PHB-based nanoparticles as drug carries for nanoscale DDS (Di Mascolo et al. 2016). Vilos et al. used the double emulsion solvent-evaporation method to encapsulate paclitaxel, an anticancer agent, into P(HB-HV) nanoparticles of an average diameter of 258.6 ± 35.7 nm (Vilos et al. 2013). The average size of these nanoparticles allows their uses in clinical studies (Peer et al. 2007). The synthesized P(HB-HV) nanoparticles were able to protect paclitaxel agent against premature degradation for at least 48 h after administration. Most importantly, due to their small size, paclitaxel-loaded nanoparticles were able to escape the blood (extravasate) and be transferred into tumor cells by the enhanced permeation and retention (EPR) phenomenon. In vitro experiments revealed a slow release of the anticancer agent from the nanoparticles and a high cytotoxic activity against cancer cell lines in primary cultures (Vilos et al. 2013).

The anticancer efficiency of the poorly soluble and high toxic agent, ellipticine (EPT), was significantly increased by its encapsulation into P(HB-HV) nanoparticles. A high loading efficiency (39.32–46.65%) of EPT into P(HB-HV) nanoparticles was achieved. The cytotoxic effect of the encapsulated- into PHA-based nanoparticles- EPT was evaluated against A549 cancer cell line, and a twofold higher inhibition percentage was observed in comparison with free EPT (Masood et al. 2013a).

Pramual et al. reported the potential use of P(HB-HV) in the enhanced photodynamic therapy (PDT) for the cancer treatment (Pramual et al. 2016). PDT uses photosensitizer (PS) compounds that when are exposed to a specific wavelength of light they produce reactive oxygen species that kills nearby cells (Dolmans et al. 2003). However, PDT suffers from some challenges, especially due to the hydrophobicity of PS molecules that leads to their aggregation and reduction of photoexcitation efficiency (Brown et al. 2004). In the study of Pramual et al. an emulsification-diffusion method was used to develop nanoparticles composed of PHB homopolymer and 2 copolymers of P(HB-HV) with 12% and 50% HV with diameters raining from 169.0 to 211.2 nm for the encapsulation of the hydrophobic photosensitizer, 5,10,15,20-Tetrakis(4-hydroxy-phenyl)-21H, 23H-porphine (pTHPP). In vitro examination revealed a continuous and slow release of pTHPP from PHA nanoparticles. In addition, the photo-cytotoxicity effect of pTHPP-loaded nanoparticles was assessed using HT-29 colon cancer cell, and a time- and concentration-dependent cell death was observed which was consistent with the release profile of pTHPP over 24 h (Pramual et al. 2016).

Similar to PHA-based microspheres, the functionalization of PHB-based nanoparticles was examinedin order to improve their drug delivery efficacy and eliminated any unwanted side effects. Shah et al. examined the development of drug delivery nanoparticles composed of P(3HB-4HB) copolymer coupled to monomethoxypoly(ethylene glycol) [P(3HV-4HB)-*b*-mPEG] for the targeted and controlled release of the chemotherapeutic agent cisplatin to tumors cells. The efficacy and bioavailability of the chemotherapeutic compound were examined in vitro. The results revealed that the amphiphilic nanoparticles containing cisplatin were effectively internalized by tumor cells. In addition, a continuous release of the therapeutic compound from the hydrophobic core of nanoparticles was observed resulting in a higher accumulation of cisplatin in tumor cells compared to the free drug in solution. Most importantly, confocal microscopy examination illustrated a suppression of tumor cell growth was suppressed while apoptosis of tumor cells was enhanced (Shah et al. 2012).

Zhang et al. reported the development of a novel targeting drug delivery system composed of the copolymer of PHB with 3-hydroxyoctanoate [P(HB-HO)] as the drug carrier and folic acid (FA) as the targeting ligand. Due to specific affinity of FA to folate receptor in HeLa cells, FA-P(HB-HO) nanoparticles have potential application for the treatment of cervical cancer. In this study doxorubicin (DOX) was used as the model anticancer drug. The DOX-loaded, folate-mediated P(HB-HO) [DOX/FA-PEG-P(HB-HO)] nanoparticles of an average size of 240 nm were prepared by W1/O/W2 solvent extraction/evaporation method. The encapsulation efficiency and drug loading were equal to 83.5% and 29.6%, and respectively. In in vitro and in vivo experiments, the DOX/FA-PEG-P(HB-HO) nanoparticles targeting showed a higher binding selectivity to cancer cells (they targeted HeLa cells more efficiently) and a higher antitumor activity compared to the non-targeted P(HB-HO) nanoparticles and free DOX. The effective binding between the folate in the PHA-based nanoparticles and folate receptor in HeLa has as results the selective binding of DOX/FA-PEG-P(HB-HO) nanoparticles to tumors and subsequent the higher released of DOX in tumors. In addition, in vivo antitumor examination in mice verified the enhanced therapeutic efficiency of the DOX/FA-PEG-P(HB-HO) nanoparticles (Zhang et al. 2016).

The conjugation of P(HB-HHx) nanoparticles with FA for the targeted delivery of etoposide, an anticancer drug, to tumors has also been reported (Kılıçay et al. 2011). P(HB-HHx) nanoparticles of different sizes ranging from 180 nm to 1.5 μm were synthesized using a modified solvent evaporation technique. FA-conjugated/etoposide-loaded P(HB-HHx) nanoparticles [etoposide/FA- P(HB-HHx) nanoparticles] exhibited the higher cytotoxic effect for HeLa cells than nanoparticles containing etoposide but not FA or free etoposide. In addition, the etoposide/FA- P(HB-HHx) nanoparticles exhibited toxicity in HeLa cells than normal fibroblast cells indicating that the presence of FA on the surface of nanoparticles promotes their selective binding to cancer cells (Kılıçay et al. 2011).

Wu et al. coated the surface of P(HB-HHx) nanoparticles with cationic polyethylenimine (PEI), at low concentration, in order improve the particle binding and uptake capacity by cells. The PEI-coated nanoparticles of an average diameter of 179 nm were loaded with rhodamine-B resulting in the (PEI)-coated rhodamine-B-loaded P(HB-HHx) nanoparticles. Subsequently, the effect of PEI-coating on the uptake and intracellular trafficking of PHA-based nanoparticles was examined in different human and animal cell types. Uncoated PHA-based nanoparticles interacted poorly with the cells, even after a 24-h incubation. On the contrary, PEI-coating improved interaction between PHA-based nanoparticles regardless the cell time, with the process was depended on both the time and concertation of nanoparticle (Wu et al. 2014).

5.8 Hydrogels

The most important characteristic of drug delivery systems is the sustained drug release from the carriers (e.g., micro- or nano-particles) which is not only accurate and but also timed controlled from few hours to month (Jogani et al. 2008). A smart and effective drug delivery system should be able to sense signals produced by disease, evaluate the level of the signal and subsequently release the appropriate amount of the encapsulated drug to fight the disease. Hydrogels are excellent candidates for the development of a smart drug delivery system as they have the characteristics of such a system (Upadhyay et al. 2017).

Hydrogels are 3-D, cross-linked networks of a water-soluble polymer, that can be molded easily in any shape and size, while they do not dissolve and most importantly they can hold a high fraction of water (thousand times of their own dry weight) (Nicolson and Vogt 2001). Usually, polymers containing hydrophobic domains naturally cross-link in aqueous solutions via reverse thermal gelation, also known as 'sol-gel' chemistry (Hoare and Kohane 2008). Physical hydrogels, formed by the self-assembly of polymers in response to environmental alterations display sol-gel transitions whereas their volume remains constant (He et al. 2008). Hydrogels composed of biodegradable copolymer that exhibit a sol-gel phase transition in response to external stimuli, i.e. alterations in temperature, pH or both, are widely used in biomedical and pharmaceutical science, such as drug delivery and tissue engineering (Nguyen and Lee 2010). The most important property of hydrogels is their highly porous structure which can be easily tuned by adjusting the density of the cross-links in the gel matrix. In addition, hydrogels are characterized by their swelling properties, i.e. the release of the loaded therapeutic agent from the swollen polymeric matrix and by their elasticity which affects the mechanical properties and therefore the stability of hydrogels (Amin et al. 2009; Simoes et al. 2012). In addition, their high porosity facilitates the loading of the drug into the gel matrix while its release profile from the polymeric gel network depends on the diffusion coefficient of the drug (Hoare and Kohane 2008). Hydrogels are also generally highly biocompatible, mainly due to their high-water content. Drug delivery hydrogels are usually developed outside of the body and loaded with therapeutic agents before placement of the hydrogel/drug complex in the body (Hoare and Kohane 2008). Another important feature of hydrogels is that they can be tuned towards stimuli while they can be tailored to be sensitive to the extracellular conditions of the body's tissue, including pH, temperature, and alterations of enzymatic activities at the diseased sited, and therefore are consider as an excellent candidate for the development of drug delivery carriers (Singh and Lee 2014; Upadhyay et al. 2017). The release of a drug from a hydrogel takes place through different mechanisms: swelling controlled, diffusion controlled, chemically controlled and environmentally-responsive release (Caló and Khutoryanskiy 2015).

A variety of natural and synthetic polymers, such as PLA, PLGA, starch hydroxyethylcellulose (HEC), NIPAAm (N-isopropylacrylamide) as well as their copolymers and triblock polymers, has been used for the synthesis of hydrogels are usually composed of hydrogels. Due to their unique physical properties, hydrogels are considered as excellent candidates for the formulations of novel DDCs in cancer therapy (Hoare and Kohane 2008; Simoes et al. 2012). However, the low encapsulation efficiency of hydrophobic drugs is probably the main disadvantage of hydrogels, as DDCs. Thus, there is a need fornew polymeric materials for the synthesis of hydrogels to improve their hydrophobic drug loading capacity.

PHA-based/stimuli-responsive materials have emerged as the most potent class of polymeric materials for the development of implantable scaffold substitute since they have the properties of polymeric hydrogels. Most importantly, PHA-based hydrogels exhibit additional properties and functionalities, including the change from reversible sol to gel, as a response to external stimuli that facilitate easy implementation during surgery. Moreover, the efficiency of these hydrogels as drug delivery carriers has been previously demonstrated (Pramual et al. 2016). Copolymers of hydrophilic biocompatible PEG with PHB, have been considered as alternative biomaterials for the development of hydrogels. Conjugations of PHB with PEG (i.e. PEGylation) of various molecular weights results in PHA-based- di-blocks and triblocks of low molecular weight and are excellent candiates for the development of injectable polymeric hydrogels for the delivery of drugs (Foster 2010).

The use of PHB in the development of hydrogels presents several advantages over the widely used alternatives. The crystallinity of PHB is higher than that of PLA or PCL while because PHB is hydrophobic, its copolymers exhibit a sol-gel transition with temperature change at relatively low concentrations (2–5% w/w) (Nguyen and Lee 2010). Loh et al. reported the synthesis of an amphiphilic and thermo-sensitive multi-block copolymer composed of PEG, poly(propylene glycol) (PPG) and PHB (PEG/PPG/PHB) as a potential candidate for the development of injectable DDCs. It has been demonstrated that in aqueous solutions, this block polymer undergoes a reversible sol-gel transition at very low concentrations upon changes in temperature. Interesting, PEG/PPG/PHB exhibit a very low critical micellization concentration a better thermal stability compared to that of PHB. The micellar packing of this novel multi-block copolymer was associated with its gelation in aqueous environments (Loh et al. 2007b). The same research group demonstrated that the PEG/PPG/PHB hydrogel is an excellent candidate for the development of drug delivery systems for the sustained release of proteins (Loh et al. 2007a). Bovine serum albumin was used as a model protein. In vitro studies revealed the controlled release of BSA from PEG/PPG/PHB hydrogel for more than 70 days. The release of BSA from hydrogel carried out in two phases: the first phase was diffusion-controlled while the second one was erosion-controlled. Li et al. reported the synthesis of a new triblock copolymer composed of poly(ethylene oxide)-poly[3-hydroxybutyrate]-poly(ethylene oxide) (PEO-PHB-PEO) in order to develop a denser and/or more stable hydrogel networks. PEO-PHB-PEO was subsequently complexed with alpha-cyclodextrin and a supramolecular hydrogel with a strong macromolecular network was formed. Fluorescein isothiocyanate (FITC) labeled dextran (FITC-dextran) was used as a model drug and loaded into the cyclodextrin/PEO-PHB-PEO hydrogel and a controlled release of the drug from hydrogel was recorded in vitro. The results of this study showed that in contrast to many simple triblock copolymer systems, this novel cyclodextrin/PEO-PHB-PEO hydrogel was suitable for the sustained and control release of macromolecular drugs for up to 1 month illustrating its potential application for the development of injectable drug delivery system (Li et al. 2006).

5.9 Non-woven Membrane

Non-woven membranes are one-dimensional (1-D) structures composed of a random array of nano- or micro-meter fibers which can act as a filter (Holowka and Bhatia 2014). Because of their unique mechanical properties and their ability to obtain different shapes and form, nanofibers are used in different medical applications such as wound dressing, tissues engineering, scaffolds and vascular prostheses (Jayakumar et al. 2010; Bizarria et al. 2014). In addition, polymeric nanofibers are excellent candidates for the development of carriers for the targeted delivery of therapeutic compounds at specific sites of application. For the use of nanofibers in medical and pharmaceutical applications, the production of fibers with diameters in the range of 100–200 nm is essential. Polymeric nanofibers are fabricated by different methods including phase separation and self-assembly and electrospinning. Among these techniques, electrospinning is the most commonly used because it is a fast, cheap and straightforward method. During this method, nanofibers are formed by electrostatic forces applying apolymeric solution of high viscosity through the application of a strong electric field between a polymer solution and grounded target. The characteristics of fibers such as composition, diameter, degradation, and mechanical properties can be modified according to the application (Dahlin et al. 2011).

Nanofibers used in medical and pharmaceutical applications must be fabricated by synthetic- or natural- biocompatible polymers. A variety of synthetic polymers including, polyurethane, poly(ethylene oxide) (PEO), poly(caprolactone) (PCL) and poly(vinyl alcohol) (PVA) as well as natural polymers such as PLA, cellulose and its derivatives and chitosan and chitin have been used to produce nanofibers for drug delivery systems and other medical applications (Sajeev et al. 2008; Pillay et al. 2013). A variety of drugs can be loaded directly into different sizes and/or types of fibers for targeted drug delivery (Yu et al. 2009). The main advantage of synthetic fibers over synthetic ones is that they exhibit excellent mechanical properties such as tensile strength a modulus (Laurencin et al. 2008; Sajeev et al. 2008). On the other hand, fibers fabricated by natural polymers have multifunctional bioactive properties and biocompatibility and thus exhibit a better interaction with the human tissues. Among natural polymers that are currently being studied for the development of novel nanofibers, the attention is focused on PHAs (Bizarria et al. 2014).

Lee et al. reported the synthesis of nanofibers composed of PHB homopolymer or poly(3-hydroxybutyrate-co-4-hydroxybutyrate) P(3HB-4HB) containing 95% 4HB. The results of this study revealed that fine fibers can be manufactured using the high molecular weight-PHB bead-free fibers at a concentration as low as 2%. However, to produce bead-free nanofibers using the low molecular weight P(3HB-4HB) copolymer, a fivefold higher concentration was required resulting in larger fibers. In addition, nanofibers composed of P(3HB-4HV) have larger diameters that nanofibers composed of PHB. Subsequently, the loading efficiency of paclitaxel a poorly water-soluble drug was on PHB and P(3HB-4HB) nanofibers was investigated. Interestingly, for various drug loadings, the different arrangements of fibers resulted in different paclitaxel release profile. The drug release rate from membranes composed of large-diameter P(3HB-4HB) nanofibers was faster probably due to the large size of pores. However, at low paclitaxel loadings, the drug release profile was more controlled and constant, compared to the higher drug loading. In addition, in this study, metal stents were uniformly coated with paclitaxel-loaded P(3HB-4HB) nanofibers revealing the potential applications of PHA drug-loaded nanofibers in the development of biocompatible drug-eluting stents (Lee et al. 2016).

Cheng et al. improved the hydrophilic, thermal, and mechanical properties of P(HB-HV) non-woven membranes using cellulose nanocrystals (CNCs) as rigid organic nanofillers. In general, P(HB-HV) non-woven membranes have weak mechanical properties, high crystallinity and hydrophobicity and thus hydrophobic drugs, such as tetracycline hydrochloride (TH) accumulate in the surface of membranes leading to the fast release of the drug. Interestingly, the hydrophilicity of nano-fibrous P(HB-HV) membranes was improved by the incorporation of the hydrophilic CNCs. In addition, in nano-fibrous P(HB-HV) membranes containing 6% (w/w) CNC a maximum drug loading of 25% and a drug loading of 98.8% were achieved while more than 86% of the loaded drug was successfully delivered within 540 h (Cheng et al. 2017).

The above studies revealed that PHA non-woven membranes are ideal candidates for the development of novel drug delivery systems. In addition, the electrospinning method has been successfully used by different research groups in order to develop PHA-based for the synthesis of non-woven membranes. Domínguez-Díaz reported the synthesis of nonwoven membranes composed of PHB and P(HB-HV) copolymer containing either 5 or 12% HV, using electrospinning and chloroform as a solvent. The diameter of fibers was at the micrometer scale while the membranes exhibited hydrophobic behavior, whereas hydrophobicity was increased in P(HB-HV) copolymer containing 12% HV (Domínguez-Díaz et al. 2012). Zhu et al. reported the development of super hydrophobic P(HB-HV) surfaces composed of micro- and nanofibers (Zhu et al. 2006). In another study, Sombatmankhong et al. developed ultrafine fiber mats from PHB, P(HB-HV), and their 50/50 (w/w) blend (Sombatmankhong et al. 2007).

5.10 Micelles

Polymeric nanocarriers are able to cross the cell walls and enter cells through the process of endocytosis and thus they are gaining increased attention for the development of carriers for the delivery of drugs to specific sites in the body (Shrivastav et al. 2013; Li and Loh 2017). Especially, nanoparticles with a diameter ranging from 50 to 150 nm are internalized by cancer cells at higher rates than normal cells due to the EPR effect (Karim et al. 2016). In addition to nanoparticles, polymeric micelles have successfully been used as nanocarriers to deliver drugs to specific targets in a controlled rate and within predefined periods of time in order to treat cancer and other diseases (Xu et al. 2009, 2013; Kedar et al. 2010). Micelles are gaining increased attention for the development of drug delivery nanocarriers due to their ability to cross many cellular barriers, carry high amounts of hydrophobic drugs and achieve site-specific delivery (Ale et al. 2012).

Micelles are aggregates of amphiphilic copolymers with sizes ranging from 50 to 100 nm. Micelles are composed of two main parts; (i) a hydrophobic core, in which the encapsulation of water-insoluble drugs can be achieved and (ii) and a hydrophilic shell (corona) that stabilizes the micelles and allows the loading of water-soluble drugs (Mikhail and Allen 2009; Owen et al. 2012). The micelles are formed spontaneously as soon as the concentration of an amphiphilic polymer exceeds the critical micelle concentration (Oerlemans et al. 2010). In general, amphiphilic copolymer blocks are di-block copolymers (A-B), tri-block copolymers (A-B-A) or grafted copolymers. Micelles not only carry therapeutic agents but also have some additional "functions" that enhance the efficient delivery to the target tissue/cell. For example, to target cancer cells more effectively, anti-cancer drug-loaded micelles maybe carry a ligand on their surface that is easily bound to the corresponding receptor of the surface of cancer cells (Nicolas et al. 2013; Liu et al. 2014b; Bahrami et al. 2015). PEG a hydrophilic polymer is widely used in the development of amphiphilic copolymers because is not toxic, flexible and electrically neutral. The copolymer of PEG [also known as poly(ethylene oxide)-PEO] with PLC, PLA or PLGA, have been used as materials for the development of micelles (Conte et al. 2014). However, PHB due its hydrophobicity is rarely used in the development of polymeric micelles. Recently, the development of micelles composed of PHA copolymers with PEG for the encapsulation the targeted delivery of therapeutic agents such as anti-cancer drugs has been reported.

Kim et al. encapsulated Doxorubicin (DOX) a chemotherapeutic against a wide range of solid tumors, in micelles composed of PEO-PHB-PEO tri-block copolymer of an average diameter of 37 nm. Due to the hydrophobic nature of PHB, the micelles were very stable while the encapsulation efficiency of DOX into the core of micelles was very high (64%). The drug release profile was examined in vitro, and only 30% of encapsulated DOX was released after 4 days. The therapeutic efficacy of DOX-loaded PEO-PHB-PEO was evaluated using a monolayer culture of SiHa human cervical carcinoma (HeLa) cells and three-dimensional (3-D) multicellular spheroids (MCS) that mimic solid tumors. In addition, PEO-PHB-PEO micelles containing DOX were conjugated with a fluorescent-labeled and subsequently injected in a tumor-bearing mouse model in order to further evaluate the antitumor activity of micelles in vivo. The results revealed that the encapsulation of DOX in PEO-PHB-PEO micelles improved penetration of the drug in 3-D MCS cultures, promote the apoptosis of cancer cells, and most importantly increased the delivery of DOX to the nucleus of cells which is the target organelle of this antitumor drug (Kim et al. 2010).

In another study, Kim et al. synthesized PHA-protein amphiphilic block copolymers composed of hydrophilic PHA-synthesis protein as the stabilized corona and PHB as the hydrophobic core, via a biosynthetic method. These novel PHA-based micelles were further functionalized by the conjugation of Cys-Asp-Cys-Arg-Gly-Asp-Cys-Phe-Cys (RGD4C), a peptide that exhibits high molecular affinity with integrin overexpressed in cancer cells, to the protein portion of micelles. RGD4C conjugated PHA-protein micelles showed an effective uptake to the targeted MDA-MB-231 breast cancer cells (Kim et al. 2009).

Loh et al. synthesized biodegradable thermal- and pH-sensitive amphiphilic triblock copolymers composed of a central hydrophobic PHB block and two hydrophilic poly(2-(di-methylamino)ethyl methacrylate) (DMAEMA) blocks. The anticancer agent was successfully encapsulated into the hydrophobic core of DMAEMA-PHB- DMAEMA micelles. The in vitro drug release studies revealed that the release rate of DOX from micelles was higher at 37 °C compared to that at 20 °C. In addition, a fast release rate of DOX was observed when micelle solution was incubated at pH 5, while a sustained drug release was observed when micelles treated at pH 7. Most importantly, these smart DMAEMA-PHB- DMAEMA micelles (i) exhibited lower toxicity towards normal cells compared with the PDMAEMA homopolymer (ii) they were effectively internalized by HeLa cells and (iii) the in vitro drug delivery efficiency was higher compared to that of commercially available Pluronics F127 micelles (Loh et al. 2013).

5.11 Bone Repair

Bone repair materials (BRM) represent an important biomaterial for human hard tissue replacement and repair in bone surgery, orthopedics, and dentistry. The development of BRM has attracted much attention due to the constantly increasing number of bone disorders induced by traumas, inflammations, and tumors (Tan et al. 2013). In general, there are three main types of BRM: the autograft bones, the allograft bones, and the artificially synthesized materials (Mankin et al. 1976; Fellah et al. 2008).

Although the autograft bones are the ideal BRM, their limited resources in combination with their abnormal morphology and dysfunction in the supply zone are the most important restrictions on their wide application. On the other hand, immune rejection reactions, contagious diseases, and tumorigenesis are the main disadvantages of autograft bones. Consequently, due to their advantage on flexibility and designability, artificially synthesized BRM become increasingly popular (Tan et al. 2013).

According to the degradation performance, BRM can be categorized into biodegradable and bio-inert materials (Sousa et al. 2003; Mano et al. 2004). Nowadays, biodegradable implanted materials have been attracting scientific attention because of their distinctive degradable and biocompatible nature, the limitation of mechanical malfunctions and the avoidance of the stress shield effect. Furthermore, since they do not require removals, secondary surgeries are also avoided resulting in a decrease of the implantation cost and surgery pain (Pietrzak et al. 1997; Maurus and Kaeding 2004; Witte et al. 2007; Wong et al. 2010). Among these biodegradable materials, biodegradable polymers represent an important category as the primary materials for tissue engineering applications and bone repair implants. According to their origin, they can be divided in natural-based materials and synthetic polymers (Tan et al. 2013). PHAs either on their own, under varying compositions, or in combination with inorganic phases, possess favorable mechanical and piezoelectric properties, degradation rates, bioactivity, and biocompatibility under favorable physiological conditions at specific time frames (Reddy et al. 2003; Chodak 2008). Their diversity renders them a family of bio-implant osteo-synthetic materials with multiple applications in hard-tissue engineering, especially bone tissue engineering (Chen and Wu 2005; Chodak 2008).

Indicatively, it has been found that the cytotoxicity of oligo-PHAs decreased with the increase of the oligo-PHAs side chain length. Being more biocompatible, medium chain length (mcl) oligo-PHAs have been proposed as PHA-based tissue engineering biomaterials due to their low toxicity of their degraded oligo- and monomers (Sun et al. 2007).

Composites of PHB, P(HB-HV), and P(HB-HHx) with either hydroxyapatite (HAp), glass-ceramic fillers, coatings or bioactive glass can be used as biodegradable bone fracture fixations or even as BRM (Chen and Wang 2002; Ni and Wang 2002), taking into account the similarity between the piezoelectric properties of PHB and those of bones (Knowles et al. 1991; Misra et al. 2006).

P(HB-HV) membranes were used to create spaces in the bone filling, as a treatment of jaw-bone defects in rats (Kostopoulos and Karring 1994). The experimental results of a 6-month study showed that P(HB-HV) membranes enhanced the development of the bone fill and successfully increased the height of the rat mandible.

Furthermore, in vivo subcutaneous implantation of P(HB-HV)-phosphate glass composites, as non-load bearing femoral implants, showed that after 4 weeks, the initial, extensive cellular activity, significantly affected by the increased concentration of the released ions from the solubilizing glass, was replaced by the entrance of cells inside the surface pores in the femoral implants formed by the glass, and the subsequent development of new bone on the surface of the implant (Knowles and Hastings 1993).

However, implantation of P(HB-HV)/HAp and P(HB-HV)/HAp-glass composites in rabbit femurs indicated better bonding behavior of the former, mainly due to a soft-tissue reaction caused by the released ions from the P(HB-HV)/HAp-glass composites (Knowles et al. 1992; Knowles 1993). In addition, studies on composites of P(HB-HV) with tri-calcium phosphate ($Ca_3(PO_4)_2$) filler in femoral and subcutaneous implants showed four times slower in vivo degradation rates compared to composites derived from poly-(D,L)-lactide (PLA) and $Ca_3(PO_4)_2$ (Jones et al. 2001).

Recently, in vivo implantation studies of porous P(HB-HV) and $Ca_3(PO_4)_2$-loaded collagen (CaP-Gelfix) foams in rat femurs showed that both cell-free and cell-loaded

P(HB-HV) matrices evoked the formation of minimal fibrous-tissue. Furthermore, macroscopic and radiological studies proved that P(HB-HV) matrices demonstrated better healing behavior, despite the establishment of the highly fibrous connective tissue and inflammation scores in the CaP-Gelfix group (Kose et al. 2004).

PHAs have also been studied as internal-bone fixation devices. In the late 1980s, PHB compression molded T-plates reinforced with carbon fibers (7%) were applied in the confrontation of tibial diaphysis osteotomies in rabbits (Vainionpaa et al. 1986). Sutures of absorbable polyglycolic acid (PGA) were used in the fixation of the implants to the tibia and compared with implants made of reinforced Vicryl. The experimental results after 12 weeks showed that the performance of reinforced PHB implants proved to be more sufficient than this of Vicryl plates, avoiding nonunion of the osteotomies, breakage and angulation.

In addition to the previous results, in vivo studies of the behavior of rivets prepared from PHB or PHB filled with HAp showed no significant difference when they were inserted into rabbit femurs (Doyle et al. 1991). The observed results also indicated favorable tissue responses and osteogenic activity of PHB.

Recent in vivo studies on the interface between PHB/HAp composites and bone after 1 month of implantation in rabbits showed bone apposition along the entire implant interface. Gradually, after 3 months of implantation, the observed developing bone formation presented an improved interlocking structure on the exposed interfacial HAp particles, followed by dense bone formation after 6 months of implantation (Luklinska and Schluckwerder 2003).

Nowadays, on the topic of nerve injury repair, new strategies on the use of PHAs as scaffolds for human bone marrow stromal cells are proposed. A P(HB-HV-HHx) composite terpolyester was compared with PLA and P(HB-HHx) in the differentiation mechanism of the human bone marrow stromal cells into nerve cells. It has been proved that the P(HB-HV-HHx) composite showed stronger cell adhesion, differentiation and proliferation (Wang et al. 2010). Following this observation, novel three-dimensional, highly porous, robust scaffolds of a composite composed of P(HB-HHx) and mesoporous bioactive glass were printed with a 3D bio-plotter (Zhao et al. 2014). The observed in vivo experimental results of the tested materials showed increased bioactivity and stimulation of both human bone marrow stromal cells adhesion and bone regeneration.

5.12 Anti-proliferative Activity

(R)-3-hydroxyalkanoic acids (RHAs), the PHA-origin, contain one chiral center and simply modified functional hydroxyl and carboxyl groups. Their structural features render them as valuable synthons or starting materials for the chemical synthesis of vitamins, antibiotics, pheromones and aromatics (Ruth et al. 2007).

In addition to RHA monomer function as structural units of chiral compounds, they are also utilized as synthons of innovative chiral polyesters, such as 3-hydroxy-3-cyclopropyl propionate, 3-hydroxy-4-chlorobutyrate and their CoA thioester derivatives (Kamachi et al. 2001). Furthermore, novel β- and γ-peptides with improved application abilities were produced through RHA modifications by replacing amino acids with 3HB residues and the chain bound oxygens in 3HB or 4HB with a NH- group (Matthews et al. 1998). These new types of peptides developed enhanced resistance to environmental microbial degradation and peptidases, long life span in mammalian serum and useful antibacterial, anti-proliferative and hemolytic properties (Park et al. 2001; Seebach et al. 2001).

In vitro experiments in murine fibroblast L929 cells, rabbit articular cartilages and human umbilical vein endothelial cells confirmed the hypothesis that properties inherent to HB act in favor of the cell growth observed on plates coated with P(HB-HHx) (Cheng et al. 2006). The results at hand showed that low HB concentrations (0.005–0.05 $g{\cdot}L^{-1}$) stimulated cell proliferation by suppressing cell death, through a signal transduction pathway dependent upon increases in $[Ca^{2+}]_I$ influx into the cells from the extracellular medium while, on the contrary, high concentrations of HB (0.05–0.1 $g{\cdot}L^{-1}$) opposed cell proliferation (Cheng et al. 2005; Xiao et al. 2007).

Cellular proliferation measurements in renal tubular cells (HK-2 cells) measured by $[^3H]$-thymidine incorporation into DNA and cell numbers, after exposure to β-HB, showed that β-HB dose-dependently (0.1–10 $mmol{\cdot}L^{-1}$) inhibited proliferation by a mechanism independent of cytotoxicity and increased collagen production (Guh et al. 2003).

In vitro studies of growth and proliferation measurements in human umbilical vein endothelial cells (HUVECs) and rabbit aorta smooth muscle cells (SMCs) on random copolyester of P(HB-HHx), with surface modifications induced by fibronectin coating and/or ammonia plasma treatment, showed that the reaction of cells to different types of polymer surface depends highly on the cell type. Moreover, it is proved that surface properties of polymers, such as: (a) chemical composition, (b) surface charge, (c) free energy, (d) hydrophilicity, and (e) crystallinity play important roles in cell proliferation (Qu et al. 2006a). As an indicative result, HUVECs proliferation rate on hydrophobic P(HB-HHx) was low compared to that of SMCs. In addition, SMCs showed a slow proliferation rate on the fibronectin-coated P(HB-HHx) (Qu et al. 2005).

Recent in vitro experimental results of the anti-proliferative activity of eighteen biotechnologically produced medium chain length (mcl)-PHA compounds consisting of predominantly (R)-3-hydroxyoctanoic acid monomers against the human lung fibroblast MRC5 cell line and melanoma cell A375, showed that almost all derivatives did not restrain the proliferation of mammalian cells even at concentrations of 3 mM, while only 3-oxooctanoic acid and (E)-oct-2-enoic had IC_{50} values of 1.6 and 1.7 mM with the human lung fibroblast cell line, respectively (Radivojevic et al. 2016).

5.13 Hemolytic Properties

The basic criteria that determine the suitability of a hemocompatible biomaterial in long-term contact with blood include strong biocompatibility and the ability to deter destruction of blood constituents and plasma proteins, thrombosis and thromboembolisms, antigenic response, etc. An important indicator of hemocompatibility is the state of the hemostasis system (Sevastianov et al. 2003). It can be estimated by various parameters such as the morphology of adhering platelets (Sevastianov et al. 2003), and the activation of the complement and coagulation cascade systems (Sevastianov and Tseytlina 1984; Titushkin et al. 2001).

Hemolytic assays of hemocompatible biomaterials play an important role in the hemolysis of red blood cells and the platelet adhesion and activation (Pramanik et al. 2015). The degree of hemolysis indicates accurately the extent of damage to erythrocytes and is a key factor in the evaluation of the biocompatibility of a biomaterial (Gao et al. 2010).

In general, biomaterials with a degree of hemolysis greater than 2% are not considered hemocompatible (Liu et al. 2014a). The lack of hemocompatibility causes the destruction of the erythrocyte membranes with the simultaneous release of hemoglobin, which induces acute renal failure (Kirkizlar et al. 2007). Therefore, biomaterials with low values of hemolysis ($\leq 2\%$) are considered suitable for wound healing applications (Ravanat et al. 2001).

Previous studies have shown that biomaterials engineering has focused on materials capable of adjusting their properties or biodegrade in response to the cellular and extracellular milieu. Such materials, specifically PHAs, have been widely considered suitable for drug delivery devices and tissue engineering (Doyle et al. 1991). PHAs, as drug carriers for medical applications, display adequate biocompatibility and biodegradability, without exhibiting detrimental interactions with blood constituents (Chen et al. 2008).

However, there are only a few results of investigations of PHA blood compatible properties presented in the current literature. In vivo studies on the hemocompatibility of poly-4-hydroxybutyrate (P4HB) implants into rabbits showed efficiently low procoagulant activity when impregnated with decellularized aortic matrix (Stamm et al. 2004). On the contrary, other PHAs, such as PHB and PHV activated the coagulation cascade systems (Sevastianov et al. 2003), whereas the copolymer P(3HB-4HB) shortened clotting time and promoted platelet adhesion (Chen et al. 2009). In general, the experimental results at hand showed that P4HB is not toxic to erythrocytes and does not influence on the fibrinolytic system and clotting factors or other plasma proteins, presenting a stronger affinity, compared to the polyvinyl chloride (PVC) group, for leukocytes and erythrocytes, specifically for neutrophils, and not for platelets (Liu et al. 2014a).

Studies on the remarkable hemocompatibility and cytocompatibility of P(HB-HHx) conjugates as blood-contact graft materials showed that alterations on the HHx content did not cause significant changes in the hemolysis assay, compared to the changes observed in the response in the platelet adhesion test, indicating that the chemical composition of P(HB-HHx) is a critical parameter for erythrocyte function, while improved biocompatibility is more dependent on the surface modifications. Additionally, P(HB-HHx) copolymer showed better hemocompatibility and cytocompatibility, reduced platelet adhesion, and less hemolysis reactivity and erythrocyte contact compared to PHB and PHBV films (Qu et al. 2006a).

Series of studies on various concentrations of PHB-PEG-PHB nanoparticles as potential drug carriers showed no observational hemolytic activities in the red blood cellular milieu confirming their suitability for intravenous administration and blood-contacting applications (Chen et al. 2008).

Additionally, recent studies on guar-gum/P(HB-HV)/curcumin wound healing blend films indicated the ability of these films to prevent damage to red blood cells mainly due to the deformation and exposure of procoagulant phospholipids on the film surface (Ravanat et al. 2001; Pramanik et al. 2015), a procedure similar to the catalytic function of activated platelets in the conversion of prothrombin to thrombin in the maturation stage of healing (Holmberg et al. 2008).

5.14 Conclusions

Drug delivery refers to technologies, approaches and systems for the adequate administration of a pharmaceutically active compound in the body to achieve its favorable therapeutic effects. Various concepts, devices, and materials for advanced drug formulations targets to healthcare applications of controlled-release technology (CRT) focusing on novel biomaterials with targeted therapeutic activity. Drug delivery systems are gaining increasing importance for human diseases treatment, through the administration of effective drug formulations via alternative to parenteral routes such as nasal and pulmonary drug delivery routes. Various drug delivery vehicles have been utilized as liposomes, nanoparticles, microspheres, hydrogels, micelles, and membranes, among others. Nowadays, nanotechnology represents an effective approach in the development of potent drug delivery systems targeted to specific diseased tissues (e.g., tumors), thereby avoiding interaction with healthy tissues. Nano-formulations composed of biodegradable synthons fulfill the crucial requirements that define an effective delivery system, such as biocompatibility and biodegradability, predictable and repeatable pharmacokinetics, low toxicity, targeting of specific cell populations or sites in an organ, increased loading capacity, controlled and tunable drug release rates. PHAs, as alternatives to synthetic polymers, are potentially emerging as next generation drug carriers of therapeutics and environmentally friendly materials with versatility and a broad range of applicability.

5.15 Opinion

Polymeric nano vehicles represent a promising type of drug delivery carriers. The development of biodegradable polymeric nanocarriers has received considerable interest recently. Functionalizing surface properties and suitable size, they can efficiently bypass physiological barriers to reach diseased tissues bearing a sufficient amount of drug. PHAs are a group of biodegradable polymers that can be synthesized by a combination of bio-chemical methods. They possess a wide range of tunable physicochemical properties that define their exceptional biocompatibility and biodegradability, while their toxic degradation products render them excellent candidates for drug delivery nanovehicles. The versatility of engineered PHAs and the consequent diversity in their physicochemical properties resulted in the development of a plethora of drug delivery systems. Undoubtedly, at the moment, PHAs are far less competitive compared to their petrochemical peers. Besides their relative poor properties, the high production cost is the major factor preventing their successful large-scale marketing. Nevertheless, the future of PHAs relies heavily on the rational design of advanced strategies to obtain more suitable micro-organisms and optimized processes for the enhancement of their production. Genetic and metabolic engineering, utilization of microbial genomes, and cloning of genes from noncultivable microorganisms can provide the potential for sufficiently inexpensive and high PHA production rates.

References

Abdul Karim A, Loh XJ (2015) Design of a micellized α-cyclodextrin based supramolecular hydrogel system. Soft Matter 11:5425–5434. https://doi.org/10.1039/C5SM00665A

Ale EC, Maggio B, Fanani ML (2012) Ordered-disordered domain coexistence in ternary lipid monolayers activates sphingomyelinase by clearing ceramide from the active phase. BBA-Biomembranes 1818:2767–2776. https://doi.org/10.1016/j.bbamem.2012.06.017

Althuri A, Mathew J, Sindhu R, Banerjee R, Pandey A, Binod P (2013) Microbial synthesis of poly-3-hydroxybutyrate and its application as targeted drug delivery vehicle. Bioresour Technol 145:290–296. https://doi.org/10.1016/j.biortech.2013.01.106

Amass W, Amass A, Tighe B (1998) A review of biodegradable polymers: uses, current developments in the synthesis and characterization of biodegradable polyesters, blends of biodegradable polymers and recent advances in biodegradation studies. Polym Int 47:89–144. https://doi.org/10.1002/(SICI)1097-0126(1998100)47:2<89::AID-PI86>3.0.CO;2-F

Amin S, Rajabnezhad S, Kohli K (2009) Hydrogels as potential drug delivery systems. Sci Res Essays 3:1175–1183

Anderson AJ, Dawes EA (1990) Occurrence, metabolism, metabolic role, and industrial uses of bacterial polyhydroxyalkanoates. Microbiol Rev 54:450–472

Ansary RH, Rahman MM, Awang MB, Katas H, Hadi H, Doolaanea AA (2016) Preparation, characterization, and in vitro release studies of insulin-loaded double-walled poly(lactide-co-glycolide) microspheres. Drug Deliv Transl Res 6:308–318. https://doi.org/10.1007/s13346-016-0278-y

Artsis MI, Bonartsev AP, Iordanskii AL, Bonartseva GA, Zaikov GE (2010) Biodegradation and medical application of microbial poly(3-hydroxybutyrate). Mol Cryst Liq Cryst 523:232–262. https://doi.org/10.1080/15421401003726519

Atkins TW (1997) Fabrication of microspheres using blends of poly(ethylene adipate) and poly(ethylene adipate)/poly(hydroxybutyrate-hydroxyvalerate) with poly(caprolactone): incorporation and release of bovine serum albumin. J Biomater Sci Polym Ed 8:833–845. https://doi.org/10.1163/156856297X00038

Avgoustakis K, Beletsi A, Panagi Z, Klepetsanis P, Karydas AG, Ithakissios DS (2002) PLGA-mPEG nanoparticles of cisplatin: in vitro nanoparticle degradation, in vitro drug release and in vivo drug residence in blood properties. J Control Release 79:123–135. https://doi.org/10.1016/S0168-3659(01)00530-2

Bahrami B, Mohammadnia-Afrouzi M, Bakhshaei P, Yazdani Y, Ghalamfarsa G, Yousefi M, Sadreddini S, Jadidi-Niaragh F, Hojjat-Farsangi M (2015) Folate-conjugated nanoparticles as a potent therapeutic approach in targeted cancer therapy. Tumour Biol 36:5727–5742. https://doi.org/10.1007/s13277-015-3706-6

Baird JA, Olayo-Valles R, Rinaldi C, Taylor LS (2010) Effect of molecular weight, temperature, and additives on the moisture sorption properties of polyethylene glycol. J Pharm Sci 99:154–168. https://doi.org/10.1002/jps.21808

Bazzo GC, Macedo ATD, Crenca JP, Silva VE, Pereira EM, Zétola M, Pezzini BR (2012) Microspheres prepared with biodegradable PHBV and PLA polymers as prolonged-release system for ibuprofen: in vitro drug release and in vivo evaluation. Braz J Pharm Sci 48:773–780. https://doi.org/10.1590/S1984-82502012000400021

Bengtsson S, Karlsson A, Alexandersson T, Quadri L, Hjort M, Johansson P, Morgan-Sagastume F, Anterrieu S, Arcos-Hernandez M, Karabegovic L, Magnusson P, Werker A (2017) A process for polyhydroxyalkanoate (PHA) production from municipal wastewater treatment with biological carbon and nitrogen removal demonstrated at pilot-scale. New Biotechnol 35:42–53. https://doi.org/10.1016/j.nbt.2016.11.005

Bizarria MTM, d'Ávila M, Innocentini-Mei L (2014) Non-woven nanofiber chitosan/PEO membranes obtained by electrospinning. Braz J Chem Eng 31:57–68. https://doi.org/10.1590/S0104-66322014000100007

Bonartsev A, Bonartseva G, Makhina T, Myshkina V, Luchinina E, Livshits V, Boskhomdzhiev A, Markin V, Iordanskii A (2006) New poly (3-hydroxybutyrate)-based systems for controlled release of dipyridamole and indomethacin. Appl Microbiol Biotechnol 42:625–630. https://doi.org/10.1134/S0003683806060159

Bonartsev A, Livshits V, Makhina T, Myshkina V, Bonartseva G, Iordanskii A (2007) Controlled release profiles of dipyridamole from biodegradable microspheres on the base of poly (3-hydroxybutyrate). Express Polym Lett 1:797–803

Bootz A, Vogel V, Schubert D, Kreuter J (2004) Comparison of scanning electron microscopy, dynamic light scattering and analytical ultracentrifugation for the sizing of poly(butyl cyanoacrylate) nanoparticles. Eur J Pharm Biopharm 57:369–375. https://doi.org/10.1016/S0939-6411(03)00193-0

Brown SB, Brown EA, Walker I (2004) The present and future role of photodynamic therapy in cancer treatment. Lancet Oncol 5:497–508. https://doi.org/10.1016/S1470-2045(04)01529-3

Bunjes H, Unruh T (2007) Characterization of lipid nanoparticles by differential scanning calorimetry, X-ray and neutron scattering. Adv Drug Deliv Rev 59:379–402. https://doi.org/10.1016/j.addr.2007.04.013

Caló E, Khutoryanskiy VV (2015) Biomedical applications of hydrogels: a review of patents and commercial products. Eur Polym J 65:252–267. https://doi.org/10.1016/j.eurpolymj.2014.11.024

Cammas S, Bear MM, Moine L, Escalup R, Ponchel G, Kataoka K, Guerin P (1999) Polymers of malic acid and 3-alkylmalic acid as synthetic PHAs in the design of biocompatible hydrolyzable devices. Int J Biol Macromol 25:273–282. https://doi.org/10.1016/S0141-8130(99)00042-2

Carino GP, Jacob JS, Mathiowitz E (2000) Nanosphere based oral insulin delivery. J Control Release 65:261–269. https://doi.org/10.1016/S0168-3659(99)00247-3

Chan Z, Zhang Z, Zhao L (2016) Folate-decorated poly(3-hydroxybutyrate-co-3-hydroxyoctanoate) nanoparticles for targeting delivery: optimization and in vivo antitumor activity. Drug Deliv 23:1830–1837. https://doi.org/10.3109/10717544.2015.1122675

Chang HM, Wang ZH, Luo HN, Xu M, Ren XY, Zheng GX, Wu BJ, Zhang XH, Lu XY, Chen F, Jing XH, Wang L (2014) Poly(3-hydroxybutyrate-co-3-hydroxyhexanoate)-based scaffolds for tissue engineering. Braz J Med Biol Res 47:533–539. https://doi.org/10.1590/1414-431X20143930

Chaturvedi K, Kulkarni AR, Aminabhavi TM (2011) Blend microspheres of poly(3-hydroxybutyrate) and cellulose acetate phthalate for colon delivery of 5-fluorouracil. Ind Eng Chem Res 50:10414–10423. https://doi.org/10.1021/ie2011005

Chee PL, Lakshmanan L, Jiang S, Ye H, Kai D, Loh XJ (2016) An injectable double-network hydrogel for cell encapsulation. Aust J Chem 69:388–393. https://doi.org/10.1071/CH15659

Chen J, Davis SS (2002) The release of diazepam from poly(hydroxybutyrate-hydroxyvalerate) microspheres. J Microencapsul 19:191–201. https://doi.org/10.1080/02652040110065431

Chen LJ, Wang M (2002) Production and evaluation of biodegradable composites based on PHB–PHV copolymer. Biomaterials 23:2631–2639. https://doi.org/10.1016/S0142-9612(01)00394-5

Chen G-Q, Wu Q (2005) The application of polyhydroxyalkanoates as tissue engineering materials. Biomaterials 26:6565–6578. https://doi.org/10.1016/j.biomaterials.2005.04.036

Chen D, Tang C, Chan K, Tsui C, Yu P, Leung M, Uskokovic P (2007) Dynamic mechanical properties and in vitro bioactivity of PHBHV/HA nanocomposite. Compos Sci Technol 67:1617–1626. https://doi.org/10.1016/j.compscitech.2006.07.034

Chen C, Cheng YC, Yu CH, Chan SW, Cheung MK, Yu PH (2008) In vitro cytotoxicity, hemolysis assay, and biodegradation behavior of biodegradable poly(3-hydroxybutyrate)-poly(ethylene glycol)-poly(3-hydroxybutyrate) nanoparticles as potential drug carriers. J Biomed Mater Res A 87:290–298. https://doi.org/10.1002/jbm.a.31719

Chen Z, Cheng S, Li Z, Xu K, Chen GQ (2009) Synthesis, characterization and cell compatibility of novel poly(ester urethane)s based on poly(3-hydroxybutyrate-co-4-hydroxybutyrate) and poly(3-hydroxybutyrate-co-3-hydroxyhexanoate) prepared by melting polymerization. J Biomater Sci Polym Ed 20:1451–1471. https://doi.org/10.1163/092050609x12457419007621

Cheng G, Cai Z, Wang L (2003) Biocompatibility and biodegradation of poly(hydroxybutyrate)/poly(ethylene glycol) blend films. J Mater Sci Mater Med 14:1073–1078. https://doi.org/10.1023/B:JMSM.0000004004.37103.f4

Cheng S, Wu Q, Yang F, Xu M, Leski M, Chen G-Q (2005) Influence of dl-β-hydroxybutyric acid on cell proliferation and calcium influx. Biomacromolecules 6:593–597. https://doi.org/10.1021/bm049465y

Cheng S, Chen G-Q, Leski M, Zou B, Wang Y, Wu Q (2006) The effect of d,l-β-hydroxybutyric acid on cell death and proliferation in L929 cells. Biomaterials 27:3758–3765. https://doi.org/10.1016/j.biomaterials.2006.02.046

Cheng M, Qin Z, Hu S, Dong S, Ren Z, Yu H (2017) Achieving long-term sustained drug delivery for electrospun biopolyester nanofibrous membranes by introducing cellulose nanocrystals. ACS Biomater Sci Eng 3:1666–1676. https://doi.org/10.1021/acsbiomaterials.7b00169

Chodak I (2008) Polyhydroxyalkanoates: origin, properties and applications. In: Belgacem M, Gandini A (eds) Monomers, polymers and composites from renewable resources. Elsevier, Amsterdam, pp 451–477. https://doi.org/10.1016/B978-0-08-045316-3.00022-3. ISBN:9780080560519

Cilurzo F, Selmin F, Minghetti P, Adami M, Bertoni E, Lauria S, Montanari L (2011) Injectability evaluation: an open issue. AAPS Pharm Sci Tech 12:604–609. https://doi.org/10.1208/s12249-011-9625-y

Clogston JD, Patri AK (2011) Zeta potential measurement. Methods Mol Biol 697:63–70. https://doi.org/10.1007/978-1-60327-198-1_6

Coats AW, Redfern JP (1963) Thermogravimetric analysis. A review. Analyst 88:906–906. https://doi.org/10.1039/an9638800906

Coimbra PA, De Sousa HC, Gil MH (2008) Preparation and characterization of flurbiprofen-loaded poly(3-hydroxybutyrate-co-3-hydroxyvalerate) microspheres. J Microencapsul 25:170–178. https://doi.org/10.1080/02652040701814140

Conte C, d'Angelo I, Miro A, Ungaro F, Quaglia F (2014) PEGylated polyester-based nanoncologicals. Curr Top Med Chem 14:1097–1114

Dahlin RL, Kasper FK, Mikos AG (2011) Polymeric nanofibers in tissue engineering. Tissue Eng 17:349–364. https://doi.org/10.1089/ten.teb.2011.0238

de Morais MG, de Morais EG, Vaz B d S, Gonçalves CF, Lisboa C, Vieira Costa JA (2016) Nanoencapsulation of the bioactive compounds of spirulina with a microalgal biopolymer coating. J Nanosci Nanotechnol 16:81–91. https://doi.org/10.1166/jnn.2016.10899

Di Mascolo D, Basnett P, Palange AL, Francardi M, Roy I, Decuzzi P (2016) Tuning core hydrophobicity of spherical polymeric nanoconstructs for docetaxel delivery. Polym Int 65:741–746. https://doi.org/10.1002/pi.5072

Díez-Pascual AM, Díez-Vicente AL (2014) Poly(3-hydroxybutyrate)/ZnO bionanocomposites with improved mechanical, barrier and antibacterial properties. Int J Mol Sci 15:10950–10973. https://doi.org/10.3390/ijms150610950

Dolmans DEJGJ, Fukumura D, Jain RK (2003) Photodynamic therapy for cancer. Nat Rev Cancer 3:380. https://doi.org/10.1038/nrc1071

Domínguez-Díaz M, Romo-Uribe A, Flores A, Cruz-Silva R (2012) Morphology-induced hydrophobic behavior of electrospun polyhydroxyalkanoate membranes. MRS Proc 1466. https://doi.org/10.1557/opl.2012.1256

Doyle C, Tanner ET, Bonfield W (1991) In vitro and in vivo evaluation of polyhydroxybutyrate and of polyhydroxybutyrate reinforced with hydroxyapatite. Biomaterials 12:841–847. https://doi.org/10.1016/0142-9612(91)90072-I

Duran N, Alvarenga MA, Da Silva EC, Melo PS, Marcato PD (2008) Microencapsulation of antibiotic rifampicin in poly(3-hydroxybutyrate-co-3-hydroxyvalerate). Arch Pharm Res 31:1509–1516. https://doi.org/10.1007/s12272-001-2137-7

Dutta RC (2007) Drug carriers in pharmaceutical design: promises and progress. Curr Pharm Des 13:761–769. https://doi.org/10.2174/138161207780249119

El-Say K (2016) Maximizing the encapsulation efficiency and the bioavailability of controlled-release cetirizine microspheres using Draper–Lin small composite design. Drug Des Dev Ther 10:825–839. https://doi.org/10.2147/DDDT.S101900

Embleton JK, Tighe BJ (1993) Polymers for biodegradable medical devices. X. Microencapsulation studies: control of poly-hydroxybutyrate-hydroxyvalerate microcapsules porosity via polycaprolactone blending. J Microencapsul 10:341–352. https://doi.org/10.3109/02652049309031524

Errico C, Bartoli C, Chiellini F, Chiellini E (2009) Poly(hydroxyalkanoates)-based polymeric nanoparticles for drug delivery. J Biomed Biotechnol 2009:571702–571702. https://doi.org/10.1155/2009/571702

Fang J, Nakamura H, Maeda H (2011) The EPR effect: unique features of tumor blood vessels for drug delivery, factors involved, and limitations and augmentation of the effect. Adv Drug Deliv Rev 63:136–151. https://doi.org/10.1016/j.addr.2010.04.009

Fellah BH, Gauthier O, Weiss P, Chappard D, Layrolle P (2008) Osteogenicity of biphasic calcium phosphate ceramics and bone autograft in a goat model. Biomaterials 29:1177–1188. https://doi.org/10.1016/j.biomaterials.2007.11.034

Fonseca AC, Serra AC, Coelho JF (2015) Bioabsorbable polymers in cancer therapy: latest developments. EPMA J 6:22. https://doi.org/10.1186/s13167-015-0045-z

Foster L (2010) PEGylation and bioPEGylation of polyhydroxyalkanoates: synthesis, characterisationand applications. In: Elnashar M (ed) Biopolymers. InTech, Bolton, pp 223–256. https://doi.org/10.5772/10265. ISBN:978-953-307-109-1

Freier T, Kunze C, Nischan C, Kramer S, Sternberg K, Saß M, HoptU T, Schmitz K-P (2002) In vitro and in vivo degradation studies for development of a biodegradable patch based on poly(3-hydroxybutyrate). Biomaterials 23:2649–2657. https://doi.org/10.1016/S0142-9612(01)00405-7

Freitas S, Merkle HP, Gander B (2005) Microencapsulation by solvent extraction/evaporation: reviewing the state of the art of microsphere preparation process technology. J Control Release 102:313–332. https://doi.org/10.1016/j.jconrel.2004.10.015

Gangrade N, Price JC (1991) Poly(hydroxybutyrate-hydroxyvalerate) microspheres containing progesterone: preparation, morphology and release properties. J Microencapsul 8:185–202. https://doi.org/10.3109/02652049109071487

Gao Y, Chen L, Zhang Z, Gu W, Li Y (2010) Linear cationic click polymer for gene delivery: synthesis, biocompatibility, and in vitro transfection. Biomacromolecules 11:3102–3111. https://doi.org/10.1021/bm100906m

Gogolewski S, Jovanovic M, Perren SM, Dillon JG, Hughes MK (1993) Tissue response and in vivo degradation of selected polyhydroxyacids: polylactides (PLA), poly(3-hydroxybutyrate) (PHB), and poly(3-hydroxybutyrate-co-3-hydroxyvalerate) (PHB/VA). J Biomed Mater Res 27:1135–1148. https://doi.org/10.1002/jbm.820270904

Gould PL, Holland SJ, Tighe BJ (1987) Polymers for biodegradable medical devices. IV. Hydroxybutyrate-valerate copolymers as non-disintegrating matrices for controlled-release oral dosage forms. Int J Pharm 38:231–237

Grage K, Jahns AC, Parlane N, Palanisamy R, Rasiah IA, Atwood JA, Rehm BHA (2009) Bacterial polyhydroxyalkanoate granules: biogenesis, structure, and potential use as nano-/micro-beads in biotechnological and biomedical applications. Biomacromolecules 10:660–669. https://doi.org/10.1021/bm801394s

Grillo R, de Melo NFS, de Lima R, Lourenço RW, Rosa AH, Fraceto LF (2010) Characterization of atrazine-loaded biodegradable poly(hydroxybutyrate-co-hydroxyvalerate) microspheres. J Polym Environ 18:26–32. https://doi.org/10.1007/s10924-009-0153-8

Guh J-Y, Chuang T-D, Chen H-C, Hung W-C, Lai Y-H, Shin S-J, Chuang L-Y (2003) β-hydroxybutyrate–induced growth inhibition and collagen production in HK-2 cells are dependent on TGF-β and Smad3. Kidney Int 64:2041–2051. https://doi.org/10.1046/j.1523-1755.2003.00330.x

Gürsel I, Hasirci V (1995) Properties and drug release behaviour of poly(3-hydroxybutyric acid) and various poly(3-hydroxybutyrate-hydroxyvalerate) copolymer microcapsules. J Microencapsul 12:185–193. https://doi.org/10.3109/02652049509015289

Gursel I, Korkusuz F, Turesin F, Alaeddinoglu NG, Hasirci V (2001) In vivo application of biodegradable controlled antibiotic release systems for the treatment of implant-related osteomyelitis. Biomaterials 22:73–80. https://doi.org/10.1016/S0142-9612(00)00170-8

Gursel I, Yagmurlu F, Korkusuz F, Hasirci V (2002) In vitro antibiotic release from poly (3-hydroxybutyrate-co-3-hydroxyvalerate) rods. J Microencapsul 19:153–164. https://doi.org/10.1080/02652040110065413

Hamoudeh M, Faraj AA, Canet-Soulas E, Bessueille F, Léonard D, Fessi H (2007) Elaboration of PLLA-based superparamagnetic nanoparticles: characterization, magnetic behaviour study and in vitro relaxivity evaluation. Int J Pharm 338:248–257. https://doi.org/10.1016/j.ijpharm.2007.01.023

Hazer DB, Kılıçay E, Hazer B (2012) Poly(3-hydroxyalkanoate)s: diversification and biomedical applications: a state of the art review. Mater Sci Eng C 32:637–647. https://doi.org/10.1016/j.msec.2012.01.021

He C, Kim SW, Lee DS (2008) In situ gelling stimuli-sensitive block copolymer hydrogels for drug delivery. J Control Release 127:189–207. https://doi.org/10.1016/j.jconrel.2008.01.005

Heathman TRJ, Webb WR, Han J, Dan Z, Chen GQ, Forsyth NR, El Haj AJ, Zhang ZR, Sun X (2014) Controlled production of poly (3-hydroxybutyrate-co-3-hydroxyhexanoate) (PHBHHx) nanoparticles for targeted and sustained drug delivery. J Pharm Sci 103:2498–2508. https://doi.org/10.1002/jps.24035

Hoare TR, Kohane DS (2008) Hydrogels in drug delivery: progress and challenges. Polymer 49:1993–2007. https://doi.org/10.1016/j.polymer.2008.01.027

Hoffman AS (2008) The origins and evolution of "controlled" drug delivery systems. J Control Release 132:153–163. https://doi.org/10.1016/j.jconrel.2008.08.012

Holmberg M, Stibius KB, Larsen NB, Hou X (2008) Competitive protein adsorption to polymer surfaces from human serum. J Mater Sci Mater Med 19:2179–2185. https://doi.org/10.1007/s10856-007-3318-9

Holowka EP, Bhatia SK (2014) Drug delivery: materials design and clinical perspective. Springer, New York, 63 p

Hsu PH, Laddu AR (1994) Analysis of adverse events in a dose titration study. J Clin Pharmacol 34:136–141. https://doi.org/10.1002/j.1552-4604.1994.tb03977.x

Hussein AS, Abdullah N, Fakru'l-razi A (2013) Optimizing the process parameters for encapsulation of linamarin into PLGA nanoparticles using double emulsion solvent evaporation technique. Adv Polym Tech 32:E486–E504. https://doi.org/10.1002/adv.21295

Imre B, Pukánszky B (2013) Compatibilization in bio-based and biodegradable polymer blends. Eur Polym J 49:1215–1233. https://doi.org/10.1016/j.eurpolymj.2013.01.019

Jain KK (2008) Drug delivery systems – an overview. Methods Mol Biol 437:1–50. https://doi.org/10.1007/978-1-59745-210-6_1

Jayakumar R, Prabaharan M, Nair SV, Tamura H (2010) Novel chitin and chitosan nanofibers in biomedical applications. Biotechnol Adv 28:142–150. https://doi.org/10.1016/j.biotechadv.2009.11.001

Jogani V, Jinturkar K, Vyas T, Misra A (2008) Recent patents review on intranasal administration for CNS drug delivery. Recent Pat Drug Deliv Formul 2:25–40. https://doi.org/10.2174/187221108783331429

Jones NL, Cooper JJ, Waters RD, Williams DF (2001) Resorption profile and biological response of calcium phosphate filled PLLA and PHB7V. In: Agarwal CM, Parr JE, Lin ST (eds) Synthetic bioabsorbable polymers for implants. Astm International, Pennsylvania, pp 69–82. https://doi.org/10.1520/STP1396-EB. ISBN:978-0-8031-5444-5

Kai D, Low ZW, Liow SS, Abdul Karim A, Ye H, Jin G, Li K, Loh XJ (2015) Development of lignin supramolecular hydrogels with mechanically responsive and self-healing properties. ACS Sustain Chem Eng 3:2160–2169. https://doi.org/10.1021/acssuschemeng.5b00405

Kamachi M, Zhang S, Goodwin S, Lenz RW (2001) Enzymatic polymerization and characterization of new poly(3-hydroxyalkanoate)s by a bacterial polymerase. Macromolecules 34:6889–6894. https://doi.org/10.1021/ma010081z

Kamaly N, Yameen B, Wu J, Farokhzad OC (2016) Degradable controlled-release polymers and polymeric nanoparticles: mechanisms of controlling drug release. Chem Rev 116:2602–2663. https://doi.org/10.1021/acs.chemrev.5b00346

Kang J-S, Lee M-H (2009) Overview of therapeutic drug monitoring. Korean J Intern Med 24:1–10. https://doi.org/10.3904/kjim.2009.24.1.1

Karim AA, Dou Q, Li Z, Loh XJ (2016) Emerging supramolecular therapeutic carriers based on host-guest interactions. Chem Asian J 11:1300–1321. https://doi.org/10.1002/asia.201501434

Kassab AC, Xu K, Denkbas EB, Dou Y, Zhao S, Piskin E (1997) Rifampicin carrying polyhydroxybutyrate microspheres as a potential chemoembolization agent. J Biomater Sci Polym Ed 8:947–961. https://doi.org/10.1163/156856297X00119

Kassab AC, Pişkin E, Bilgic S, Denkbaş E, Xu K (1999) Embolization with polyhydroxybutyrate (PHB) microspheres: in-vivo studies. J Bioact Compat Polym 14:291–303

Kawaguchi T, Tsugane A, Higashide K, Endoh H, Hasegawa T, Kanno H, Seki T, Juni K, Fukushima S, Nakano M (1992) Control of drug release with a combination of prodrug and polymer matrix: antitumor activity and release profiles of 2′, 3′-diacyl-5-fluoro-2′-deoxyuridine from poly (3-hydroxybutyrate) microspheres. J Pharm Sci 81:508–512. https://doi.org/10.1002/jps.2600810606

Kazuhiko J, Masahiro N, Miho K (1986) Controlled release of aclarubicim, an anticancer antibiotic, from poly-β-hydroxybutyric acid microspheres. J Control Release 4:25–32. https://doi.org/10.1016/0168-3659(86)90030-1

Kedar U, Phutane P, Shidhaye S, Kadam V (2010) Advances in polymeric micelles for drug delivery and tumor targeting. Nanomedicine 6:714–729. https://doi.org/10.1016/j.nano.2010.05.005

Khang G, Kim SW, Cho JC, Rhee JM, Yoon SC, Lee HB (2001) Preparation and characterization of poly(3-hydroxybutyrate-co-3-hydroxyvalerate) microspheres for the sustained release of 5-fluorouracil. Biomed Mater Eng 11:89–103

Kılıçay E, Demirbilek M, Türk M, Güven E, Hazer B, Denkbas EB (2011) Preparation and characterization of poly(3-hydroxybutyrate-co-3-hydroxyhexanoate) (PHBHHX) based nanoparticles for targeted cancer therapy. Eur J Pharm Sci 44:310–320. https://doi.org/10.1016/j.ejps.2011.08.013

Kilicay E, Karahaliloglu Z, Hazer B, Tekin IÖ, Denkbas EB (2016) Concanavaline A conjugated bacterial polyester-based PHBHHx nanoparticles loaded with curcumin for breast cancer therapy. J Microencapsul 33:274–285. https://doi.org/10.3109/02652048.2016.1169325

Kim H-N, Lee J, Kim H-Y, Kim Y-R (2009) Enzymatic synthesis of a drug delivery system based on polyhydroxyalkanoate-proteinblock copolymers. Chem Commun 46:7104–7106. https://doi.org/10.1039/B912871A

Kim TH, Mount CW, Gombotz WR, Pun SH (2010) The delivery of doxorubicin to 3-D multicellular spheroids and tumors in a murine xenograft model using tumor-penetrating triblock polymeric micelles. Biomaterials 31:7386–7397. https://doi.org/10.1016/j.biomaterials.2010.06.004

Kirkizlar O, Kendir M, Karaali Z, Ure U, Ozbay G, Selcuk D, Kazancioglu R (2007) Acute renal failure in a patient with severe hemolysis. Int Urol Nephrol 39:651–654. https://doi.org/10.1007/s11255-006-9096-3

Knowles JC (1993) Development of a natural degradable polymer for orthopaedic use. J Med Eng Technol 17:129–137. https://doi.org/10.3109/03091909309087593
Knowles JC, Hastings GW (1993) In vitro and in vivo investigation of a range of phosphate glass-reinforced polyhydroxybutyrate-based degradable composites. J Mater Sci Mater Med 4:102–106. https://doi.org/10.1007/bf00120377
Knowles JC, Mahmud FA, Hastings GW (1991) Piezoelectric characteristics of a polyhydroxybutyrate-based composite. Clin Mater 8:155–158. https://doi.org/10.1016/0267-6605(91)90024-A
Knowles JC, Hastings GW, Ohta H, Niwa S, Boeree N (1992) Development of a degradable composite for orthopaedic use: in vivo biomechanical and histological evaluation of two bioactive degradable composites based on the polyhydroxybutyrate polymer. Biomaterials 13:491–496. https://doi.org/10.1016/0142-9612(92)90099-A
Korsatko W, Wabnegg B, Tillian H, Braunegg G, Lafferty R (1983) Poly-D-hydroxybutyric acid-a biologically degradable vehicle to regard release of a drug. Pharm Ind 45:1004–1007
Kose GT, Kenar H, Hasirci N, Hasirci V (2003) Macroporous poly(3-hydroxybutyrate-co-3-hydroxyvalerate) matrices for bone tissue engineering. Biomaterials 24:1949–1958. https://doi.org/10.1016/S0142-9612(02)00613-0
Kose G, Korkusuz F, Korkusuz P, Hasirci V (2004) In vivo tissue engineering of bone using poly(3-hydroxybutyric acid-co-3-hydroxyvaleric acid) and collagen scaffolds. Tissue Eng 10:1234–1250. https://doi.org/10.1089/ten.2004.10.1234
Kost J, Langer R (2001) Responsive polymeric delivery systems. Adv Drug Deliv Rev 46:125–148. https://doi.org/10.1016/S0169-409X(00)00136-8
Kostopoulos L, Karring T (1994) Guided bone regeneration in mandibular defects in rats using a bioresorbable polymer. Clin Oral Implants Res 5:66–74. https://doi.org/10.1034/j.1600-0501.1994.050202.x
Kreuter J (1978) Nanoparticles and nanocapsules – new dosage forms in the nanometer size range. Pharm Acta Helv 53:33–39
Laurencin CT, Kumbar SG, Nukavarapu SP, James R, Hogan MV (2008) Recent patents on electrospun biomedical nanostructures: an overview. Recent Pat Biomed Eng 1:68–78. https://doi.org/10.2174/1874764710801010068
Le Ray AM, Chiffoleau S, Iooss P, Grimandi G, Gouyette A, Daculsi G, Merle C (2003) Vancomycin encapsulation in biodegradable poly(epsilon-caprolactone) microparticles for bone implantation. Influence of the formulation process on size, drug loading, in vitro release and cytocompatibility. Biomaterials 24:443–449. https://doi.org/10.1016/S0142-9612(02)00357-5
Lee Y-F, Sridewi N, Ramanathan S, Sudesh K (2016) The influence of electrospinning parameters and drug loading on polyhydroxyalkanoate (PHA) nanofibers for drug delivery. Int J Biotechnol Wellness Ind 4:103–113. https://doi.org/10.6000/1927-3037.2015.04.04.1
Lenz RW, Marchessault RH (2005) Bacterial polyesters: biosynthesis, biodegradable plastics and biotechnology. Biomacromolecules 6:1–8. https://doi.org/10.1021/bm049700c
Li Z, Loh XJ (2017) Recent advances of using polyhydroxyalkanoate-based nanovehicles as therapeutic delivery carriers. Wiley Interdiscip Rev Nanomed Nanobiotechnol 9:e1429. https://doi.org/10.1002/wnan.1429
Li J, Li X, Ni X, Wang X, Li H, Leong KW (2006) Self-assembled supramolecular hydrogels formed by biodegradable PEO-PHB-PEO triblock copolymers and alpha-cyclodextrin for controlled drug delivery. Biomaterials 27:4132–4140. https://doi.org/10.1016/j.biomaterials.2006.03.025
Li W, Ding Y, Rai R, Roether JA, Schubert DW, Boccaccini AR (2014) Preparation and characterization of PHBV microsphere/45S5 bioactive glass composite scaffolds with vancomycin releasing function. Mater Sci Eng C Mater Biol Appl 41:320–328. https://doi.org/10.1016/j.msec.2014.04.052
Li X, Chang H, Luo H, Wang Z, Zheng G, Lu X, He X, Chen F, Wang T, Liang J, Xu M (2015) Poly (3-hydroxybutyrate-co-3-hydroxyhexanoate) scaffolds coated with PhaP-RGD fusion protein promotes the proliferation and chondrogenic differentiation of human umbilical cord mesenchymal stem cells in vitro. J Biomed Mater Res A 103:1169–1175. https://doi.org/10.1002/jbm.a.35265

Li W, Jan Z, Ding Y, Liu Y, Janko C, Pischetsrieder M, Alexiou C, Boccaccini AR (2016) Facile preparation of multifunctional superparamagnetic PHBV microspheres containing SPIONs for biomedical applications. Sci Rep 6:23140. https://doi.org/10.1038/srep23140

Liechty WB, Kryscio DR, Slaughter BV, Peppas NA (2010) Polymers for drug delivery systems. Annu Rev Chem Biomol Eng 1:149–173. https://doi.org/10.1146/annurev-chembioeng-073009-100847

Liu Y, Cai D, Yang J, Wang Y, Zhang X, Yin S (2014a) In vitro hemocompatibility evaluation of poly (4-hydroxybutyrate) scaffold. Int J Clin Exp Med 7:1233–1243

Liu Y, Zhu YH, Mao CQ, Dou S, Shen S, Tan ZB, Wang J (2014b) Triple negative breast cancer therapy with CDK1 siRNA delivered by cationic lipid assisted PEG-PLA nanoparticles. J Control Release 192:114–121. https://doi.org/10.1016/j.jconrel.2014.07.001

Loh XJ, Goh SH, Li J (2007a) Hydrolytic degradation and protein release studies of thermogelling polyurethane copolymers consisting of poly[(R)-3-hydroxybutyrate], poly(ethylene glycol), and poly(propylene glycol). Biomaterials 28:4113–4123. https://doi.org/10.1016/j.biomaterials.2007.05.016

Loh XJ, Goh SH, Li J (2007b) New biodegradable thermogelling copolymers having very low gelation concentrations. Biomacromolecules 8:585–593. https://doi.org/10.1021/bm0607933

Loh XJ, Ong SJ, Tung YT, Choo HT (2013) Dual responsive micelles based on poly[(R)-3-hydroxybutyrate] and poly(2-(di-methylamino)ethyl methacrylate) for effective doxorubicin delivery. Polym Chem 4:2564–2574. https://doi.org/10.1039/C3PY00096F

Lu X-Y, Zhang Y, Wang L (2010) Preparation and in vitro drug-release behavior of 5-fluorouracil-loaded poly(hydroxybutyrate- co -hydroxyhexanoate) nanoparticles and microparticles. J Appl Polym Sci 8:2944–2950. https://doi.org/10.1002/app.31806

Lu XY, Ciraolo E, Stefenia R, Chen GQ, Zhang Y, Hirsch E (2011) Sustained release of PI3K inhibitor from PHA nanoparticles and in vitro growth inhibition of cancer cell lines. Appl Microbiol Biotechnol 89:1423–1433. https://doi.org/10.1007/s00253-011-3101-1

Lü L-X, Zhang X-F, Wang Y-Y, Ortiz L, Mao X, Jiang Z-L, Xiao Z-D, Huang N-P (2013) Effects of hydroxyapatite-containing composite nanofibers on osteogenesis of mesenchymal stem cells in vitro and bone regeneration in vivo. ACS Appl Mater Interfaces 5:319–330. https://doi.org/10.1021/am302146w

Lu X-Y, Li M-C, Zhu X-L, Fan F, Wang L-L, Ma J-G (2014) Microbial synthesized biodegradable PHBHHxPEG hybrid copolymer as an efficient intracellular delivery nanocarrier for kinase inhibitor. BMC Biotechnol 14:4–4. https://doi.org/10.1186/1472-6750-14-4

Luef KP, Stelzer F, Wiesbrock F (2015) Poly(hydroxy alkanoate)s in medical applications. Chem Biochem Eng Q 29:287–297. https://doi.org/10.15255/CABEQ.2014.2261

Luklinska ZB, Schluckwerder H (2003) In vivo response to HA-polyhydroxybutyrate/polyhydroxyvalerate composite. J Microsc 211:121–129. https://doi.org/10.1046/j.1365-2818.2003.01204.x

Mader JS, Hoskin DW (2006) Cationic antimicrobial peptides as novel cytotoxic agents for cancer treatment. Expert Opin Investig Drugs 15:933–946. https://doi.org/10.1517/13543784.15.8.933

Mahapatro A, Singh DK (2011) Biodegradable nanoparticles are excellent vehicle for site directed in-vivo delivery of drugs and vaccines. J Nanobiotechnol 9:55. (51–59). https://doi.org/10.1186/1477-3155-9-55

Makadia HK, Siegel SJ (2011) Poly lactic-co-glycolic acid (PLGA) as biodegradable controlled drug delivery carrier. Polymers 3:1377–1397. https://doi.org/10.3390/polym3031377

Mankin HJ, Fogelson FS, Thrasher AZ, Jaffer F (1976) Massive resection and allograft transplantation in the treatment of malignant bone tumors. N Engl J Med 294:1247–1255. https://doi.org/10.1056/nejm197606032942301

Mano JF, Sousa RA, Boesel LF, Neves NM, Reis RL (2004) Bioinert, biodegradable and injectable polymeric matrix composites for hard tissue replacement: state of the art and recent developments. Compos Sci Technol 64:789–817. https://doi.org/10.1016/j.compscitech.2003.09.001

Martínez Rivas CJ, Tarhini M, Badri W, Miladi K, Greige-Gerges H, Nazari QA, Galindo Rodríguez SA, Román RÁ, Fessi H, Elaissari A (2017) Nanoprecipitation process: from encapsulation to drug delivery. Int J Pharm 532:66–81. https://doi.org/10.1016/j.ijpharm.2017.08.064

Masood F, Chen P, Yasin T, Fatima N, Hasan F, Hameed A (2013a) Encapsulation of Ellipticine in poly-(3-hydroxybutyrate-co-3-hydroxyvalerate) based nanoparticles and its in vitro application. Mater Sci Eng C Mater Biol Appl 33:1054–1060. https://doi.org/10.1016/j.msec.2012.11.025
Masood F, Chen P, Yasin T, Hasan F, Ahmad B, Hameed A (2013b) Synthesis of poly-(3-hydroxybutyrate-co-12 mol % 3-hydroxyvalerate) by *Bacillus cereus* FB11: its characterization and application as a drug carrier. J Mater Sci Mater Med 24:1927–1937. https://doi.org/10.1007/s10856-013-4946-x
Matthews J, Gademann K, Jaun B, Seebach D (1998) Linear and cyclic [small beta]3-oligopeptides with functionalised side-chains (-CH2OBn, -CO2Bn, -CH2CH2CO2Bn) derived from serine and from aspartic and glutamic acid. J Chem Soc Perkin Trans 1:3331–3340. https://doi.org/10.1039/A805478I
Maurus PB, Kaeding CC (2004) Bioabsorbable implant material review. Oper Tech Sports Med 12:158–160. https://doi.org/10.1053/j.otsm.2004.07.015
Mayya KS, Bhattacharyya A, Argillier JF (2003) Micro-encapsulation by complex coacervation: influence of surfactant. Polym Int 52:644–647. https://doi.org/10.1002/pi.1125
McGinity JW, O'Donnell PB (1997) Preparation of microspheres by the solvent evaporation technique. Adv Drug Deliv Rev 28:25–42. https://doi.org/10.1016/S0169-409X(97)00049-5
Mikhail AS, Allen C (2009) Block copolymer micelles for delivery of cancer therapy: transport at the whole body, tissue and cellular levels. J Control Release 138:214–223. https://doi.org/10.1016/j.jconrel.2009.04.010
Minost A, Delaveau J, Bolzinger M-A, Fessi H, Elaissari A (2012) Nanoparticles via nanoprecipitation process. Recent Pat Drug Deliv Formul 6:250–258. https://doi.org/10.2174/187221112802652615
Misra SK, Valappil SP, Roy I, Boccaccini AR (2006) Polyhydroxyalkanoate (PHA)/inorganic phase composites for tissue engineering applications. Biomacromolecules 7:2249–2258. https://doi.org/10.1021/bm060317c
Modi S, Koelling K, Vodovotz Y (2011) Assessment of PHB with varying hydroxyvalerate content for potential packaging applications. Eur Polym J 47:179–186. https://doi.org/10.1016/j.eurpolymj.2010.11.010
Monnier A, Rombouts C, Kouider D, About I, Fessi H, Sheibat-Othman N (2016) Preparation and characterization of biodegradable polyhydroxybutyrate-co-hydroxyvalerate/polyethylene glycol-based microspheres. Int J Pharm 513:49–61. https://doi.org/10.1016/j.ijpharm.2016.08.066
Mousavioun P, Doherty WOS, George G (2010) Thermal stability and miscibility of poly(hydroxybutyrate) and soda lignin blends. Ind Crop Prod 32:656–661. https://doi.org/10.1016/j.indcrop.2010.08.001
Müller DJ, Dufrêne YF (2008) Atomic force microscopy as a multifunctional molecular toolbox in nanobiotechnology. Nat Nanotechnol 3:261–269. https://doi.org/10.1038/nnano.2008.100
Murueva AV, Shishatskaya EI, Kuzmina AM, Volova TG, Sinskey AJ (2013) Microparticles prepared from biodegradable polyhydroxyalkanoates as matrix for encapsulation of cytostatic drug. J Mater Sci Mater Med 24:1905–1915. https://doi.org/10.1007/s10856-013-4941-2
Murugesan S, Ganesan S, Averineni RK, Nahar M, Mishra P, Jain NK (2007) PEGylated poly(lactide-co-glycolide) (PLGA) nanoparticulate delivery of docetaxel: synthesis of diblock copolymers, optimization of preparation variables on formulation characteristics and in vitro release studies. J Biomed Nanotechnol 3:52–60. https://doi.org/10.1166/jbn.2007.012
Narayani R, Panduranga Rao K (1996) Gelatin microsphere cocktails of different sizes for the controlled release of anticancer drugs. Int J Pharm 143:255–258. https://doi.org/10.1016/S0378-5173(96)04685-6
Nguyen MK, Lee DS (2010) Injectable biodegradable hydrogels. Macromol Biosci 10:563–579. https://doi.org/10.1002/mabi.200900402
Ni J, Wang M (2002) In vitro evaluation of hydroxyapatite reinforced polyhydroxybutyrate composite. Mater Sci Eng C 20:101–109. https://doi.org/10.1016/S0928-4931(02)00019-X

Nicolas J, Mura S, Brambilla D, Mackiewicz N, Couvreur P (2013) Design, functionalization strategies and biomedical applications of targeted biodegradable/biocompatible polymer-based nanocarriers for drug delivery. Chem Soc Rev 42:1147–1235. https://doi.org/10.1039/c2cs35265f

Nicolson PC, Vogt J (2001) Soft contact lens polymers: an evolution. Biomaterials 22:3273–3283. https://doi.org/10.1016/S0142-9612(01)00165-X

Nobes G, Maysinger D, Marchessault R (1998) Polyhydroxyalkanoates: materials for delivery systems. Drug Deliv 5:167–177. https://doi.org/10.3109/10717549809052032

O'Connor S, Szwej E, Nikodinovic-Runic J, O'Connor A, Byrne AT, Devocelle M, O'Donovan N, Gallagher WM, Babu R, Kenny ST, Zinn M, Zulian QR, O'Connor KE (2013) The anti-cancer activity of a cationic anti-microbial peptide derived from monomers of polyhydroxyalkanoate. Biomaterials 34:2710–2718. https://doi.org/10.1016/j.biomaterials.2012.12.032

Oerlemans C, Bult W, Bos M, Storm G, Nijsen JF, Hennink WE (2010) Polymeric micelles in anticancer therapy: targeting, imaging and triggered release. Pharm Res 27:2569–2589. https://doi.org/10.1007/s11095-010-0233-4

Orts WJ, Nobes GA, Kawada J, Nguyen S, Yu G-E, Ravenelle F (2008) Poly (hydroxyalkanoates): biorefinery polymers with a whole range of applications. The work of robert H. Marchessault. Can J Chem 86:628–640. https://doi.org/10.1139/v08-050

Owen S, Chan DPY, Shoichet MS (2012) Polymeric micelle stability. Nano Today 7:53–65. https://doi.org/10.1016/j.nantod.2012.01.002

Panith N, Assavanig A, Lertsiri S, Bergkvist M, Surarit R, Niamsiri N (2016) Development of tunable biodegradable polyhydroxyalkanoates microspheres for controlled delivery of tetracycline for treating periodontal disease. J Appl Polym Sci 133:44128. https://doi.org/10.1002/app.44128

Park K (2014) Controlled drug delivery systems: past forward and future back. J Control Release 190:3–8. https://doi.org/10.1016/j.jconrel.2014.03.054

Park S, Lee SH, Lee SY (2001) Preparation of optically active β-amino acids from microbial polyester polyhydroxyalkanoates. J Chem Res 2001:498–499. https://doi.org/10.3184/030823401103168640

Peer D, Karp JM, Hong S, Farokhzad OC, Margalit R, Langer R (2007) Nanocarriers as an emerging platform for cancer therapy. Nat Nanotechnol 2:751–760. https://doi.org/10.1038/nnano.2007.387

Peng Q, Zhang Z-R, Sun X, Zuo J, Zhao D, Gong T (2010) Mechanisms of phospholipid complex loaded nanoparticles enhancing the oral bioavailability. Mol Pharm 7:565–575. https://doi.org/10.1021/mp900274u

Peng X, Chen Y, Li Y, Wang Y, Zhang X (2016) A long-acting BMP-2 release system based on poly(3-hydroxybutyrate) nanoparticles modified by amphiphilic phospholipid for osteogenic differentiation. Biomed Res Int 2016:5878645. https://doi.org/10.1155/2016/5878645

Pich A, Schiemenz N, Corten C, Adler H-JP (2006) Preparation of poly(3-hydroxybutyrate-co-3-hydroxyvalerate) (PHBV) particles in O/W emulsion. Polymer 47:1912–1920. https://doi.org/10.1016/j.polymer.2006.01.038

Pietrzak WS, Sarver DR, Verstynen ML (1997) Bioabsorbable polymer science for the practicing surgeon. J Craniofac Surg 8:87–91. https://doi.org/10.1097/00001665-199703000-00004

Pillay V, Dott C, Choonara YE, Tyagi C, Tomar L, Kumar P, du Toit LC, Ndesendo VMK (2013) A review of the effect of processing variables on the fabrication of electrospun nanofibers for drug delivery applications. J Nanomater 2013:1–22. https://doi.org/10.1155/2013/789289

Poletto FS, Fiel LA, Donida B, Ré MI, Guterres SS, Pohlmann AR (2008) Controlling the size of poly(hydroxybutyrate-co-hydroxyvalerate) nanoparticles prepared by emulsification–diffusion technique using ethanol as surface agent. Colloids Surf A Physicochem Eng Asp 324:105–112. https://doi.org/10.1016/j.colsurfa.2008.04.003

Poste G, Kirsh R (1983) Site-specific (targeted) drug delivery in cancer therapy. Nat Biotechnol 1:869–878. https://doi.org/10.1038/nbt1283-869

Pouton CW, Akhtar S (1996) Biosynthetic polyhydroxyalkanoates and their potential in drug delivery. Adv Drug Deliv Rev 18:133–162

Pramanik N, Mitra T, Khamrai M, Bhattacharyya A, Mukhopadhyay P, Gnanamani A, Basu RK, Kundu PP (2015) Characterization and evaluation of curcumin loaded guar gum/polyhydroxyalkanoates blend films for wound healing applications. RSC Adv 5:63489–63501. https://doi.org/10.1039/C5RA10114J

Pramual S, Assavanig A, Bergkvist M, Batt CA, Sunintaboon P, Lirdprapamongkol K, Svasti J, Niamsiri N (2016) Development and characterization of bio-derived polyhydroxyalkanoate nanoparticles as a delivery system for hydrophobic photodynamic therapy agents. J Mater Sci Mater Med 27:40. https://doi.org/10.1007/s10856-015-5655-4

Pushpamalar J, Veeramachineni AK, Owh C, Loh XJ (2016) Biodegradable polysaccharides for controlled drug delivery. ChemPlusChem 81:504–514. https://doi.org/10.1002/cplu.201600112

Qi X, Liu Y, Ding ZY, Cao JQ, Huang JH, Zhang JY, Jia WT, Wang J, Liu CS, Li XL (2017) Synergistic effects of dimethyloxallyl glycine and recombinant human bone morphogenetic protein-2 on repair of critical-sized bone defects in rats. Sci Rep 7:1–13. https://doi.org/10.1038/srep42820

Qu X-H, Wu Q, Liang J, Qu X, Wang S-G, Chen G-Q (2005) Enhanced vascular-related cellular affinity on surface modified copolyesters of 3-hydroxybutyrate and 3-hydroxyhexanoate (PHBHHx). Biomaterials 26:6991–7001. https://doi.org/10.1016/j.biomaterials.2005.05.034

Qu X-H, Wu Q, Chen G-Q (2006a) In vitro study on hemocompatibility and cytocompatibility of poly(3-hydroxybutyrate-co-3-hydroxyhexanoate). J Biomater Sci Polym Ed 17:1107–1121. https://doi.org/10.1163/156856206778530704

Qu X-H, Wu Q, Zhang K-Y, Chen GQ (2006b) In vivo studies of poly(3-hydroxybutyrate-co-3-hydroxyhexanoate) based polymers: biodegradation and tissue reactions. Biomaterials 27:3540–3548. https://doi.org/10.1016/j.biomaterials.2006.02.015

Radivojevic J, Skaro S, Senerovic L, Vasiljevic B, Guzik M, Kenny ST, Maslak V, Nikodinovic-Runic J, O'Connor KE (2016) Polyhydroxyalkanoate-based 3-hydroxyoctanoic acid and its derivatives as a platform of bioactive compounds. Appl Microbiol Biotechnol 100:161–172. https://doi.org/10.1007/s00253-015-6984-4

Ravanat J-L, Douki T, Cadet J (2001) Direct and indirect effects of UV radiation on DNA and its components. J Photochem Photobiol B 63:88–102. https://doi.org/10.1016/S1011-1344(01)00206-8

Ravi S, Peh KK, Darwis Y, Murthy BK, Singh TRR, Mallikarjun C (2008) Development and characterization of polymeric microspheres for controlled release protein loaded drug delivery system. Indian J Pharm Sci 70:303–309. https://doi.org/10.4103/0250-474X.42978

Ray S, Kalia VC (2017) Biomedical applications of polyhydroxyalkanoates. Indian J Microbiol 57:261–269. https://doi.org/10.1007/s12088-017-0651-7

Reddy CSK, Ghai R, Rashmi, Kalia VC (2003) Polyhydroxyalkanoates: an overview. Bioresour Technol 87:137–146. https://doi.org/10.1016/S0960-8524(02)00212-2

Ruth K, Grubelnik A, Hartmann R, Egli T, Zinn M, Ren Q (2007) Efficient production of (R)-3-hydroxycarboxylic acids by biotechnological conversion of polyhydroxyalkanoates and their purification. Biomacromolecules 8:279–286. https://doi.org/10.1021/bm060585a

Saito T, Tomita K, Juni K, Ooba K (1991) In vivo and in vitro degradation of poly(3-hydroxybutyrate) in rat. Biomaterials 12:309–312. https://doi.org/10.1016/0142-9612(91)90039-D

Sajeev US, Anoop Anand K, Menon D, Nair S (2008) Control of nanostructures in PVA, PVA/chitosan blends and PCL through electrospinning. Bull Mater Sci 31:343–351. https://doi.org/10.1007/s12034-008-0054-9

Salman M, Sahin A, Onur M, Öge K, Kassab A, Aypar Ü (2003) Tramadol encapsulated into polyhydroxybutyrate microspheres: in vitro release and epidural analgesic effect in rats. Acta Anaesthesiol Scand 47:1006–1012. https://doi.org/10.1034/j.1399-6576.2003.00180.x

Sanathanan LP, Peck CC (1991) The randomized concentration-controlled trial: an evaluation of its sample size efficiency. Control Clin Trials 12:780–794. https://doi.org/10.1016/0197-2456(91)90041-J

Scheithauer EC, Li W, Ding Y, Harhaus L, Roether JA, Boccaccini AR (2015) Preparation and characterization of electrosprayed daidzein–loaded PHBV microspheres. Mater Lett 158:66–69. https://doi.org/10.1016/j.matlet.2015.05.133

Seebach D, Albert M, Arvidsson PI, Rueping M, Schreiber JV (2001) From the biopolymer PHB to biological investigations of unnatural β- and γ-peptides. Chimia 55:345–353

Sendil D, Gursel I, Wise DL, Hasirci V (1999) Antibiotic release from biodegradable PHBV microparticles. J Control Release 59:207–217. https://doi.org/10.1016/S0168-3659(98)00195-3

Sevastianov VI, Tseytlina EA (1984) The activation of the complement system by polymer materials and their blood compatibility. J Biomed Mater Res 18:969–978. https://doi.org/10.1002/jbm.820180902

Sevastianov VI, Perova NV, Shishatskaya EI, Kalacheva GS, Volova TG (2003) Production of purified polyhydroxyalkanoates (PHAs) for applications in contact with blood. J Biomater Sci Polym Ed 14:1029–1042. https://doi.org/10.1163/156856203769231547

Shah M, Choi MH, Ullah N, Kim MO, Yoon SC (2011) Synthesis and characterization of PHV-block-mPEG diblock copolymer and its formation of amphiphilic nanoparticles for drug delivery. J Nanosci Nanotechnol 11:5702–5710. https://doi.org/10.1166/jnn.2011.4493

Shah M, Ullah N, Choi MH, Kim MO, Yoon SC (2012) Amorphous amphiphilic P(3HV-co-4HB)-b-mPEG block copolymer synthesized from bacterial copolyester via melt transesterification: nanoparticle preparation, cisplatin-loading for cancer therapy and in vitro evaluation. Eur J Pharm Biopharm 80:518–527. https://doi.org/10.1016/j.ejpb.2011.11.014

Shah M, Ullah N, Choi MH, Yoon SC (2014) Nanoscale poly(4-hydroxybutyrate)-mPEG carriers for anticancer drugs delivery. J Nanosci Nanotechnol 14:8416–8421. https://doi.org/10.1166/jnn.2014.9924

Sharma PK, Munir RI, de Kievit T, Levin DB (2017) Synthesis of polyhydroxyalkanoates (PHAs) from vegetable oils and free fatty acids by wild-type and mutant strains of Pseudomonas chlororaphis. Can J Microbiol 63:1009–1024. https://doi.org/10.1139/cjm-2017-0412

Shin SY, Lee SH, Yang ST, Park EJ, Lee DG, Lee MK, Eom SH, Song WK, Kim Y, Hahm KS, Kim JI (2001) Antibacterial, antitumor and hemolytic activities of alpha-helical antibiotic peptide, P18 and its analogs. J Pept Res 58:504–514. https://doi.org/10.1034/j.1399-3011.2001.00934.x

Shishatskaya EI, Goreva AV, Voinova ON, Inzhevatkin EV, Khlebopros RG, Volova TG (2008a) Evaluation of antitumor activity of rubomycin deposited in absorbable polymeric microparticles. Bull Exp Biol Med 145:358–361. https://doi.org/10.1007/s10517-008-0091-9

Shishatskaya EI, Voinova ON, Goreva AV, Mogilnaya OA, Volova TG (2008b) Biocompatibility of polyhydroxybutyrate microspheres: in vitro and in vivo evaluation. J Mater Sci Mater Med 19:2493–2502. https://doi.org/10.1007/s10856-007-3345-6

Shrivastav A, Kim H-Y, Kim Y-R (2013) Advances in the applications of polyhydroxyalkanoate nanoparticles for novel drug delivery system. Biomed Res Int 2013:12. https://doi.org/10.1155/2013/581684

Simoes S, Figueiras A, Veiga F (2012) Modular hydrogels for drug delivery. J Biomater Nanobiotechnol 3:185. https://doi.org/10.4236/jbnb.2012.32025

Singh NK, Lee DS (2014) In situ gelling pH- and temperature-sensitive biodegradable block copolymer hydrogels for drug delivery. J Control Release 193:214–227. https://doi.org/10.1016/j.jconrel.2014.04.056

Singh R, Lillard JW Jr (2009) Nanoparticle-based targeted drug delivery. Exp Mol Pathol 86:215–223. https://doi.org/10.1016/j.yexmp.2008.12.004

Sombatmankhong K, Sanchavanakit N, Pavasant P, Supaphol P (2007) Bone scaffolds from electrospun fiber mats of poly(3-hydroxybutyrate), poly(3-hydroxybutyrate-co-3-hydroxyvalerate) and their blend. Polymer 48:1419–1427. https://doi.org/10.1016/j.polymer.2007.01.014

Sousa RA, Reis RL, Cunha AM, Bevis MJ (2003) Processing and properties of bone-analogue biodegradable and bioinert polymeric composites. Compos Sci Technol 63:389–402. https://doi.org/10.1016/S0266-3538(02)00213-0

Stamm C, Khosravi A, Grabow N, Schmohl K, Treckmann N, Drechsel A, Nan M, Schmitz KP, Haubold A, Steinhoff G (2004) Biomatrix/polymer composite material for heart valve tissue engineering. Ann Thorac Surg 78:2084–2092. https://doi.org/10.1016/j.athoracsur.2004.03.106

Suarez-Jimenez G-M, Burgos-Hernandez A, Ezquerra-Brauer J-M (2012) Bioactive peptides and depsipeptides with anticancer potential: sources from marine animals. Mar Drugs 10:963–986. https://doi.org/10.3390/md10050963

Sun J, Dai Z, Zhao Y, Chen G-Q (2007) In vitro effect of oligo-hydroxyalkanoates on the growth of mouse fibroblast cell line L929. Biomaterials 28:3896–3903. https://doi.org/10.1016/j.biomaterials.2007.05.011

Sur S, Fries AC, Kinzler KW, Zhou S, Vogelstein B (2014) Remote loading of preencapsulated drugs into stealth liposomes. Proc Natl Acad Sci U S A 111:2283–2288. https://doi.org/10.1073/pnas.1324135111

Tan L, Yu X, Wan P, Yang K (2013) Biodegradable materials for bone repairs: a review. J Mater Sci Technol 29:503–513. https://doi.org/10.1016/j.jmst.2013.03.002

Tan Y, Hu Y, Liu X, Yin Z, Chen XW, Liu M (2016) Improving drug safety: from adverse drug reaction knowledge discovery to clinical implementation. Methods 110:14–25. https://doi.org/10.1016/j.ymeth.2016.07.023

Tian H, Tang Z, Zhuang X, Chen X, Jing X (2012) Biodegradable synthetic polymers: preparation, functionalization and biomedical application. Prog Polym Sci 37:237–280. https://doi.org/10.1016/j.progpolymsci.2011.06.004

Titushkin IA, Vasin SL, Rozanova IB, Pokidysheva EN, Alekhin AP, Sevastianov VI (2001) Carbon coated polyethylene: effect of surface energetics and topography on human platelet adhesion. ASAIO J 47:11–17. https://doi.org/10.1097/00002480-200101000-00005

Tiwari G, Tiwari R, Sriwastawa B, Bhati L, Pandey S, Pandey P, Bannerjee SK (2012) Drug delivery systems: an updated review. Int J Pharma Investig 2:2–11. https://doi.org/10.4103/2230-973X.96920

Upadhyay M, Adena SKR, Mishra B (2017) Hydrogels: an introduction to a controlled drug delivery device, synthesis and application in drug delivery and tissue engineering. Austin J Biomed Eng 4:1037

Vainionpaa S, Vihtonen K, Mero M, Patiala H, Rokkanen P, Kilpikari J, Tormala P (1986) Biodegradable fixation of rabbit osteotomies. Acta Orthop Scand 57:237–239. https://doi.org/10.3109/17453678608994386

Valappil SP, Misra SK, Boccaccini A, Roy I (2006) Biomedical applications of polyhydroxyalkanoates, an overview of animal testing and in vivo responses. Expert Rev Med Devices 3:853–868. https://doi.org/10.1586/17434440.3.6.853

Vardhan H, Mittal P, Adena SKR, Upadhyay M, Mishra B (2017) Development of long-circulating docetaxel loaded poly (3-hydroxybutyrate-co-3-hydroxyvalerate) nanoparticles: optimization, pharmacokinetic, cytotoxicity and in vivo assessments. Int J Biol Macromol 103:791–801. https://doi.org/10.1016/j.ijbiomac.2017.05.125

Vega-Castro O, Contreras-Calderon J, León E, Segura A, Arias M, Pérez L, Sobral PJA (2016) Characterization of a polyhydroxyalkanoate obtained from pineapple peel waste using *Ralsthonia eutropha*. J Biotechnol 231:232–238. https://doi.org/10.1016/j.jbiotec.2016.06.018

Vilar G, Tulla-Puche J, Albericio F (2012) Polymers and drug delivery systems. Curr Drug Deliv 9:367–394

Vilos, Gutiérrez Cutiño M, Escobar J, Morales Zavala F, Denardin J, Velasquez L, Altbir (2013) Superparamagnetic poly (3-hydroxybutyrate-co-3 hydroxyvalerate) (PHBV) nanoparticles for biomedical applications. Electron J Biotechnol 16. https://doi.org/10.1016/j.ejbt.v16n5-8

Vitezić D, Božina N, Mršić-Pelčić J, Turk VE, Francetić I (2016) Personalized medicine in clinical pharmacology. In: Bodiroga-Vukobrat N, Rukavina D, Pavelić K, Sander GG (eds) Personalized medicine: a new medical and social challenge. Springer, Cham, pp 265–278. https://doi.org/10.1007/978-3-319-39349-0. ISBN:978-3-319-39349-0

Wang L, Wang Z-H, Shen C-Y, You M-L, Xiao J-F, Chen G-Q (2010) Differentiation of human bone marrow mesenchymal stem cells grown in terpolyesters of 3-hydroxyalkanoates scaffolds into nerve cells. Biomaterials 31:1691–1698. https://doi.org/10.1016/j.biomaterials.2009.11.053

Witte F, Ulrich H, Rudert M, Willbold E (2007) Biodegradable magnesium scaffolds: part 1: appropriate inflammatory response. J Biomed Mater Res A 81:748–756. https://doi.org/10.1002/jbm.a.31170

Wong HM, Yeung KWK, Lam KO, Tam V, Chu PK, Luk KDK, Cheung KMC (2010) A biodegradable polymer-based coating to control the performance of magnesium alloy orthopaedic implants. Biomaterials 31:2084–2096. https://doi.org/10.1016/j.biomaterials.2009.11.111

Wu LP, Wang D, Parhamifar L, Hall A, Chen G-Q, Moghimi SM (2014) Poly(3-hydroxybutyrate-co-R-3-hydroxyhexanoate) nanoparticles with polyethylenimine coat as simple, safe, and versatile vehicles for cell targeting: population characteristics, cell uptake, and intracellular trafficking. Adv Healthc Mater 3:817–824. https://doi.org/10.1002/adhm.201300533

Xiao X-Q, Zhao Y, Chen G-Q (2007) The effect of 3-hydroxybutyrate and its derivatives on the growth of glial cells. Biomaterials 28:3608–3616. https://doi.org/10.1016/j.biomaterials.2007.04.046

Xiong YC, Yao YC, Zhan XY, Chen GQ (2010) Application of polyhydroxyalkanoates nanoparticles as intracellular sustained drug-release vectors. J Biomater Sci Polym Ed 21:127–140. https://doi.org/10.1163/156856209x410283

Xu JX, Tang JB, Zhao LH, Shen YQ (2009) Advances in the study of tumor pH-responsive polymeric micelles for cancer drug targeting delivery. Acta Pharm Sin 44:1328–1335

Xu W, Ling P, Zhang T (2013) Polymeric micelles, a promising drug delivery system to enhance bioavailability of poorly water-soluble drugs. J Drug Deliv 2013:1–15. https://doi.org/10.1155/2013/340315

Ye H, Owh C, Loh XJ (2015) A thixotropic polyglycerol sebacate-based supramolecular hydrogel showing UCST behavior. RSC Adv 5:48720–48728. https://doi.org/10.1039/C5RA08222F

Yu H-Y, Qin Z-Y (2014) Surface grafting of cellulose nanocrystals with poly(3-hydroxybutyrate-co-3-hydroxyvalerate). Carbohydr Polym 101:471–478. https://doi.org/10.1016/j.carbpol.2013.09.048

Yu D-G, Zhu L-M, White K, Branford-White C (2009) Electrospun nanofiber-based drug delivery systems. Health 01:67–75. https://doi.org/10.4236/health.2009.12012

Yu X, Trase I, Ren M, Duval K, Guo X, Chen Z (2016) Design of nanoparticle-based carriers for targeted drug delivery. J Nanomater 2016:1–15. https://doi.org/10.1155/2016/1087250

Zawidlak B, Kawalec M, Bosek I, Łuczyk-Juzwa M, Adamus G, Rusin A, Filipczak P, Głowala-Kosińska M, Wolanska K, Krawczyk Z, Kurcok P (2010) Synthesis and antiproliferative properties of ibuprofen-oligo(3-hydroxybutyrate) conjugates. Eur J Med Chem 45:1833–1842. https://doi.org/10.1016/j.ejmech.2010.01.020

Zhang C, Zhang Z, Zhao L (2016) Folate-decorated poly(3-hydroxybutyrate-co-3-hydroxyoctanoate) nanoparticles for targeting delivery: optimization and in vivo antitumor activity. Drug Deliv 23:1830–1837. https://doi.org/10.3109/10717544.2015.1122675

Zhao X (2011) Bioactive materials in drug delivery systems. In: Zhao X, Courtney JM, Qian H (eds) Bioactive materials in medicine. Woodhead Publishing, Cambridge, pp 247–265. https://doi.org/10.1533/9780857092939.2.247. ISBN:978-1-84569-624-5

Zhao S, Zhu M, Zhang J, Zhang Y, Liu Z, Zhu Y, Zhang C (2014) Three dimensionally printed mesoporous bioactive glass and poly(3-hydroxybutyrate-co-3-hydroxyhexanoate) composite scaffolds for bone regeneration. J Mater Chem B 2:6106–6118. https://doi.org/10.1039/C4TB00838C

Zhu M, Zuo W, Yu H, Yang W, Chen Y (2006) Superhydrophobic surface directly created by electrospinning based on hydrophilic material. J Mater Sci 41:3793–3797. https://doi.org/10.1007/s10853-005-5910-z

Zhu XH, Wang CH, Tong YW (2009) In vitro characterization of hepatocyte growth factor release from PHBV/PLGA microsphere scaffold. J Biomed Mater Res A 89:411–423. https://doi.org/10.1002/jbm.a.31978

Applications of Polyhydroxyalkanoates Based Nanovehicles as Drug Carriers

6

Mohanasundaram Sugappriya, Dorairaj Sudarsanam, Jerrine Joseph, Mudasir A. Mir, and Chandrabose Selvaraj

Abstract

Polyhydroxyalkanoates (PHAs) are the biodegradable polyesters synthesized from the bacterial origin with high therapeutic importance due to their major biopolymer properties like biodegradability, biocompatibility, non-toxicity and thermo-plasticity. These hypotheses make the PHAs as one of the drug carrier or nanovehicles for therapeutic drug delivery. Drug delivery is one of the emerging fields in biological science, which gives application of delivering the promising drug molecules in the desired drug target, with increasing dissolution rate and bioavailability. Understanding the drug-polymer interactions and controlled release makes the combo of the perfect drug molecule and the biopolymer, and in this, the PHA is reported to be a better natural origin drug carrier. Several reports suggest that, chemical entities of PHA show enhanced activity and used as one of the promising nanocarrier. Experimental test by several assays shows PHA's as one of the compatible biopolymers in tissues. This chapter comes up with various applications related to PHAs like medical implants, adhesion

M. Sugappriya
Research and Development Centre, Bharathiar University, Coimbatore, Tamil Nadu, India

D. Sudarsanam
Department of Zoology and Advanced Biotechnology, Loyola College, Chennai, Tamil Nadu, India

J. Joseph
Centre for Drug Discovery and Development, Jeppiaar Research Park, Sathyabama Deemed to be University, Chennai, Tamil Nadu, India

M. A. Mir
Centre for Plant Biotechnology, Division of Biotechnology, Sher-e-Kashmir University of Agricultural Sciences and Technology of Kashmir, Shalimar, India

C. Selvaraj (✉)
School of Basic Science, Indian Institute of Technology Mandi, Kamand, Himachal Pradesh, India

V. C. Kalia (ed.), *Biotechnological Applications of Polyhydroxyalkanoates*, https://doi.org/10.1007/978-981-13-3759-8_6

barriers, drug carriers, tissue engineering, Nano based targeted delivery. Overall, this chapter will provide a perfect base for studies on PHA related to drug carrier studies.

Keywords
Biopolymer · Polyhydroxyalkanotes · Carriers · Nanovehicles · Implants · Scaffolds · Suture · Cancer · Wound healing · Bioplastics · Drug delivery

Abbreviations

APTS	*para*-toluene sulfonic acid monohydrate
DTXL	Docetaxel
HA	hydroxyapatite
HDI	1,6-hexamethylene di-isocyanate
P3	poly(esterurethane)s
P3HB4HB	poly(R-3-hydroxybutyrate-co-4-hydroxybutyrate)
P4HB	poly-4-hydroxybutyrate;
PHA(s)	Polyhydroxyalkanoate(s)
PhaP	amphiphilic PHA granule binding protein
PhaR	PHA synthesis repressor protein
PHB	poly-(R)-3-hydroxybutyrate;
PHB	Polyhydroxybutyrate
PHBHHx	poly(R-3-hydroxybutyrate-co-R-3-hydroxyhexanoate)
PHBHx	Poly(hydroxybutyrate-co-hydroxyhexanoate)
PHBO	Poly(hydroxybutyrate-co-hydroxyoctanoate)
PHBV	Poly(hydroxybutyrate-co-hydroxyvalerate)
PHBV	poly(R-3-hydroxybutyrate-co-R-3-hydroxyvalerate)
PHBVHHx	poly(R-3-hydroxy-butyrate-co-R-3-hydroxyvalerate-co-R-3-hydroxyhexanoate)
PHHx	Poly(hydroxyhexanoate)
PHHx	Poly(hydroxyhexanoate)
PHO	Poly(hydroxyoctanoate)
PHO	poly-(R)-3-hydroxyoctanoate;
PPC	Poly propylene carbonate

6.1 Introduction

As far with trending biological science, the Bioplastics are playing the core importance in multiple applications, holds the specialty of the pollution free product and environment friendly materials. In this, one of the notable bioplastic called "Polyhydroxyalkanoates"(PHAs) has introduced a several applications which replace the conventional plastics with various physical attributes. Polymers are the renewable resources among which the PHA is a biodegradable polymer accumulated as storage materials in the cytoplasm of the bacterial cells (Troschl et al. 2017). PHA is produced in the lack of nutrients, namely nitrogen and phosphorous, with carbon source found plethora. The first discovery of Polyhydroxybutyrate (PHB), which is a member of PHA family by Lemoine in 1927. PHA is very easily produced by the microorganisms in nature which shows its biodegradability (Kourmentza et al. 2017). It acts as an alternative for the traditional polymers where PHA exhibits mechanical and thermal properties similar to the plastics. The side chain length plays a vital role in maintaining these properties which give them stability. There are more than 150 different types of hydroxy alkanoic acids known for the bacterial storage polyesters. Ecological niches namely the marine habitats, estuaries, rhizosphere, microbial mats and marine sediments are the most convenient places for the microbes to perform its role and interact with all other microorganisms to constitute the ecosystem. The PHA accumulation will be occurred by microorganisms, with the higher stress lever in the environmental factors and considering this factor, the researchers have focused the industrial waste and effluents from textile industry, distilleries, dairy farms, oil mills, paper industry, animal residues, agricultural wastes and biodiesel waste products for PHA production. Since PHAs are produced under scarcity of nutrients, but with excess carbon, they are stored as inclusion bodies in polymerized form. The PHA has been previously produced from the sludge by aerobic dynamic feeding and anaerobic system in fermented effluents and sequencing batch reactors. The production of PHA from mixed cultures enriched the glycogen accumulating organisms, phosphate accumulating organisms and Archae.

PHA as a stored energy source makes the bacterium *Bacillus megaterium* in water a biocontrol agent (Defoirdt et al. 2009). The product which contains PHA breaks down completely into carbon dioxide and water, which are the major significant molecules for the photosynthesis. This could be a reason; PHA fits into the carbon cycle similar to petroleum-based plastics. The nanoparticles have a wide range of applications in various fields, as for the medical fields is concerned, they are used as the covering material for the drugs, which can be peculiarly designed and altered with respect to the target sites without altering the durability of the living cells. The PHA overcomes the commercial bioplastics like poly lactic acid (PLA), s-poly-glycolic acid (PGA) and poly (dl-lactide-co-glycolide) (PLGA),

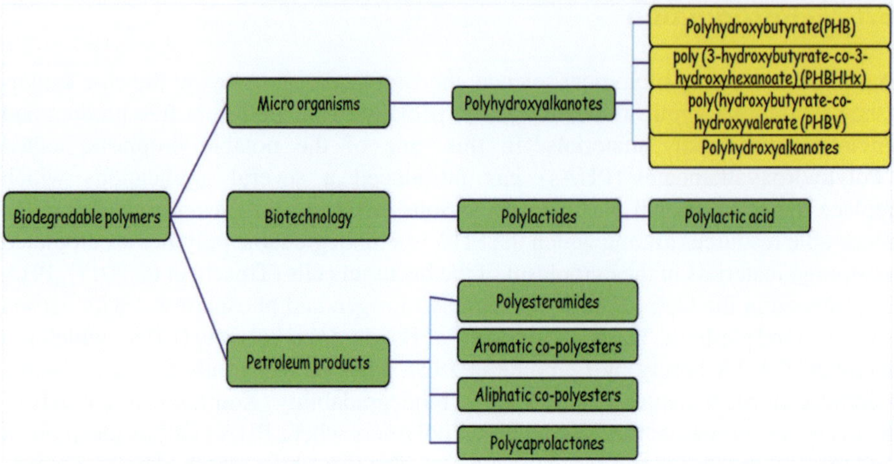

Fig. 6.1 Different biodegradable polymers

since the monomers of 3-hydroxybutyric acid and 4-hydroxybutyric acid are removed from the body in less than an hour since they are less acidic when compared to other bioplastics. The elimination or the eradication of PHA depends on the dispensation methods like nanofibers, films, scaffolds and also varies from one tissue to the other. The PHA degradation products like 3HB (3-hydroxy butyrate) and 3 HV (3-hydroxy valerate) are less acidic with pKa values of 4.7 than the glycolic and lactic acid with 4.72 (Fig. 6.1).

The drug deliveries are highly significant tools for the delivery of pharmacological compounds to the tissues, organs and cells. They are done by the polymers which make the drugs reach the target site safer with reduced side effects. One of the traditionally used polymers is the silicone which has been uncertain to cause cancers. To overcome these drawbacks, there is a growing need to find nontoxic, biocompatible and biodegradable polymers (Michalak et al. 2016). The poly(3-hydroxybutyrate) and the poly(3-hydroxybutyrate-*co*-3-hydroxyvalerate (PHBV) are the commonly used drug carriers and has more merits than the other polymers in tissue engineering. The polylactate, polyglycolate and poly(lactide-co-glycolide) are the chemically synthesized polymers which are biodegradable and biocompatible. The merit of PHA over other polymers is that they are impregnated into the human body and need not be removed, which are degraded by the system without any adverse effects. Hence it is aid to be a better bet against synthesized polymers (Errico et al. 2009). On hydrolytic degradation PHB produces 3 hydroxy butyrate, a metabolite in the human blood, were 3HB itself contains pharmaceutical properties. In order to improve the efficacy, the copolymers are blended with hexanoate, 1,4-butanediol, propionate and valerate. Poly(3-hyroxybutyrate-co-3-hydroxyhexanoate) belonging to the PHA family is physically blended with PHB (Misra et al. 2010). The glycosaminoglycans are used along with PHAs to improve the mechanical strength and biological properties of PHA biopolymer. There are certain inorganic compounds like bioactive glass, nano-HA (hydroxyapatite), calcium silicate and zinconium oxides added to PHB and PHBV for enhancing the cell interactions in tissue engineering and bone implants (Pascu et al. 2013). For example, calcium silicate enhances the bone bioactivity; cell adhesion of the composite, the bioglass grades 45S5 (45 wt% SiO2, 24.5 wt% Na2O, 24.5 wt% CaO, and 6.0 wt% P2O5) interconnects PHB with the formation of hydroxyalkanote within a short period, making it an excellent material for bone regeneration. The PHB is soluble in few solvents like dimethyl formamide, dichloromethane and chloroform and hence mixing of polymers with PHB is a challenging task. More care is taken during thermal molding, which makes the polymers break (Vroman and Tighzert 2009; Manavitehrani et al. 2016).

6.2 Structure of PHA

PHA's are crystalline in nature, hydrophobic with chirally pure configuration monomers. The PHAs inside the bacterial cells as refractile bodies and as environmental friendly degradable plastics was initially found in early 1988 (Shrivastav et al. 2013). The PHA composition has been established by Griebel in 1968, in bacterium *Bacillus megaterium*. These have a huge array of applications in the fields of marine, medical and agriculture industry. The PHAs are stored as granules in the bacterial cells (Shively 2011). These granules contain proteins (1.87%), lipids (0.46%) and polyesters (97.7%). The region depicted in brackets is one monomer and R is the residual group, which varies based on the length of the carbon chains (Fig. 6.2). There are different types of PHA, which are differentiated based on the length of the side chain and the main chains. The types are based on the number of the carbon atoms in the monomeric units, as the short chain length PHA's (scl-PHAs) with three to five carbons and medium chain length PHA's (mcl-PHAs) with six to fourteen carbon atoms (Aslan et al. 2016). The enzyme involved in PHA biosynthesis is the PHA synthase, which polymerizes the monomeric hydroxyalkanote substrates.

There are around 14 pathways involved in the PHA biosynthesis and are classified based on the primary sequences.

Class I: One type- PhaC which forms homodimer, Scl monomer C3-C5 carbon chain lengths.

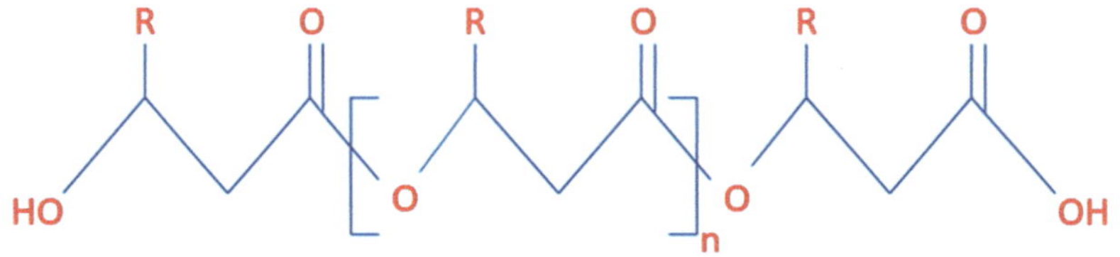

Fig. 6.2 Structure of polyhydroxyalkanotes

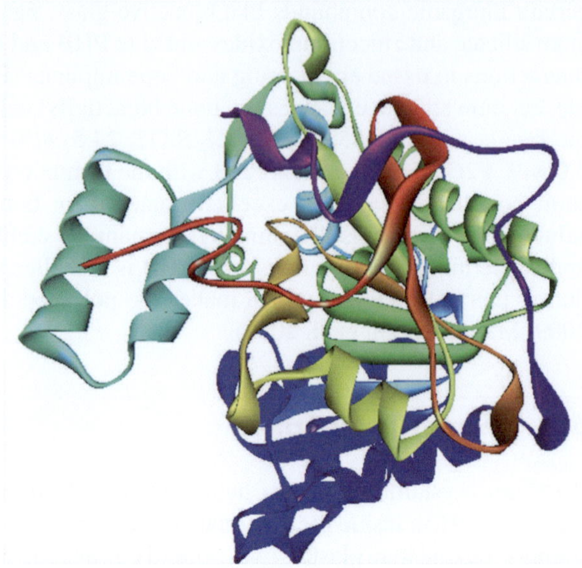

Fig. 6.3 Structure of class I polyhydroxybutyrate synthase – Cupriavidus necator, 3D visualization using DS Visualizer

Class II: Two types- PhaC1 and PhaC2, Mcl monomer C6-C14 carbon chain lengths.

Class III: PhaC-PhaE, PhaC and PhaR, and forms heterodimer, Scl monomer C3-C5 carbon chain lengths

Class IV: Similar to class III, Scl monomer C3-C5 carbon chain lengths

In spite of the discovery three decades ago, there is no structural information of PHB synthase, the first structure of the catalytic domain of a PHB synthase is the PhaC from *C. necator* with PDB ID: 5T6O (Structure of the catalytic domain of the class I polyhydroxybutyrate synthase from *Cupriavidus necator*) (Fig. 6.3) (Chek et al. 2017). The scl-PHAs are used for the production of packing materials and disposable items, whereas the mcl-PHAs are of highly important applications like medicinal implants, matrix for drug delivery, surgical sutures etc., (Chen 2009a, b). Due to the diverse properties and different strategies used in the production of PHA, the functional group modifications like –COOH, OH, halogens and epoxy may enhance the thermal and mechanical properties. As the Scl and Mcl possess different monomeric units, their mixture possesses greater physical and thermal properties which are through the enzyme engineering (Sharma et al. 2017).

6.3 Bacterial PHA

PH3B, a homopolymer of 3-hydroxybutyrate is the most widely used member of PHA. The molecular weight of PHA depends upon the organism's growth conditions and the extraction methods. The organisms *Bacillus megaterium, Ralstonia eutrophus* is said to produce PHB due to the physiological stress in a mixed or pure culture. The microbial synthesis of PHB involves condensation of two molecules of acetyl CoA to produce acetoacetyl CoA which is reduced to hydroxybutyryl CoA which is the monomer to polymerize PHB (Laycock et al. 2013).

6.4 PHA Production

The commercial production of PHA is by the fermentation of microorganisms namely *Ralstonia eutropha, Pseudomonas putida, Escherichia coli, Aeromonas caviae, Azotobacter sp*, *Bacillus sp,* and *Methylobacterium sp*. Though various microorganisms produce PHA, only a few are selected for the commercial production. Those organisms which have the metabolic activity of synthesizing PHA molecules are specifically selected. They play a vital role in the ion channel complexes, where degradation of PHB polymer yields 3-hydroxybutyrate which is a natural metabolite responsible for the formation of ketone bodies in animals. The ketone body is biosynthesized in the mitochondria of liver and used by brain (Roy 2015). One of the commonly used strains is *Cupriavidus necator* for the industrial production of PHB, P3HB4HB and PHBV. The recombinant *Escherichia coli* are used in the production of PHA, PHV and PHBV were PHAs are also produced in plants (Fig. 6.4).

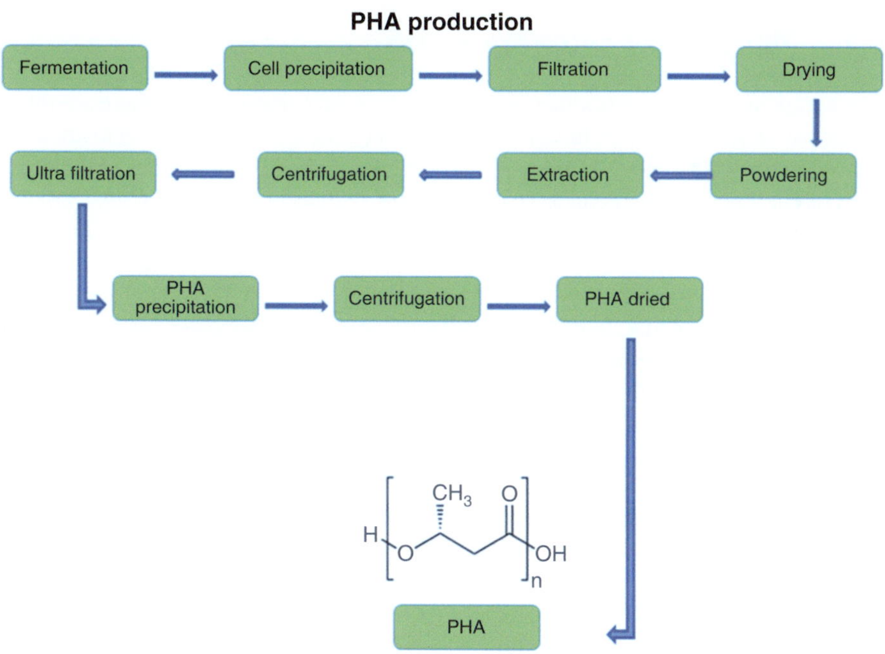

Fig. 6.4 Production and extraction of PHA

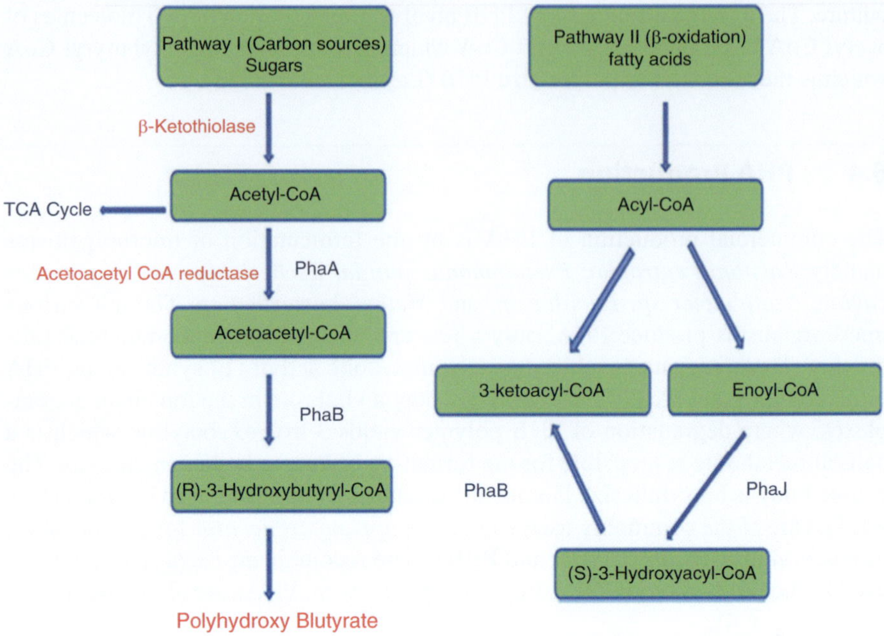

Fig. 6.5 Pathways in PHA production

6.5 Biosynthetic Pathways of PHA

The composition of the monomer is based on the used carbon source; there are three major pathways of PHA. Pathway 1 is two acetyl CoA from the TCA cycle are condensed to form aceto-acetyl CoA with the generation of 3HB monomers in the presence of enzyme β-keto-thiolase. The acetyl CoA reductase and acetyl CoA form the 3-hydroxybutyryl-CoA, finally yielding poly (3-hydroxybutyrate) P(3HB). Pathway 2 is the fatty acid metabolism which generates different hydroxyalkanote monomers, where the trans-2-enoyl-CoA is converted to (R)-hydroxyacyl-CoA by a (R)-specific enoyl-CoA hydratase. Pathway generates monomers form carbon sources like sucrose, glucose and fructose, which involves the conversion of acyl carrier protein (ACP) form to the CoA form by the enzyme acyl-ACP-CoA transacylase (Fig. 6.5).

6.6 Sources and Microorganisms Used for PHB Production

6.6.1 Eubacteria

In the year 1925, PHB accumulation was reported in *Bacillus megaterium* in the form of cytoplasmic inclusions, the PHA synthesizing bacteria are most commonly

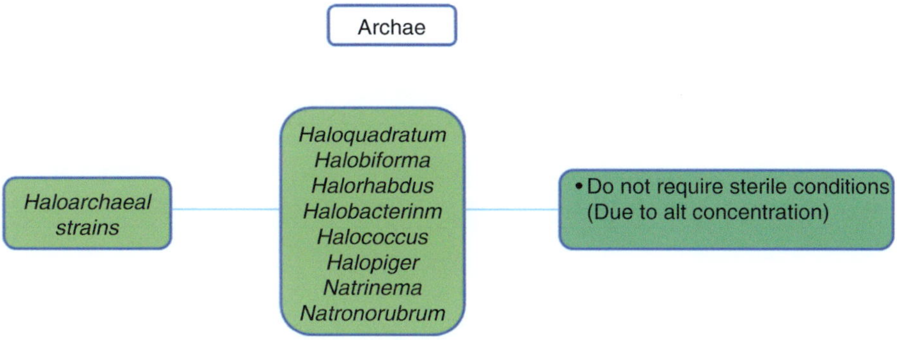

Fig. 6.6 Archaea involved in PHB production

seen in nitrogen fixing organisms. They are classified into two groups based on the culture conditions, with the first group containing bacteria which requires excess carbon and limited nitrogen and oxygen, and the second group does not require limited nutrients. For the large-scale production, the culture conditions are the most important factor to be taken on account. The recombinant *E. coli* was shown to have high PHA accumulation due to the presence of PHA biosynthesis genes (Attia et al. 2010; Shukla 2017).

6.6.2 Archaea

Haloarchaea are the class of Euryarchaeota, which requires high salt concentrations to grow, and they are saturated with salt in order to maintain the cell wall stability, in the absence of the salt the cell wall breaks (Fig. 6.6).

6.6.3 Cyanobacteria

Cyanobacteria accumulated PHA by making use of the energy sources like CO_2 and sunlight. These bacteria naturally possess PHA synthase enzyme. *Synechocystis sp.* PCC6803, *spirulinaplatensis* UMACC 161, accumulates 10% PHB as of the cell dry weight. Some of the cyanobacteria require phosphate limited conditions for the accumulation of PHB (Sudhesh et al. 2006).

6.6.4 Agro-industrial Residue

To reduce the production costs of the bioplastics the industrial and agricultural wastes as nitrogen and carbon sources are utilized for the biosynthesis of degradable bioplastics.

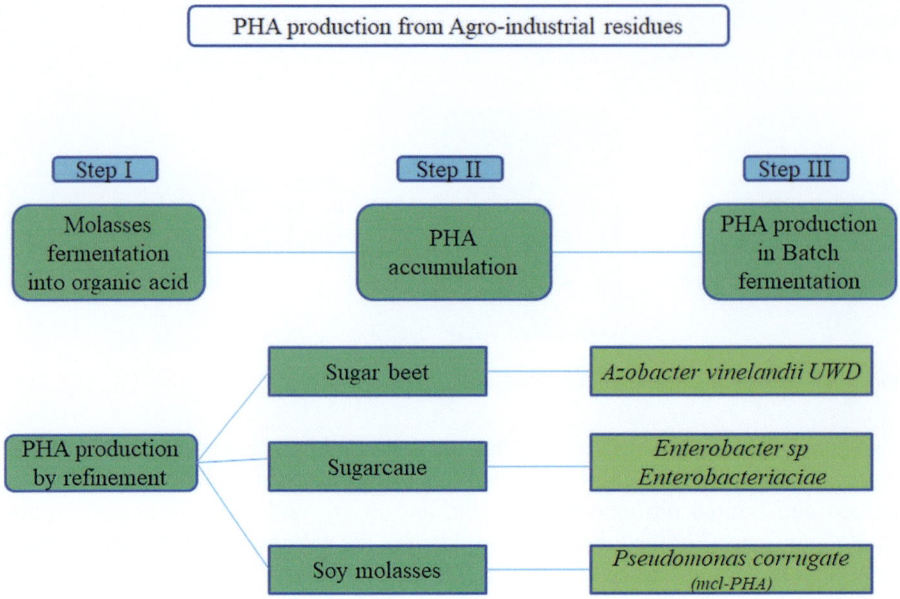

Fig. 6.7 PHA production from Agro-industrial wastes

6.6.5 Molasses

The by-product obtained during the refining process of grapes, sugar beet and sugarcane. The accumulation of PHB was reported in the year 1992, in *Azobactervinelandii* UWD species. PHA production involves a three-step process, with step I, the fermentation of molasses into organic acids, followed by the Step II with the accumulation of PHA, and the step III, with the production of PHA in batches. On the refinement of the sugar beet, sugarcane and soy molasses PHA accumulation was observed in certain organisms (Fig. 6.7).

6.6.6 Whey Hydrolysates

Whey is the by-product of cheese and dairy industry obtained from the separation of casein from whole milk. The cheese whey equals the amount of milk used during processing, and the disposal of these waste causes pollution to the environment. PHA production has been reported in the cheese whey, in *Paracoccusdenitrificans* DSM 413 (Pantazaki et al. 2009; Koller et al. 2011) (Fig. 6.8).

6.6.7 Lignocellulose Material

Lignocelluloses consist of 30–50% cellulose, 15–35% lignin, and 20–50% hemicelluloses. The wheat straw is an agricultural residue, a most abundant cheap carbon source in bioplastics production. The *Halomonasboliviensis* from the wheat bran

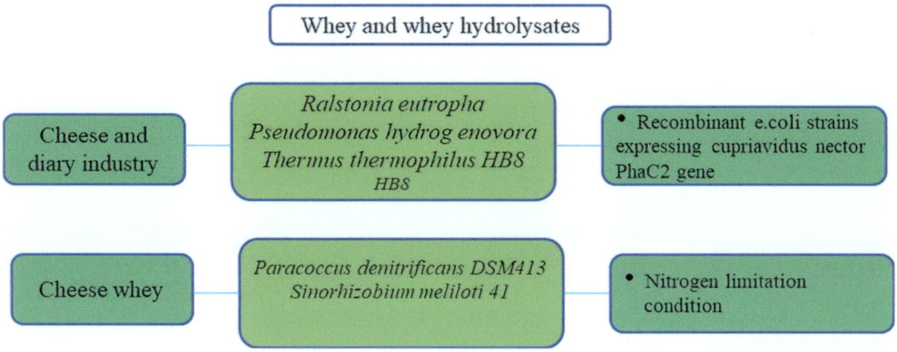

Fig. 6.8 PHA production from whey

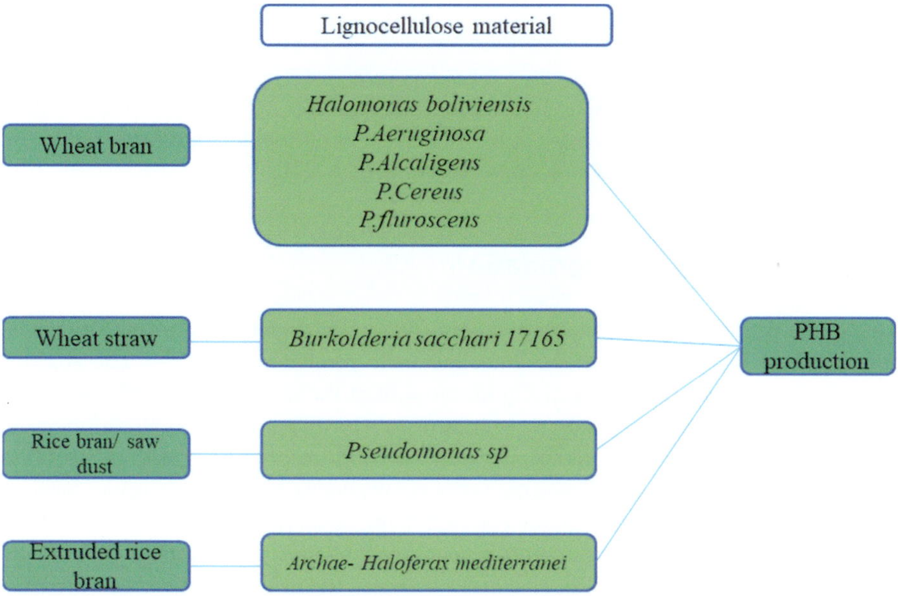

Fig. 6.9 PHB production from lignocellulose

was reported for the production of PHB, were the wheat bran hydrolyzed by the enzyme to provide mixture of sugars. Similarly, the wheat straw, rice bran and extruded rice bran are the sources, for the organisms in the production of PHB (Fig. 6.9).

6.6.8 Vegetable Oils and Fats

The triacylglycerols (TAG) are an important feedstock for the production of PHA, the *Aeromonascaviae* biosynthesize PHA from the triacylglycerol. The type of the substrate and the monomer type have a relationship, were the high saturated fats

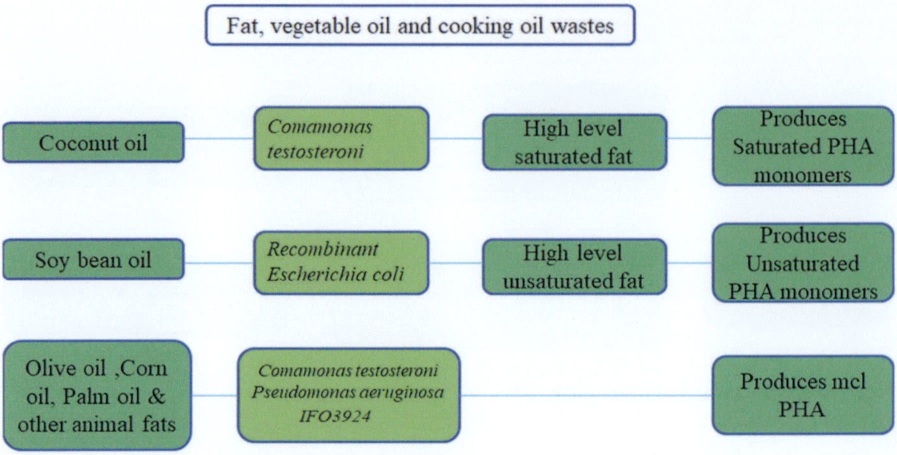

Fig. 6.10 PHA production from oil contaminants

produce high saturated PHA monomers and the high unsaturated fats produce unsaturated monomers. Olive oil, palm oil and other animal fats are responsible for the production of medium chain length PHAs (Fig. 6.10).

6.7 Microbial Biodegradation

Due to the presence of xenobiotic compounds the microorganisms face the environmental stress which in turn makes the chances to produce more PHA. The amount of carbon is found to be seen in higher amount at the oil contaminated sites with lesser nitrogen, which makes a prerequisite for the microbes *Pseudomonas, Acinetobacter, brochothrix, Ralstonia,* and*Yokenella* to produce PHA. It has been reported that the organism *Ralstoniaeutropha* produces PHB nearly up to 50% of cell dry weight due to the presence of sodium benzoate and phenol as carbon sources but limiting nutrients. The waste products comprising toluene as constituent enables various microbes like *Rhodococcusaetherivorans* to produce copolymer P(3HB-*co*-HV) as it makes more carbon source with limiting nutrients. Hence these wastes can be made as useful valuable products like PHA (Hori et al. 2009) (Fig. 6.11).

6.7.1 Extremophile Organisms

Halophiles are organisms which live in high salt concentrations, also known as salt loving organisms. They are classified under Archae, still there are some bacterial halophiles and eukaryota. They live in salt concentrations nearly five times higher than the salt concentrations in oceans, islands and salt lakes. For the first time the *Halobacterium marismortui* was found to have PHB accumulation identified using the freeze fracture technique (Robenek and Severs 2008). The bacterium

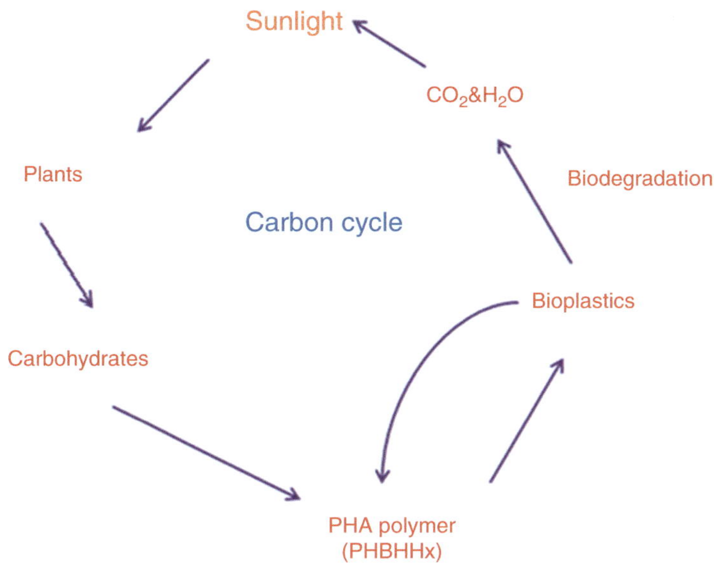

Fig. 6.11 Carbon cycle

Haloferaxmediterranei is said to produce around 65% of PHA of cells dry weight, in case were the glucose and sucrose acts as the carbon source, limitation of the oxygen enhances the PHA production. *H.boliviensis*, are said to produce higher amount of PHB nearly up to 88% of cell dry weight with sodium acetate and butyric acid as carbon sources (Kourmentza et al. 2017).

6.7.2 Phosynthetic Bacteria

Cyanobacteria are reported to produce PHA by oxygenic photosynthesis, as they can produce PHA naturally and have the tendency to store PHA. The PHA producing organisms are found to be species specific, and mostly produce PHB.

6.7.3 Rhizospheres

The rhizospheres are the place for the microbes to harbor which enhances the growth of roots in plants, and it's the place of exchange of nutrients for interaction of microbes hence the rhizospheres are the reservoirs of PHA. *Pseudomonas extrem-australis, Pseudomonas orientalis, Pseudomonas brassicacearum, Burkholderiaterricola* and *Lysobactergummosus*are reported to be the PHA producers. The microbe *Azospirrillumbrasilense*is a growth promoting rhizobacterium in plants (Muller-Santos et al. 2015). The need for the utilization and storage of poly--hydroxyalkanoate is for the maintenance of shelf-life of bacteria. There are various merits in the production of PHA like growth promotion in plants, cell multiplication, motility and colonization. Hence the oilseed rape, sugar beet and wheat rhizospheres have more PHA production, which means those plants which are rich in oily and carbohydrates products are said to produce more PHA.

6.7.4 Antibiotic Producers

Streptomyces, gram positive, aerobic bacteria are said to produce PHA as intracellular granules with rich carbon source for the antibiotics synthesis. Various other strains of *Streptomyces* produce PHB and play's an important role in antibiotic production.

6.8 Different Types of PHAs

Based on the number of carbons, the types of PHAs are classified. Among various types PHB and PHBV are the extensively used commercial polymeric biomaterials. The molecular weight and the structure are manipulated by changing the growth conditions.

6.8.1 PHB

The Poly-β-hydroxybutyrate (PHB) is a homopolymer of (R) – β-hydroxybutyric acid, a head to tail homopolymer, naturally found in bacteria (Chen and Wu 2005; Philip et al. 2007). It's the biodegradable and biocompatible polymer when used as implants gets degraded gradually by the temperature and produces non-toxic metabolites which are excreted through urine. Polyhydroxy butyrate and polyvinyl alcohol are used as scaffolds, prepared by the electro spinning in the field of tissue engineering. The collagen coated Nano fibrous PHBV loaded with somatic stem cells showed better healing property in rat models (Asran et al. 2010).

6.8.2 PHV, PHHx, PHHp, PHO

The polyhydroxyvalerate obtained from the *P.oleovorans*, in the activated sludge, few other monomers like 3HV, 3- hydroxyheptanoate (3HHp), and 3-hydroxyhexanoate (3HHx) existed as major components (Chen et al. 2013) (Fig. 6.12). The Poly (hydroxyhexanoate) (PHHx) from *P. putida KTHH03,* polyhydroxyheptanoate (PHHp) from *P. putida KTHH03,* PHO Poly(hydroxyoctanoate) from *Streptmoyceslividans*and PHN from *Alcaligenessp.* The therapeutic applications are mentioned in the Table 6.1 and medical applications are mentioned in the Fig. 6.13.

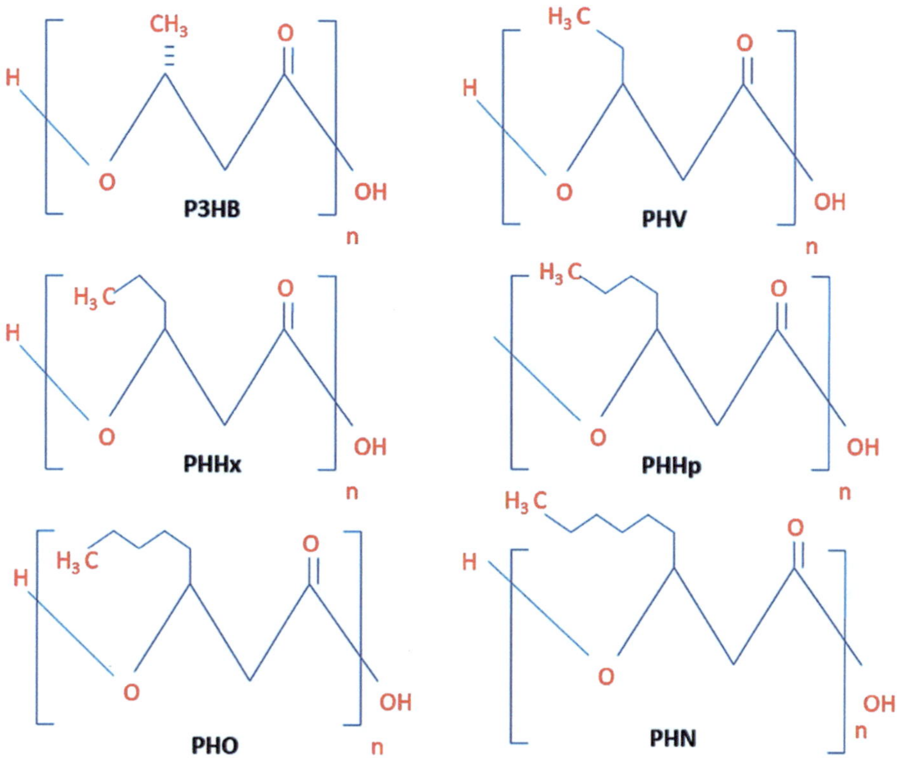

Fig. 6.12 Different kinds of PHAs

6.9 Biodegradability of PHA

An important role which makes the PHAs compatible is its biocompatibility. Since the PHAs are of the biological origin they are catabolized into water and carbon dioxide by microbes found in water, soil and sewage (Jaglan et al. 2017). These microbes excrete extracellular enzymes to hydrolyze solid PHA into water soluble monomers and oligomers, and use the end products as nutrients. The detriment of PHB seems to be faster in vivo than the in vitro because of the enzymes which catalyze the degradation at body temperature. The PHA polymers are degraded at the site of implantation by natural nonspecific esterase and lipases (Shrivastav et al. 2013). Lobler reported that there could be some impact of lipases on the PHA implants which has been shown by the lipase activities observed near the PHA implants in rat gastrointestinein vivo. For the use in medical applications PHA must be very specific and biocompatible, which should not cause any serious reactions as it must be introduced into the host organisms. The PHA must not evoke any immune response, when impregnated with the drug as the carrier vehicle. Over a period the PHA releases the drug which acts as the dosing agent. The dosage of the compounds

Table 6.1 Type of PHAs and their therapeutic applications

Polyhydroxyalkanoates	Properties	References
PHB, PHBHHx, PLA	Biodegradation and Biocompatible, PHB exhibits better properties than PLA	Chen and Zhang (2017)
	Tendon tissue regeneration, high affinity towards bone marrow cells, chondrocytes	
PHBVHHx	Tissue engineering – Skin and bones, proliferation of WJ-MSCs, PHBVHHx, better than all other biopolymers	Ji et al. (2009)
PHB-PLA-PCL (Triblock copolymers)	Supports cell growth and anti-blood clotting	Chen and Zhang (2017)
PHB, Ma-PHBHHx	Cell growth and fast biodegradable	Hong et al. (2008)
P3HB4HB-PEG, PHB-PEG	Drug delivery system (Injection), hydrophobic wound healing property	Loh et al. (2007)
PHB-PMLA, PHB-PEG	Tissue compatibility, regeneration of tissues	Bonartsev et al. (2016)
PhaP-RGD, PhaP-IKVAV, PHB, PHBHHx	Proliferation, cell adhesion and differentiation of the chondrocytes in rats.	Xie et al. (2013)
PHBHHx,PHBV,PLA coated with PhaP-RGD	Cell adhesion, accelerated growth on the coated films.	Chen et al. (2009a, b)
PHBHHx, PHB modified by ammonia plasma	Cell growth of umbilical endothelial cells in human, and aorta muscle cells-rabbit	Pompe et al. (2007)
PHBHHx-Insulin-Phospholipid, PHB-PMLA	Nanoparticle drug delivery matrices, hydrophilic drug delivery, insulin (long lasting)	Peng et al. (2012)
PHBHHx, PHB blend with NaOH	Proliferation of chondrocytes and elongation property	Yang et al. (2002)
PHBHHx, PHB blended with HAP, composites	Cell adhesion: MG-63 cells, PHBHHx-Hap: elastic modulus, cell growth, in vivo and in vitro alkaline phosphatase activity	Ding et al. (2016)
PHBHHx	Biodegradable, biocompatible, mechanical and can be altered with varying molecular weight	Shangguan et al. (2006)
PHBV, PHB	Anti-inflammatory, biodegradable	Shangguan et al. (2006)
PHBHHx-PhaP-ligand	In vivo and in vitro drug delivery matrices, specificity with colon cancer cells.	Yao et al. (2008) and Kwon et al. (2014)
P4HB nanoparticles	Sustained release of cisplatin from PHBHHx, inhibiting cancer growth	Lu et al. (2011) and Shah et al. (2014)
PHA nanoparticles	Anticancer drug release	Masood (2016)
PHB- Scaffolds –Nano bio glass	Tissue engineering: Bones, improves cellular proliferation	Hajiali et al, (2012)

(continued)

Table 6.1 (continued)

Polyhydroxyalkanoates	Properties	References
Medium chain length PHA	PHBHHx blend with PPC improves mechanical properties, Artificial heart and blood vessels	Zhang et al. (2007)
PHBHHx	Lithography (mimic of the cellular environment)	Dai et al. (2009)
PHBHHx	Formation of films on the PHBHHx, mixed with nontoxic solvents, artificial nerve conduits	Zhou et al. (2010)
PLA, PHB, P3HB4HB	Proliferation of HaCaT cell line, as nanofibre	Li et al. (2008a, b)
PHB-TCP	3D printing	Wang et al. (2006)
PHB,PHO blend	Growth and differentiation, mechanical properties of nerves	Lizarraga-Valderrama et al. (2015)
PHB	Nerve regeneration, scaffold attachment, proliferation and survival	Young et al. (2002)
PHBHHx coated porcine aortic valve	Artificial heart valve and blood vessels, promoting repopulation of cells	Sodian et al. (2000a, b) and Wu et al. (2007a, b)
P3HB4HB based polyurethane block copolymer, PHBV and bacterial cellulose	The 4HB content in P3HB4HB enhanced elastin formation and artificial blood vessels, PHBV+ cellulose (same as P3HB4HB)	Li et al. (2010)
HA, PHB	Hydroxyapatite or PHB coated with collagen enhanced osteogenic growth and differentiation	Rentsch et al. (2010)
PHBHHx	Islet transplantation, insulin gene expression	Yang et al. (2009) and Wang et al. (2006)
	Tarsal repair	
PHBVHHx	Nerve conduits, differentiation of human bone marrow mesenchymal cells	Wang et al. (2010)
PHBHHx coated PhaP-RGD	Tissue engineering: bone, repair spinal cord injury, cartilage tissue engineering	You et al. (2011) and Lomas et al. (2013)
P(3HB), P(4HB)- Tube form	Fibrosis reduction	Lobler et al. (2011)
P(3HB), P(4HB)- Microspheres	Bone disorders, anticancer drug delivery	Francis et al. (2011)
P(3HB-co-3 HV), magnetite composite	Intestinal tissue – targeted drug delivery	Vilos et al. (2013)
P(3HB-co-3 HV) nanoparticle	Folate mediated drug delivery (Doxorubicin)	Zhang et al. (2010)
P(3HB-co-3HHx)	Folate mediated anticancer drug delivery	Kiliçay et al. (2011))
P(3HB),P(4HB)/HA, P(3HO)	Skin tissue engineering	Peschel et al. (2008)

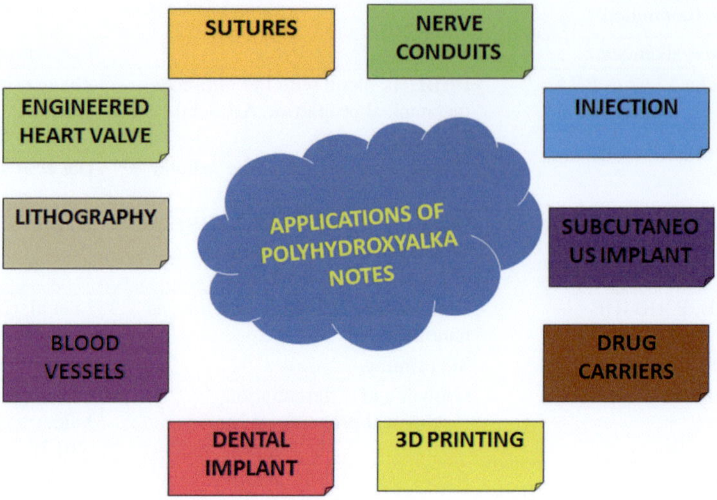

Fig. 6.13 Biomedical applications of PHA

can be altered by changing the properties of the biopolymer, like different side chains for PHA. As the PHA are biocompatible and does not elicit any immune response they can be used in the implants and sutures.

6.10 Biocompatibility of PHA

The implant material though being biodegradable should also be biocompatible with the host organisms. The polymers should not produce any toxic responses in the body, its supports the cell adhesion and growth which are the key factors for the commercialization.

6.10.1 *In Vitro* Cell and Tissue Compatibility

Basically, the monomers as they are not harmful for the tissues and cells, the biodegradable products are preferred for the medicinal applications. The PHB polymer when used as a pericardial patch in sheep, did not elicit any adverse effects. The PLAs are used as the standard polymer approved by FDA for implantation, the PHBHHx elicited a mild tissue response and found superior than PLA, as they exhibit thermo and mechanical properties. The PHBVHHx studied using human bone marrow mesenchymal cells showed better adhesion property than the PLA film and also enhanced osteogenic differentiation. Similarly, they showed better adhesion and proliferation of the Wharton's jelly derived mesenchymal stem cells. On comparison with other PHA materials, P3HB4HBHHx and PHBVHHx exhibited better adhesion and cellular proliferation than tricalcium phosphate.

6.10.2 *In Vivo* Medical Implants

The PHAs are used as drug delivery matrices for encapsulating the hydrophobic drugs in the form of nanoparticles. The PHA nanoparticles, proteins or polypeptides fused with PhaP are developed, were PhaP fused with ligands and the overexpression of the genes of *E.Coli* attaches to the hydrophobic particle. These ligands pull the PhaP and PHA particles to the targets; this ligand-PhaP-PHA complex has been proved for specific drug delivery in both in vivo and in vitro. The insulin phospholipid complex (INS-PLC) loaded with PHBHHx was tested with the rats and found that the 20% of insulin released in the first 8 h within 31 days. The hypoglycemic effect sustained for more than 3 days and the subcutaneous injection of INS-PLC-NPs prolonged the therapeutic effect than administering with insulin solution indicating the challenging role of INS-PLC-NPs complex, with long term insulin release formulation. For targeting the microcapillary endothelial cells in the pancreas, the PHBHHx microspheres were prepared. Adenoviruses carrying a green fluorescence protein gene complexed with PHBHHx microspheres and injected in the pancreas of mice. About 40% of the cells were labeled after 5 days from surgery and the islet cells labeled occasionally. The insulin entrapment in PHBHHx nanoparticles, enhanced the efficiency of insulin in INS-PLC-NPs system. These paved the way for targeting capillary related disorders using PHBHHx microsphere-adenoviral complex (Zhang et al. 2012).

6.11 Biomedical Implants

The PHAs like PHO, PHB, PHBHHx, PHBV, and P4HB are used in biomedical applications since they are in sufficient quantities. For past 20 years PHA and composites have been used for many biomedical applications like sutures, staples, screws, plates(bone), slings, patches, adhesion barriers, nerve guides, filling agents, scaffolds, dressings, substitutes for skin injury, ligaments, cell implants, hemostats, orthopedic pins, tissue regeneration, meniscus regeneration, done dowels and spinal fusion cages (Chen et al. 2001; Chen and Wu 2005; Wang et al. 2008; Bian et al. 2009). The composition of PHA is altered to enhance the biocompatibility, degradation time under physiological conditions. 3-hydroxybutyrate promoted cell proliferation in highly dense cultures which prevented the cell death making the biopolymer PHBHHx a good candidate for tissue engineering, which may regenerate large number of cells. PHBHHx microparticles also promote the L929cell proliferations with increase in the intracellular calcium concentration. The P4HB is the successfully used implant material approved by FDA (Cheng et al. 2006). For the disease osteomyelitis, the poly(3HB-co-3 HV) rods with the sulperazone antibiotic were used. The rods loaded with 50% of the drug in the form of granules to the poly(3HB-co-3HV) solvent solution and the resulting paste was molded as rods. The animal studies were performed using *staphylococcus aureus* infected implants in rabbit tibias and replaced with the biopolymeric rods. The infection was found to be totally eradicated in 15 days. Similarly using poly(3HB-co-3HV) rods with the drug sulbactam-cefoper for the treatment of osteomyelitis has been reported (Gursel et al. 2002). (Korkusuz et al. 2001) used the poly(3HB-co-3HV) and poly(3HB-co-4HB) as antibiotic carriers for the osteomyelitis, the poly(3HB-co-4HB) was less rigid and easier to handle.

6.12 Subcutaneous Implants

The lipophilic drugs when administered orally are normally destroyed by the liver, so the transdermal delivery makes an option of delivering lipophilic drugs. The skin is one of the toughest paths for the diffusion of the drugs, which consist of a lipid bilayer; hence more focus is towards the transdermal delivery (Metcalfe and Ferguson 2006). The gonadotropin releasing hormone is entrapped with the P(3HB) and placed in sheep subcutaneously. The hormone induces ovulation and promotes the pre-ovulatory follicular growth. The release of the hormone was observed for about 2–4 h, and the incidence of ovulation was observed.

6.13 Ceramic Scaffolds

The ceramic scaffolds and biodegradable polymers are the drug delivery vehicles, and certain synthetic polymers could not be used for drug delivery as it produces adverse responses due to acidic degradation product. Therefore, the biocompatible polymer like PHAs supports cellular growth. The PHAs when in combinations with the bioactive ceramics improves the mechanical properties of the scaffolds.

6.14 Urology

The poly(3HB) and poly(3HB-*co*-3HV) material used in the repair and reconstruction of ureter. The interlaced tubes were coated with poly(3HB-*co*-3HV) and implanted in dogs for urethra replacement. The urethra tissue has been reconstructed in all tested animals after a period of 6–9 months, showing the efficiency of the poly(3HB) in reconstruction without any adverse effects (Jiménez and Kenny 2014).

6.15 Blood Vessels and Heart Valve

Due to the flexible property of PHAs they are used for the biomedical applications in making artificial blood vessels and heart valves. The P3HB4HB containing 4 hydroxy butyrate showed the formation of the elastin and its strength, which has the ability to induce formation of elastins and hence P3HB4HB polymers are used for making the artificial blood vessels (Cheng et al. 2008). As P3HB4HB is more fragile and less strong, copolymers with strong structures using poly (propylene glycol)- poly(ethylene glycol)-poly(propylene glycol) (PPG-PEG-PPG), were synthesized using

1,6-hexamethylene di-isocyanate (HDI).The stronger blood vessel materials are made using the materials like PHBHV and cellulose. The blending was made by the solvent casting method into films with varying ratios based on the blood vessels, when blended with other materials makes the PHBHHx a more suitable for engineering blood vessels. Blending of PHBHHx with poly propylene carbonate (PPC) increases the mechanical properties of the biopolymer making an ideal model. The porcine aortic valves decellularized and coated with the PHBHHx and implanted in the sheep without cardiopulmonary bypass. Then explanted valve is studied after 16 weeks for its histological and biochemical examination. It was observed that the valves maintained the original shape with less calcification than the control. The in vitro assay revealed the PHBHHx increases the tensile strength and enhanced repopulation of the cells making it similar to the native for valve replacement. The P4HB is absorbable and more flexible and found to be superior to the traditional polymeric materials. It finds use in wound healing, drug delivery and orthopedic surgeries (Wu et al. 2007a, b). The vascular cells from the ovine carotid arteries were harvested in vitro and seeded on the P4HB valve scaffold by cardiopulmonary bypass. The native valves were cut out and replaced using the 2 cm segments of pulmonary artery by cell seeded construct. After the observations all the animals survived with no thrombus formation. Such scaffold using P4HB can be used as tissue engineered heart valves and functions for a period of 120 days (Sodian et al. 2000a, b).

6.16 Hard and Soft Tissue Engineering

As the PHAs are biocompatible and long term degradability make it favorable biopolymer for the tissue engineering. The commonly used polymers in hard tissue engineering are the poly lactide-co-glycolide and polyethylene glycol. The P(3HB) strengthens the natural bone and induces osteogenesis and healing (Yang 2001). The P(3HB) are used in recovering the damaged gastrointestinal tract, pericardial patches, repair of peripheral nerves (Novikov et al. 2002). The medium chain length biopolymers are favorable for the soft tissue engineering due to their elastomeric and soft properties. The P(3HO-co-3HHx) is used to replace soft tissues like blood vessels and valves and also in hard tissue engineering like bones (Sodian et al. 2000a, b; Stock et al. 2000).

6.17 Hydrogels

The hydrophilic polymers with large water content are the hydrogels, which forms three-dimensional macro molecular networks. It is advantageous in tissue engineering as because of its permeability to nutrients. The 3D printing, molding and micro patterning has the impact on the hydrogels. There are natural and synthetic hydrogels based on the polymeric material as its components. The natural hydrogel forming polymers are agarose, chitosan, alginate and dextran and the synthetic hydrogel forming polymers are the polyethylene oxide, polyacrylic acid and polyvinyl alcohol. The hybrid hydrogels consist the peptide and the synthetic domains. The macromolecular polymers made of cross linked chains are used in the drug development and in various biomedical applications. The most favorable property for the hydrogels is its porous morphology which allows easier diffusion of drugs and nutrients within and outside the matrix. Hence these hydrogels are widely used as the drug carrier vehicle. The drug release rate can be altered or modified based on the mesh size of the network which enhances diffusion mediated release. The drugs impregnated hydrogels are used to induce therapeutic applications and the controlled release of the small molecule drugs, proteins or peptides. These gels are grafted onto biomaterials by polymerization, entrapment and graft coupling. Hydrogels are used as cardiovascular implants, urinary tracts, dressing of wounds and as controlled drug release devices. The coating with the help of hydrogels makes the biomaterials more compatible. The role of hydrogels in antibacterial activity using curcumin loaded nanoparticles into hydrogel was studied by (Vimala et al. 2014). The thin films were fabricated by the adsorption of dextran sulphate (DS) and poly (allylamine hydrochloride) (PAH) to drug delivery. The leaf extracts of Hybanthus enneasermus were used for the silver nanoparticles production and prepared as thin films and these silver nanoparticles along with the films showed the release of the drug more efficiently. The drug release of the silver nanoparticles coated polyelectrolyte thin films showed high release of moxifloxacin hydrochloride than the silver nanoparticles alone (Fig. 6.14).

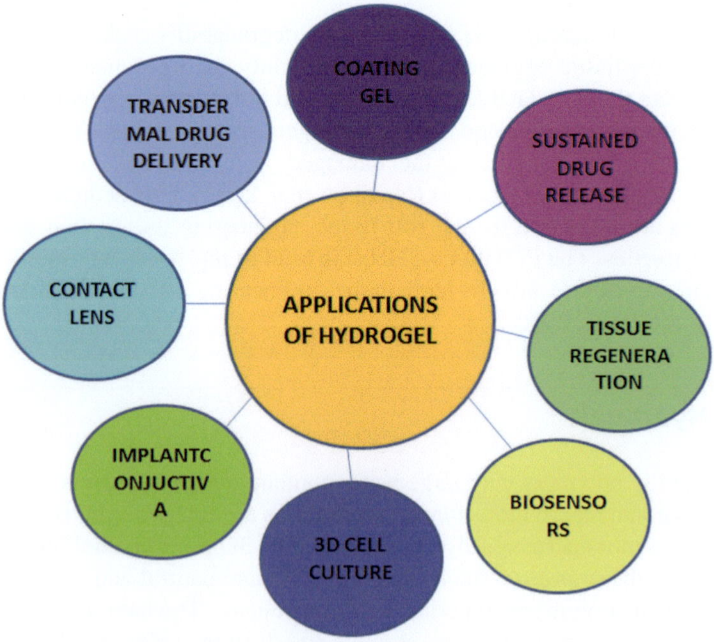

Fig. 6.14 Biomedical applications of hydrogels

6.18 Wound Healing

The largest organ in the human body is the skin and those damages caused might be due to the injury, burn or diseases. Skin act as a barrier in the human body which protects from the environment, skin substitutes are done mainly for the purpose as the grafts for healing wounds and for in vitro assay for skin testing of products. The skin replacements are through tissue engineered substitutes, like biomaterials and these biomaterials are designed for the skin wounds, and must enhance the growth of skin cells like fibroblast, keratinocytes and melanocytes. The biomaterials act as an artificial extracellular matrix for the attachment of the cells. Both the natural and synthetic polymers like glycosaminoglycans, polypeptides, hyaluronan and polylactide are used as substitutes were there are few demerits for the materials. The main demerit includes the inability of molecular signaling making it a tough task during surgery. Hence to overcome these issues PHAs are used for tissue regeneration. The seeding of melanocytes on the polymer scaffolds and in glands namely sebaceous gland and sweat glands. The drawback of the synthetic materials is lack of cell recognition as the heparin binding domains present on the binding sites of many proteins. The P(3HO) films supports cell differentiation and maturation of the HaCaT cell line. The PHA material terpolyster P(3HB-co-3 HV-co-HHx), used as a scaffold material in skin substitutes. This complex as it contains the 3HB, 3 HV, 3HHx polymers they have high mechanical and thermal properties.

6.19 Nerve Conduits

The PHB is brittle and the high elastic properties make it an ideal material as elastic conduit with tube structures. The PHB scaffold enhances proliferation and attachment of Schwann cells and its regeneration. The PHB conduits were tested for nerve injury, the peripheral nerve regeneration was studied using rabbits and showed these PHBs would be suitable for long term nerve injuries. The PHBHHx promotes central nervous system growth and differentiation on 2D and 3D matrices beneficial for the CNS repair. On comparison with 2D, the 3D nanofibers were more suitable for the central nervous system attachment, synaptogenesis and synaptic outgrowth. The PHA with varying ratios of PHB was found to be rough which makes the stiffness in the polymer when PHB is increased. Though they are blended with many ratios, the one with 25:75makes it significant for the neuronal cells in growth and differentiation. The PHAs would serve as the base for the manufacture of nerve conduits (Lizarraga-Valderrama et al. 2015).

6.20 Biomedical 3D Printing

The 3D porous cubes were made based on the laser sintering technology and the cubes are altered as per the virtual model. On comparison of the clinically used PLA, PGA and PLGA, PHB implants showed unaltered pH which makes them tolerated by cells. The PHB are used as coating materials for the dissolving implants (Meischel et al. 2016).

6.21 Biomedical Lithography

The two lithographic methods micro molding and hot embossing are used for PHBHHx arrays of microstructures in culturing cells in local environment in desired shapes, with different configurations like octagons, rectangles and squares. The microstructures of PHBHHx are used to impersonate cellular microenvironment for cell culture (Zhou et al. 2012; Li et al. 2014).

6.22 Injections

The PHBHHx dissolved in nontoxic solvents like dimethylacetamide, N-methylpyrrolidone, 1,4-dioxane, dimethyl sulphoxide, and 1,4-butanolide, which is the injectable implant method and forms a film around the site of injection in animals. The films were formed due to the interaction between the body fluids and the PHBHHx solvents. Such implant system was used for the tissue adhesion preventing film formation in surgeries (Dai et al. 2009).

6.23 Eyelid Reconstruction

The eyelid reconstruction involves the tarsal repair, were the traditional methods did not show satisfactory effects. The PHBHHx material is used as the tarsal substitute tested on the Sprague Dawley rats. Both the commercial scaffolds and the PHBHHx showed better effects though it showed inflammation for 2 weeks but still better than the blank. Since it's a strong material with high elastic proprieties and biodegradability it's proved that PHBHHx would be a good candidate for tarsal repair (Zhou et al. 2010). The transplantation of Islets is the alternative for the treatment of diabetes. The spherical islet like structures were observed on the PHBHHx films on cultivation, and cells when assayed by MTT showed high metabolic activity. After a growth on the PHBHHx films the extracellular secretion and the insulin gene expression were upregulated making it ideal for islet transplantation (Yang et al. 2009).

6.24 Osteogenic and Chondrogenic Effects

The biomaterial initiates the changes on the cell behavior which has specific surface features. The planar and grooved films exhibit osteogenic and chondrogenic effects on mesenchymal stem cells (Li et al. 2014). The presence of the biomaterials impacts the increase and decrease of specific genes for the variation in cellular response, which shows that the surface modifications are the major factor for the cellular response. The cells sense the material through adhesion formation or the integrin clusters. The strategy behind the proliferation of the cell lines are through the enhanced combination of a PHA binding protein with a mcl-PHA (Geiger et al. 2009).

6.25 Tissue Engineered Heart Valve

Sodian and his team made an attempt in tissue engineering using the flexible properties of the PHA, in implanting a trileaflet tissue engineered heart valve in a lamb with P(3HHx-co-3HO) polymer. It was found to work very efficiently for more than 3 months without any clots but had slight contraction (Sodian et al. 2000a, b; Levine et al. 2015). The poly glycolic acid when used as a blend with Poly (3hydroxyhexanoate-co-3hydroxyoctanoate), the heart valve regurgitation was seen after implantation in 6 months. The surface modification has a high impact when designing the PHA biopolymer (Qu et al. 2006).

6.26 Medical Scaffolding Material

In the medical terms the scaffolding is made of polymer fibers used for bone and tissue regrowth in various organs. The scaffolds are inserted into the host for the repair of the damaged tissue, once the tissue grows the scaffolds, are absorbed in the body and completely disappears.

Few studies report the hydroxyapatite as scaffolds used in the treatment of bone disorders. The PHB and polyglycolic acid were used as scaffolds for the pulmonary arteries in the sheep, where the PHA based scaffold are inserted surgically into the host. Also, the PHB were used in the rats with the spinal cord injury by neuronal generation (Novikov et al. 2002).

6.27 Surgical Material (Sutures)

The sterile fibers may be absorbable or non-absorbable used in the medical applications for tissue fixation, ligation and wound healing. During the implantation, absorbable sutures are made which loses their tensile strength and gets degraded as nontoxic products (Chen et al. 2014). The polymeric materials are more specific for making sutures, were the material must exhibit high tensile strength for the effectiveness of the sutures made for closure of wounds. The sutures must exhibit high tensile strength, flexibility, resistance to bacterial growth, easily to tie, grip, good absorbance and eventually disappearing property. The polymers Poly(3hydroxybutyrate-co-3hydroxyhexanoate) and PLA exhibited such properties and have been used as biomedical application. The TephaFLEX suture is the most well-known first approved suture by US FDA, which are more familiar due to its mechanical properties.

6.28 Drug Release

PHA are used as devices in drug delivery, an in vitro assay using P(HB-*co*-HV) polymer, impregnated with drug gentamicin showed a slow release of the drug into the aqueous solution for 2 weeks. The release of the drug can be altered using the amount of drug loaded. The sustained release using biopolymers makes the drugs reach the target site at a given concentrations (Gursel et al. 2002).

6.29 Drug Carriers

As PHAs are hydrophobic and biocompatible they are turned into porous matrices, microspheres, nanoparticles, films and micro capsules. So, the drugs like antibiotics, anticancer, anesthetic agents, anti-inflammatory agents, vaccines and steroids can be trapped in the polymer and delivered to the host organisms. The PHA microspheres are used as carriers of steroids, were PHB is used in the preparation of microspheres containing the progesterone as drug. PHB microspheres showed the higher rate of release of the drug when compared with other polymers. The anticancer drug lomustine incorporated in PHB and PLA microspheres as drug carriers and found the release of drug faster from the PHB microspheres than PLA. The increased rate of drug release was observed with the incorporation of butyl esters in PHB microspheres (Chen et al. 2009a, b; Shrivastav et al. 2013). The PHBV polymers incorporated with tetracycline acts as the carriers for the periodontal diseases and were found to be effective for the controlled release of the drug. The tetracycline are loaded in PHBV microspheres and in micro capsules in both the acidic and neutral forms, showed the complete release of the drug before degradation (Chang and Sultana 2017). The PHB microspheres were incorporated with rubomycin an antitumor drug, in mice for its proliferative activity in Ehrlich's carcinoma. (Lu et al. 2011) reported PHA nanoparticles as drug carriers for ant cancerous studies. The sustained release of the PI3K(TGX221) for blocking the proliferation of cancer cell lines. The TGX221 slowly released from the PHA based nanoparticles, gradually inhibited the growth of cancerous cells. The chemotherapeutic drugs which are highly toxic and poorly water soluble are encapsulated with PHA nanoparticles. The chemotherapeutic molecule Docetaxel (DTXL) used in treatment of tumor types. DTXL formulated with PHB nanoparticles increased the therapeutic index and blood longevity. The PLGA is the traditionally used nanoparticles for DTXL whereas the product of PHB is the nontoxic compound, 3-hydroxybutyric acid. In terms of loading, release and cytotoxicity the PHB polymers exhibited high loading capacity.

Cisplatin, a familiar chemotherapeutic agent used against different tumors, loaded in self assembled copolymer nanoparticles, showed the sustained the release of the cisplatin from the hydrophobic domain enhancing the apoptosis of the tumor cells (Shah et al. 2012). The PHA polymers releases 3HB monomers, which increases the calcium level in the cells inturn protecting the mitochondria, hence suggesting these biopolymers as a drug for nervous and bone disorders. For obtaining a purified R10 monomer the medium chain length PHA (R10) was depolymerized and conjugated with a D-peptide (DP18) and derivatives, which enhanced the anticancer activities. The anticancer activity was more superior with the R10-DP18 than the peptides combined with decanoic acid (O'Connor et al. 2013).The PHA nano particles can be used for targeted drug delivery (Shrivastav et al. 2013). The carcinogenicity of PHAs were analyzed using deoxyribonucleic acid (DNA) aneuploid and telomerase activity, which showed that the PHAs supports cell growth and hence not susceptible to tumor induction, which implies PHAs as safe biopolymers (Peng et al. 2012). The PhaP called as Phasins, the PHA binding protein is seen immobilized on the surface of the nanoparticles through hydrophobic interactions between PHA and PhaP. The overexpression of the recombinant genes in microbes produces PhaP fused ligands. The ligand-PhaP-PHA nano based drug delivery has been proved in vitro and in vivo (Yao et al. 2008). The costimulation of T-cells by B7(costimulatory molecules), showed hetrologous expression of the B7-2 molecule and construction of an immobilized molecule using PHA nano particles as immune activation agent. The B7-2 molecule fused at the N terminal with PhaP and expressed in *E.coli* strain DE3. The purified B7-2-PhaP was immobilized on the surface of nanoparticle. The immobilized particle provide the signals to induce activation of T-cell and in vitro proliferation (Li et al. 2012).

6.30 Microparticulate Carriers

The drug delivery methods with controlled and sustained release of the drugs with maximum activity and minimized side effects are the microparticulate drug delivery. The drugs in the form of solids or liquids covered by the polymers of different thickness depending upon the permeability of the cell membrane. The poly 3HB and poly (3HB-co-3HV) are the examples of microparticulate carriers produced by solvent evaporation method, which varies by the composition of polymers, particle size, crystallization rate and molecular weight. The valerate content usually slows down the drug release rate due to less crystalline and hence valerate incorporated with poly 3HB tends to decrease the physical damage. Before the degradation of the polymers the drugs are completely released, which is diffusion based.

6.31 Tablets

The tablets were prepared using poly(3HB) containing diltiazem, the vasodilator and tested in vitro. The complete release of the drug was seen when drug loaded with the higher concentration of biopolymer and the release rate was slower with lower concentrations (Korkusuz et al. 2001). The contraceptive pills are the commonly used steroids; estrogens are readily absorbed through mucous membranes, gastrointestinal tract and through skin. There are certain adverse effects on steroids, but still due to the missing steroid synthesis, they are taken endogenously. Estrogens are used for the treatment of ovarian development failure, prevention of heart attacks and osteoporosis. The steroids encapsulated with biopolymer PHA are used to minimize the side effects. The PHB and P(3HB-3HV) are used to prepare microspheres containing the progesterone. The surface of the microspheres is based on the composition and temperature, when there is a change in the solvents the microspheres becomes smoother. The in vitro release was seen slowest with the other polymers were as found to be efficient with PHB (Chen and Davis 2002).

6.32 Drugs

The derivatives of 3HB are the ketone body produced in animals and also human, which is the degradation product of PHA. The derivatives of 3HB plays a role in the cell apoptosis and the in the cytosolic Ca^{2+}concentration in the mouse glial cells, in the presence of 3HB the apoptosis decreased significantly (Xiao et al. 2007). The in vitro assay showed that elevation of cytosolic Ca^{2+} concentration is because of the 3HB derivatives. The intracellular and the extracellular calcium contributed for the increase in calcium level mediating the signaling of cellular pathways. The 3HBME which is also one of the derivative of 3HB shown to be efficient in cellular permeation than the other derivatives D-3HB and DL-3HB. The brain which is the most sensitive organ requires more energy for the activity, which is supplemented with metabolic substrates. The metabolic activity of the neuroglial cells was seen elevated in the presence of the PHB derivatives. An increase in the ERK2and connexin 36 protein in brain tissues was observed with the treatment using 3HBME. Through the signaling pathways 3HB enhances the memory and the learning skills (Zou et al. 2009).

6.33 Prodrug

Prodrugs are those inactive compounds metabolized inside the human system as drugs with the action of enzymes. The poly(4HB) is the prodrug of 4-hydroxybutyrate, which were tested using the rats given with low molecular weight polymer poly(4HB). The serum concentration was elevated within a short period of time. The prolonged release of monomers had an importance for the narcolepsy (Sudesh et al. 2000). The properties of the prodrug depends on the susceptibility, physio-chemical properties and the enzymatic regeneration were as the chemical modifications alone does not enhance the activity of prodrug, in turn the combinations with the polymeric conjugates (Williams and Martin 2005). The controlled release of 5-fluoro-2′-deoxyuridine (FUdR) ester prodrugs as microspheres using PHB were investigated. The PHB-450 microspheres containing the drug showed rapid complete release containing FUdR within 24 h. The release profiles were observed with the poly lactic acid and the FUdR prodrug and the lower content spheres showed significant faster release. To identify the drug release rate for anti-tumor activity an in vivo assay using P388 mouse leukemia cells through intraperitoneal injections and the single administration containing FUdR was effective with the use of high molecular weight PHB. The combinations of the prodrug and the biopolymer PHB showed a controlled and sustained release with increased therapeutic applications as the anti-tumor agent FUdR. The molecular weight along with the physio chemical properties of PHB microspheres was biocompatible and with low toxicity.

6.34 Drug Delivery of PHAs

Drug delivery is the reach of the therapeutic drug at the given concentration to the target, which triggers the pharmacological response. The dosage of the drugs and its frequency has to be increased to overcome the side effects of the drugs. The biopolymers designed, and its composition is based on the target place of the drug release. An in vitro and in vivo assay of the drug 7- hydroxyetheophylline in P(3HB) tablets, implanted on the neck fold subcutaneously showed a high drug release rate for 80% drug loading, and the complete release was observed within 24 h (Brigham and Sinskey 2012; Ali and Jamil 2016; Ferre-Guell and Winterburn 2018). On comparison with the P(3HB) and PHV polymers the release rate was slower with the P(3HB-co-3 HV) compressed biopolymers. The main difference between the drug releases was the structure, composition and porosity.

6.35 Mechanism of Controlled Release

6.35.1 Diffusion Release Mechanism

- Monolithic device
- Reservoir device

The active pharmaceutical ingredient is added to the polymeric solution or dispersed within the polymer. The drug is slowly released into the physiological fluid by dissolution, followed by diffusion. Though the drug release may not be 100% but it's the simplest method of drug delivery, is the monolithic device.

In reservoir device, active pharmaceutical ingredient is placed on the film, based on the porosity and the thickness the drug released. The drug with low solubility makes a possibility to stop the process, and those with high solubility makes the drug burst increasing the concentration of the drug to the surrounding parts (Vasant and John 2011).

6.36 Osmotic Release

The drug delivery which employs the osmotic pressure for delivering the active pharmaceutical ingredient, were the drug is coated with the semi permeable material. The orally administered drug involves this type of release. The permeability is based on the degree of acetylation, were the permeability increases with increase in acetyl content. When the drug immersed in the aqueous medium, due to the hydrostatic pressure within the drug, it diffuses through the membrane. The thickness ranges from 200 to 300 μm. Around 80% of the drugs are expelled completely beyond which the residual release rate is slower (Kwon and Furgeson 2007).

6.37 Polymer Erosion

The importance of the polymer is highly noted only by the biodegradability. The drug releases through three different mechanisms.

- The drug is attached with the polymer backbone by covalent linkage, and the drug release is observed when the backbone is hydrolytically cleaved.
- The drug is encapsulated with the biopolymer and released when the polymer undergoes degradation. The drug release rate is proportional to the polymer degradation rate.
- This mechanism involves the diffusion and erosion of the polymer.

The small chain length PHA is ideal for drug delivery than medium chain length as they are degraded by surface erosion. The rate of drug release is based on the crystallinity of the polymer. On the other hand, medium chain length polymer provides controlled drug release because of their low melting point. The drawback of Mcl PHAs is the inability of production in large quantity.

6.38 Oral Drug Delivery

The larger molecules like proteins and peptide are challenging task in reaching the targets when taken orally (Yang 2001). The orally taken drugs must maintain its stability and structure until reaching the site of action. The hindrances are due to the enzymes in the stomach which prevents the drugs bioactivity and the pharmacokinetics. The mucosal lining prevents the penetration of the drugs to the site of action (Shishatskaya et al. 2011). Hence P (3HB) is used for the sustained release of the drug to be delivered, which is based on the molecular weight of the polymer. For example when the molecular weight increased from 3000 to 6000 kDa, the release rate also increases. The P(3HB) were used as compressed compact discs for the treatment of periodontal diseases for the delivery of tetracycline and the reduction of the gingival infections has been observed over a period of 10 days (Volova 2004). Similarly P(3HB) microspheres encapsulation sulphamethizole was taken orally in greyhound dogs. The absorption rate was good in both the in vivo and in vitro profiles (Koller 2018). The vasodilator, Diltrazem compressed with the P(3HB) tablets was used in studying the drug release rate in vitro. When there is 45% drug loading complete release of the drug has been observed.

6.39 Intravenous Administration

The anticancer drug Imoustine, entrapped with the P (3HB) microspheres, intravenously injected on lung carcinoma mice, were the drug release was better with the P(3HB) than the poly lactic acid (Makadia and Siegel 2011; Panchal et al. 2012).

6.40 Nanoparticles-Based Targeted Drug Delivery

The drug when reaches the target becomes an efficient molecule against the disease, were there are no side effects when the drug reaches the target. Cancer is one of the dreadful diseases and focus towards the targeted drug delivery is based on the delivery particles its accumulation within the target. The new approach to target the cancer cells for drug delivery and imaging based on the oil-water emulsion method, using the PHA synthase as catalyst. The coupling of hydrophobic surface of PHB nanoparticle and the PHB chain, in the presence of the synthase linked with a ligand, organizes the nanoparticle in the aqueous environment. RGD4C fused with PHA synthase and the tumor specific ligand, organizes the surface of the nanoparticle, which showed affinity towards the MDA-MB-231 breast cancer cells. Hence showing the capability of the drug carrier molecule especially bound towards the PHB nanoparticle (Lee et al. 2011). The development in the field of nanotechnology has been a promising aspect in the drug delivery. The delivery of drugs to various parts of the body, to the targets and the sustained release of the drug. Based on the target the surface of the nanoparticles can be modified. The hydrophobic and hydrophilic drugs, vaccines, proteins, macromolecules and vaccines are delivered using the nano particles. Based on the target of the drugs, targeted based drug delivery system plays a vital role in delivering the exact dosage of drugs for the specified time without any adverse effects on other parts. More focus towards the chemotherapeutic drugs and proteins and peptides which are delicate (Yamamoto et al. 2005). The modification in the nanoparticles with the ligand enhances the efficacy and the cellular uptake of nanoparticles through cell mediated endocytosis, an increase in the intracellular concentration of the drug increases the therapeutic activity. The PHB microspheres used for the delivery of formalinized vaccine of Staphylococcal Enterotoxin B for the lymphoid tissue (Fig. 6.15) In Payer's patches the microspheres exhibited good absorption (Shi et al. 2010).

6.41 Folate Mediated Targeting

Folate is important in the biosynthesis of bases for DNA synthesis and hence required in large amounts. Since because of its role in DNA synthesis, the folate receptor is regulated in various cancer types. There are various reports which shows the over expression of folate receptors in ovarian carcinomas. The density of the folate receptor gradually increases with the progression of the disease, and hence focusing folate in targeted drug delivery for cancers.

Zhang et al. proposed targeting a drug delivery system using [P(3HB-co-3HO)] as carrier, folic acid as ligand and doxorubicin as anticancer drug. The folate conjugated particles were endocytosed by HeLa cells than the non-folate conjugate. The cytotoxicity was high, and the in vitro assay has shown the intracellular uptake of the nano particles were efficient with the HeLa cells. The in vivo studies on mice having the HeLa tumors xenograft exhibited similar activity with minimal side effects (Zhang et al. 2010). In a study, the chemotherapeutic drug etoposide (ETO)

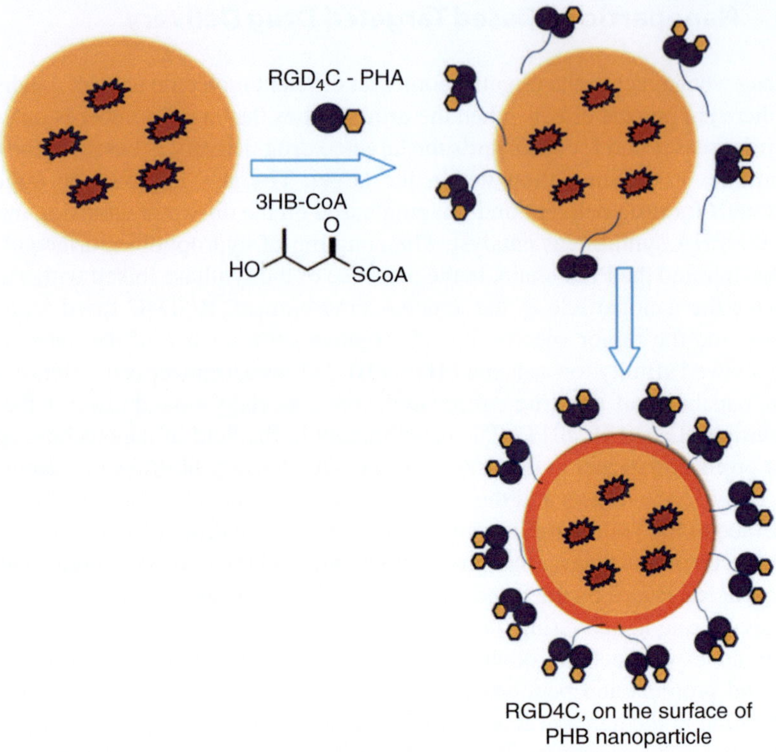

Fig. 6.15 Drug delivery with nano based biopolymers

was delivered with the P(3HB-co-3HHx) conjugated with the folate(FA). From the treatment it was observed that the complex ETO/FA- P(3HB-co-3HHx) showed higher cytotoxicity than the non-conjugated particles. The L929 cells, which have low folate receptor expression, and this may be the reason behind the reduced effect. On comparison DOX/FA-P(3HB-co-3HHx) was more cytotoxic than complex with etoposide which may be due to the difference in the drugs. Though both the drugs are topoisomerase II inhibitors, doxorubicin generated reactive oxygen species (ROS), due to which they exhibit more cytotoxicity. So, targeting folate receptor with the folate conjugated PHA increases the selectivity and cytotoxicity for the cancer therapeutics (Andronescu and Grumezescu 2017).

6.42 Targeting Ligands by PHA Binding Proteins

The targeting method based on the ligand conjugates using PhaP and PhaC, were PhaP is the amphiphilic protein with low molecular weight produced during the PHA biosynthesis, and these forms a layer on the granules and stabilizes the intracellular environment in the bacteria. And PhaC is covalently linked to the granules

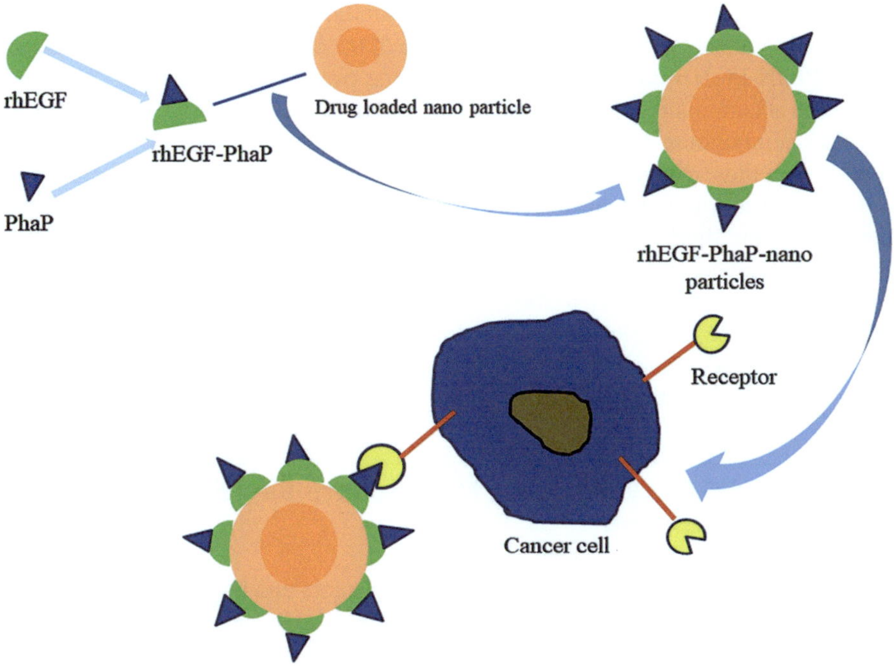

Fig. 6.16 Ligand targeting by PHA binding proteins

by the cysteine residues, and these PhaC and PhaP genes produces fusion proteins. As PhaP has higher affinity towards PHAs, they are linked to the PHA particles, and the fusion proteins seem to maintain the enzyme activity (Chen 2009a, b). Based on the ligand targeting, the PhaP fused with the human epidermal growth factor (rhEGF), the upregulation of the EGF is an important consideration in cancer types mainly hepatocellular carcinoma. The rhEGF-PhaP fusion proteins were linked with the drug loaded nanoparticles, the P(3HB-co-3HHx) (Fig. 6.16). The drug loaded is the rhodamine B isothiocynate (RBITC) used for analysis of tissue deposition. The murine hepatoma22 (H22) model cells were implanted in mice and then treated with the drug RBITC. The fluorescence assay showed the accumulation of RBITC loaded rhEGF-PhaP-PHBHHX nanoparticles within tumors (York et al. 2002; Potter 2005; Yao et al. 2008; Kwon et al. 2014).

6.43 Targeting Using Superparamagnetic Iron Oxide Nanoparticles

The superparamagnetic iron oxide nanoparticles encapsulated with P(3HB-co-3 HV) forms a theragnostic particles, which means the combination of therapeutic and diagnostic properties in a single agent (Jain et al. 2008; Vilos et al. 2013). The SPION-P(3HB-co-3HV) are biocompatible and inhibits the generation of ROS. This superparamagnetic property facilitates the accumulation of the drug in the targeted site. The in vivo assay, with the administration of the drug Alexa-Fluor encapsulated with SPION-P(3HB-co-3 HV) to the rats in the presence or absence of magnetic field. The accumulation of the nanoparticles measured through fluorescence showed for the accumulation of SPION-P(3HB-co-3 HV nanoparticles in the target was higher with those particles exposed to magnetism. Hence these PHA biomaterials are compatible with the controlled drug release and the nanoparticles can be altered based targeting ligands in order to have increased effects within the tissues. If the drugs of higher doses are administered, drug localization plays a vital role to reduce the risk of side effects (Allen 2004). An example of such therapies is the cancer therapeutics as small molecule kinase inhibitors and monoclonal antibodies (Grillo-Lopez et al. 2000). Based on the target antigen and the agents, PHAs act as a better nano carrier for targeted drug delivery.

6.44 Polyhydroxyalkanoate Microsphere-Based Protein Encapsulation

Proteins are used as vaccines, enzymes, growth factors and hormones for biomedical applications. The major disadvantage of proteins is the short half lives in vivo. Due to the high molecular weight of proteins, they are not completely absorbed by the system. The protein stability is maintained by the microencapsulated capsules. Biopolymer microspheres provide an opportunity in decreasing the recurrent injections, which is based on the size of the microspheres (Yang 2001).

6.45 Chemical Modification of PHA

The PHA contains the methyl, bromine, hydroxyl and nitrophenyl derivatives produced from the microorganisms. These functional groups are said to enhance the activity and the biomedical applications of the PHAs. Such modifications are still under progress and challenging to modulate the structure of the polyesters by making alterations in the functionality and the molecular weight. Introducing functional groups to the PHA by chemical reactions makes a possibility to widely use the PHA as biocompatible and biodegradable component in the application of medicinal therapeutics. There are various chemical modifications like the hydroxylation, carboxylation, epoxidation and the chlorination.

6.46 Modification Using Hydroxyl Group

6.46.1 Monohydroxylated Macromonomer

The acid or base catalyzed reactions with *para*-toluene sulfonic acid monohydrate (APTS) and methanolysis produced oligomers with terminal hydroxyl group. The ester bonds of PHO are stable at pH 12 and after hydrolysis the pH was 14 with the formation of hydroxyl group at one end of the chain. The decrease in the molar mass is based on the nature of the solvent; the chain cleavage occurred more rapid in toluene than dichloroethane. From the spectral analysis the acid catalyzed condition using APTS ended in the formation of hydroxyl group in one end and the carboxylic group on the other. The cyclic structures have been observed and the high molecular weight polyhydroxyalkanotes are exposed to methanolysis in the presence of a catalyst sulphuric acid resulted in the formation of monohydroxylated macromonomers with carboxylic acid were formed. Thus, many such monohydroxylated PHAs can be formed (Nguyen et al. 2002; Sun et al. 2007; Timbart et al. 2007; Xiao et al. 2007; Wu et al. 2008; Zhou et al. 2012).

6.46.2 Dihydroxylated Macromonomer

The transesterification with 1,4-butanediol in the presence of catalyst *p*-toluenesulfonicacid. The hydroxyl group in the molecule cleaves the ester bond forming a telechelic polymer, which is a prepolymer and enters into various reactions with the end groups, as the cleavage is an important mechanism the molecular weight decreases gradually. By the combinations of PHAs with different side chain length the P3 poly(esterurethane)s, 4HB poly(esterurethane)s and PHBHHx poly(esterurethane)s are obtained. These possess more cell compatibility than the pure forms, and hence used as biomaterials for would healing. During synthesis the PHBHHx replaced by PHHxHO, the polymer obtained becomes more hydrophobic, hence modifying the physical properties enhances the biomedical applications of the polymers (Saad et al. 2001a, b; Chen et al. 2009a, b).

6.46.3 Modification Using the Carboxylic Group

The functional groups with carboxylic acids play an important role for the binding of targeting enzymes, bioactive compounds and the hydrophilic components. These carboxylic groups when incorporated in the polymers enhance the hydrophilicity (Renard et al. 2007). The PHAs with the unsaturated groups can be converted to carboxyl groups by oxidation process, theses polymers are used as graft copolymers preparation. In another approach the PHA oligomers terminated with carboxyl group forms methacrylic PHAs (Lee and Park 2000; Kurth et al. 2002).

6.46.4 Modifications Using Epoxy Group

The reaction by which alkenes is converted to epoxide; the epoxidation involves the chemical modifications to enhance the physical property like the elasticity of the compounds. The conditions like cross-linking and attachment of ionizable groups plays a role in the epoxidation process. In the acidic conditions there is no much change in the chains of poly(3-hydroxyoctanoate-co-3-hydroxyundec-10-enoate) whereas cross linked PHOUS are formed from the epoxidation reactions in the presence of succinic anhydride, and these cross linking enhances the elasticity of the polymers without disturbing the biodegradability (Hazer 2010; Kai and Loh 2013; Li et al. 2016).

6.46.5 Modification Using Chlorination

The process of passing of chlorine gas through the unsaturated PHAs gives the PHA-Cl complex with the chlorine content of about 5–40%. And these thermal properties of the PHA-Cl complex were changed in terms of the transition temperature. Due to the chlorination the temperature increased, and the mobility of the molecules were dependent on the chlorine content. From the PHA chlorinated derivatives the quaternary ammonium salts, and phenyl derivatives were obtained, the benzene and the PHA-Cl complex forms the cross linked polymers (Arkin and Hazer 2002).

6.47 Conclusion

Biopolymers like polyhydroxyalkanoates are produced as energy- and carbon-storage materials by many bacteria. The PHAs produced from the bacteria are with varying monomer based on the substrates, due to the physical properties from flexible to brittle and elastic makes better replacements for the petroleum based plastics. The PHA production from various sources like industrial waste, molasses, cheese and whey products, lignocellulosic materials, oils from cooking wastes which are the cheapest sources for the PHA production. As they are more peculiar due to the biocompatibility, biodegradability, non-toxicity, and chemical diversity they are with various biomedical applications like sutures, surgical implants, and as drug carriers which has increased the interest in the use of PHAs. The tissue engineering is one of the research areas growing exponentially and the scaffold materials play a vital role. PHA based targeted drug delivery in cancer therapy, and in the delivery of biological molecules, vaccines and steroids. The unmodified PHAs are highly hydrophobic and crystalline, due to poor compatibility between polymers, and the different chemical modifications may come up with novel class of biopolymers as water soluble PHAs. They are considered favorable as they are reproducible and are synthesized from natural renewable resources. The applications of PHAs make the replacement with the commercial plastics as better improvements and as environmental friendly materials in the formulating into useful biomedical products.

6.48 Prospects and Challenges

Despite various advantages of PHA, all PHAs produced to date are mainly from the bacterial origin, which is a very costly process, and this is a main reason for the exploitation of PHA from research. The harmful effects of the plastics from petroleum-based materials and the hike in petroleum prices will create a big challenge for the production of bioplastics. The PHAs have a wide application as drug carrier for cancer therapy and tumor treatment, with the formulation as micro capsules, microspheres and Nano particles, corresponds to the speedy release and to reach the target. PHAs without any modifications chemically leads to high crystallinity, and porous formation on the cell surface which leads to uncontrolled release of the drugs. The development of PHAs with respect to related molecular weight, its physico-chemical and surface properties are main considerations for the distribution of the drug.

Few PHA types have shown to exhibit antimicrobial properties, which reduce the amount of drug used for the treatment and free from side effects.

- Through the metabolic engineering methods PHAs with unique monomers can be obtained.
- Novel bacterial strains of bacteria which can produce unique monomers with enhanced properties.
- Advancements can be made in the production of bioplastics with peculiar properties like nontoxic and biocompatibility to prevent from immune response.
- Production of PHA using low cost substrates, to improve the economy.
- Genetic modifications of bacteria to improve the yield of PHA.
- So far, no studies have been proved for the carcinogenesis implanting PHA in vivo, though PHA promotes cell proliferation.
- Blending of smaller and larger PHA helps maintaining the stability and increases biodegradation. Blending PHA with PLGA increases the rate of degradation.
- The poly-4-hydroxybutyrate is the only approved material for the sutures by FDA which took more than a decade for approval.
- The PHAs are more focused towards the packaging industries, which has to be directed towards the biomedical applications.
- The purification of the PHAs for the complete removal of lipopolysaccharides, bacterial impurities and residual proteins to make it suitable for the biomedical implants.
- A great demand in the field of plastic surgery, the 3D printing material plays a vital role.
- Treatment and diagnosis using single particle administration, production of multifunctional theragnostics particles.
- Targeting using superparamagnetic approach for specificity and localization of the cells within tissues.

References

Ali I, Jamil N (2016) Polyhydroxyalkanoates: current applications in the medical field. Front Biol 11:19–27. https://doi.org/10.1007/s11515-016-1389-z

Allen TM (2004) Drug delivery systems: entering the mainstream. Science 303:1818–1822. https://doi.org/10.1126/science.1095833

Andronescu E, Grumezescu AM (2017) Nanostructures for oral medicine, 1st edn. Elsevier, Amsterdam. ISBN: 9780323461498

Arkin AH, Hazer B (2002) Chemical modification of chlorinated microbial polyesters. Biomacromolecules 3:1327–1335. https://doi.org/10.1021/bm020079v

Aslan AK, Ali MD, Morad NA, Tamunaidu P (2016) Polyhydroxyalkanoates production from waste biomass. IOP Conf Ser: Earth Environ Sci 36:012040. https://doi.org/10.1088/1755-1315/36/1/012040

Asran AS, Razghandi K, Aggarwal N, Michler GH, Groth T (2010) Nanofibers from blends of polyvinyl alcohol and polyhydroxy butyrate as potential scaffold material for tissue engineering of skin. Biomacromolecules 11:3413–3421. https://doi.org/10.1021/bm100912v

Attia R, Nazia J, Nighat N, Shahida H (2010) Bacteria from contaminated urban and hilly areas as a source of polyhydroxyalkanoates production. Afr J Biotechnol 9:1919–1925. https://doi.org/10.5897/ajb09.1003

Bian Y, Wang Y, Aibaidoula G, Chen G, Wu Q (2009) Evaluation of poly(3-hydroxybutyrate-co-3-hydroxyhexanoate) conduits for peripheral nerve regeneration. Biomaterials 30:217–225. https://doi.org/10.1016/j.biomaterials.2008.09.036

Bonartsev AP, Zharkova II, Yakovlev SG, Myshkina VL, Makhina TK, Zernov AL, Kirpichnikov MP (2016) 3D-Scaffolds from Poly(3-hydroxybutyrate) Poly(ethylene glycol) Copolymer for Tissue Engineering. J Biomater Tissue Eng 6:42–52. https://doi.org/10.1166/jbt.2016.1414

Brigham CJ, Sinskey AJ (2012) Applications of polyhydroxyalkanoates in the medical industry. Int J Biotech Well Ind 1:53–60. https://doi.org/10.6000/1927-3037.2012.01.01.03

Chang HC, Sultana N (2017) PLA/PHBV electrospun membrane: fabrication, coating with conductive PEDOT:PSS and antibacterial activity of drug loaded membrane. Cogent Eng, 4. https://doi.org/10.1080/23311916.2017.1322479

Chek MF, Kim S, Mori T, Arsad H, Samian MR, Sudesh K, Hakoshima T (2017) Structure of polyhydroxyalkanoate (PHA) synthase PhaC from *Chromobacterium* sp. USM2, producing biodegradable plastics. Sci Rep 7. https://doi.org/10.1038/s41598-017-05509-4

Chen G (2009a) ChemInform abstract: a mcrobial polyhydroxyalkanoates (PHA) based bio- and materials industry. ChemInform, 40. https://doi.org/10.1002/chin.200948268

Chen G (2009b) A microbial polyhydroxyalkanoates (PHA) based bio- and materials industry. Chem Soc Rev 38:2434. https://doi.org/10.1039/b812677c

Chen J, Davis SS (2002) The release of diazepam from poly(hydroxybutyrate- hydroxyvalerate) microspheres. J Microencapsul 19:191–201. https://doi.org/10.1080/02652040110065431

Chen G, Wu Q (2005) The application of polyhydroxyalkanoates as tissue engineering materials. Biomaterials 26:6565–6578. https://doi.org/10.1016/j.biomaterials.2005.04.036

Chen G, Zhang J (2017) Microbial polyhydroxyalkanoates as medical implant biomaterials. Artif Cells Nanomed Biotechnol 46:1–18. https://doi.org/10.1080/21691401.2017.1371185

Chen GQ, Zhang G, Park SJ, Lee SJ (2001) Industrial scale production of poly(3-hydroxybutyrate-co-3-hydroxyhexanoate). Appl Microbiol Biotechnol 57:50–55. https://doi.org/10.1007/s002530100755

Chen Z, Cheng S, Xu K (2009a) Block poly(ester-urethane)s based on poly(3-hydroxybutyrate-co-4-hydroxybutyrate) and poly(3-hydroxyhexanoate-co-3-hydroxyoctanoate). Biomaterials 30:2219–2230. https://doi.org/10.1016/j.biomaterials.2008.12.078

Chen Z, Cheng S, Li Z, Xu K, Chen G (2009b) Synthesis, characterization and cell compatibility of novel poly(ester urethane)s based on poly(3-hydroxybutyrate-co-4-hydroxybutyrate) and poly(3-hydroxybutyrate-co-3-hydroxyhexanoate) prepared by melting polymerization. J Biomater Sci Polym Ed 20:1451–1471. https://doi.org/10.1163/092050609x12457419007621

Chen D, Fang J, Shao Q, Ye J, Ouyang D, Chen J (2013) Biodegradation of tetrahydrofuran by Pseudomonas oleovorans DT4 immobilized in calcium alginate beads impregnated with activated carbon fiber: mass transfer effect and continuous treatment. Bioresour Technol 139:87–93. https://doi.org/10.1016/j.biortech.2013.04.037

Chen Y, Tsai Y, Chou I, Tseng S, Wu H (2014) Application of biodegradable polyhydroxyalkanoates as surgical films for ventral hernia repair in mice. Int J Polym Sci 1–11. https://doi.org/10.1155/2014/789681

Cheng S, Wu Q, Zhao Y, Zou B, Chen G (2006) Effect of poly(hydroxybutyrate-co-hydroxyhexanoate) microparticles on growth of murine fibroblast L929 cells. Polym Degrad Stab 91:3191–3196. https://doi.org/10.1016/j.polymdegradstab.2006.07.010

Cheng S, Chen Z, Chen G (2008) The expression of cross-linked elastin by rabbit blood vessel smooth muscle cells cultured in polyhydroxyalkanoate scaffolds. Biomaterials 29:4187–4194. https://doi.org/10.1016/j.biomaterials.2008.07.022

Dai Z, Zou X, Chen G (2009) Poly(3-hydroxybutyrate-co-3-hydroxyhexanoate) as an injectable implant system for prevention of post-surgical tissue adhesion. Biomaterials 30:3075–3083. https://doi.org/10.1016/j.biomaterials.2009.02.015

Defoirdt T, Boon N, Sorgeloos P, Verstraete W, Bossier P (2009) Short-chain fatty acids and poly-β-hydroxyalkanoates: (New) Biocontrol agents for a sustainable animal production. Biotechnol Adv 27:680–685. https://doi.org/10.1016/j.biotechadv.2009.04.026

Ding Y, Li W, Muller T, Schubert DW, Boccaccini AR, Yao Q, Roether JA (2016) Electrospun polyhydroxybutyrate/Poly(ε-caprolactone)/58S sol–gel bioactive glass hybrid scaffolds with highly improved osteogenic potential for bone tissue engineering. ACS Appl Mater Interfaces 8:17098–17108. https://doi.org/10.1021/acsami.6b03997

Errico C, Bartoli C, Chiellini F, Chiellini E (2009) Poly(hydroxyalkanoates)-based polymeric nanoparticles for drug delivery. J Biomed Biotechnol 2009:1–10. https://doi.org/10.1155/2009/571702

Ferre-Guell A, Winterburn J (2018) Biosynthesis and characterization of polyhydroxyalkanoates with controlled composition and microstructure. Biomacromolecules 19:996–1005. https://doi.org/10.1021/acs.biomac.7b01788

Francis L, Meng D, Knowles J, Keshavarz T, Boccaccini A, Roy I (2011) Controlled delivery of gentamicin using poly(3-hydroxybutyrate) microspheres. Int J Mol Sci 12:4294–4314. https://doi.org/10.3390/ijms12074294

Geiger B, Spatz JP, Bershadsky AD (2009) Environmental sensing through focal adhesions. Nat Rev Mol Cell Biol 10:21–33. https://doi.org/10.1038/nrm2593

Grillo-Lopez A, White C, Dallaire B, Varns C, Shen C, Wei A, Rosenberg J (2000) Rituximab the first monoclonal antibody approved for the treatment of lymphoma. Curr Pharm Biotechnol 1:1–9. https://doi.org/10.2174/1389201003379059

Gursel I, Yagmurlu F, Korkusuz F, Hasirci V (2002) In vitro antibiotic release from poly(3-hydroxybutyrate-co-3-hydroxyvalerate) rods. J Microencapsul 19:153–164. https://doi.org/10.1080/02652040110065413

Hajiali H, Hosseinalipour M, Karbasi S, Shokrgozar MA (2012) The influence of bioglass nanoparticles on the biodegradation and biocompatibility of poly (3-hydroxybutyrate) scaffolds. Int J Artif Organs 35:1015–1024. https://doi.org/10.5301/ijao.5000119

Hazer B (2010) Amphiphilic poly (3-Hydroxy Alkanoate)s: potential candidates for medical applications. Energy Power Eng 02:31–38. https://doi.org/10.4236/epe.2010.21006

Hong S, Lin Y, Lin C (2008) Crystallization and degradation behaviors of treated polyhydroxybutyrates. React Funct Polym 68:1516–1511. https://doi.org/10.1016/j.reactfunctpolym.2008.08.003

Hori K, Abe M, Unno H (2009) Production of triacylglycerol and poly(3-hydroxybutyrate-co-3-hydroxyvalerate) by the toluene-degrading bacterium Rhodococcusaetherivorans IAR1. J Biosci Bioeng 108:319–324. https://doi.org/10.1016/j.jbiosc.2009.04.020

Jaglan S, Yadav R, Siwach P, Singh N (2017) Recent updates on molecular biotechnological intervention in isabgol. Plant Biotechnol: Recent Adv Dev 233–253. https://doi.org/10.1007/978-981-10-4732-9_12

Jain TK, Reddy MK, Morales MA, Leslie-Pelecky DL, Labhasetwar V (2008) Biodistribution, clearance, and biocompatibility of iron oxide magnetic nanoparticles in rats. Mol Pharm 5:316–327. https://doi.org/10.1021/mp7001285

Ji G, Wei X, Chen G (2009) Growth of human umbilical cord wharton's jelly-derived mesenchymal stem cells on the terpolyester poly(3-hydroxybutyrate-co-3-hydroxyvalerate-co-3-hydroxyhexanoate). J Biomater Sci Polym Ed 20:325–339. https://doi.org/10.1163/156856209x412191

Jimenez A, Kenny JM (2014) 4th international conference on biodegradable and bio-based polymers (BIOPOL-2013). Polym Degrad Stab 108:139. https://doi.org/10.1016/j.polymdegradstab.2014.07.019

Kai D, Loh XJ (2013) Polyhydroxyalkanoates: chemical modifications toward biomedical applications. ACS Sustain Chem Eng 2:106–119. https://doi.org/10.1021/sc400340p

Kilicay E, Demirbilek M, Turk M, Guven E, Hazer B, Denkbas EB (2011) Preparation and characterization of poly(3-hydroxybutyrate-co-3-hydroxyhexanoate) (PHBHHX) based nanoparticles for targeted cancer therapy. Eur J Pharm Sci 44:310–320. https://doi.org/10.1016/j.ejps.2011.08.013

Koller M (2018) Biodegradable and biocompatible Polyhydroxyalkanoates (PHA): Auspicious microbial macromolecules for pharmaceutical and therapeutic applications. Molecules 23:E362. https://doi.org/10.20944/preprints201801.0218.v1

Koller M, Hesse P, Salerno A, Reiterer A, Braunegg G (2011) A viable antibiotic strategy against microbial contamination in biotechnological production of polyhydroxyalkanoates from surplus whey. Biomass Bioenergy 35:748–753. https://doi.org/10.1016/j.biombioe.2010.10.008

Korkusuz F, Korkusuz P, Eksioglu F, Gursel I, Hasırci V (2001) In vivo response to biodegradable controlled antibiotic release systems. J Biomed Mater Res A 55:217–228. https://doi.org/10.1002/1097-4636(200105)55:23.3.co;2-p

Kourmentza C, Placido J, Venetsaneas N, Burniol-Figols A, Varrone C, Gavala HN, Reis MA (2017) Recent advances and challenges towards sustainable polyhydroxyalkanoate (PHA) production. Bioengineering 4:55. https://doi.org/10.3390/bioengineering4020055

Kurth N, Renard E, Brachet F, Robic D, Guerin P, Bourbouze R (2002) Poly(3-hydroxyoctanoate) containing pendant carboxylic groups for the preparation of nanoparticles aimed at drug transport and release. Polymer 43:1095–1101. https://doi.org/10.1016/s0032-3861(01)00692-9

Kwon G, Furgeson D (2007) Biodegradable polymers for drug delivery systems. Biomed Polym. https://doi.org/10.1201/9781439823972.ch4

Kwon H, Jung S, Kim H, Parker SA, Batt CA, Kim Y (2014) A multi-functional polyhydroxybutyrate nanoparticle for theranostic applications. J Mater Chem B 2:3965–3971. https://doi.org/10.1039/c4tb00304g

Laycock B, Halley P, Pratt S, Werker A, Lant P (2013) The chemomechanical properties of microbial polyhydroxyalkanoates. Prog Polym Sci 38:536–583. https://doi.org/10.1016/j.progpolymsci.2012.06.003

Lee MY, Park WH (2000) Preparation of bacterial copolyesters with improved hydrophilicity by carboxylation. Macromol Chem Phys 201:2771–2774. https://doi.org/10.1002/1521-3935(20001201)201:183.0.co;2-v

Lee J, Jung S, Park C, Kim H, Batt CA, Kim Y (2011) Tumor-specific hybrid polyhydroxybutyrate nanoparticle: surface modification of nanoparticle by enzymatically synthesized functional block copolymer. Bioorg Med Chem 21:2941–2944. https://doi.org/10.1016/j.bmcl.2011.03.058

Levine AC, Sparano A, Twigg FF, Numata K, Nomura CT (2015) Influence of cross-linking on the physical properties and cytotoxicity of polyhydroxyalkanoate (PHA) scaffolds for tissue engineering. ACS Biomater Sci Eng 1:567–576. https://doi.org/10.1021/acsbiomaterials.5b00052

Li X, Zhang Y, Chen G (2008a) Nanofibrous polyhydroxyalkanoate matrices as cell growth supporting materials. Biomaterials 29:3720–3728. https://doi.org/10.1016/j.biomaterials.2008.06.004

Li ZB, Kollias HD, Wagner KR (2008b) Myostatin directly regulates skeletal muscle fibrosis. J Biol Chem 283:19371–19378. https://doi.org/10.1074/jbc.m802585200

Li SY, Dong CL, Wang SY, Ye HM, Chen G (2010) Microbial production of polyhydroxyalkanoate block copolymer by recombinant Pseudomonas putida. Appl Microbiol Biotechnol 90:659–669. https://doi.org/10.1007/s00253-010-3069-2

Li M, Liu Q, Lu X, Zhang Y, Wang L (2012) Heterologous expression of human costimulatory molecule B7-2 and construction of B7-2 immobilized polyhydroxyalkanoate nanoparticles for use as an immune activation agent. BMC Biotechnol 12:43. https://doi.org/10.1186/1472-6750-12-43

Li X, Chang H, Luo H, Wang Z, Zheng G, Lu X, Xu M (2014) Poly (3-hydroxybutyrate-co-3-hydroxyhexanoate) scaffolds coated with PhaP-RGD fusion protein promotes the proliferation and chondrogenic differentiation of human umbilical cord mesenchymal stem cells in vitro. J Biomed Mater Res B 103:1169–1175. https://doi.org/10.1002/jbm.a.35265

Li Z, Yang J, Loh XJ (2016) Polyhydroxyalkanoates: opening doors for a sustainable future. NPG Asia Mater 8:e265–e265. https://doi.org/10.1038/am.2016.48

Lizarraga-Valderrama LR, Nigmatullin R, Taylor C, Haycock JW, Claeyssens F, Knowles JC, Roy I (2015) Nerve tissue engineering using blends of poly(3-hydroxyalkanoates) for peripheral nerve regeneration. Eng Life Sci 15:612–621. https://doi.org/10.1002/elsc.201400151

Lobler M, Sternberg K, Stachs O, Allemann R, Grabow N, Roock A, Guthoff R (2011) Polymers and drugs suitable for the development of a drug delivery drainage system in glaucoma surgery. J Biomed Mater Res B Appl Biomater 97B:388–395. https://doi.org/10.1002/jbm.b.31826

Loh XJ, Goh SH, Li J (2007) New biodegradable thermogelling copolymers having very low gelation concentrations. Biomacromolecules 8:585–593. https://doi.org/10.1021/bm0607933

Lomas AJ, Webb WR, Han J, Chen G, Sun X, Zhang Z, Forsyth NR (2013) Poly (3-hydroxybut yrate-co-3-hydroxyhexanoate)/collagen hybrid scaffolds for tissue engineering applications. Tissue Eng Part C Methods 19:577–585. https://doi.org/10.1089/ten.tec.2012.0457

Lu X, Ciraolo E, Stefenia R, Chen G, Zhang Y, Hirsch E (2011) Sustained release of PI3K inhibitor from PHA nanoparticles and in vitro growth inhibition of cancer cell lines. Appl Microbiol Biotechnol 89:1423–1433. https://doi.org/10.1007/s00253-011-3101-1

Makadia HK, Siegel SJ (2011) Poly lactic-co-glycolic acid (PLGA) as biodegradable controlled drug delivery carrier. Polymers 3:1377–1397. https://doi.org/10.3390/polym3031377

Manavitehrani I, Fathi A, Badr H, Daly S, Negahi Shirazi A, Dehghani F (2016) Biomedical applications of biodegradable polyesters. Polymers 8:20. https://doi.org/10.3390/polym8010020

Masood F (2016) Polymeric nanoparticles for targeted drug delivery system for cancer therapy. Mater Sci Eng C 60:569–578. https://doi.org/10.1016/j.msec.2015.11.067

Meischel M, Eichler J, Martinelli E, Karr U, Weigel J, Schmöller G, Stanzl-Tschegg S (2016) Adhesive strength of bone-implant interfaces and in-vivo degradation of PHB composites for load-bearing applications. J Mech Behav Biomed Mater 53:104–118. https://doi.org/10.1016/j.jmbbm.2015.08.004

Metcalfe AD, Ferguson MW (2006) Tissue engineering of replacement skin: the crossroads of biomaterials, wound healing, embryonic development, stem cells and regeneration. J R Soc Interface 4:413–437. https://doi.org/10.1098/rsif.2006.0179

Michalak M, Kurcok P, Hakkarainen M (2016) Polyhydroxyalkanoate-based drug delivery systems. Polym Int 66:617–622. https://doi.org/10.1002/pi.5282

Misra SK, Ansari TI, Valappil SP, Mohn D, Philip SE, Stark WJ, Boccaccini AR (2010) Poly(3-hydroxybutyrate) multifunctional composite scaffolds for tissue engineering applications. Biomaterials 31:2806–2815. https://doi.org/10.1016/j.biomaterials.2009.12.045

Muller-Santos M, Maltempi de Souza E, De Oliveira Pedrosa F, Chubatsu LS (2015) Polyhydroxybutyrate in Azospirillumbrasilense. Handbook for Azospirillum 241–250. https://doi.org/10.1007/978-3-319-06542-7_13

Nguyen S, Yu G, Marchessault RH (2002) Thermal degradation of poly(3-hydroxyalkanoates): preparation of well-defined oligomers. Biomacromolecules 3:219–224. https://doi.org/10.1021/bm0156274

Novikov LN, Novikova LN, Mosahebi A, Wiberg M, Terenghi G, Kellerth J (2002) A novel biodegradable implant for neuronal rescue and regeneration after spinal cord injury. Biomaterials 23:3369–3376. https://doi.org/10.1016/s0142-9612(02)00037-6

O'Connor S, Szwej E, Nikodinovic-Runic J, O'Connor A, Byrne AT, Devocelle M, O'Connor KE (2013) The anti-cancer activity of a cationic anti-microbial peptide derived from monomers of polyhydroxyalkanoate. Biomaterials 34:2710–2718. https://doi.org/10.1016/j.biomaterials.2012.12.032

Panchal B, Bagdadi A, Roy I (2012) Polyhydroxyalkanoates: the natural polymers produced by bacterial fermentation. Adv Natur Polym 397–421. https://doi.org/10.1007/978-3-642-20940-6_12

Pantazaki AA, Papaneophytou CP, Pritsa AG, Liakopoulou-Kyriakides M, Kyriakidis DA (2009) Production of polyhydroxyalkanoates from whey by thermus thermophilus HB8. Process Biochem 44:847–853. https://doi.org/10.1016/j.procbio.2009.04.002

Pascu EI, Stokes J, McGuinness GB (2013) Electrospun composites of PHBV, silk fibroin and nano-hydroxyapatite for bone tissue engineering. Mater Sci Eng C 33:4905–4916. https://doi.org/10.1016/j.msec.2013.08.012

Peng Q, Zhang Z, Gong T, Chen G, Sun X (2012) A rapid-acting, long-acting insulin formulation based on a phospholipid complex loaded PHBHHx nanoparticles. Biomaterials 33:1583–1588. https://doi.org/10.1016/j.biomaterials.2011.10.072

Peschel G, Dahse H, Konrad A, Wieland GD, Mueller P, Martin DP, Roth M (2008) Growth of keratinocytes on porous films of poly(3-hydroxybutyrate) and poly(4-hydroxybutyrate) blended with hyaluronic acid and chitosan. J Biomed Mater Res A 85A:1072–1081. https://doi.org/10.1002/jbm.a.31666

Philip S, Keshavarz T, Roy I (2007) Polyhydroxyalkanoates: biodegradable polymers with a range of applications. J Chem Technol Biotechnol 82:233–247. https://doi.org/10.1002/jctb.1667

Pompe T, Keller K, Mothes G, Nitschke M, Teese M, Zimmermann R, Werner C (2007) Surface modification of poly(hydroxybutyrate) films to control cell–matrix adhesion. Biomaterials 28:28–37. https://doi.org/10.1016/j.biomaterials.2006.08.028

Potter M (2005) Influence of homologous phasins (PhaP) on PHA accumulation and regulation of their expression by the transcriptional repressor PhaR in Ralstoniaeutropha H16. Microbiology 151:825–833. https://doi.org/10.1099/mic.0.27613-0

Qu X, Wu Q, Zhang K, Chen G (2006) In vivo studies of poly(3-hydroxybutyrate-co-3-hydroxyhexanoate) based polymers: biodegradation and tissue reactions. Biomaterials 27:3540–3548. https://doi.org/10.1016/j.biomaterials.2006.02.015

Renard E, Langlois V, Guerin P (2007) Chemical modifications of bacterial polyesters: from stability to controlled degradation of resulting polymers. Corros Eng Sci Technol 42:300–311. https://doi.org/10.1179/174327807x238918

Rentsch C, Rentsch B, Breier A, Hofmann A, Manthey S, Scharnweber D, Zwipp H (2010) Evaluation of the osteogenic potential and vascularization of 3D poly(3)hydroxybutyrate scaffolds subcutaneously implanted in nude rats. J Biomed Mater Res A 92A:185–195. https://doi.org/10.1002/jbm.a.32314

Robenek H, Severs NJ (2008) Recent advances in freeze-fracture electron microscop: the replica immunolabeling technique. Biol Proced Online 10:9–19. https://doi.org/10.1251/bpo138

Roy I (2015) Polyhydroxyalkanoate (PHA) based blends composites and nanocomposites. Royal Society of Chemistry, Cambridge, UK

Saad GR, Lee YJ, Seliger H (2001a) Synthesis and characterization of biodegradable poly(ester-urethanes) based on bacterial poly(R-3-hydroxybutyrate). J Appl Polym Sci 83:703–718. https://doi.org/10.1002/app.2265

Saad GR, Lee YJ, Seliger H (2001b) Synthesis and thermal properties of biodegradable poly(ester-urethane)s based on chemo-synthetic poly[(R,S)-3-hydroxybutyrate]. Macromol Biosci 1:91–99. https://doi.org/10.1002/1616-5195(20010301)1:33.0.co;2-8

Shah M, Ullah N, Choi MH, Kim MO, Yoon SC (2012) Amorphous amphiphilic P(3HV-co-4HB)-b-mPEG block copolymer synthesized from bacterial copolyester via melt transesterification: nanoparticle preparation, cisplatin-loading for cancer therapy and in vitro evaluation. Eur J Pharm Biopharm 80:518–527. https://doi.org/10.1016/j.ejpb.2011.11.014

Shah M, Ullah N, Choi MH, Yoon SC (2014) Nanoscale poly(4-hydroxybutyrate)-mPEG carriers for anticancer drugs delivery. J Nanosci Nanotechnol 14:8416–8421. https://doi.org/10.1166/jnn.2014.9924

Shangguan Y, Wang Y, Wu Q, Chen G (2006) The mechanical properties and in vitro biodegradation and biocompatibility of UV-treated poly(3-hydroxybutyrate-co-3-hydroxyhexanoate). Biomaterials 27:2349–2357. https://doi.org/10.1016/j.biomaterials.2005.11.024

Sharma P, Munir R, Blunt W, Dartiailh C, Cheng J, Charles T, Levin D (2017) Synthesis and physical properties of polyhydroxyalkanoate polymers with different monomer compositions by recombinant pseudomonas putida LS46 expressing a novel PHA SYNTHASE (PhaC116) enzyme. Appl Sci 7:242. https://doi.org/10.3390/app7030242

Shi J, Votruba AR, Farokhzad OC, Langer R (2010) Nanotechnology in drug delivery and tissue engineering: from discovery to applications. Nano Lett 10:3223–3230. https://doi.org/10.1021/nl102184c

Shishatskaya E, Goreva A, Kalacheva G, Volova T (2011) Biocompatibility and resorption of intravenously administered polymer microparticles in tissues of internal organs of laboratory animals. J Biomater Sci Polym Ed 22:2185–2203. https://doi.org/10.1163/092050610x537138

Shively JM (2011) Inclusions in prokaryotes. Springer, Berlin/Heidelberg. https://doi.org/10.1007/3-540-33774-1

Shrivastav A, Kim HY, Kim YR (2013) Advances in the applications of polyhydroxyalkanoate nanoparticles for novel drug delivery system. Biomed Res Int 2013:581684. https://doi.org/10.1155/2013/581684

Shukla P (2017) Recent advances in applied microbiology, Springer Nature Singapore Ltd, ISBN 978-981-10-5274-3

Sodian R, Hoerstrup SP, Sperling JS, Daebritz S, Martin DP, Moran AM, Mayer JE (2000a) Early in vivo experience with tissue-engineered trileaflet heart valves. Circulation 102(Supplement 3):III-22–III-29. https://doi.org/10.1161/01.cir.102.suppl_3.iii-22

Sodian R, Sperling JS, Martin DP, Egozy A, Stock U, Mayer JE, Vacanti JP (2000b) Technical report: fabrication of a trileaflet heart valve scaffold from a polyhydroxyalkanoate biopolyester for use in tissue engineering. Tissue Eng 6:183–188. https://doi.org/10.1089/107632700320793

Stock UA, Sakamoto T, Hatsuoka S, Martin DP, Nagashima M, Moran AM, Mayer JE (2000) Patch augmentation of the pulmonary artery with bioabsorbable polymers and autologous cell seeding. J Thorac Cardiovasc Surg 120:1158–1167. https://doi.org/10.1067/mtc.2000.109539

Sudesh K, Abe H, Doi Y (2000) Synthesis, structure and properties of polyhydroxyalkanoates: biological polyesters. Prog Polym Sci 25:1503–1555. https://doi.org/10.1016/s0079-6700(00)00035-6

Sudhesh K, Taguchi K, Doi Y, Yoshiharu (2006) Can cyanobacteria be a potential PHA producer?

Sun J, Dai Z, Zhao Y, Chen G (2007) In vitro effect of oligo-hydroxyalkanoates on the growth of mouse fibroblast cell line L929. Biomaterials 28:3896–3903. https://doi.org/10.1016/j.biomaterials.2007.05.011

Timbart L, Renard E, Tessier M, Langlois V (2007) Monohydroxylated poly(3-hydroxyoctanoate) oligomers and its functionalized derivatives used as macroinitiators in the synthesis of degradable diblock copolyesters. Biomacromolecules 8:1255–1265. https://doi.org/10.1021/bm060981t

Troschl C, Meixner K, Drosg B (2017) Cyanobacterial PHA production—review of recent advances and a summary of three years' working experience running a pilot plant. Bioengineering 4:26. https://doi.org/10.3390/bioengineering4020026

Vasant Ranade V, John Cannon B (2011) Role of polymers in drug delivery. Drug Delivery Systems, 3rd 79–136. https://doi.org/10.1201/b10846-4

Vilos C, Gutierrez M, Escobar RA, Morales F, Denardin J, Velasquez L, Altbir D (2013) Superparamagnetic Poly (3-hydroxybutyrate-co-3 hydroxyvalerate) (PHBV) nanoparticles for biomedical applications. Electron J Biotechnol 16. https://doi.org/10.2225/vol16-issue5-fulltext-8

Vimala K, Varaprasad K, Sadiku R, Ramam K, Kanny K (2014) Development of novel protein–Ag nanocomposite for drug delivery and inactivation of bacterial applications. Int J Biol Macromol 63:75–82. https://doi.org/10.1016/j.ijbiomac.2013.10.021

Volova TG (2004) Polyhydroxyalkanoates – plastic materials of the 21st century: production, properties, applications. Nova Science Publishers, Hauppauge

Vroman I, Tighzert L (2009) Biodegradable polymers. Materials 2:307–344. https://doi.org/10.3390/ma2020307

Wang Z, Hu H, Wang Y, Wang Y, Wu Q, Liu L, Chen G (2006) Fabrication of poly(3-hydroxybutyrate-co-3-hydroxyhexanoate) (PHBHHx) microstructures using soft lithography for scaffold applications. Biomaterials 27:2550–2557. https://doi.org/10.1016/j.biomaterials.2005.11.026

Wang Y, Bian Y, Wu Q, Chen G (2008) Evaluation of three-dimensional scaffolds prepared from poly(3-hydroxybutyrate-co-3-hydroxyhexanoate) for growth of allogeneic chondrocytes for cartilage repair in rabbits. Biomaterials 29:2858–2868. https://doi.org/10.1016/j.biomaterials.2008.03.021

Wang L, Wang Z, Shen C, You M, Xiao J, Chen G (2010) Differentiation of human bone marrow mesenchymal stem cells grown in terpolyesters of 3-hydroxyalkanoates scaffolds into nerve cells. Biomaterials 31:1691–1698. https://doi.org/10.1016/j.biomaterials.2009.11.053

Williams SF, Martin DP (2005) Applications of polyhydroxyalkanoates (PHA) in medicine and pharmacy. Biopolymers Online. https://doi.org/10.1002/3527600035.bpol4004

Wu LP, Wei X, Ling JQ (2007a) Expression of matrix extracellular phosphoglycoprotein mRNA in human periodontal ligament cell osteogenic differentiation. Chin J Stomatol 43:362–366

Wu S, Liu Y, Cui B, Qu X, Chen G (2007b) Study on decellularized porcine aortic valve/poly (3-hydroxybutyrate-co-3-hydroxyhexanoate) hybrid heart valve in sheep model. Artif Organs 31:689–697. https://doi.org/10.1111/j.1525-1594.2007.00442.x

Wu L, Chen S, Li Z, Xu K, Chen G (2008) Synthesis, characterization and biocompatibility of novel biodegradable poly[((R)-3-hydroxybutyrate)-block-(D,L-lactide)-block-(ε-caprolactone)] triblock copolymers. Polym Int 57:939–949. https://doi.org/10.1002/pi.2431

Xiao X, Zhao Y, Chen G (2007) The effect of 3-hydroxybutyrate and its derivatives on the growth of glial cells. Biomaterials 28:3608–3616. https://doi.org/10.1016/j.biomaterials.2007.04.046

Xie H, Li J, Li L, Dong Y, Chen G, Chen KC (2013) Enhanced proliferation and differentiation of neural stem cells grown on PHA films coated with recombinant fusion proteins. Acta Biomater 9:7845–7854. https://doi.org/10.1016/j.actbio.2013.04.038

Yamamoto H, Kuno Y, Sugimoto S, Takeuchi H, Kawashima Y (2005) Surface-modified PLGA nanosphere with chitosan improved pulmonary delivery of calcitonin by mucoadhesion and opening of the intercellular tight junctions. J Control Release 102:373–381. https://doi.org/10.1016/j.jconrel.2004.10.010

Yang Y (2001) Morphology, drug distribution, and in vitro release profiles of biodegradable polymeric microspheres containing protein fabricated by double-emulsion solvent extraction/evaporation method. Biomaterials 22:231–241. https://doi.org/10.1016/s0142-9612(00)00178-2

Yang X, Zhao K, Chen G (2002) Effect of surface treatment on the biocompatibility of microbial polyhydroxyalkanoates. Biomaterials 23:1391–1397. https://doi.org/10.1016/s0142-9612(01)00260-5

Yang X, Li, H, Chen M, Zou X, Zhu L, Wei C, Chen G (2009) Enhanced insulin production from murine islet beta cells incubated on poly(3-hydroxybutyrate-co-3-hydroxyhexanoate). J Biomed Mater Res A9999A, NA-NA. https://doi.org/10.1002/jbm.a.32379

Yao Y, Zhan X, Zhang J, Zou X, Wang Z, Xiong Y, Chen G (2008) A specific drug targeting system based on polyhydroxyalkanoate granule binding protein PhaP fused with targeted cell ligands. Biomaterials 29:4823–4830. https://doi.org/10.1016/j.biomaterials.2008.09.008

York GM, Stubbe J, Sinskey AJ (2002) The ralstoniaeutrophaPhaR protein couples synthesis of the PhaPPhasin to the presence of polyhydroxybutyrate in cells and promotes polyhydroxybutyrate production. J Bacteriol 184:59–66. https://doi.org/10.1128/jb.184.1.59-66.2002

You M, Peng G, Li J, Ma P, Wang Z, Shu W, Chen G (2011) Chondrogenic differentiation of human bone marrow mesenchymal stem cells on polyhydroxyalkanoate (PHA) scaffolds coated with PHA granule binding protein PhaP fused with RGD peptide. Biomaterials 32:2305–2313. https://doi.org/10.1016/j.biomaterials.2010.12.009

Young R, Terenghi G, Wiberg M (2002) Poly-3-hydroxybutyrate (PHB): a resorbable conduit for long-gap repair in peripheral nerves. Br J Plast Surg 55:235–240. https://doi.org/10.1054/bjps.2002.3798

Zhang L, Zheng Z, Xi J, Gao Y, Ao Q, Gong Y, Zhang X (2007) Improved mechanical property and biocompatibility of poly(3-hydroxybutyrate-co-3-hydroxyhexanoate) for blood vessel tissue engineering by blending with poly(propylene carbonate). Eur Polym J 43:2975–2986. https://doi.org/10.1016/j.eurpolymj.2007.04.007

Zhang C, Zhao L, Dong Y, Zhang X, Lin J, Chen Z (2010) Folate-mediated poly(3-hydroxybutyrate-co-3-hydroxyoctanoate) nanoparticles for targeting drug delivery. Eur J Pharmacol Biopharm 76:10–16. https://doi.org/10.1016/j.ejpb.2010.05.005

Zhang S, Zheng D, Fan W, Wei D, Peng S, Tang M, Wei C (2012) Transient embolization with microspheres of polyhydroxyalkanoate renders efficient adenoviral transduction of pancreatic capillary in vivo. J Gene Med 14:530–539. https://doi.org/10.1002/jgm.2650

Zhou J, Peng S, Wang Y, Zheng S, Wang Y, Chen G (2010) The use of poly(3-hydroxybutyrate-co-3-hydroxyhexanoate) scaffolds for tarsal repair in eyelid reconstruction in the rat. Biomaterials 31:7512–7518. https://doi.org/10.1016/j.biomaterials.2010.06.044

Zhou L, Chen Z, Chi W, Yang X, Wang W, Zhang B (2012) Mono-methoxy-poly(3-hydroxybutyrate-co-4-hydroxybutyrate)-graft-hyper-branched polyethylenimine copolymers for siRNA delivery. Biomaterials 33:2334–2344. https://doi.org/10.1016/j.biomaterials.2011.11.060

Zou X, Li H, Wang S, Leski M, Yao Y, Yang X, Chen G (2009) The effect of 3-hydroxybutyrate methyl ester on learning and memory in mice. Biomaterials 30:1532–1541. https://doi.org/10.1016/j.biomaterials.2008.12.012

Memory Enhancers

7

Eleftherios Halevas, Georgios K. Katsipis, and Anastasia A. Pantazaki

Abstract

Polyhydroxyalkanoates (PHAs) constitute a family of naturally-occurring microbial polymers possessing excellent physicochemical properties, non-toxic behavior, biocompatibility and biodegradability which render them distinctive candidates for industrial applications especially in medicinal and pharmaceutical areas. In this review, we deepened in various cellular processes to collect and record information highlighting the biotechnological applications of PHA monomers or derivatives as memory amplifiers. Neurological conditions under the umbrella term of "dementia" are concerning millions of people worldwide, and their prevalence rises exponentially. These diseases are generally defined by gradual loss of cognitive and physical abilities, due to severe dysfunction of important central nervous system (CNS) areas. Neuronal and synaptic degeneration, though not specified in detail, are most probably multi-etiological events, caused by abnormal protein aggregation, neuro-inflammation, oxidative stress, and dys-regulated extracellular or intracellular signaling and energy supply. Therefore, multifunctional polymeric formulations without side effects, such as PHAs (polymers of hydroxy-organic acids), offer numerous applications in the prevention and treatment of various diseases. PHA monomers or derivatives (e.g. 3-hydroxybutyrate, 3-hydroxybutyrate methyl ester, 3-hydroxybutyl-3-hydroxybutyrate) are now proven to act as artificial ketogenic compounds and memory enhancers administered in ketogenic diets. Ketogenic diet (KD) is a well-known alternative for the treatment of neurological conditions, as the produced ketones affect protein modifications, attenuate oxidative and inflammatory stress and modulate signaling pathways contributing to neurogenesis. Collectively, PHAs can ameliorate brain and neuronal activity, improve memory

E. Halevas · G. K. Katsipis · A. A. Pantazaki (✉)
Laboratory of Biochemistry, Department of Chemistry, Aristotle University, Thessaloniki, Greece
e-mail: natasa@chem.auth.gr

V. C. Kalia (ed.), *Biotechnological Applications of Polyhydroxyalkanoates*,
https://doi.org/10.1007/978-981-13-3759-8_7

recall and even alleviate important pathological features of neurodegenerative problems, such as Alzheimer's disease (AD) – related amyloid plaques.

Keywords
Biopolymers · Polyhydroxyalkanoates · Memory enhancers

7.1 Introduction

PHAs have been established as a complex class of biodegradable homo- and co-polymers since the late 1950s. They consist of various hydroxyalkanoic acids (HAs) as monomers whose alteration leads to the development of biopolymers with desirable mechanical and structural properties, biocompatible behavior and favorable degradation kinetics under controlled physiological conditions. PHAs are synthesized by bacteria and gathered in PHA granules as energy and carbon storage composites, in the presence of high carbon concentrations and very low concentrations of growth nutrients, such as sulphur, nitrogen, oxygen or phosphorus (Kumar et al. 2015a, b, c; Patel et al. 2015a, b, 2016; Kalia et al. 2016; Koller et al. 2017). Bacteria are able instead of metabolizing available carbon sources through the Tricarboxylic acid cycle (TCA), to generate energy utilizing the acetylCoA by diverting it towards the polyhydroxyalknoate (PHA) biosynthetic pathway (Singh et al. 2015; Ray and Kalia 2016, 2017). Their high degree of polymerization in combination with their biodegradability, insolubility in water, nontoxic behavior, thermoplastic, piezoelectric and elastomeric properties, render them functional materials for various biotechnological applications in the food and packaging industries, agriculture, pharmacy, medicine, and as basic synthons in chemical processes (Anderson and Dawes 1990).

Poly(3-hydroxybutyric acid) (P(3HB)), one of the most extensively studied constituent of bacterial PHAs, was initially reported to be produced by the bacterium *Bacillus megaterium* in 1926 (Valappil et al. 2006). Two novel HAs, 3-hydroxyhexanoate (3HHx) and 3-hydroxyvalerate (3HV), were identified in the 1960s and 1970s (Valappil et al. 2006). Future research efforts led to the growth of small amounts of 3-hydroxyoctanoate (3HO), and 3HHx units in *Pseudomonas oleovorans* on feed containing *n*-octane (Valappil et al. 2006). By the end of the 1980s, various HA constituents, such as 4-hydroxyalkanoates, and 5-hydroxyalkanoates were discovered, whereas after the late 1980s research was focused on the development of PHA polymers with tailored properties for medical use, through the involvement of cloning and clarification of genes in the biosynthetic procedure (Valappil et al. 2006).

At present, scientists have focused their interest on the specification of tertiary and quaternary structures of the key enzymes in PHA biosynthesis, the PHA synthases. These multifunctional enzymes define the: (i) substrate specificity, (ii) catalytic mechanisms, and (iii) molecular weight of each PHA produced (Valappil et al. 2006).

PHAs are categorized, according to the carbon atoms within a PHA monomer, into the: (i) short-chain length PHAs with 3-5 carbon atoms as *scl*-PHAs, (ii) medium-chain length PHAs with 6-14 carbon atoms as *mcl*-PHAs, and (iii) long-chain length PHAs with $\geq$15 carbon atoms as *lcl*-PHAs (Khanna and Srivastava 2005). Until nowadays, more than 150 kinds of PHA monomers have been characterized. New species of PHAs arise through the physical or chemical modification of naturally-occurring PHAs (Zinn and Hany 2005), or through recombinant organisms that produce PHAs with specific properties (Escapa et al. 2011).

The discovery of various monomers with numerous useful properties increased interest in PHAs. The first industrial copolymer consisting of 3HB and 3HV was developed by Imperial Chemical Industries (ICI) and sold under the trade name of Biopol® as biodegradable and renewable replacements instead of the petrochemically derived plastics (Valappil et al. 2006). The increasing commercial impact of P(3HB) and P(3HB-*co*-3HV) copolymer, enabled their future assessment as medical biomaterials in biotechnological applications. HAs are responsible for the stimulation of Ca^{2+} channels which in turn act as an aid in amplifying memory especially in AD patients (Xiao et al. 2007; Zou et al. 2009; Chen. 2010b; Magdouli et al. 2015).

Dementia, as a debilitating neurodegenerative disorder, affects memory, orientation, thinking, calculation, learning capacity, comprehension, judgment, and language (Jotheeswaran et al. 2010; Gross et al. 2012). It affects around 35 million people worlwide, and the observed global prevalence for individuals over 60 years old is 4.7%. AD is the most common type of dementia and contributes to 60–70% of the dementia cases (Honjo et al. 2012). Another prevalent type of dementia is vascular dementia, which represents 17% of dementia cases (Ross et al. 2006). Other types of dementia occurring along with AD pathology are Parkinson's disease (PD) and Lewy body dementia (LBD). Since the distinction between these types of dementias is not clear, scientists tend to believe that their main difference lies in the duration of each type (Langa et al. 2004; Plassman et al. 2007).

Dementia is an age-related disease. This fact deposits an enormous social burden in respect of economics and human inflictions, especially in the context of an outstretched aging population. Currently, there is no cure, however, it has been estimated that successful preventive interventions by delaying the symptoms of late-onset AD, will reduce the predicted prevalence and incidence rates of the disease (Jorm et al. 2005).

AD is notably high among patients who are above 65 years of age and it is characterized by a progressive deterioration of cognitive function, ending with severe brain damage (Mattson 2008; Braak and Del Tredici 2012; Revett et al. 2013). Pathological data on the disease imply that oxidative stress is closely associated with tissue damage such as advanced glycation end products (Smith et al. 1994a, b), nitration (Smith et al. 1997), carbonyl-modified neurofilament proteins and free carbonyls (Smith et al. 1991), and products of lipid peroxidation (Montine et al. 1996; Sayre et al. 1997). Moreover, Aβ (amyloid-beta plaque) aggregation, hyperphosphorylation of tau (τ) protein, and reduced synthesis of the neurotransmitter acetylcholine are some of the most prevalent hypotheses on the cause and progression of AD (Zhang et al. 2013).

In general, oxidative stress is affected by a lack of balance in the antioxidant defence systems and the production of reactive oxygen species (ROS) (Jung et al. 2009). The cytopathological significance of the damage induced by oxidative stress is observed through the up-regulation of the antioxidant enzyme heme oxygenase-1 (Smith et al. 1994a, b; Premkumar et al. 1995), associated with the alteration of amyloidogenic deposits of tau (τ) (Takeda et al. 2000), found in the neurofibrillary tangles of mild AD patient brains (Nday et al. 2015; Halevas et al. 2016). AD symptoms often commence with loss of ability to create novel memories, leading to confusion. Eventually, the inability for self-care often leads patients to their committal in institutionalization (Allen et al. 2013).

The ketogenic diet (KD) is considered a putative preventive treatment for various types of brain diseases, such as AD or epilepsy. During KD, carbohydrates are replaced by fats resulting in the increase of blood-borne ketone bodies (KBs) levels. Since drug development for AD is based on the acetylcholinesterase inhibition or the antibody of Aβ with no satisfactory results on the prevention or reversion of the disease progress, scientists have focused their research interest on the development of drugs that confront the dysfunction of mitochondria (Colell et al. 2009; Swerdlow et al. 2010; Du et al. 2010; Zhu et al. 2013), and the impairment of energy metabolism (Liang et al. 2008; O'Connor et al. 2008).

KBs have been reported as the only alternative to energy supplement for the brain. The major component of KBs, 3-hydroxybutyrate (3HB), and a degradation product of microbial, natural and biocompatible biopolymer P(3HB) (Williams 2008, 2009), have been proved to possess neuroprotective properties. However, due to its charged nature and acidity, 3HB is not considered an ideal drug candidate (Zhang et al. 2013). On the other hand, 3-hydroxybutyrate methyl ester (HBME), the esterification product of 3HB, has a lower polarity with a neutral pH, providing better evidence on its bioavailability and ability to go through the blood-brain barrier (BBB) more efficiently than 3HB (Zhang et al. 2013). Consequently, HBME constitutes a more effective drug candidate for developing CNS biotechnology, especially anti-AD and neuroprotective pharmaceutical prevention and treatment.

7.2 Production of PHA Monomers, Their Biotechnological Applications, and Economic Impacts

PHAs are a category of polyesters consisting of a great number of chiral R-hydroxyalkanoic acids (R-HAs), constituting an abundant source of optically pure chiral compounds, bi-functional hydroxyl acids of R(−) configuration that can be utilized as starting materials for several biotechnological processes such as the synthesis of many pharmacological bioactive compounds including drugs, vitamins, antibiotics and antifungal agents, hormones, pheromones, β-peptides, enzymes inhibitors, siderophores, biofuels, aromatics as well as chiral synthons for several fine chemicals (Ren et al. 2010). Therefore, an emphasis has been given on the utilization of PHA degradative pathway products and their chemical modifications, which provide unique properties for biomedical applications (Hazer et al. 2012; Martinez et al. 2014; Ke et al. 2017).

PHA monomers can be obtained by several strategies such as chemical de novo synthesis of R-HAs, or chemical degradation of PHAs, biotransformation, microbial de novo biosynthesis, in vitro enzymatic degradation of purified PHAs, enzymatic in vivo depolymerization attributed to intracellular PHA depolymerases of wild-type bacteria or R-HA production by metabolic pathway engineering by use of recombinant microorganisms (Ren et al. 2010).

It is remarkable that all these follow-up compounds possess higher economic value in the commercial market than PHAs themselves. However, PHA degradation and recovery of the yielded monomers or oligomers are highly energy-consuming processes. To resolve this problem research studies were directed towards the purification of PHA hydrolyzing enzymes such as depolymerases (Jendrossek et al. 1996; Jendrossek and Handrick 2002; Kadouri et al. 2005; Kim et al. 2007; Jendrossek 2007; Papaneophytou et al. 2009, Papaneophytou and Pantazaki, 2011), and the development of effective methods to activate these enzymes in vivo within the cells (Cheng et al. 2005; Ren et al. 2010) or extracellularly (Papaneophytou et al. 2009, Papaneophytou and Pantazaki, 2011). Additionally, other strategies aimed at the decrease of the market price of the initial PHAs products by using low-value sources such as industrial wastes (Pantazaki et al. 2009) or by the simultaneous production of other useful biotechnological products (Pantazaki et al. 2011), in order to benefit from multipotent advantageous applications such as the production of monomers and oligomers.

7.3 Ketone Bodies, Ketosis, Ketogenic Diet

In order to understand in depth, the basis and the extent of PHAs and their monomers biomedical applications, it is important to elucidate the involvement of the principal of them, 3HB, their natural counterpart, in various cellular processes. Analytically, 3HB is one of the main natural endogenous circulating KBs, as acetoacetate and acetone.

In the mitochondrial milieu, 3HB constitutes the precursor of acetoacetate, which is produced during a catabolic reaction catalyzed by the enzyme 3HB dehydrogenase 1 (BDH1) and directly linked to the reduction of nicotinamide adenine dinucleotide (NAD) to $NADH^+$. Subsequently, acetoacetate is converted to acetoacetyl-CoA and subsequently to acetyl-CoA, which is a component of the citric acid (TCA) cycle (Liu et al. 2014), as presented in Fig. 7.1.

KBs are products of the catabolism of fatty acids in the liver and are used as an alternative energy source (fuel) instead of blood glucose, when glucose availability is insufficient, by enhancing the fatty acid-beta oxidation levels in the brain cells (Panov et al. 2014). The maximal concentrations of circulating KB can be up

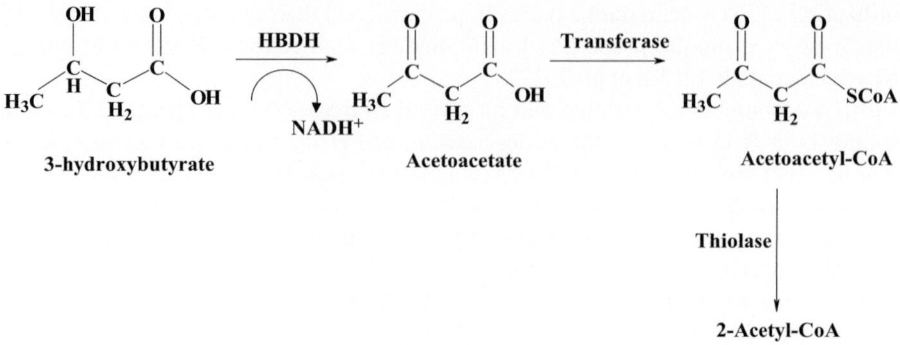

Fig. 7.1 Enzymatic steps involved in the 3HB catabolic process to TCA cycle entry as acetyl-CoA

regulated by adhering to a ketogenic diet (KD). It is considered that a high-fat content and adequate-protein diet, in which carbohydrate intake is drastically reduced and replaced with body dietary sources of glucose, exhorts a metabolic transition towards the consumption of fatty acids (Hammami 1997; LaManna et al. 2009).

Accumulation of KBs in the blood circulation leads to a metabolic situation known as ketosis. It is a survival mechanism functioning as a metabolic state activated during starvation, or due to lack of carbohydrate digestion (Brownlow et al. 2013). It should be noted that the plasma concentration of 3HB is estimated in the range of 0.02–0.09 $\mu mol \cdot mL^{-1}$ (Blomqvist et al. 2002). In addition, the metabolism of KBs mimics some actions of insulin and overcomes insulin resistance indicating a useful therapeutic potential in the alleviation of the pathological hallmarks of cognitive impairment (Sato et al. 1995).

A key feature that increases the lifespan of many organisms is dietary restriction (Greer and Brunet 2009). The ketone moiety 3HB functions as a Dietary Restriction (DR) mimetic compound (Newman and Verdin 2014), by increasing in the plasma during DR, and additionally by decreasing the levels of oxidative stress after exogenous administration (Shimazu et al. 2013).

In general, the KD is regarded as an effective non-pharmacological treatment providing symptomatic avail, and effective disease-ameliorating activity of various neurological disorders typified/provoked by the death of neurons, including epilepsy, AD, Huntington's disease (HD), PD, stroke and brain injury (Gasior et al. 2006). Studies on 3HB functionality have proved its utilization as an energy substrate in hemorrhage shock rats (Katayama et al. 1994), and head trauma patients (Hiraide et al. 1991).

7.3.1 Ketotherapeutic Intervention

KBs serve as alternative energy sources, and they can be increased as liver glycogen supplies get consumed after prolonged fasting (Cotter et al. 2013; Newman and

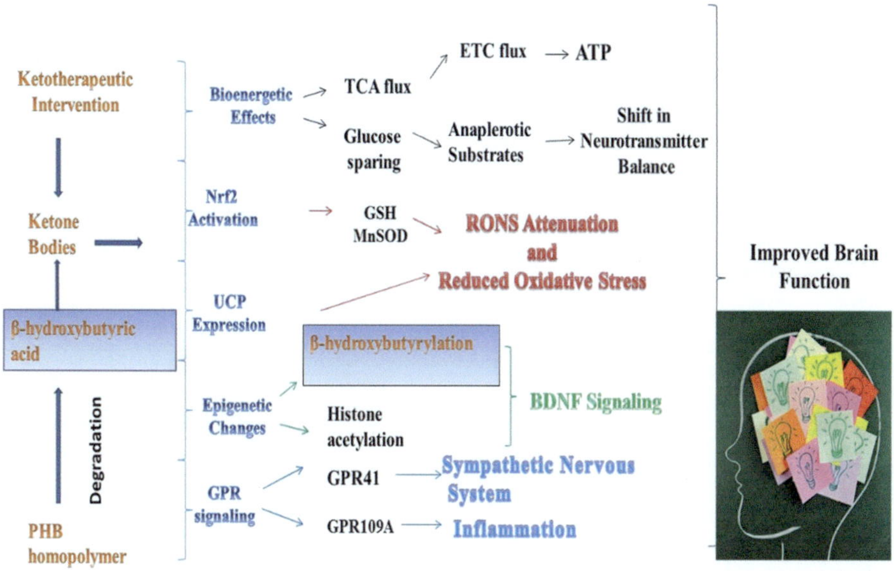

Fig. 7.2 Ketotherapeutic interventions through KB influence on brain function. *Nrf2* nuclear factor erythroid 2-related factor, *UCP* uncoupling protein, *GPR* G-coupled protein receptor, *TCA* tricarboxylic acid, *GSH* glutathione, *MnSOD* manganese superoxide dismutase, *ETC* electron transport chain, *RONS* reactive oxygen-nitrogen species, *ATP* adenosine triphosphate, *BDNF* brain derived neurotrophic factor (Koppel and Swerdlow 2017)

Verdin 2014). It has been proved that recent ketotherapeutic approaches influence the CNS through a variety of mechanisms (Fig. 7.2). More specifically, the action of KBs includes: (a) post-translational modifications of proteins, (b) attenuation of RONS production, (c) enhancement of the anti-oxidant stress response pathway expressions, (d) modulation of G-coupled protein receptor (GPR) and the signaling pathways brain derived neurotrophic factor (BDNF), (e) contribution to substrates replacement and (f) exhibition of anti-inflammatory effects (Brownlow et al. 2013; Koppel and Swerdlow 2017).

7.4 The Case of the Ubiquitous (or Omnipresent) 3-Hydroxybutyrate

Hydrolysis of PHB yields 3HB and its oligomers. 3HB is a physiological constituent of tissues and blood (Reusch et al. 1992; Malm et al. 1992; Peng et al. 1996), primarily originated by the long chain fatty acids catabolism in the liver and liberated to the peripheral tissues and plasma, where it diffuses rapidly, and subsequently, penetrates cell membranes serving as a lipogenic precursor and an oxidative combustible material (fuel) (Kesl et al. 2016). Sodium derivatives of D-3-hydroxybutyrate (D-3HB), DL-3-hydroxybutyrate (DL-3HB), and Methyl 3-hydroxybutyrate

(3HBME), ethyl (±)-3-hydroxybutyrate, 3-hydroxybutyl-(R)-3-hydroxybutyrate/ketone monoester (Table 7.1) are derivatives of 3HB, some are KBs produced in vivo in animals and humans, significantly useful for applications associated with tissue engineering (Chen and Wu, 2005), mainly due to the inhibition of cell death and the increase in cytosolic Ca^{2+} levels (Chen et al. 2010).

Table 7.1 Structures of (i) Sodium (S)-3-hydroxybutyrate (D-3HB), (ii) Sodium DL-3-hydroxybutyrate (DL-3HB), (iii) Methyl 3-hydroxybutyrate (3HBME), (iv) Ethyl (±)-3-hydroxybutyrate, and (v) 3-hydroxybutyl-(R)-3-hydroxybutyrate/ketone monoester

Structures	Compound
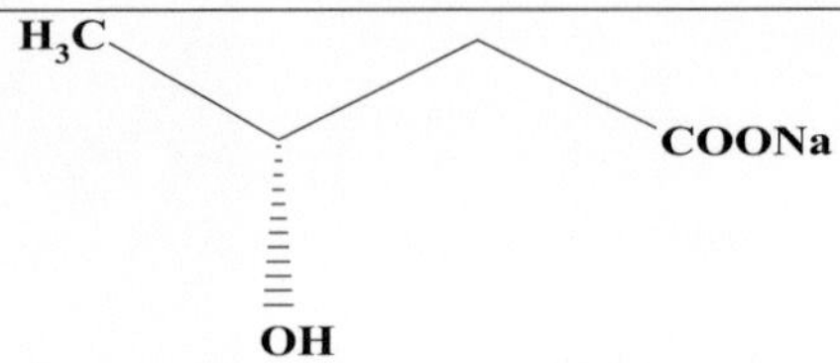	**Sodium (S)-3-hydroxybutyrate (D-3HB)**
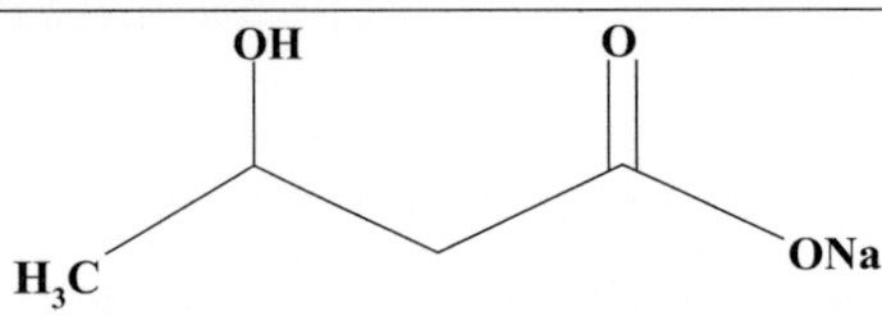	**Sodium DL-3-hydroxybutyrate (DL-3HB)**
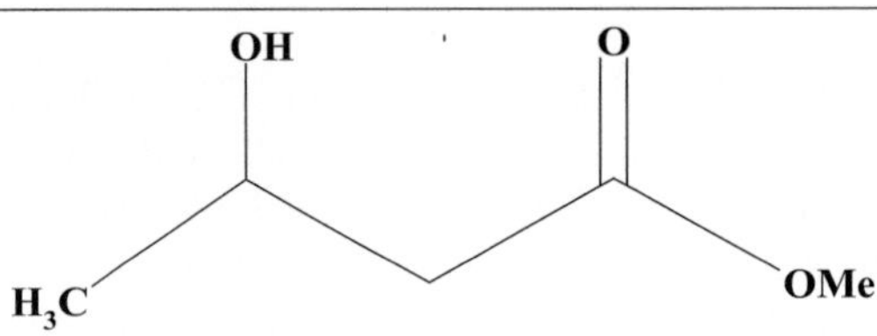	**Methyl 3-hydroxybutyrate (3HBME)**
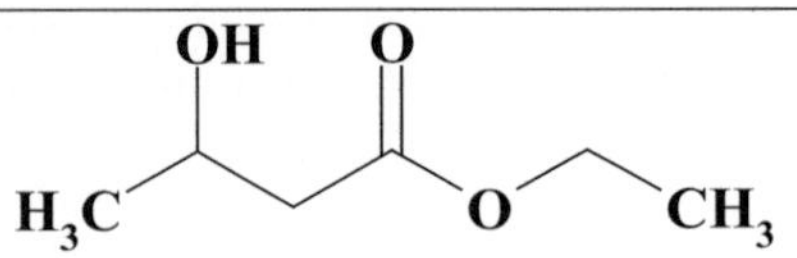	**Ethyl (±)-3-hydroxybutyrate**
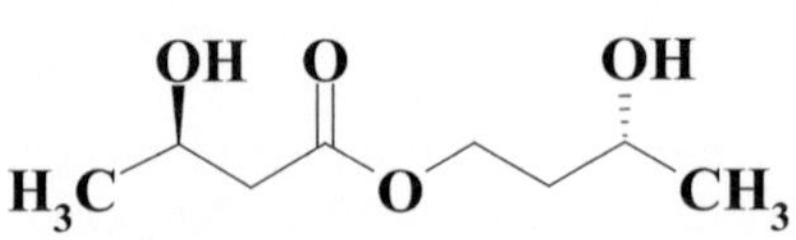	**3-hydroxybutyl-(R)-3-hydroxybutyrate/ketone monoester**

7.4.1 Therapeutic Applications of 3-Hydroxybutyrate as Memory Enhancer

7.4.1.1 3-Hydroxybutyrate as an Anti-aging Agent Against Cognitive Impairment

Aging is one of the primary risk factors that progressively deteriorate the decline of cell and tissue function in neurodegenerative diseases such as PD and AD. The development of premature aging is directly related to mitochondrial dysfunction (Braeckman et al. 1999; Hansen et al. 2005; Dell'agnello et al. 2007; Copeland et al. 2009), whilst the increase in the mitochondrial reactive oxygen species (ROS) generation provokes the progressive damage of cellular macromolecules (Harman 2009). AD specific characteristics also include early and region-specific impairments of cerebral glucose metabolism. KBs are considered intra-corporeal glucose deprivation products and are metabolized by the brain. Recent scientific research results in animal models and human trials have indicated that 3HB, as a DR mimetic compound, delays the onset and progression of age-related neurodegenerative diseases, such as AD (Pasinetti et al. 2007; Mercken et al. 2012), by protecting cultured hippocampal neurons from $A\beta_{1-42}$ mediated toxicity and increased inflammation due to increased ROS levels (Kashiwaya et al. 2000; Van der Auwera et al. 2005; Tamagno et al. 2008). Research results on 3HB supplementation in *C. elegans* nematodes led to the increase of worm thermotolerance, whilst it partially hindered glucose toxicity, delayed AD Aβ toxicity and decreased PD alpha-synuclein aggregation. Additionally, 3HB extended the lifespan of *C. elegans* nematodes through histone deacetylases (HDACs) inhibition and via the activation of conserved stress response pathways (Edwards et al. 2014).

Furthermore, the role of oxidative stress in AD is recently well-accepted (Chen and Zhong 2014; Huang et al. 2016; Liu et al. 2017). In this frame, 3HB represents a potent protective agent against oxidative stress conditions, through its inhibitory activity of both classes of HDACs (I and IIa), in the increase of the expression of genes that encode antioxidant stress response factors, such as the forkhead box O3 (FOXO3A), the transcriptional factor DAF16, the mammalian orthologue of the stress-responsive lifespan regulator in worms, and the MT2 (Shimazu et al. 2013). Additionally, the 3HB-mediated inhibition of HDACs and the activation of the well-conserved antioxidant stress response mechanisms, result in lifespan extension and protection against metabolic, proteotoxic and thermal stress conditions in the cellular milieu (Edwards et al. 2014). Recently, new research orientations are based on the thinking that acute elevation of serum 3HB concentration via the oral administration of medium chain triglycerides (MCTs) meliorates memory and attention in AD individuals setting the stage for the development of an innovative therapeutic approach towards the effective confrontation of neurodegenerative disorders (Reger et al. 2004).

7.4.1.2 The Protective Role of 3-Hydroxybutyrate Against Parkinson's and Huntington's Diseases

PD main clinical characteristics are bradykinesia, muscle rigidity, and tremor of the distal extremities, whilst its pathological features include eosinophilic cytoplasmic Lewy-body inclusions consisted of: (i) ubiquitin, (ii) α-synuclein nucleoprotein, and (iii) death of essential nigral dopaminergic neurons (Dunnett and Bjorklund 1999). Most important causes of PD are infections, environmental toxins or genetic irregularities, and it can be temporarily treated by L-3,4-dihydroxyphenylalanine (L-dopa) administration (Dunnett and Bjorklund 1999). Experimental results on the protective role of ketones against amyloid $A\beta_{1-42}$ or 1-methyl-4-phenylpyridinium (MPP^+) exposure of either hippocampal or mesencephalic neurons, respectively, suggest that mitochondrial dys-operation is the key factor in both of these common neurological disorders (Dunnett and Bjorklund 1999; Kashiwaya et al. 2000), despite the pathophysiological and genetic diversity of AD and PD etiology. AD and PD share a common dysfunction in the protein degradation process, possibly due to the defective mitochondrial energy generation. Thus, increase of ketones may promote protection of neurons by contributing in prevention and in treatment against memory decline related diseases AD and PD, in cases where cure with L-dopa must be period restricted. However, alternative biotechnologically produced dietary sources of ketones are beneficial without displaying the harmful side effects of contemporary KDs, although the high-fat KD is not considered appropriate for adults due to its atherogenic capacity (Gerngross 1999).

Taken into account all these biochemical experimental data, it is noteworthy to add that regulation of 3HB amounts can constitute a targeted neuroprotective approach for the cure against PD. More specifically, 3HB has been proved to confer protection through oxidative phosphorylation enhancement, in a stereo-specific and dose-dependent way, against the structural and functional toxic activities of the parkinsonian toxin 1-Methyl-4-phenyl-1,2,3,6-tetrahydropyridine (MPTP), such as degeneration of dopaminergic substantia nigra pars compacta (SNpc) striatal fibers and neurons, PD-like motor abnormalities, and loss of striatal dopamine (Duan and Mattson 1999; Przedborski et al. 2003; Tieu et al. 2003).

HD is a genomic abnormalities-derived neurological disease provoked by the expansion of CAG (cytosine-adenine-guanine) trinucleotide repeats in the gene coding for polyglutamine domain for the protein huntingtin. HD main characteristics are protrusive perdition of medium-size spiny neurons and constitution of protein assemblages in the striatum and its cortical connections and the cerebral cortex (Vonsattel et al. 2011). Epigenetic modifications and bioenergetic defects play also a decisive role in the onset of the disease. To that end, in vitro and in vivo experimental results have demonstrated that the bioenergetic and epigenetic effects of 3HB confer neuroprotection in two animal models, a genetic one of HD and a murine toxic one of striatal neuronal perdition, by hindering the histone deacetylation stimulated by an erroneous form of huntingtin, the mutant huntingtin (mhtt) (Lim et al. 2011).

7.4.1.3 The Beneficial Role of 3-Hydroxybutyrate in Cell Proliferation and Prevention of Cell Death

Apart from the various potential therapeutic applications of 3HB in the amelioration of tissue damage, protein catabolism, and metabolic dysfunction (Cheng et al. 2006), experimental studies have proved that 3HB promotes cell proliferation in cell cultures (Cheng et al. 2005). The results at hand indicated that 3HB either speeds up cell cycle progression by stimulating impermanent increases in intracellular calcium (Ca^{2+}) levels, which subsequently activate signal transduction pathways responsible for the regulation of the cell cycle (Berridge 1995), or decreases cell apoptosis, by enhancement of the mitochondrial respiration, bypassing glycolysis, while they are also introduced in The Citric Acid (TCA) cycle (Zou et al. 2002; Nakamura et al. 2003). In general, 3HB could supplement Adenosine Triphosphate (ATP) production, thereby hindering necrosis at high levels of cell growth, intervened by cell-to-cell contacts.

Recent experimental results on new drug formulations with various benefits such as enhanced bio-distribution and bioavailability, prolonged drug release and decreased cytotoxicity, have shown that conjugation of ibuprofen, a Non-steroidal anti-inflammatory drug (NSAID), linked to PHA oligomers, and particularly oligo(3-hydroxybutyrate) (O3HB), exhibited significantly increased inhibitory ability against proliferation of HT-29 and HCT 116 colon cancer cells in comparison with unconjugated ibuprofen. Collectively, the observed results indicated that conjugation enhances the cellular uptake of the conjugate, decreases cytotoxicity and improves the biopharmaceutical properties of the drugs (Zawidlak-Węgrzyńska et al. 2010).

7.4.1.4 The Role of 3-Hydroxybutyrate in Neuronal Death

Macro-autophagy is considered an intracellular destructive mechanism exclusively associated with the hydrolytic potency of lysosomes and responsible for the disassembly of damaged or dysfunctional organelles and proteins (Mizushima et al. 2008; Tanida 2011). It is considered a highly conserved process stimulated under various types of stress, such as nutritional stress, providing energy and sustaining cell survival (Ogata et al. 2006; Kroemer et al. 2010; Alirezaei et al. 2010). The autophagic flux is a process initiated by the recruitment of a multi-component proteinic complex, important for the formation of autophagosomes or double membrane vesicles, which then join with lysosomes to produce autophagolysosomes, where degradation of damaged proteins and intracellular components occur by hydrolytic enzymes. A dysfunction of the autophagic flux can result in increase in formation of autophagosomes leading to neuronal cell death (Kulbe et al. 2014; Sarkar et al. 2014). It has been shown that 3HB stimulates the autophagic flux by decreasing the formation rate of autophagosomes under conditions of energy deficiency and glucose withdrawal and minimizes neuronal death stimulated by glucose restriction (GD) (Camberos-Luna et al. 2016).

Glutamate is an important neurotransmitter which acts excitatory in the brain of mammals. Nevertheless, increase in extracellular levels activates neuronal death through the excitotoxicity process (Olney 1971; Choi 1992). Further scientific studies have estimated the contribution of oxidative damage in the promotion of excitotoxicity derived from inhibition of glycolytic metabolism, highlighting the neuroprotective potentiality of 3HB in the confrontation of the in vivo excitotoxic oxidative impairment (Mejía-Tober et al. 2006).

The results at hand indicate that systemic administration of 3HB prevents neuronal damage and lipoperoxidation, and stimulates two cellular processes occurred in isolated mitochondria, ATP generation and oxygen depletion (Tieu et al. 2003), implying that administration of KBs, such as 3HB, in the mitochondrial metabolism, provides in the cells the ability to confront the neurotoxic effect originated from the activation of glutamate receptor. In accordance are studies performed in vitro, demonstrated the benefit of 3HB administration during hypoxia promoting the mitochondrial membrane durability and survival of hippocampal neuronal cell cultures, by preventing the generation of free radicals and the liberation of apoptotic molecules (Masuda et al. 2005). Furthermore, various in vitro studies indicate that 3HB opposes MPP^+ (the heroin analogue 1-methyl-4-phenylpyridinium), and β-amyloid toxicity (Kashiwaya et al. 2000) affecting memory, sustains creatine-phosphate and ATP at the initial amounts during glucose deprivation (Wada et al. 1997), and maintains synaptic communication as well as neuronal integrity (Izumi et al. 1998). In addition, evaluation of the in vitro effect of 3HB and its derivatives, DL-3HB and methyl D-3-hydroxybutyrate (M-D-3HB) on cellular apoptotic death and cytosolic Ca2+ levels of mouse glial cells indicated that the proportion of apoptotic cells significantly decreased when they are exposed to the aforementioned compounds (Xiao et al. 2007).

7.4.1.5 The Role of 3-Hydroxybutyrate as a Blocker of NLRP3 Inflammasome Activation in AD

The NLRP3 inflammasome is a significant multiprotein assemblage acting as a sensor of the innate immunity system found in macrophages with a fundamental role in the control of caspase-1 activation and the secretion of pro-inflammatory cytokines IL-1β and IL-18 (Martinon et al. 2009; Lamkanfi and Dixit 2014; Wen et al. 2013). The activation of inflammasome derives as a response to various damage-linked structural features, including among them amyloids, called DAMPs (damage-associated molecular patterns) (Masters et al. 2010; Heneka et al. 2013). Ablation of NLRP3 attenuates AD and age-related functional decline (Heneka et al. 2013; Youm et al. 2013).

Since the particular nutritional conditions are related to the modified functions of immune cells, recent experimental studies have proved that 3HB specifically affects the activated state of inflammasome that is also ATP-stimulated as well ordinary mechanisms of signal transduction in response to different structurally NLRP3 activators (Youm et al. 2015). Thus, under DR conditions, the role of 3HB as a metabolic signal can reduce the responses of the innate immune system, consuming sparingly ATP to assure the proper operation of organelles, which are ketone-dependent, like the heart and brain. Thus, the use of 3HB, in this case, is to set the stage for new pharmacological approaches to confronting chronic inflammatory diseases, where NLRP3 is involved (Youm et al. 2015).

7.4.1.6 Cerebro-Protective Effect of 3-Hydroxybutyrate on Anoxia, Hypoxia and Ischemia

The brain is an active tissue, whose basic source of energy supplies is glucose oxidation. Although glucose provision limitation and metabolism under the situations of ischemia and hypoxia, lead to brain injuries such as infarct and cerebral edema, it have been shown in epidemiological studies that hyperglycemia was linked with greater dangers of both cerebral ischemia and strokes occurrence (Asplund et al.

1980; Riddle and Hart 1982; Pulsinelli et al. 1983). Moreover, studies in several animal models have verified that the brain injury, provoked by cerebral ischemia, was accelerated by glucose administration (Siemkowicz and Hansen 1978; Ibayashi et al. 1986; Natale et al. 1990). The elevated levels of anaerobic glucose metabolism, during such conditions as hypoxia and ischemia, cause in tissues an overproduction of lactate leading to augmented intracellular quantities of proton H^+ (Rehncrona et al. 1981; Raichle 1983; Siesjo 1988). It has been reported that fasting under such conditions as hypoglycemia and ketosis, increased 3HB levels in the blood, preventing rats from displaying brain infarction after ischemia and hypoxia (Go et al. 1988).

Furthermore, studies carried out for elucidating the effect of 3HB administration showed the cerebroprotective activity and the notable protraction of the survival time under hypoxia, anoxia, and ischemia in rats and mice, subjected to bilateral common carotid artery ligation (BLCL). These results were obtained by reducing cerebral edema formation, maintaining higher ATP levels and inhibiting lipid peroxidation and anaerobic lactate accumulation (Suzuki et al. 2001). These research data indicate that 3HB has the potential to be utilized as a clinical nutritional treatment of acute cerebrovascular disorders (Suzuki et al. 2002).

7.4.1.7 The Boosting Role of 3-Hydroxybutyrate in the Brain Synthesis of Kynurenic Acid

Kynurenic acid (KYNA) is produced endogenously along the kynurenine pathway and constitutes a neuroactive intermediate product of tryptophan metabolism. Its formulation is involved in a process catalyzed by the enzymes kynurenine aminotransferases (KAT I–III) (Guidetti et al. 1997, 2007). Brain KYNA is synthesized intracellularly in astrocytes and its secreted extracellular amounts reach the nanomolar levels (Moroni et al. 1988; Turski et al. 1989; Németh et al. 2006). KYNA possesses enhanced anticonvulsant and neuroprotective activities, especially this related to the inhibition of strychnine-insensitive site of N-methyl-D-aspartate receptors (NMDA) (Urbanska et al. 1991; Stone et al. 2001; Schwarcz and Pellicciari 2002; Németh et al. 2006), the sufficiency in the noncompetitive reduction of $\alpha7$ nicotinic receptors and the enhancement of the expression of $\alpha4\beta2$ nicotinic receptors by increasing the presynaptic liberation of glutamate (Carpenedo et al. 2001; Hilmas et al. 2001; Luccini et al. 2007).

Experimental data have proved that relatively low concentrations of a racemic mixture of 3HB stimulate the brain biosynthetic pathway of KYNA in primary glial cells as well as in cortical slices as effectively as during a KD or ketosis in diabetes (Gilbert et al. 2000; Fukao et al. 2004). Moreover, in vitro studies confirmed that high levels of circulating 3HB accompanied by acidosis or mild to severe hypoglycemic conditions inhibited KYNA synthesis in rat cortical tissue pieces. Furthermore, data from cultures of glial cells demonstrated that 3HB enhances the KYNA biosynthetic enzymes (KAT I and II) expression (Chmiel-Perzyńska et al. 2011). In general, the action of 3HB is to augment the levels of KYNA synthesis in the brain that via interaction with presynaptic NMDA receptors, may cause a reduction of the glutamate liberation and minimize the effects of the postsynaptic glutamate receptors contributing to their neuroprotective potency (Chmiel-Perzyńska et al. 2011).

7.4.1.8 Accumulation of 3HB Induced by Physical Exercise Promotes Brain-Derived Neurotrophic Factor Production

Exercise is beneficial for both physical and mental health. Corporeal exercise may influence the production of valuable proteins in the brain, particularly the levels of a significant protein called brain-derived neurotrophic factor (BDNF). Recent experimental studies demonstrated that 3HB accumulation induced by the physical exercise in the hippocampal milieu acts as an energy source and as an inhibitor of HDACs (class I) in the induction of BDNF expression (Fig. 7.3).

It has been proved that BDNF causes enhancement of mental competences, counteracting depression and anxiety as effectively in mice as in humans in a similar manner (Sleiman et al. 2016). In general, BDNF is necessary in order to survive susceptible populations of neurons. Furthermore, it plays an important role in dendritic and axonal neuronal cell growth as well as in synaptogenesis (Bibel and Barde 2000; Alsina et al. 2001). It has been proved that low levels of BDNF provoke depression, whereas they are elevated after administration of antidepressant treatment (Duman and Monteggia 2006; Martinowich et al. 2007).

In addition, exercise increases BDNF levels in the CNS promoting amelioration of cognitive ability and of comportment resembling depression (Russo-Neustadt et al. 2000; Marais et al. 2009), as it has been shown in animals and in patients suffering from mental disorders with depressive behavior (Sleiman et al. 2016), and neurodegenerative disorders such as PD (Frazzitta et al. 2014) or AD (Crabb et al. 2014).

Previous work indicated the neuroprotective effect of 3HB in HD (Lim et al. 2011) and PD (Kashiwaya et al. 2000; Tieu et al. 2003), on dopaminergic and striatal neurons, respectively (Autry and Monteggia 2012). Moreover, recent studies proved that the induction of 3HB in the brain after treatment of AD mice with structural analogs of glucose (e.g. 2-deoxy-D-glucose), delayed the occurrence/or advancement of bioenergetic deficiencies and the correlated β-amyloid

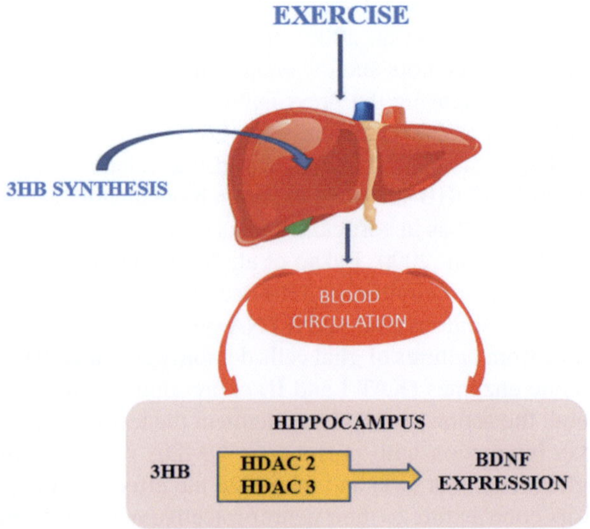

Fig. 7.3 A model mechanism indicating the effect of exercise on the induction of BDNF expression in the hippocampal milieu. Exercise provokes 3HB synthesis within the liver. Subsequently, 3HB is transferred through blood circulation to the hippocampus, where it induces BDNF expression through HDAC inhibition

encumbrance (Ralser et al. 2008; Yao et al. 2011). Therefore, since 3HB represents an endogenous molecule with the ability to cross the BBB, the experimental results at hand suggest that through physical exercise and peripheral metabolism 3HB upregulates BDNF transcription in the hippocampus, affecting the gene expression and epigenetic monitoring in the brain and the synaptic transmission (Sleiman et al. 2016).

7.5 Modified PHA Monomers. The Cases of 3-Hydroxybutyrate Methyl Ester and 3-Hydroxyalkanoate Methyl Esters

7.5.1 The 3-Hydroxybutyrate Methyl Ester

The 3-hydroxybutyrate methyl ester (3HBME) or other 3-hydroxyalkanoate methyl esters (3HAME) (Fig. 7.4) derive as products of the methyl esterification of *scl*-PHAs such as PHB (and PHV for 3HVME), or *mcl*-PHAs, including all HAs with equal and higher than six carbon atoms (Fig. 7.2) (Pantazaki et al., 2003). They have been considered as a basic source for PHA-based biofuels production (Zhang et al. 2009; Chen et al. 2010a). Modified PHA monomers such as 3HBME can also be employed as drugs against mitochondrial damage (Zhang et al. 2013).

Ex vivo production of 3HBME can be easily achieved by acidic methanolysis of PHB in a chloroform solution, at 67 °C for 60 h (de Roo et al. 2002; Wang et al. 2010; Zhang et al. 2009). Direct degradations of PHB to 3HB and 3HBME have also been reported (Seebach et al. 2003). To our knowledge, no in vivo data have been reported regarding the existence of 3HBME in biological systems. However, several microbial species, plants and possibly mammals are capable of producing both 3HB and PHB, with the latter proved to be carried by blood lipoproteins and

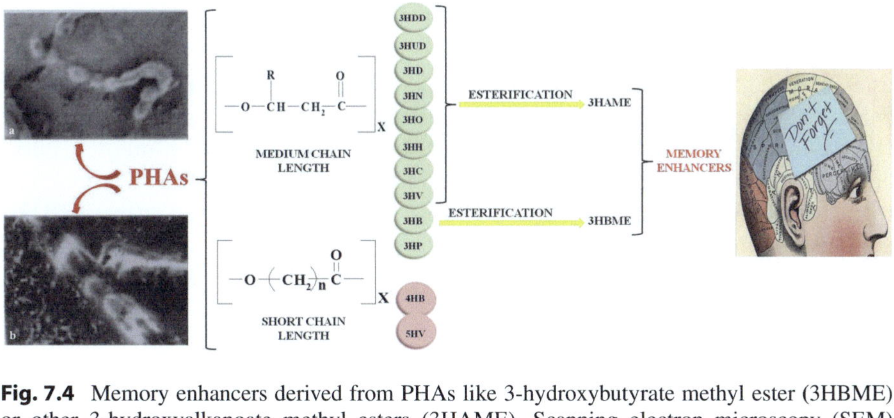

Fig. 7.4 Memory enhancers derived from PHAs like 3-hydroxybutyrate methyl ester (3HBME) or other 3-hydroxyalkanoate methyl esters (3HAME). Scanning electron microscopy (SEM) images of PHAs produced by the thermophilic bacterium *Thermus thermophilus* HB8. (Pantazaki et al. 2003)

albumin and intracellularly located in mitochondria and microsomes (Reusch et al. 1992; Reusch 2015). So, it is possible that some esterases or special de-esterases may yield 3HBME.

7.5.2 Bioavailability and Pharmacokinetics of 3-Hydroxybutyrate Methyl Ester

As it has been previously stated extensively, 3HB has been reported to possess strong neuroprotective properties (Kashiwaya et al. 2000; Reger et al. 2004; Cheng et al. 2013). 3HB is transported through the BBB via the monocarboxylate transporter family SLC16 and the sodium-dependent transporters SMCT and SLC5A with a relatively good rate (K_M of 6.03 mM in cultured rat astrocytes) (Tildon et al. 1994; Achanta and Rae 2017). Additionally, it has been proved that the bioavailability of 3HB in the brain is related to various factors including diet, age and species (Regen et al. 1983; Leino et al. 2001; Ito et al. 2011).

It is interesting that the bioavailability of 3HB in animals treated with 3HBME is higher than the ones treated with analogous shots of 3HB (7.5 times higher than basal levels compared to 1.6 times, respectively) (Uchida et al. 2011). Although the particular behavior has yet to be clarified, researchers have speculated that the amphiphilicity and small size of 3HBME could assist its easy transport through the BBB and its better distribution and transformation to 3HB in the brain (Zhang et al. 2013).

7.5.3 3-Hydroxybutyrate Methyl Ester and 3-Hydroxybutyrate Ethyl Ester as Neuroprotective Agents and Memory Enhancers

Memory function and learning action are highly energy-demanding processes. In so, if the brain is supplied with an excess of suitable metabolic substrates, an enhancement of these processes should be expected. Research results on neuroglial cells cultures supplemented with 3HB or its derivatives [(3-hydroxybutyrate ethyl ester (HBEE) and HBME] demonstrated that the neuroglial cell metabolic activity was remarkably elevated ($p < 0.05$) after 1 and 2 days of cultivation. In all cases, HBME-treated cells were more active than both HBEE and 3HB. Moreover, intracerebral expression of the receptor for the 3HB uptake into the cells, a protein upregulated in macrophages by interferon-g (PUMA-G), occurred after 3HBME administration in mice (Zou et al. 2009).

Furthermore, it was also affirmed that cerebral proteins such as phosphorylated extracellular-signal-regulated kinase 2 (ERK2) and connexin 36 were upregulated after 3HBME administration, but in statistically not significant way. Moreover, mice performed significantly better in the Morris water-maze than non-treted controls or mice administered with the neuronal stimulator acetyl-l-carnitine.

Additionally, 3HBME could increase the gap junctional intercellular communication between neurons (Zou et al. 2009).

In vivo studies were also performed on double transgenic mice models of AD that exhibited age-dependent amyloidal plaques aggregation in the cerebral cortex and the hippocampus and declining learning and memory ability. After training and daily treatment with different doses of 3HBME for 2.5 months, animals scored better in water-maze than non-treated controls and animals treated with AXONA– an FDA approved supplement for AD (Zhang et al. 2009). Significant alleviation of amyloidal burden was also demonstrated in HBME treated mice, accompanied by inhibition of brain atrophy around ventricles. Additionally, transcription of two genes encoding apolipoprotein E (ApoE) that promotes amyloidal aggregation, and caspase-3, an important protease in cell death, were down-regulated in mice treated with 3HBME (Zhang et al. 2013).

In the same study, PC12 cells were treated with NaN_3 to inhibit mitochondrial electron chain, ATP production and to induce oxidative stress. Interestingly, cells supplemented with 3HBME significantly increased ATP levels and inhibited ROS production. Additionally, 3HBME treatment restored $NADH/NAD^+$ ratio to the normal levels, after Rotenone treatment. Furthermore, 3HBME proved capable of protecting PC-12 cells from starvation-induced apoptosis, and carbonyl cyanide m-chlorophenyl hydrazine (CCCP)-induced dissociation of mitochondrial membrane potential (Zhang et al. 2013).

7.5.4 (R)-3-Hydroxybutyl (R)-3-Hydroxybutyrate/Ketone Monoester

(R)-3-hydroxybutyl (R)-3-hydroxybutyrate (D-3HB), also known as ketone monoester (KME), is a synthetic oil consisted mainly from (R)-3-hydroxybutyl (R)-3-hydroxybutyrate (94%) and other small-chain organic esters or non-esters. The purpose of this product, firstly synthesized at the University of Oxford, is to induce artificial mild ketosis, by its enzymatic hydrolysis to 3HB and butane-1,3-diol, with the latter being enzymatically turned also to 3HB (Clarke et al. 2012a, b).

7.5.4.1 Bioavailability and Pharmacokinetics of Ketone Monoester

Experimental studies have proved that KME is metabolized quickly, shortly after injection of a single dose of 714 $mg{\cdot}Kg^{-1}$ in adult humans and mice, as it cannot be traced in the systemic circulation. At the same time, blood concentrations of 3HB and acetoacetate increased, reaching a plateau after 1.5–2.5 h. In addition, it was verified that KME is hydrolyzed to its components (i.e. D-3HB and R-1,3-butanediol) and by that an increase in the levels of ketones in systemic circulation is achieved, as R-1,3-butanediol is hepatically metabolized to D-β-hydroxybutyrate and acetoacetate. Also, ingestion of the KME over a period of 5 days did not impose any significant adverse effects, even in repeated doses of 2142 mg·Kg^{-1}bw/day. There were also no effects noticed on the development of mice embryos (Clarke et al. 2012a, b).

7.5.4.2 Ketone Monoester as a Possible Treatment for Alzheimer's Disease

Human studies on the employment of KME on a 51-aged male patient suffering from the early-onset AD, after three doses of 21.5 g daily for 3 days and three doses of 28.7 g daily thereafter, showed a remarkable improvement in the mood even after the first dose. Moreover, an improvement in the patient's impaired linguistic skills was noticed. After increased dosages, the patient also began to perform everyday activities without prompting or assistance, activities that he could not perform prior to the medication. After 6 weeks, memory retrieval has also been enhanced. It is interesting that the patient's physician noted that elevated cognitive performance paralleled plasma ketone levels and induced a decline in interaction skills and mood as ketone levels fell gradually toward basal levels. Additionally, a decline in total LDL and HDL cholesterol was recorded after 20 months of supplementation. No other noteworthy biochemical alteration was noticed in blood tests (Newport et al. 2015).

A synoptic presentation of all the in vitro, ex vivo, in vivo and clinical studies of the memory- enhancing activity of PHAs monomers was recorded in Table 7.2.

7.6 Conclusions

Microbes, as cellular factories exploiting low-cost carbon sources, can produce PHAs. PHA derivatives constitute high-value products, important curative agents and memory amplifiers. Memory decline, on the cognitive level, is a major issue of AD and related neurodegenerative diseases, from which millions of people suffer in the world and is set to grow rapidly in the coming decades. Although PHAs possess unique properties such as biodegradability and biocompatibility, their high cost at the moment, acts as a deterrent for their large-scale utilization. The newly developed field of PHA-based drugs functioning as memory enhancing moieties will help in the development of novel treatments against neurodegenerative disorders. The applicability of PHAs in those pharmaceutical and medical areas is more than plausible in the near future. However, at present only 3HB and its related products have been demonstrated for therapeutic activities, but more PHA monomers need to be evaluated for their medical efficacy in the near future. These memory enhancing monomers could be then utilized on their own, or even as encapsulating agents of multi-functional therapeutic compounds for their effective delivery through the BBB against AD.

Table 7.2 Synoptic presentation of the in vitro, ex vivo, in vivo and clinical studies of the memory enhancing activity of PHAs monomers

Compound	In vitro studies	Ex vivo studies	In vivo studies	Clinical studies	Results-conclusions	References
3HB			Male ddY mice and Wistar rats		Remarkable protective effects against hypoxia, anoxia and ischemia-induced metabolic change	Suzuki et al. (2001)
					Decrease of cerebral water and sodium contents	
					Maintenance of high ATP and low lactate levels	
3HB			Male Wistar rats with permanent (p)-occlusion and transient (t)-occlusion of middle cerebral artery (MCA)		Decrease of cerebral edema formation and infarct area	Suzuki et al. (2002)
					Improvement of cerebral energy metabolism during ischemia	
					Inhibition of lipid peroxidation after reperfusion	
3HB		Rat brain slices from adult male rats			Stimulation of brain CO_2 production by reduced oxygen availability in the presence of ketones	Kirsch and D'Alecy (1984)
3HB	*C. elegans* nematodes				Extension of mean lifespan by approximately 20% through inhibiting HDACs and through the activation of conserved stress response pathways	Edwards et al. (2014)
					Increase of worm thermotolerance and partial prevention of glucose toxicity	
					Upregulation of 3HB dehydrogenase activity and increase of oxygen consumption in the worms	
					Delay of AD amyloid-beta toxicity	
					Decrease of PD alpha-synuclein aggregation	

(continued)

Table 7.2 (continued)

Compound	In vitro studies	Ex vivo studies	In vivo studies	Clinical studies	Results-conclusions	References
3HB				20 individuals with probable AD or amnestic mild cognitive impairment	Oral doses of medium chain triglycerides (MCTs) increase serum 3HB levels	Reger et al. (2004)
					Improvement of memory and attention in individuals with AD or mild cognitive impairment	
3HB	Lipopolysaccharide (LPS)-primed mouse bone marrow–derived macrophages (BMDMs)		*Oxct1*$^{fl/fl28}$ and *Atg5*$^{fl/fl}$ mice crossed with *LysM*-Cre (B6.129P2-Lyz2tm1(cre)Ifo/J) animals and breeded with Tamoxifen-inducible Cre mice (B6.Cg-Tg(CAG-cre/Esr1*)5Amc/J)		3HB affects ATP-induced inflammasome activation or common signaling mechanisms in response to structurally diverse NLRP3 activators	Youm et al. (2015)
	CD14^{+} monocytes sorted from cryopreserved peripheral blood mononuclear cells				3HB as a metabolic signal reduces the innate immune responses	
3HB	Immature primary cortical neurons from C57BL/6 mice	Hippocampal slices	Male mice		High BDNF levels in the CNS promote improvement in cognitive ability and depressive-like behavior	Sleiman et al. (2016)
					Through physical exercise and peripheral metabolism 3HB upregulates BDNF transcription in the hippocampus, affecting the epigenetic control and gene expression in the brain and the synaptic transmission	

3HB	Glial cultures from Wistar rats of both sexes	Cortical slices of brain tissue obtained from adult male Wistar rats			3HB enhances brain synthesis of kynurenic acid (KYNA)	Chmiel-Perzyńska et al. (2011)
					BHB reduces the activity of postsynaptic glutamate receptors and contributes to neuroprotective effects of KYNA	
D-3HB	Cortical cultured neurons from Wistar rat embryos				D-3HB stimulates the autophagic flux preventing autophagosome accumulation and neuronal death induced by glucose deprivation (GD)	Camberos-Luna et al. (2016)
D-3HB			Male Wistar rats		D-3HB supplementation of mitochondrial metabolism prevents neuronal damage and lipoperoxidation	Mejía-Tober et al. (2006)
D-3HB	Human embryonic kidney cells (HEK293)	Mouse serum after a 24-h fast	Mice on calorie restriction (CR)		D-3HB is an endogenous and specific inhibitor of class I histone deacetylases (HDACs)	Shimazu et al. (2013)
					D-3HB induced histone acetylation and gene expression promote stress resistance in the kidney	
D-3HB	Brains of male C57BL mice		Male C57BL mice		Infusion of D-3HB confers partial protection against dopaminergic neurodegeneration and motor deficits induced by 1-Methyl-4-phenyl-1,2,3,6-tetrahydropyridine (MPTP) neurotoxin	Tieu et al. (2003)
					D-3HB improves mitochondrial respiration and ATP production	

(continued)

Table 7.2 (continued)

Compound	In vitro studies	Ex vivo studies	In vivo studies	Clinical studies	Results-conclusions	References
D-3HB	Brains of male C57BL mice		Male C57Bl/6 mice		D-3HB attenuates motor deficits, striatal lesions, and microgliosis in mice treated with 3-nitropropionic acid (3-NP) neurodegenerative toxin	Lim et al. (2011)
	Pheochromocytoma cells (PC12 cells)		Female hemizygous R6/2 breeders bred with male B6CBAFI/J mice		D-3HB extents life span, attenuates motor deficits, and prevents striatal histone deacetylation in transgenic R6/2 mice	
					D-3HB prevents histone deacetylation in PC12 cells	
D-3HB	Primary serum-free culture of 14-day embryonic mesencephalic cells				D-3HB protects cultured mesencephalic neurons from MPP1 toxicity and hippocampal neurons from Ab_{1-42} toxicity	Kashiwaya et al. (2000)
	Hippocampal cells were dissected from embryonic rats				D-3HB corrects defects in mitochondrial energy generation	
Sodium D-3-hydroxybutyrate			Bal/c mouse newborn primary dissociated glial cell cultures		Cell apoptosis decreased significantly in the presence of 3HB and its derivatives	Xiao et al. (2007)
Sodium DL-3-hydroxybutyrate					3HB derivatives enhanced cytosolic Ca^{2+} concentration	
Methyl-D-3-hydroxybutyrate						

DL-3HB	Murine fibroblast L929 cells				3HB stimulates cell cycle progression mediated by a signaling pathway dependent upon increases in $[Ca^{2+}]_i$	Cheng et al. (2005, 2006)
	Human umbilical vein endothelial cells					
	Rabbit articular cartilages				3HB promotes proliferation in high-density cultures of L929 cells by preventing apoptotic and necrotic cell death	
Ibuprofen-oligo(3-hydroxybutyrate) conjugates (Ibu-OHB)	HT-29 and HCT 116 colon cancer cells		WIST rats		More effective cellular uptake of Ibu-OHB than free Ibu	Zawidlak-Węgrzyńska et al. (2010)
					Conjugation increases anticancer potential of Ibu	
					Conjugation improves antiproliferative activity, bioavailability and toxicity compared to free Ibu	
3-hydroxybutyrate methyl ester (3HBME)	Neuroglial cells of BALB/c mice				Metabolic activity was remarkably elevated	Zou et al. (2009)
					Enhanced gap junctional intercellular communication between neurons	
3-hydroxybutyrate methyl ester (3HBME)			BALB/c mice		Brain 3HB receptor PUMA-G, connexin 36 protein and phosphorylated ERK2 were upregulated	Zou et al. (2009)
					Mice performed significantly better in the Morris water maze than either the negative controls (no treatment) or positive controls	
3-hydroxybutyrate methyl ester (3HBME)	PC12 pheochromocytoma cell line				HBME inhibited cell apoptosis under glucose deprivation	Zhang et al. (2013)
					HBME rescued activities of mitochondrial respiratory chain complexes and stabilized the mitochondrial membrane potential	
					HBME decreased the generation of ROS	

(continued)

Table 7.2 (continued)

Compound	In vitro studies	Ex vivo studies	In vivo studies	Clinical studies	Results-conclusions	References
			C57/BL6 mice		HBME crossed the blood brain barrier easier compared with charged 3HB	Zhang et al. (2013)
					AD mice treated with HBME performed significantly better in the Morris water maze compared with other groups	
					Reduced amyloid-β deposition in mouse brains after intragastrical administration of	
(R)-3-hydroxybutyl (R)-3-hydroxybutyrate (D-3HB) or Ketone monoester (KME)				A 51-year old AD patient	Remarkable improvement in patients's mood and his ability to participate in conversations	Newport et al. (2015)
					Improvement in cognitive and everyday skills like writing, memory retrieval, personal care	
					Total cholesterol fell from 244 to 163 mg/dl, HDL cholesterol fell from 85 to 68 mg/dl, and LDL cholesterol from 145 to 81 mg/dl	

References

Achanta LB, Rae CD (2017) β-Hydroxybutyrate in the brain: one molecule, multiple mechanisms. Neurochem Res 42:35–49. https://doi.org/10.1007/s11064-016-2099-2

Alirezaei M, Kembal CC, Flynn CT, Wood MR, Whitton JL, Kiosses WB (2010) Short-term fasting induces profound neuronal autophagy. Autophagy 6:702–710. https://doi.org/10.4161/auto.6.6.12376

Allen SJ, Watson JJ, Shoemark DK, Barua NU, Patel NK (2013) GDNF, NGF and BDNF as therapeutic options for neurodegeneration. Pharmacol Ther 138:155–175. https://doi.org/10.1016/j.pharmthera.2013.01.004

Alsina B, Vu T, Cohen-Cory S (2001) Visualizing synapse formation in arborizing optic axons in vivo: dynamics and modulation by BDNF. Nat Neurosci 4:1093–1101. https://doi.org/10.1038/nn735

Anderson AJ, Dawes EA (1990) Occurrence, metabolism, metabolic role, and industrial uses of bacterial polyhydroxyalkanoates. Microbiol Rev 54:450–472

Asplund K, Hagg E, Helmers C, Lithner F, Strand T, Wester PO (1980) The natural history of stroke in diabetic patients. Acta Med Scand 207:417–424. https://doi.org/10.1111/j.0954-6820.1980.tb09749.x

Autry AE, Monteggia LM (2012) Brain-derived neurotrophic factor and neuropsychiatric disorders. Pharmacol Rev 64:238–258. https://doi.org/10.1124/pr.111.005108

Berridge MJ (1995) Calcium signalling and cell proliferation. BioEssays 17:491–500. https://doi.org/10.1002/bies.950170605

Bibel M, Barde YA (2000) Neurotrophins: key regulators of cell fate and cell shape in the vertebrate nervous system. Genes Dev 14:2919–2937. https://doi.org/10.1101/gad.841400

Blomqvist G, Alvarsson M, Grill V, Von Heijne G, Ingvar M, Thorell JO, Stone-Elander S, Widén L, Ekberg K (2002) Effect of acute hyperketonemia on the cerebral uptake of ketone bodies in non diabetic subjects and IDDM patients. Am J Physiol – Endocrinol Metab 283:E20–E28. https://doi.org/10.1152/ajpendo.00294.2001

Braak H, Del Tredici K (2012) Where, when, and in what form does sporadic Alzheimer's disease begin? Curr Opin Neurol 25:708–714. https://doi.org/10.1097/WCO.0b013e32835a3432

Braeckman BP, Houthoofd K, De Vreese A, Vanfleteren JR (1999) Apparent uncoupling of energy production and consumption in long-lived Clk mutants of *Caenorhabditis elegans*. Curr Biol 9:493–496. https://doi.org/10.1016/S0960-9822(99)80216-4

Brownlow ML, Benner L, D'Agostino D, Gordon MN, Morgan D (2013) Ketogenic diet improves motor performance but not cognition in two mouse models of Alzheimer's pathology. PLoS One 12:e75713. https://doi.org/10.1371/journal.pone.0075713

Camberos-Luna L, Gerónimo-Olvera C, Montiel T, Rincon-Heredia R, Massieu L (2016) The ketone body, β-hydroxybutyrate stimulates the autophagic flux and prevents neuronal death induced by glucose deprivation in cortical cultured neurons. Neurochem Res 41:600–609. https://doi.org/10.1007/s11064-015-1700-4

Carpenedo R, Pittaluga A, Cozzi A, Attucci S, Galli A, Raiteri M, Moroni F (2001) Presynaptic kynurenate-sensitive receptors inhibit glutamate release. Eur J Neurosci 13:2141–2147. https://doi.org/10.1046/j.0953-816x.2001.01592.x

Chen GQ (2010a) Biofunctionalization of polymers and their applications. Adv Biochem Eng/Biotechnol 125:29–45. https://doi.org/10.1007/10_2010_89

Chen GQ (2010b) Plastics completely synthesized by bacteria: polyhydroxyalkanoates. In: Plastics from bacteria. Natural functions and applications, microbiology monographs, vol. 14. Springer-Verlag, Heidelberg, p 17–37. ISBN 978-3-642-03286-8. https://doi.org/10.1007/978-3-642-03287-5

Chen GQ, Wu Q (2005) The application of polyhydroxyalkanoates as tissue engineering materials. Biomaterials 26:6565–6578. https://doi.org/10.1016/j.biomaterials.2005.04.036

Chen Z, Zhong C (2014) Oxidative stress in Alzheimer's disease. Neurosci Bull 30:271–281. https://doi.org/10.1007/s12264-013-1423-y

Cheng S, Wu Q, Yang F, Xu M, Leski M, Chen GQ (2005) Influence of DL-beta-hydroxybutyric acid on cell proliferation and calcium influx. Biomacromolecules 6:593–597. https://doi.org/10.1021/bm049465y

Cheng S, Chen GQ, Leski M, Zou B, Wang Y, Qiong W (2006) The effect of D,L-β-hydroxybutyric acid on cell death and proliferationin L929 cells. Biomaterials 27:3758–3765. https://doi.org/10.1016/j.biomaterials.2006.02.046

Cheng B, Lu H, Bai B, Chen J (2013) d-β-Hydroxybutyrate inhibited the apoptosis of PC12 cells induced by H2O2 via inhibiting oxidative stress. Neurochem Int 62:620–625. https://doi.org/10.1016/j.neuint.2012.09.011

Chmiel-Perzyńska I, Kloc R, Perzyński A, Rudzki S, Urbańska EM (2011) Novel aspect of ketone action: b-Hydroxybutyrate increases brain synthesis of kynurenic acid in vitro. Neurotox Res 20:40–50. https://doi.org/10.1007/s12640-010-9220-0

Choi D (1992) Excitotoxic cell death. J Neurobiol 23:1261–1276. https://doi.org/10.1002/neu.480230915

Clarke K, Tchabanenko K, Pawlosky R, Carter E, King MT, Musa-Veloso K, Ho M, Roberts A, Robertson J, Vanltallie TB, Veech RL (2012a) Kinetics, safety and tolerability of (R)-3-hydroxybutyl (R)-3-hydroxybutyrate in healthy adult subjects. Regul Toxicol Pharmacol 63:401–408. https://doi.org/10.1016/j.yrtph.2012.04.008

Clarke K, Tchabanenko K, Pawlosky R, Carter E, Knight NS, Murray AJ, Cochlin LE, King MT, Wong AW, Roberts A, Robertson J, Veech RL (2012b) Oral 28-day and developmental toxicity studies of (R)-3-hydroxybutyl (R)-3 hydroxybutyrate. Regul Toxicol Pharmacol 63:196–208. https://doi.org/10.1016/j.yrtph.2012.04.001

Colell A, Fernandez A, Fernandez-Checa JC (2009) Mitochondria, cholesterol and amyloid beta peptide: a dangerous trio in Alzheimer disease. J Bioenerg Biomembr 41:417–423. https://doi.org/10.1007/s10863-009-9242-6

Copeland JM, Cho J, Lo T Jr, Hur JH, Bahadorani S, Arabyan T, Rabie J, Soh J, Walker DW (2009) Extension of Drosophila life span by RNAi of the mitochondrial respiratory chain. Curr Biol 19:1591–1598. https://doi.org/10.1016/j.cub.2009.08.016

Cotter DG, Schugar RC, Crawford PA (2013) Ketone body metabolism and cardiovascular disease. Am J Physiol Heart Circ Physiol 304:H1060–H1076. https://doi.org/10.1152/ajpheart.00646.2012

Crabb DP, Smith ND, Zhu H, Nielson KA, Woodard JL, Seidenberg M, Durgerian S, Hazlett KE, Figueroa CM, Kandah CC, Kay CD, Matthews MA (2014) What's on TV? Detecting age-related neurodegenerative eye disease using eye movement scanpaths. Front Aging Neurosci 6. https://doi.org/10.3389/fnagi.2014.00312

de Roo G, Kellerhals MB, Ren Q, Witholt B, Kessler B (2002) Production of chiral R-3-hydroxyalkanoic acids and R-3-hydroxyalkanoic acid methylesters via hydrolytic degradation of polyhydroxyalkanoate synthesized by pseudomonads. Biotechnol Bioeng 77:717–722. https://doi.org/10.1002/bit.10139

Dell'agnello C, Leo S, Agostino A, Szabadkai G, Tiveron C, Zulian A, Prelle A, Roubertoux P, Rizzuto R, Zeviani M (2007) Increased longevity and refractoriness to Ca(2+)-dependent neurodegeneration in Surf1 knockout mice. Hum Mol Genet 16:431–444. https://doi.org/10.1093/hmg/ddl477

Du H, Guo L, Yan S, Sosunov AA, McKhann GM, Yan SS (2010) Early deficits in synaptic mitochondria in an Alzheimer's disease mouse model. Proc Natl Acad Sci U S A 107:18670–18675. https://doi.org/10.1073/pnas.1006586107

Duan W, Mattson MP (1999) Dietary restriction and 2-deoxyglucose administration improve behavioral outcome and reduce degeneration of dopaminergic neurons in models of Parkinson's disease. J Neurosci Res 57:195–206. https://doi.org/10.1002/(SICI)1097-4547(19990715)57:2<195::AID-JNR5>3.0.CO;2-P

Duman RS, Monteggia LM (2006) A neurotrophic model for stress-related mood disorders. Biol Psychiatry 59:1116–1127. https://doi.org/10.1016/j.biopsych.2006.02.013

Dunnett SB, Bjorklund A (1999) Prospects for new restorative and neuroprotective treatments in Parkinson's disease. Nature (London) 399:A32–A39

Edwards C, Canfield J, Copes N, Rehan M, Lipps D, Bradshaw PC (2014) D-beta-hydroxybutyrate extends lifespan in *C. elegans*. Aging (Albany NY) 6:621–644. https://doi.org/10.18632/aging.100683

Escapa IF, Morales V, Martino VP, Pollet E, Avérous L, García JL, Prieto MA (2011) Disruption of β-oxidation pathway in *Pseudomonas putida* KT2442 to produce new functionalized PHAs with thioester groups. Appl Microbiol Biotechnol 89:1583–1598. https://doi.org/10.1007/s00253-011-3099-4

Frazzitta G, Maestri R, Ghilardi MF, Riboldazzi G, Perini M, Bertotti G, Boveri N, Buttini S, Lombino FL, Uccellini D, Turla M, Pezzoli G, Comi C (2014) Intensive rehabilitation increases BDNF serum levels in parkinsonian patients: a randomized study. Neurorehabil Neural Repair 28:163–168. https://doi.org/10.1177/1545968313508474

Fukao T, Lopaschuk GD, Mitchell GA (2004) Pathways and control of ketone body metabolism: on the fringe of lipid biochemistry. Prostaglandins Leukot Essent Fat Acids 70:243–251. https://doi.org/10.1016/j.plefa.2003.11.001

Gasior M, Rogawski MA, Hartman AL (2006) Neuroprotective and disease-modifying effects of the ketogenic diet. Behav Pharmacol 17:431–439

Gerngross TU (1999) Can biotechnology move us toward a sustainable society? Nat Biotechnol 17:541–542. https://doi.org/10.1038/9843

Gilbert DL, Pyzik PL, Freeman JM (2000) The ketogenic diet: seizure control correlates better with serum b-hydroxybutyrate than with urine ketones. J Child Neurol 15:787–790. https://doi.org/10.1177/088307380001501203

Go KG, Prenen GHM, Korf J (1988) Protective effect of fasting upon cerebral hypoxic-ischemic injury. Metab Brain Dis 3:257–263

Greer EL, Brunet A (2009) Different dietary restriction regimens extend lifespan by both independent and overlapping genetic pathways in *C. elegans*. Aging Cell 8:113–127. https://doi.org/10.1111/j.1474-9726.2009.00459.x

Gross AL, Jones RN, Habtemariam DA, Fong TG, Tommet D, Quach L, Schmitt E, Yap L, Inouye SK (2012) Delirium and long-term cognitive trajectory among persons with dementia. Arch Intern Med 172:1324–1331. https://doi.org/10.1001/archinternmed.2012.3203

Guidetti P, Okuno E, Schwarcz R (1997) Characterization of rat brain kynurenine aminotransferases I and II. J Neurosci Res 50:457–465. https://doi.org/10.1002/(SICI)1097-4547(19971101)50:3<457::AID-JNR12>3.0.CO;2-3

Guidetti P, Amori L, Sapko MT, Okuno E, Schwarcz R (2007) Mitochondrial aspartate aminotransferase: a third kynurenateproducing enzyme in the mammalian brain. Neurochemistry 102:103–111. https://doi.org/10.1111/j.1471-4159.2007.04556.x

Halevas E, Nday CM, Salifoglou A (2016) Hybrid catechin silica nanoparticle influence on Cu(II) toxicity and morphological lesions in primary neuronal cells. J Inorg Biochem 163:240–249. https://doi.org/10.1016/j.jinorgbio.2016.04.017

Hammami MM (1997) Book review: diabetes mellitus: a fundamental and clinical text. Ann Saudi Med 17:264

Hansen M, Hsu AL, Dillin A, Kenyon C (2005) New genes tied to endocrine, metabolic, and dietary regulation of lifespan from a *Caenorhabditis elegans* genomic RNAi screen. PLoS Genet 1:119–128. https://doi.org/10.1371/journal.pgen.0010017

Harman D (2009) Origin and evolution of the free radical theory of aging: a brief personal history, 1954–2009. Biogerontology 10:773–781. https://doi.org/10.1007/s10522-009-9234-2

Hazer DB, Kılıçay E, Hazer B (2012) Poly(3-hydroxyalkanoate)s: Diversification and biomedical applications: a state of the art review. Mater Sci Eng C32:637–647. https://doi.org/10.1016/j.msec.2012.01.021

Heneka MT, Kummer MP, Stutz A, Delekate A, Schwartz S, Saecker A, Griep A, Axt D, Remus A, Tzeng T, Gelpi E, Halle A, Korte M, Latz E, Golenbock D (2013) NLRP3 is activated in Alzheimer's disease and contributes to pathology in APP/PS1 mice. Nature 493:674–678. https://doi.org/10.1038/nature11729

Hilmas C, Pereira EF, Alkondon M, Rassoulpour A, Schwarcz R, Albuquerque EX (2001) The brain metabolite kynurenic acid inhibits alpha7 nicotinic receptor activity and increase non-alph 7 nicotinic receptor expression; physiopathological implications. J Neurosci 21:7463–7473

Hiraide A, Katayama M, Sugimoto H, Yoshioka T, Sugimoto T (1991) Effect of 3-hydroxybutyrate on posttraumatic metabolism in man. Surgery 109:176–181

Honjo K, Black SE, Verhoeff NP (2012) Alzheimer's disease, cerebrovascular disease, and the β-amyloid cascade. Can J Neurol Sci 39:712–728

Huang WJ, Zhang X, Chen WW (2016) Role of oxidative stress in Alzheimer's disease. Biomed Rep 4:519–522. https://doi.org/10.3892/br.2016.630

Ibayashi S, Fujishima M, Sadoshima S, Yoshida F, Shiokawa O, Ogata J, Omae T (1986) Cerebral blood flow and tissue metabolism in experimental cerebral ischemia of spontaneously hypertensive rats with hyper-, normo-, and hypoglycemia. Stroke 17:261–266. https://doi.org/10.1161/01.STR.17.2.261

Ito K, Uchida Y, Ohtsuki S, Aizawa S, Kawakami H, Katsukura Y, Kamiie J, Terasaki T (2011) Quantitative membrane protein expression at the blood-brain barrier of adult and younger cynomolgus monkeys. J Pharm Sci 100:3939–3950. https://doi.org/10.1002/jps.22487

Izumi Y, Ishii K, Katsuki H, Benz AM, Zorumski CF (1998) Beta-hydroxybutyrate fuels synaptic function during development. Histological and physiological evidence in rat hippocampal slices. J Clin Invest 101:1121–1132. https://doi.org/10.1172/JCI1009

Jendrossek D (2007) Peculiarities of PHA granules preparation and PHA depolymerase activity determination. Appl Microbiol Biotechnol 74:1186–1196. https://doi.org/10.1007/s00253-007-0860-9

Jendrossek D, Handrick R (2002) Microbial degradation of polyhydroxyalkanoates. Annu Rev Microbiol 56:403–432. https://doi.org/10.1146/annurev.micro.56.012302.160838

Jendrossek D, Schirmer A, Schlegel HG (1996) Biodegradation of polyhydroxyalkanoic acids. Appl Microbiol Biotechnol 46:451–463

Jorm AF, Dear KB, Burgess NM (2005) Projections of future numbers of dementia cases in Australia with and without prevention. Aust N Z J Psychiatry 39:959–963. https://doi.org/10.1080/j.1440-1614.2005.01713.x

Jotheeswaran AT, Williams JD, Prince MJ (2010) The predictive validity of the 10/66 dementia diagnosis in Chennai, India: a 3-year follow-up study of cases identified at baseline. Alzheimer Dis Assoc Dis 24:296–302. https://doi.org/10.1097/WAD.0b013e3181d5e540

Jung HA, Min BS, Yokozawa T, Lee JH, Kim YS, Choi JS (2009) Anti-Alzheimer and antioxidant activities of *Coptidis Rhizoma* alkaloids. Biol Pharm Bull 32:1433–1438. https://doi.org/10.1248/bpb.32.1433

Kadouri D, Jurkevitch E, Okon Y, Castro-Sowinski S (2005) Ecological and agricultural significance of bacterial polyhydroxyalkanoates. Crit Rev Microbiol 31:55–67. https://doi.org/10.1080/10408410590899228

Kalia VC, Prakash J, Koul S (2016) Biorefinery for glycerol rich biodiesel industry waste. Indian J Microbiol 56:113–125. https://doi.org/10.1007/s12088-016-0583-7

Kashiwaya Y, Takeshima T, Mori N, Nakashima K, Clarke K, Veech RL (2000) D-beta-hydroxybutyrate protects neurons in models of Alzheimer's and Parkinson's disease. Proc Natl Acad Sci U S A 97:5440–5444. https://doi.org/10.1073/pnas.97.10.5440

Katayama M, Hiraide A, Sugimoto H, Yoshioka T, Sugimoto T (1994) Effect of ketone bodies on hyperglycemia and lactic acidemia in hemorrhagic stress. JPEN J Parenter Enteral Nutr 18:442–446. https://doi.org/10.1177/0148607194018005442

Ke Y, Zhang XY, Ramakrishna S, He LM, Wu G (2017) Reactive blends based on polyhydroxyalkanoates: preparation and biomedical application. Mater Sci Eng C Mater Biol Appl 70:1107–1119. https://doi.org/10.1016/j.msec.2016.03.114

Kesl SL, Poff AM, Ward NP, Fiorelli TN, Ari C, Van Putten AJ, Sherwood JW, Arnold P, D'Agostino DP (2016) Effects of exogenous ketone supplementation on blood ketone, glucose, triglyceride, and lipoprotein levels in Sprague-Dawley rats. Nutr Metab (Lond) 13:9. https://doi.org/10.1186/s12986-016-0069-y

Khanna S, Srivastava AK (2005) Recent advances in microbial polyhydroxyalkanoates. Process Biochem 40:607–619. https://doi.org/10.1016/j.procbio.2004.01.053

Kim DY, Kim HW, Chung MG, Rhee YH (2007) Biosynthesis, modification, and biodegradation of bacterial medium-chain-length polyhydroxyalkanoates. J Microbiol 45:87–97

Kirsch JR, D'Alecy LG (1984) Hypoxia induced preferential ketone utilization by Rat Brain slices. Stroke 15:319–323. https://doi.org/10.1161/01.STR.15.2.319

Koller M, Marsalek L, de Sousa Dias MM, Braunegg G (2017) Producing microbial polyhydroxyalkanoate (PHA) biopolyesters in a sustainable manner. New Biotechnol 37:24–38. https://doi.org/10.1016/j.nbt.2016.05.001

Koppel SJ, Swerdlow RH (2017) Neuroketotherapeutics: a modern review of a century-old therapy. Neurochem Int pii: S0197-0186(17)30227-9. https://doi.org/10.1016/j.neuint.2017.05.019

Kroemer G, Mariño G, Levine B (2010) Autophagy and the integrated stress response. Mol Cell 40:280–293. https://doi.org/10.1016/j.molcel.2010.09.023

Kulbe JR, Levy JMM, Coultrap SJ, Thorburn A, Baye KU (2014) Excitotoxic glutamate insults block autophagic flux in hippocampal neurons. Brain Res 1542:12–19. https://doi.org/10.1016/j.brainres.2013.10.032

Kumar P, Mehariya S, Ray S, Mishra A, Kalia VC (2015a) Biodiesel industry waste: a potential source of bioenergy and biopolymers. Indian J Microbiol 55:1–7. https://doi.org/10.1007/s12088-014-0509-1

Kumar P, Mehariya S, Ray S, Mishra A, Kalia VC (2015b) Biotechnology in aid of biodiesel industry effluent (glycerol): biofuels and bioplastics. In: Kalia VC (ed) Microbial factories. Springer, New Delhi, pp 105–119. https://doi.org/10.1007/978-81-322-2598-0

Kumar P, Ray S, Patel SKS, Lee JK, Kalia VC (2015c) Bioconversion of crude glycerol to polyhydroxyalkanoate by *Bacillus thuringiensis* under non-limiting nitrogen conditions. Int J Biol Macromol 78:9–16. https://doi.org/10.1016/j.ijbiomac.2015.03.046

LaManna JC, Salem N, Puchowicz M, Erokwu B, Koppaka S, Flask C, Lee Z (2009) Ketones suppress brain glucose consumption. Adv Exp Med Biol 645:301–306. https://doi.org/10.1007/978-0-387-85998-9_45

Lamkanfi M, Dixit VM (2014) Mechanisms and functions of inflammasomes. Cell 157:1013–1022. https://doi.org/10.1016/j.cell.2014.04.007

Langa KM, Foster NL, Larson EB (2004) Mixed dementia: emerging concepts and therapeutic implications. JAMA 292:2901–2908. https://doi.org/10.1001/jama.292.23.2901

Leino RL, Gerhart DZ, Duelli R, Enerson BE, Drewes LR (2001) Diet-induced ketosis increases monocarboxylate transporter (MCT1) levels in rat brain. Neurochem Int 38:519–527

Liang WS, Reiman EM, Valla J, Dunckley T, Beach TG, Grover A, Niedzielko TL, Schneider LE, Mastroeni D, Caselli R, Kukull W, Morris JC, Hulette CM, Schmechel D, Rogers J, Stephan DA (2008) Alzheimer's disease is associated with reduced expression of energy metabolism genes in posterior cingulate neurons. Proc Natl Acad Sci U S A 105:4441–4446. https://doi.org/10.1073/pnas.0709259105

Lim S, Chesser AS, Grima JC, Rappold PM, Blum D, Przedborski S, Tieu K (2011) D-β-hydroxybutyrate is protective in mouse models of Huntington's disease. PLoS One 6:e24620. https://doi.org/10.1371/journal.pone.0024620

Liu Z, Ciocea A, Devireddy L (2014) Endogenous siderophore 2,5-dihydroxybenzoic acid deficiency promotes anemia and splenic iron overload in mice. Mol Cell Biol 34:2533–2546. https://doi.org/10.1128/MCB.00231-14

Liu Z, Zhou T, Ziegler AC, Dimitrion P, Zuo L (2017) Oxidative stress in neurodegenerative diseases: from molecular mechanisms to clinical applications. Oxidative Med Cell Longev 2017:2525967. https://doi.org/10.1155/2017/2525967

Luccini E, Musante V, Neri E, Raiteri M, Pittaluga A (2007) N-methyl-D-aspartate autoreceptors respond to low and high agonist concentrations by facilitating, respectively, exocytosis and carrier-mediated release of glutamate in rat hippocampus. J Neurosci Res 85:3657–3665. https://doi.org/10.1002/jnr.21446

Magdouli S, Brar SK, Blais JF, Tyagi RD (2015) How to direct the fatty acid biosynthesis towards polyhydroxyalkanoates production? Biomass Bioenergy 74:268–279. https://doi.org/10.1016/j.biombioe.2014.12.017

Malm T, Bowald S, Bylock A, Saldeen T, Busch C (1992) Regeneration of pericardial tissue on absorbable polymer patches implanted into the pericardial sac. An immunohistochemical, ultrastructural and biochemical study in the sheep. Scand J Thorac Cardiovasc Surg 26:15–21. https://doi.org/10.3109/14017439209099048

Marais L, Stein DJ, Daniels WM (2009) Exercise increases BDNF levels in the striatum and decreases depressive-like behavior in chronically stressed rats. Metab Brain Dis 24:587–597. https://doi.org/10.1007/s11011-009-9157-2

Martinez V, Dinjaski N, De Eugenio LI, De la Pena F, Prieto MA (2014) Cell system engineering to produce extracellular polyhydroxyalkanoate depolymerase with targeted applications. Int J Biol Macromol 71:28–33. https://doi.org/10.1016/j.ijbiomac.2014.04.013

Martinon F, Mayor A, Tschopp J (2009) The inflammasomes: guardians of the body. Annu Rev Immunol 27:229–265. https://doi.org/10.1146/annurev.immunol.021908.132715

Martinowich K, Manji H, Lu B (2007) New insights into BDNF function in depression and anxiety. Nat Neurosci 10:1089–1093. https://doi.org/10.1038/nn1971

Masters SL, Dunne A, Subramanian SL, Hull RL, Tannahill GM, Sharp FA, Becker C, Franchi L, Yoshihara E, Chen Z, Mullooly N, Mielke LA, Harris J, Coll RC, Mills KH, Mok KH, Newsholme P, Nuñez G, Yodoi J, Kahn SE, Lavelle EC, O'Neill LA (2010) Activation of the NLRP3 inflammasome by islet amyloid polypeptide provides a mechanism for enhanced IL-1β in type 2 diabetes. Nat Immunol 11:897–904. https://doi.org/10.1038/ni.1935

Masuda R, Monahan J, Kashiwaya Y (2005) D-beta-Hydroxybutyrate is neuroprotective against hypoxia in serum-free hippocampal primary cultures. J Neurosci Res 80:501–509

Mattson MP (2008) Glutamate and neurotrophic factors in neuronal plasticity and disease. Ann N Y Acad Sci 1144:97–112. https://doi.org/10.1196/annals.1418.005

Mejía-Tober J, Montiel T, Massieu L (2006) D-β-hydroxybutyrate prevents glutamate-mediated lipoperoxidation and neuronal damage elicited during glycolysis inhibition in vivo. Neurochem Res 31:1399–1408. https://doi.org/10.1007/s11064-006-9189-5

Mercken EM, Carboneau BA, Krzysik-Walker SM, de Cabo R (2012) Of mice and men: the benefits of caloric restriction, exercise, and mimetics. Ageing Res Rev 11:390–398. https://doi.org/10.1016/j.arr.2011.11.005

Mizushima N, Levine B, Cuervo AM, Klionsky D (2008) Autophagy fights disease through cellular self-digestion. Nature 451:1069–1075. https://doi.org/10.1038/nature06639

Montine TJ, Amarnath V, Martin ME, Strittmatter WJ, Graham DG (1996) E-4-hydroxy-2-nonenal is cytotoxic and cross-links cytoskeletal proteins in P19 neuroglial cultures. Am J Pathol 148:89–93

Moroni F, Russi P, Lombardi G, Beni M, Carlà V (1988) Presence of kynurenic acid in the mammalian brain. J Neurochem 51:177–180. https://doi.org/10.1111/j.1471-4159.1988.tb04852.x

Nakamura S, Shibuya M, Saito Y, Nakashima H, Saito F, Higuchi A, Tsubota K (2003) Protective effect of D-beta-hydroxybutyrate on corneal epithelia in dry eye conditions through suppression of apoptosis. Invest Ophthalmol Vis Sci 44:4682–4688. https://doi.org/10.1167/iovs.03-0198

Natale JE, Stante SM, D'Alecy LG (1990) Elevated brain lactate accumulation and increased neurologic deficit are associated with modest hyperglycemia in global ischemia. Resuscitation 19:271–289. https://doi.org/10.1016/0300-9572(90)90107-P

Nday CM, Halevas E, Jackson GE, Salifoglou A (2015) Quercetin encapsulation in modified silica nanoparticles: potential use against Cu(II)-induced oxidative stress in neurodegeneration. J Inorg Biochem 145:51–64. https://doi.org/10.1016/j.jinorgbio.2015.01.001

Németh H, Toldi J, Vécsei L (2006) Kynurenines, Parkinson's disease and other neurodegenerative disorders: preclinical and clinical studies. J Neural Transm Suppl 70:285–304

Newman JC, Verdin E (2014) Ketone bodies as signaling metabolites. Trends Endocrinol Metab 25:42–52. https://doi.org/10.1016/j.tem.2013.09.002

Newport MT, VanItallie TB, Kashiwaya Y, King MT, Veech RL (2015) A new way to produce hyperketonemia: use of ketone ester in a case of Alzheimer's disease. Alzheimers Dement 11:99–103. https://doi.org/10.1016/j.jalz.2014.01.006

O'Connor T, Sadleir KR, Maus E, Velliquette RA, Zhao J, Cole SL, Eimer WA, Hitt B, Bembinster LA, Lammich S, Lichtenthaler SF, Hébert SS, De Strooper B, Haass C, Bennett DA, Vassar R (2008) Phosphorylation of the translation initiation factor eIF2 alpha increases BACE1 levels and promotes amyloidogenesis. Neuron 60:988–1009. https://doi.org/10.1016/j.neuron.2008.10.047

Ogata M, Hino S, Saito A, Morikawa K, Kondo S, Kanemoto S, Murakami T, Taniguchi M, Tanii I, Yoshinaga K, Shiosaka S, Hammarback JA, Urano F, Imaizumi K (2006) Autophagy is activated for cell survival after endoplasmic reticulum stress. Mol Cell Biol 26:9220–9231. https://doi.org/10.1128/MCB.01453-06

Olney JW (1971) Glutamate-induced neuronal necrosis in the infant mouse hypothalamus. An electron microscopic study. J Neuropathol Exp Neurol 30:75–90

Panov A, Orynbayeva Z, Vavilin V, Lyakhovich V (2014) Fatty acids in energy metabolism of the central nervous system. Biomed Res Int 2014:472–459. https://doi.org/10.1155/2014/472459

Pantazaki AA, Tambaka MG, Langlois V, Guerin P, Kyriakidis DA (2003) Polyhydroxyalkanoate (PHA) biosynthesis in *Thermus thermophilus*: purification and biochemical properties of PHA synthase. Mol Cell Biochem 254:173–183

Pantazaki AA, Papaneophytou CP, Pritsa AG, Liakopoulou-Kyriakides M, Kyriakidis DA (2009) Production of polyhydroxyalkanoates from whey by *Thermus thermophilus* HB8. Process Biochem 44:847–853. https://doi.org/10.1016/j.procbio.2009.04.002

Pantazaki AA, Papaneophytou CP, Lambropoulou DA (2011) Simultaneous polyhydroxyalkanoates and rhamnolipids production by *Thermus thermophilus* HB8. AMB Express 1:17. https://doi.org/10.1186/2191-0855-1-17

Papaneophytou CP, Pantazaki AA (2011) A novel affinity chromatographic material for the purification of extracellular polyhydroxybutyrate depolymerases. J Polym Environ 19:876–886. https://doi.org/10.1007/s10924-011-0345-x

Papaneophytou CP, Pantazaki AA, Kyriakidis DA (2009) An extracellular polyhydroxybutyrate depolymerase in *Thermus thermophilus* HB8. Appl Microbiol Biotechnol 83:659–668. https://doi.org/10.1007/s00253-008-1842-2

Papaneophytou CP, Velali EE, Pantazaki AA (2011) Purification and characterization of an extracellular medium-chain length polyhydroxyalkanoate depolymerase from *Thermus thermophilus* HB8. Polym Degrad Stab 96:670–678. https://doi.org/10.1016/j.polymdegradstab.2010.12.015

Pasinetti GM, Zhao Z, Qin W, Ho L, Shrishailam Y, Macgrogan D, Ressmann W, Humala N, Liu X, Romero C, Stetka B, Chen L, Ksiezak-Reding H, Wang J (2007) Caloric intake and Alzheimer's disease. Experimental approaches and therapeutic implications. Interdiscip Top Gerontol 35:159–175. https://doi.org/10.1159/000096561

Patel SKS, Kumar P, Singh S, Lee JK, Kalia VC (2015a) Integrative approach for hydrogen and polyhydroxybutyrate production. In: Kalia VC (ed) Microbial factories: waste treatment. Springer, New Delhi, pp 73–85. https://doi.org/10.1007/978-81-322-2598-0_5

Patel SKS, Kumar P, Singh S, Lee JK, Kalia VC (2015b) Integrative approach to produce hydrogen and polyhydroxybutyrate from biowaste using defined bacterial cultures. Bioresour Technol 176:136–141. https://doi.org/10.1016/j.biortech.2014.11.029

Patel SKS, Lee JK, Kalia VC (2016) Integrative approach for producing hydrogen and polyhydroxyalkanoate from mixed wastes of biological origin. Indian J Microbiol 56:293–300. https://doi.org/10.1007/s12088-016-0595-3

Peng T, Gibula P, Yao KD, Goosen MF (1996) Role of polymers in improving the results of stenting in coronary arteries. Biomaterials 17:685–694. https://doi.org/10.1016/0142-9612(96)86738-X

Plassman BL, Langa KM, Fisher GG, Heeringa SG, Weir DR, Ofstedal MB, Burke JR, Hurd MD, Potter GG, Rodgers WL, Steffens DC, Willis RJ, Wallace RB (2007) Prevalence of dementia in the United States: the aging, demographics, and memory study. Neuroepidemiology 29:125–132. https://doi.org/10.1159/000109998

Premkumar DR, Smith MA, Richey PL, Petersen RB, Castellani R, Kutty RK, Wiggert B, Perry G, Kalaria RN (1995) Induction of heme oxygenase-1 mRNA and protein in neocortex and cerebral vessels in Alzheimer's disease. J Neurochem 65:1399–1402. https://doi.org/10.1046/j.1471-4159.1995.65031399.x

Przedborski S, Kostic V, Giladi N, Eidelberg D (2003) Dopaminergic system in Parkinson's disease. In: Sidhu A, Laruelle M, Vernier P (eds) Dopamine receptors and transporters. Marcel Dekker, New York, pp 363–402

Pulsinelli WA, Levy DE, Sigsbee B, Scherer P, Plum F (1983) Increased damage after ischemic stroke in patients with hyperglycemia with or without established diabetes mellitus. Am J Med 74:540–544. https://doi.org/10.1016/0002-9343(83)91007-0

Raichle ME (1983) The pathophysiology of brain ischemia. Ann Neurol 13:2–10. https://doi.org/10.1002/ana.410130103

Ralser M, Wamelink MM, Struys EA, Joppich C, Krobitsch S, Jakobs C, Lehrach H (2008) A catabolic block does not sufficiently explain how 2-deoxy-D-glucose inhibits cell growth. Proc Natl Acad Sci U S A 105:17807–17811. https://doi.org/10.1073/pnas.0803090105

Ray S, Kalia VC (2016) Microbial co-metabolism and polyhydroxyalkanoate co-polymers. Indian J Microbiol 57:39–47. https://doi.org/10.1007/s12088-016-0622-4

Ray S, Kalia VC (2017) Co-metabolism of substrates by *Bacillus thuringiensis* regulates polyhydroxyalkanoate co-polymer composition. Bioresour Technol 224:743–747. https://doi.org/10.1016/j.biortech.2016.11.089

Regen DM, Callis JT, Sugden MC (1983) Studies of cerebral beta-hydroxybutyrate transport by carotid injection; effects of age, diet and injectant composition. Brain Res 271:289–299

Reger MA, Henderson ST, Hale C, Cholerton B, Baker LD, Watson GS, Hyde K, Chapman D, Craft S (2004) Effects of β-hydroxybutyrate on cognition in memory-impaired adults. Neurobiol Aging 25:311–314. https://doi.org/10.1016/S0197-4580(03)00087-3

Rehncrona S, Rosen I, Siesjo BK (1981) Brain lactic acidosis and ischemic cell damage: 1. Biochemistry and neurophysiology. J Cereb Blood Flow Metabol 1:297–311. https://doi.org/10.1038/jcbfm.1981.34

Ren Q, Ruth K, Thöny-Meyer L, Zinn M (2010) Enatiomerically pure hydroxycarboxylic acids: current approaches and future perspectives. Appl Microbiol Biotechnol 87:41–52. https://doi.org/10.1007/s00253-010-2530-6

Reusch RN (2015) Poly-(R)-3-hydroxybutyrates (PHB) are atherogenic components of lipoprotein Lp(a). Med Hypotheses 85:1041–1043. https://doi.org/10.1016/j.mehy.2015.07.027

Reusch RN, Sparrow AW, Gardiner J (1992) Transport of poly betahydroxybutyrate in human plasma. Biochim Biophys Acta 1123:33–40. https://doi.org/10.1016/0005-2760(92)90168-U

Revett TJ, Baker GB, Jhamandas J, Kar S (2013) Glutamate system, amyloid β peptides and tau protein: functional interrelationships and relevance to Alzheimer disease pathology. J Psychiatry Neurosci 38:6–23. https://doi.org/10.1503/jpn.110190

Riddle MC, Hart J (1982) Hyperglycemia, recognized and unrecognized, as a risk factor for stroke and transient ischemic attacks. Stroke 13:356–359. https://doi.org/10.1161/01.STR.13.3.356

Ross ED, Shah SN, Prodan CI, Monnot M (2006) Changing relative prevalence of Alzheimer disease versus Non-Alzheimer disease dementias: have we underestimated the looming dementia epidemic? Dement Geriatr Cogn Disord 22:273–277. https://doi.org/10.1159/000095127

Russo-Neustadt AA, Beard RC, Huang YM, Cotman CW (2000) Physical activity and antidepressant treatment potentiate the expression of specific brain-derived neurotrophic factor transcripts in the rat hippocampus. Neuroscience 101:305–312. https://doi.org/10.1016/S0306-4522(00)00349-3

Sarkar C, Zhao Z, Aungst S, Sabirzhanov B, Faden AI, Lipinski MM (2014) Impaired autophagy flux is associated with neuronal cell death after traumatic brain injury. Autophagy 10:2208–2222. https://doi.org/10.4161/15548627.2014.981787

Sato K, Kashiwaya Y, Keon CA, Tsuchiya N, King MT, Radda GK, Chance B, Clarke K, Veech RL (1995) Insulin, ketone bodies, and mitochondrial energy transduction. FASEB J 9:651–658

Sayre LM, Zelasko DA, Harris PL, Perry G, Salomon RG, Smith MA (1997) 4-Hydroxynonenal-derived advanced lipid peroxidation end products are increased in Alzheimer's disease. J Neurochem 68:2092–2097. https://doi.org/10.1046/j.1471-4159.1997.68052092.x
Schwarcz R, Pellicciari R (2002) Manipulation of brain kynurenines: glial targets, neuronal effects, and clinical opportunities. J Pharmacol Exp Ther 303:1–10. https://doi.org/10.1124/jpet.102.034439
Seebach D, Beck AK, Breitschuh R, Job K (2003) Direct degradation of the biopolymer poly[(R)-3-hydroxybutyric acid] to (R)-3-hydroxybutanoic acid and its methyl ester. In: Org. Synth. Wiley, Hoboken, p 39–39. https://doi.org/10.1002/0471264180.os071.05
Shimazu T, Hirschey MD, Newman J, He W, Shirakawa K, Le Moan N, Grueter CA, Lim H, Saunders LR, Stevens RD, Newgard CB, Farese RV Jr, de Cabo R, Ulrich S, Akassoglou K, Verdin E (2013) Suppression of oxidative stress by beta-hydroxybutyrate, an endogenous histone deacetylase inhibitor. Science 339:211–214. https://doi.org/10.1126/science.1227166
Siemkowicz E, Hansen AJ (1978) Clinical restitution following cerebral ischemia in hypo-, normo-and hyperglycemic rats. Acta Neurol Scand 58:1–8. https://doi.org/10.1111/j.1600-0404.1978.tb02855.x
Siesjo BK (1988) Acidosis and ischemic brain damage. Neurochem Pathol 9:31–88
Singh M, Kumar P, Ray S, Kalia VC (2015) Challenges and opportunities for the customizing polyhydroxyalkanoates. Indian J Microbiol 55:235–249. https://doi.org/10.1007/s12088-015-0528-6
Sleiman SF, Henry J, Al-Haddad R, El Hayek L, AbouHaidar E, Stringer T, Ulja D, Karuppagounder SS, Holson EB, Ratan RR, Ninan I, Chao MV (2016) Exercise promotes the expression of brain derived neurotrophic factor (BDNF) through the action of the ketone body *β*-hydroxybutyrate. eLife 5:e15092. https://doi.org/10.7554/eLife.15092
Smith CD, Carney JM, Starke-Reed PE (1991) Excess brain protein oxidation and enzyme dysfunction in normal aging and in Alzheimer disease. Proc Natl Acad Sci U S A 88:10540–10543
Smith MA, Kutty RK, Richey PL (1994a) Heme oxygenase-1 is associated with the neurofibrillary pathology of Alzheimer's disease. Am J Pathol 145:42–47
Smith MA, Taneda S, Richey PL (1994b) Advanced Maillard reaction end products are associated with Alzheimer disease pathology. Proc Natl Acad Sci U S A 91:5710–5714
Smith MA, Harris PLR, Sayre LM, Beckman JS, Perry G (1997) Widespread peroxynitrite-mediated damage in Alzheimer's disease. J Neurosci 17:2653–6537
Stone TW, Behan WM, Jones PA, Darlington LG, Smith RA (2001) The role of kynurenines in the production of neuronal death, and the neuroprotective effect of purines. J Alzheimers Dis 3:355–366. https://doi.org/10.3233/JAD-2001-3401
Suzuki M, Suzuki M, Sato K, Dohi S, Sato T, Matsuura A, Hiraide A (2001) Effect of *β*-Hydroxybutyrate, a cerebral function improving agent on cerebral hypoxia, anoxia and ischemia in mice and rats. Jpn J Pharmacol 87:143–150. https://doi.org/10.1254/jjp.87.143
Suzuki M, Suzuki M, Kitamura Y, Mori S, Sato K, Dohi S, Sato T, Matsuura A, Hiraide A (2002) *β*-Hydroxybutyrate, a cerebral function improving agent, protects rat brain against ischemic damage caused by permanent and transietn focal cerebral ischemia. Jpn J Pharmacol 89:36–43. https://doi.org/10.1254/jjp.89.36
Swerdlow RH, Burns JM, Khan SM (2010) The Alzheimer's disease mitochondrial cascade hypothesis. J Alzheimers Dis 20:S265–S279. https://doi.org/10.3233/JAD-2010-100339
Takeda A, Smith MA, Avila J, Nunomura A, Siedlak SL, Zhu X, Perry G, Sayre LM (2000) In Alzheimer's disease, heme oxygenase is coincident with Alz50, an epitope of tau induced by 4-hydroxy-2-nonenal modification. J Neurochem 75:1234–1241. https://doi.org/10.1046/j.1471-4159.2000.0751234.x
Tamagno E, Guglielmotto M, Aragno M, Borghi R, Autelli R, Giliberto L, Muraca G, Danni O, Zhu X, Smith MA, Perry G, Jo DG, Mattson MP, Tabaton M (2008) Oxidative stress activates a positive feedback between the gamma- and beta-secretase cleavages of the beta-amyloid precursor protein. J Neurochem 104:683–695. https://doi.org/10.1111/j.1471-4159.2007.05072.x

Tanida I (2011) Autophagy basics. Microbiol Immunol 55:1–11. https://doi.org/10.1111/j.1348-0421.2010.00271.x

Tieu K, Perier C, Caspersen C, Teismann P, Wu DC, Yan SD, Naini A, Vila M, Jackson-Lewis V, Ramasamy R, Przedborski S (2003) D-β-Hydroxybutyrate rescues mitochondrial respiration and mitigates features of Parkinson disease. J Clin Invest 112:892–901. https://doi.org/10.1172/JCI200318797

Tildon JT, McKenna MC, Stevenson JH (1994) Transport of 3-hydroxybutyrate by cultured rat brain astrocytes. Neurochem Res 19:1237–1242

Turski WA, Gramsbergen JBP, Traitler H, Schwarcz R (1989) Rat brain slices produce and liberate kynurenic acid upon exposure to L-kynurenine. J Neurochem 52:1629–1636. https://doi.org/10.1111/j.1471-4159.1989.tb09218.x

Uchida Y, Ohtsuki S, Katsukura Y, Ikeda C, Suzuki T, Kamiie J, Terasaki T (2011) Quantitative targeted absolute proteomics of human blood-brain barrier transporters and receptors. J Neurochem 117:333–345. https://doi.org/10.1111/j.1471-4159.2011.07208.x

Urbanska E, Ikonomidou C, Sieklucka M, Turski WA (1991) Aminooxyacetic acid produces excitotoxic lesions in the rat striatum. Synapse 9:129–135. https://doi.org/10.1002/syn.890090207

Valappil SP, Misra SK, Boccaccini AR, Roy I (2006) Expert Rev Med Devices 3:853–868. https://doi.org/10.1586/17434440.3.6.853

Van der Auwera I, Wera S, Van Leuven F, Henderson ST (2005) A ketogenic diet reduces amyloid beta 40 and 42 in a mouse model of Alzheimer's disease. Nutr Metab (Lond) 2:28. https://doi.org/10.1186/1743-7075-2-28

Vonsattel JP, Keller C, Cortes Ramirez EP (2011) Huntington's disease-neuropathology. Handb Clin Neurol 100:83–100. https://doi.org/10.1016/B978-0-444-52014-2.00004-5

Wada H, Okada Y, Nabetani M, Nakamura H (1997) The effects of lactate and b-hydroxybutyrate on the energy metabolism and neural activity of hippocampal slices from adult and immature rat. Brain Res Dev Brain Res 101:1–7

Wang SY, Wang Z, Liu MM, Xu Y, Zhang XJ, Chen GQ (2010) Properties of a new gasoline oxygenate blend component: 3-Hydroxybutyrate methyl ester produced from bacterial poly-3-hydroxybutyrate. Biomass Bioenergy 34:1216–1222. https://doi.org/10.1016/j.biombioe.2010.03.020

Wen H, Miao EA, Ting JP (2013) Mechanisms of NOD-like receptor-associated inflammasome activation. Immunity 39:432–441. https://doi.org/10.1016/j.immuni.2013.08.037

Williams DF (2008) On the mechanisms of biocompatibility. Biomaterials 29:2941–2953

Williams DF (2009) On the nature of biomaterials. Biomaterials 30:5897–5909

Xiao XQ, Zhao Y, Chen GQ (2007) The effect of 3-hydroxybutyrate and its derivatives on the growth of glial cells. Biomaterials 28:3608–3616. https://doi.org/10.1016/j.biomaterials.2007.04.046

Yao J, Chen S, Mao Z, Cadenas E, Brinton RD (2011) 2-Deoxy-D-glucose treatment induces ketogenesis, sustains mitochondrial function, and reduces pathology in female mouse model of Alzheimer's disease. PLoS One 6:e21788. https://doi.org/10.1371/journal.pone.0021788

Youm YH, Grant RW, McCabe LR, Albarado DC, Nguyen KY, Ravussin A, Pistell P, Newman S, Carter R, Laque A, Münzberg H, Rosen CJ, Ingram DK, Salbaum JM, Dixit VD (2013) Canonical Nlrp3 inflammasome links systemic low-grade inflammation to functional decline in aging. Cell Metab 18:519–532. https://doi.org/10.1016/j.cmet.2013.09.010

Youm YH, Nguyen KY, Grant RW, Goldberg EL, Bodogai M, Kim D, D'Agostino D, Planavsky N, Lupfer C, Kanneganti TD, Kang S, Horvath TL, Fahmy TM, Crawford PA, Biragyn A, Alnemri E, Dixit VD (2015) The ketone metabolite β-hydroxybutyrate blocks NLRP3 inflammasome–mediated inflammatory disease. Nat Med 21:263–269. https://doi.org/10.1038/nm.3804

Zawidlak-Węgrzyńska B, Kawalec M, Bosek I, Łuczyk-Juzwa M, Adamus G, Rusin A, Filipczak P, Glowala-Kosińska M, Wolańska K, Krawczyk Z, Kurcok P (2010) Synthesis and antiproliferative properties of ibuprofen–oligo(3-hydroxybutyrate) conjugates. Eur J Med Chem 45:1833–1842. https://doi.org/10.1016/j.ejmech.2010.01.020

Zhang XJ, Luo RC, Wang Z, Deng Y, Chen GQ (2009) Applications of (*R*)-3-hydroxyalkanoatemethyl esters derived from microbial polyhydroxyalkanoates as novel biofuels. Biomacromolecules 10:707–711. https://doi.org/10.1021/bm801424e

Zhang J, Cao Q, Li S, Lu X, Zhao Y, Guan JS, Chen JC, Wu Q, Chen GQ (2013) 3-Hydroxybutyrate methyl ester as a potential drug against Alzheimer's disease via mitochondria protection mechanism. Biomaterials 34:7552–7562. https://doi.org/10.1016/j.biomaterials.2013.06.043

Zhu X, Perry G, Smith MA, Wang X (2013) Abnormal mitochondrial dynamics in the pathogenesis of Alzheimer's disease. J Alzheimers Dis 33:S253–S262. https://doi.org/10.3233/JAD-2012-129005

Zinn M, Hany R (2005) Tailored material properties of polyhydroxyalkanoates through biosynthesis and chemical modification. Adv Eng Mater 7:408–411. https://doi.org/10.1002/adem.200500053

Zou Z, Sasaguri S, Rajesh KG, Suzuki R (2002) dl-3-Hydroxybutyrate administration prevents myocardial damage after coronary occlusion in rat hearts. Am J Physiol Heart Circ Physiol 283:H 1968–H 1974. https://doi.org/10.1152/ajpheart.00250.2002

Zou XH, Li HM, Wang S, Leski M, Yao YC, Yang XD, Huang QJ, Chen GQ (2009) The effect of 3-hydroxybutyrate methyl ester on learning and memory in mice. Biomaterials 30:1532–1541. https://doi.org/10.1016/j.biomaterials.2008.12.012

Biotechnological Application of Polyhydroxyalkanoates and Their Composites as Anti-microbials Agents

8

Sanjay K. S. Patel, Kumar Sandeep, Mamtesh Singh, Gajendra P. Singh, Jung-Kul Lee, Shashi K. Bhatia, and Vipin C. Kalia

Abstract

Bio-polymers are widely synthesized by living organisms to assist several biological functions. Bacterial biopolymers are produced as a polyhydroxyalkanoates (PHA) in the form intracellular cellular storage material under the nutrient limiting or stress conditions in the excess of carbon source and recognized as a suitable alternative to petrochemical-based synthetic plastic due to their biodegradable and eco-friendly nature. Primarily, the synthesis of PHA is considered as costly process due to the high price of feed and its poor grade of thermal and mechanical properties of homopolymers such as polyhydroxybutyrate. The modification of PHA through chemical or biological synthesis and as composite materials through the incorporation of nanoparticles or other substrate and materials seems viable approaches to improve the properties of PHA for the novel application in the area of biomedical applications including anti-microbial agents. In this book chapter, we are evaluating the potential application of PHA, derivatives of PHA and modified with composite materials as anti-microbial agents.

S. K. S. Patel (✉) · J.-K. Lee · V. C. Kalia
Department of Chemical Engineering, Konkuk University, Seoul, Republic of Korea

K. Sandeep
Dr. B. R. Ambedkar Institute Rotary Cancer Hospital, All India Institute of Medical Sciences, New Delhi, India

M. Singh
Department of Zoology, Gargi College, University of Delhi, Delhi, India

G. P. Singh
Mathematical Sciences and Interdisciplinary Research Lab (MathSciIntR-Lab), School of Computational and Integrative Sciences, Jawaharlal Nehru University, New Delhi, India

S. K. Bhatia
Department of Biological Engineering, College of Engineering, Konkuk University, Seoul, Republic of Korea

V. C. Kalia (ed.), *Biotechnological Applications of Polyhydroxyalkanoates*,
https://doi.org/10.1007/978-981-13-3759-8_8

Keywords
Antimicrobials · Biocontrol · Biodegration · Biopolymers · Composite materials · Polyhydroxyalkanoates

8.1 Introduction

The significant development in area of energy, nanotechnology and polymers science, including biological processes has received worldwide attention (Kalia et al. 2003, 2016; Kalia and Purohit 2008; Patel et al. 2010, 2012a, 2014a, 2016a, b, c, d, e, 2017a, b, c, 2018a, b, c, d; Patel and Kalia 2013; Kumar et al. 2013, 2015a, b, c; Bhatia et al. 2016, 2018; Mardina et al. 2016; Kiran et al. 2017). Synthetic polymers such as plastics are produced more than 200 million tonnes, annually. These substances have recognized as major solid waste as an environmental pollutants due to their extensive uses a large scale and non-degradable properties. Polyhydroxyalkanoates (PHAs) are a natural polymer and suggested as a biocompatible alternative to the synthetic plastics (Singh et al. 2009, 2015; Patel et al. 2012b; Ray and Kalia 2016, 2017a; Zhang et al. 2018). Currently, PHA production on an industrial scale is still very costly in comparison to petrochemical-based synthetic plastics (Porwal et al. 2008; Singh et al. 2009; Rehm 2010). World-widely, various efforts are running to develop suitable biodegradable polymers for their environmental waste management and biotechnological applications. PHA has been recognized as the best candidates as a biodegradable polymer. Although, numbers of biodegradable plastics have been known, including bacterial polyesters cellulose-derived polymers, starch-based materials, and synthetic polymers. The degradation abilities of degradable polymers are highly dependent on their chemical, physical and/or mechanical properties. Biodegradation of polymeric compounds is occurred through either aerobic or anaerobic processes involving microbes, including bacteria such as *Actinomycetes*, *Alcaligenes*, *Bacillus*, *Pseudomonas*, *Klebsiella*, *Nocardia*, *Streptomyces* and *Thermoactinomycetes*, and fungi such as *Candida*, *Cladosporium*, *Ganoderma*, *Penicillium*, *Phanerochaete*, *Sporotrichum*, *Thermomyces* and *Trametes* (Defoirdt et al. 2009; Singh et al. 2009; Leja and Lewandowicz 2010; Rehm 2010; Ray et al. 2018).

Biological approaches are appeared most promising as it is non-polluting and can exploit biological wastes as substrate for the cost-effective biotransformation, including biopolymers such as PHA over carbohydrates, specially pure sugars (Singh et al. 2009; Patel et al. 2010, 2011, 2015a, b, 2016f, 2017d; Kumar et al. 2016; Prakash et al. 2018a, b; Ray and Kalia 2017b, c; Ray et al. 2018). The synthesis of PHA as co-polymers showed better thermal and mechanical properties than the polyhydroxybutyrate (PHB). Further, the properties of PHA can be tailored with interesting physical features through the selection of suitable producing strains, substrates and cosubstrates. Recently, PHA has shown a broad potential applications as agriculture, biocontrol agents, biofuels, horticulture, packaging and biomedical such as anti-osteoporosis agents, biodegradable implants, drugs carriers, medical devices, memory enhancer and tissue engineering (Defoirdt et al. 2009; Radivojevic et al. 2016; Ray and Kalia 2017a; Zhang et al. 2018). Also, the various strategies have been suggested to improve the properties of PHA through their derivatization or modification as composites materials by chemical and biological synthesis, solvent-casting, electrospinning, chemical cross-linking, physical mixing, dip-coating, melting and mixing processes for the suitable biotechnological and biomedical applications. In this book chapter, the anti-microbial properties of PHA and their derivatives or composite with other materials are discussed against bacteria, fungi, and virus for their potential applications.

8.2 Polyhydroxyalkanoate (PHA) Production

PHAs are a kind of aliphatic polystres and produced by many prokaryotic organisms as reserve material up to 90% of their dry cell mass (DCM) (Singh et al. 2009, 2013; Porwal et al. 2008; Kumar et al. 2013, 2014; Bhatia et al. 2015). Primarily, the synthesis of PHAs has occurred under physiological stress conditions, including excess of carbon (C) source in the environment and limitations of essential nutrients such as magnesium, potassium, nitrogen, phosphorus, sulphur or iron during their growth period (Kumar et al. 2009; Singh et al. 2009). Under stress environmental conditions, it also serves as a food source which is mobilized by PHA depolymerase. PHAs are synthesized either in homo- or co-polymers forms. PHB is the most studied homo-polymer of PHA and produced by a wide range of bacteria. For PHB biosynthesis, three distinct enzymes are involved: (i) The first step is catalyzed by the enzyme β-ketothiolase (encoded by phaA, 2 acetyl-CoA molecules combine to form acetoacetyl- CoA), (ii) Acetoacetyl-CoA is then reduced by another enzyme, NADPHdependent acetoacetyl-CoA dehydrogenase (encoded by phaB) and (iii) in the final reaction, the monomers [(R)-3-hydroxybutyryl-CoA] are polymerized into PHB by PHB synthase (encoded by phaC) (Reddy et al. 2003; Kalia et al. 2007; Singh et al. 2009; Rehm 2010; Kumar et al. 2013; Ray and Kalia 2016). The variability of PHB produced by different organisms is regulated by the type of phaC gene present in a particular organism. In addition, some PHA producers use secondary pathways: (i) methylmalonyl CoA pathway, (ii) the de novo fatty acid synthetic pathway and (iii) via a five-step metabolic pathway assisted by two stereospecific 2-enoyl-CoA hydratases prior to polymerization (Kumar et al. 2013; Ray and Kalia 2016).

The large-scale production of PHA is limited by the high cost of feed. High crystalline nature and very low strength of PHB has resulted in basic additional drawbacks for applications (Singh et al. 2009; Rehm 2010; Kumar et al. 2013). Microorganisms including, *Actinobacillus, Aeromonas, Alcaligenes*, *Azotobactor, Bacillus*, *Brevibacterium*, *Clostridium*, *Comamonas, Cupriavidus, Escherichia, Holoferax, Klebsiella, Lactobacillus, Pseudomonas*, *Ralstonia, Rhizobium, Rhodococcus*, *Streptococcus* and *Streptomyces* have been demonstrated for PHA production using sugars and biowaste materials (Porwal et al. 2008; Singh et al. 2009, 2015; Kumar et al. 2013; Ray and Kalia 2016). Microbial PHAs classified into two major sub-groups based on the C chain length of their repeating monomeric units as short chain length (scl) PHAs with a 3–5 C atoms and medium chain length (mcl) PHAs with a 6–18 C atoms (Radivojevic et al. 2016). The various approaches involving a suitable complementation of organism and biowaste(s) as feed are required for producing PHAs with different properties such as high molecular weight and compositions for largescale biotechnological applications. In the comparison of homopolymers (PHB), copolymers have better film forming and mechanical properties quite similar to low-density polyethylene. These features improve their strength and processability. Thus, efforts are needed towards improving product quality (physical properties) and efficiency for the biotechnological applications of biopolymers, including biomedical (Defoirdt et al. 2009; Singh et al. 2009, 2015; Ray and Kalia 2017a; Zhang et al. 2018).

8.3 PHA as Anti-microbial Agents

Various protective approaches have been established to control microbial diseases through uses of anti-microbial peptides, immune-stimulants, probionts or quorum sensing-interference (Defoirdt et al. 2009; Kalia 2013, 2014; Kalia and Purohit 2011; Kumar et al. 2015d; Koul et al. 2016; Kalia et al. 2017, 2019). Recently, due to the potential risk of antibiotic resistance development through their excessive uses, there is growing awareness towards antibiotics application such as animal production and aquaculture (Allen et al. 2012). Therefore, alternative approaches to control pathogenic microbes has been suggested as short-chain fatty acids (SCFAs) as anti-microbial agents in animal and aquaculture production (Van Immerseel et al. 2006). Generally, SCFAs, including formate, acetate, propionate, butyrate, and valerate have been known for anti-bacterial activity towards enterobacteria such as *Escherichia* and *Salmonella* (Defoirdt et al. 2009). Here, SCFAs anti-bacterial properties are highly dependent on the physiological conditions of microorganisms and external environment characteristics. The PHAs degradation has occurred through numerous mechanisms, including chemical decomposition and enzymatic hydrolysis or hydrolysis. The extracellular degradation of PHAs is widely demonstrated among bacteria and fungi. It is generally believed that PHB can be chemically hydrolysed in both acidic and alkaline solutions. The microbial biodegradation of PHAs is highly dependent on the physiochemical characteristics, including crystallinity, stereoregularity, composition and surface accessibility. PHAs can be used directly or modify by chemical or biological synthesis, chemical cross-linking or as nanomaterials hybrids for the effective anti-microbial agents (Jendrossek and Handrick 2002; Defoirdt et al. 2009; Zhang et al. 2018). The schematic represetation of PHA as an anti-microbial agent is given in Fig. 8.1.

8.3.1 Microbial Synthesized PHAs as Anti-microbial Agents

The application of microbial synthesized PHAs as anti-microbial agents is presented in Table 8.1. The chiral enantiomer of naturally produced bacterial PHAs

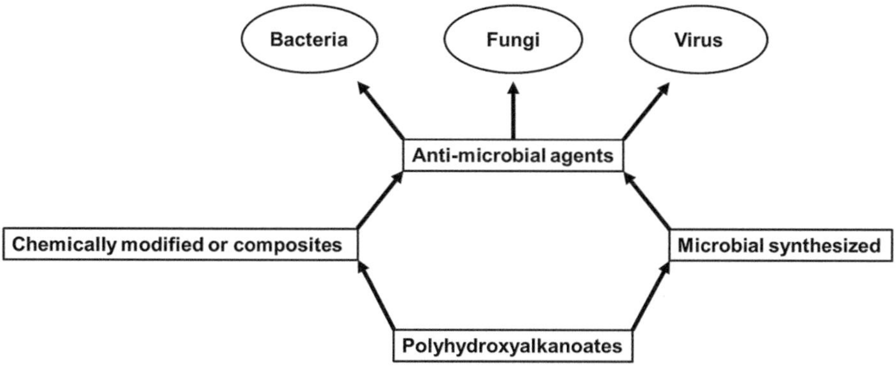

Fig. 8.1 Schematic presentation of polyhydroxyalkanoates as an anti-microbial agents

such as (R)-3-hydroxycarboxylic acids (3HAs) contains both hydroxyl and carboxylic groups, which has lead to their suitable industrial applications in the synthesis of fungicides, vitaminsa and anti-microbial through chemical modifications or introducing new chiral centers. Also, mcl 3HAs has suggested as better bactericidal effects over scl 3HAs (Van Immerseel et al. 2003; Allen et al. 2012). Here, the toxicity of HAs has been associated with its membrane uncoupling capabilities, which may interfere with cytoplasmic membrane and membrane proteins, including electron transport system (adenosine triphosphate production). Allen et al. have synthesized mcl (R)-3HAs synthons by *Streptomyces* sp. JM3 (JN166713) from *Jatropha curcas* oil as C source and used its crude form for effective anti-microbial properties towards various pathogens, including *Escherichia coli* O157:H7, *Listeria monocytogenes* ATCC 7644 and *Salmonella typhimurium* ATCC 14028 (Allen et al. 2012). The crude (R)-3HAs synthons exhibited significant decrease in viability of pathogenic cultures within first 2–4 h of incubation over controls with a minimum inhibitory concentration (MIC) of 25.0, 12.5 and 25.0 mg/ml of culture solution for *E. coli* O157:H7, *L. monocytogenes* ATCC 7644 and *S. typhimurium* ATCC 14028, respectively. Similarly, copolymers poly(3-hydroxybutyrate-co-70%4-hydroxybutyrate) [P(3HB-co-70%4HB)] produced by *Cupriavidus* sp. USMAA1020 exhibited anti-microbial agents against *Staphylococcus aureus* (Hema et al., 2013). *Pseudomonas oleovorans* ATCC 29347 synthesized poly(3-hydroxyoctanoate) (PHO) grafted along with vinylimidazole exhibited significant high up to 90% reduction within 12 h incubation of bacterial (*Escherichia coli* ATCC 25922 as a Gram-positive and *Staphylococcus aureus* ATCC 6538P as a Gram-negative) and fungal (*Candida albicans* ATCC 11651) pathogens at concentration of 2.0% (Chung et al. 2012). In contrast, the PHO as a control did not exhibit significant anti-microbial activity. These results suggested that vinylimidazole grafted PHO can be used effectively as an anti-bacterial and anti-fungal agent.

Vibrio as a Gram-negative bacteria has been recognized as a common dreadful pathogenic organism to the aquaculture system (Kiran et al. 2014). The biofilms formation potential of these pathogens has resulted in a major threat to the living

Table 8.1 Anti-microbial application of microbial synthesized polyhydroxyalkanoates (PHA)

PHA		Antimicrobial properties		
Type/ supplementation	Organism	Organism	Inhibition (%)/duration (h)	References
Bacteria				
3HAs	*Streptomyces* sp. JM3	*Escherichia coli* O157:H7	25.0[a]/24	Allen et al. (2012)
		Listeria monocytogenes ATCC 7644	12.5[a]/24	
		Salmonella typhimurium ATCC 14028	25.0[a]/24	
PHO/ vinylimidazole	*Pseudomonas oleovorans* ATCC 29347	*Escherichia coli* ATCC 25922	90.0/12	Chung et al. (2012)
		Staphylococcus aureus ATCC 6538P	90.0/12	
PHB	*Brevibacterium casei* MSI04	*Vibrio alginolyticus*	92.0/2	Kiran et al. (2014)
		Vibrio harveyi	88.0/2	
		Vibrio fischeri	96.0/2	
		Vibrio parahaemolyticus	92.0/2	
		Vibrio vulnificus	96.0/2	
PHB/eugenol	*Bacillus mycoides* DFC1	*Bacillus cereus* MTCC 1272	70.0/48	Narayanan et al. (2013)
		Escherichia coli NCIM 23058	81.1/48	
		Staphylococcus aureus MTCC 737	73.6/48	
		Salmonella typhimurium MTCC 98	86.8/48	
PHB/vanillin	*B. mycoides* DFC1	*Escherichia coli* MTCC 23058	100[a]/24	Xavier et al. (2015)
		Salmonella typhimurium MTCC 98	80.0[a]/24	
		Shigella flexneri MTCC 1457	80.0[a]/24	
		Staphylococcus aureus MTCC 737	80.0[a]/24	
PHB/melanin	*B. casei* MSI04	*Staphyloccoccus aureus* (MDRSA)	100/48	Kiran et al. (2017)
P(3HB-co-4HB)	*Cupriavidus* sp. USMAA1020	*Staphylococcus aureus*	-[b]/24	Hema et al. (2013)
P(HB-co-HHx)	*Cupriavidus necator* Re2001/ pCB81	*Rhodococcus opacus* PD63	42.0/24	Kehail and Brigham 2018)

(continued)

Table 8.1 (continued)

PHA		Antimicrobial properties		
Type/ supplementation	Organism	Organism	Inhibition (%)/duration (h)	References
Fungi				
PHB/eugenol	*B. mycoides* DFC1	*Aspergillus flavus* MTCC 277	91.6/192	Narayanan et al. (2013)
		Aspergillus niger MTCC 281	93.3/192	
		Penicillium sp. MTCC 4610	93.3/192	
		Rhizopus stolonifera MTCC 162	93.8/192	
PHB/vanillin	*B. mycoides* DFC 1	*A. flavus* MTCC 277	50.0[a]/72	Xavier et al. (2015)
		Aspergillus fumigatus MTCC 15066	50.0[a]/72	
		Aspergillus niger MTCC 478	50.0[a]/72	
		Aspergillus parasiticus MTCC 8189	50.0[a]/72	
		Aspergillus ochraceus MTCC 1877	50.0[a]/72	
		Penicillium clavigerum MTCC 9182	50.0[a]/72	
		Penicillium viridicatum MTCC 4973	50.0[a]/72	
PHA/ vinylimidazole	*P. oleovorans* ATCC 29347	*Candida albicans* ATCC 11651	90.0/12	Chung et al. (2012)

3-HAs 3-hydroxycarboxylic acids, *PHO* poly(3-hydroxyoctanoate), *PHB* polyhydroxybutyrate, *P(3HB-co-4HB)* poly(3-hydroxybutyrate-co-4-hydroxybutyrate), *P(HB-co-HHx)* poly(3-hydroxybutyrate-*co*-3-hydroxyhexanoate)

[a]Minimum inhibitory concentration (μg g^{-1})

[b]Not available

organisms and also exhibited high resistance towards anti-biotic compounds due to the formation of persistent non-multiplying cells. The application of natural PHAs as an anti-biofilms and anti-adhesive against the *Vibrio* in shrimp aquaculture could be an affective approach to control these pathogens. The PHB produced by marine bacterium *Brevibacterium* casei MSI04 exhibited 80–95% anti-adhesive effect towards *Vibrio* spp., including *V. alginolyticus, V. fischeri, V. harveyi, V. parahaemolyticus and V. vulnificus* on the different surfaces of glass and polystyrene (Kiran et al. 2014). On the other hand, melanin-PHB-glycerol synthesised hybrid with melanine and PHB produced by *Pseudomonas* sp. WH001 55 and *B. casei* MSI04 was very effective against biofilm-forming multi-drug resistant *Staphylococcus aureus* (MDRSA) (Kiran et al. 2017). Similarly, lysozyme immobilized on poly(3-hydroxybutyrate-*co*-3-hydroxyhexanoate) [P(HB-*co*-HHx)] produced by *Cupriavidus necator* Re2001/pCB81 showed significant anti-biofilm forming potential against *Rhodococcus opacus* PD63 (Kehail and Brigham 2018).

Narayanan et al. synthesized eugenol incorporated PHB produced by *Bacillus mycoides* DFC1 hybrid for effective anti-microbial applications against food spoilage bacterial and fungal pathogens, including *Bacillus cereus* MTCC 1272, *Escherichia coli* NCIM 23058, *Staphylococcus aureus* MTCC 737, *Salmonella typhimurium* MTCC 98, *Aspergillus flavus* MTCC 277, *Aspergillus niger* MTCC 281, *Penicillium* sp. MTCC 4610 and *Rhizopus stolonifera* MTCC 162 (Narayanan et al. 2013). The anti-microbial eugenol/PHB hybrid containing eugenol 80 μg/g of PHB exhibited growth reduction up to 86.8% and 93.8% of bacterial and fungal pathogens for the incubation of 1 and 7 days, respectively. On the other hand, vanillin incorporation in the PHB synthesized by *B. acillus mycoides* DFC1 showed significant anti-microbial activity against bacterial (*Escherichia coli* MTCC 23058, *S. typhimurium* MTCC 98, *Shigella flexneri* MTCC 1457 and *S. aureus* MTCC 737) and fungal (*A. flavus* MTCC 277, *Aspergillus fumigatus* MTCC 15066, *Aspergillus niger* MTCC 478, *Aspergillus parasiticus* MTCC 8189 and *Aspergillus ochraceus* MTCC 1877, *Penicillium clavigerum* MTCC 9182 and *Penicillium viridicatum* MTCC 4953) at vanillin concentrations of ≥80 and ≥50 μg/g of PHB, respectively (Xavier et al. 2015). These results suggested that eugenol or vanillin incorporated PHB can be effectively used as ani-bactrial and anti-fungal agents for its applications in the food processing industry (Narayanan et al. 2013; Xavier et al. 2015).

8.3.2 Modified PHAs as a Composite Materials

PHAs are mostly (R)-enantiomer of 3-HAs with the optically active center. (R)-3HAs constituted PHA with diverse functional side groups, including unsaturated bonds, ester, cyano, amino, halogen, and phenyl (Radivojevic et al. 2016). These compound can be successfully synthesized using chemical of enzymatic depolymerisation of microbial PHAs with potential applications in chemical and pharmaceutical industries. The anti-microbial properties of composite PHA materials are present in Table 8.2. Radivojevic et al. synthesized 18 (R)-3-hydroxyoctanoic acids derivatives through the chemical modification of PHA produced by *P. putida* GPo1, including halo, unsaturated methyl, and benzyl esters as anti-microbial agents (Radivojevic et al. 2016). These compounds exhibited anti-microbial activity against bacteria (Gram-positive and Gram-negative), including *Escherichia coli* NCTC 9001, *Salmonella typhimurium* NCTC 12023, *Staphylococcus aureus* NCTC 6571 and *Listeria monocytogenes* NCTC 11994, and fungi (*Candida albicans* ATCC 10231 and *Micrococcus gypseum* ATCC24102). Here, the presence of carboxylic groups in the derivatized compounds was essential for the anti-microbial activity and these compounds exhibited higher ant-microbial potential towards fungi over the bacterial strains. Also, (R)-HOAs derivative founded very effective towards opportunistic pathogen *Pseudomonas aeruginosa* PAO1 by inhibiting the

Table 8.2 Anti-microbial application of modified polyhydroxyalkanoates (PHA) as a composite material

Modified PHA or composite	Synthesis method	Antimicrobial agent against	References
Bacteria			
3-HOA derivatives	Chemical synthesis	*Escherichia coli* NCTC 9001	Radivojevic et al. (2016)
		Listeria monocytogenes NCTC 11994	
		Micrococcus gypseum ATCC 24102	
		Pseudomonas aeruginosa PAO1 NCTC 10332	
		Salmonella typhimurium NCTC 12023	
		Staphylococcus aureus NCTC 6571	
Ag/chitosan/P(3HB-co-4HB)	Solvent-casting and chemical cross-linking	*Bacillus subtilis* ATCC 6633	Rennukka et al. (2014)
		Escherichia coli ATCC 11303	
		Pseudomonas aeruginosa USM-AR2	
		Staphylococcus aureus ATCC 12600	
Ag/PHA	Electrospinning	*Listeria monocytogenes* CECT 7467	Castro-Mayorga et al. (2017a)
		Salmonella enterica CECT 4300	
Ag/PHB	Biological synthesis	*L. mnocytogenes* CECT 7467	Castro-Mayorga et al. (2018a)
		S. enterica CECT 4300	
Ag/PHBV	Physical mixing	*L. monocytogenes* CECT7467	Castro-Mayorga et al. (2014)
		S. enterica CECT4300	
Cellulose-ZnO/P(3-HB-co-3-HV)	Electrospinning	*Escherichia coli*	Abdalkarim et al. (2017)
		Staphylococcus aureus	

(continued)

Table 8.2 (continued)

Modified PHA or composite	Synthesis method	Antimicrobial agent against	References
Chitosan-polycaprolactone/P(3-HB-co-3-PHBHHx)	Electrospinning	*Escherichia coli* ATCC 25922	Diez-Pascual and Diez-Vicente (2016)
		Pseudomonas aeruginosa ATCC 27853	
		Staphylococcus epidermidis ATCC 12228	
		Staphylococcus aureus ATCC 12600	
Cloisite 30B/P(3HB-co-4HB)	Solvent-casting	*S. aureus*	Hema et al. (2013)
Claytone APA/P(3HB-co-4HB)	Solvent-casting	*S. aureus*	
Cinnamaldehyde/PHBV + zein	Electrospinning	*Listeria monocytogenes*	Cerqueira et al. (2016)
Cinnamaldehyde + PHBV/PHBV + zein	Electrospinning	*L. monocytogenes*	
Cinnamaldehyde + alginate/PHBV + zein	Electrospinning	*L. monocytogenes*	
CuO/PHBV	Electrospinning	*L. monocytogenes* CECT 7467	Castro-Mayorga et al. (2018b)
		S. enterica CECT 4300	
Ti/PHBV	Dip-coating process	*E. coli*	Rodríguez-Contreras et al. (2017)
		S. aureus	
ZnO/PHBV	Solvent-casting	*E. coli* ATCC 25922	Diez-Pascual and Diez-Vicente (2014)
		S. aureus ATCC 12600	
	Coating	*L. monocytogenes* CECT 7467	Castro-Mayorga et al. (2017b)
	Electrospinning	*L. monocytogenes* CECT 7467	
	Melting and mixing	*L. monocytogenes* CECT 7467	
Fungi			
3-HOA derivatives	Chemical synthesis	*Candida albicans* ATCC 10231	Radivojevic et al. (2016)
Virus			
Ag/PHBV3/PHBV18	Electrospinning	Feline calicivirus	Castro-Mayorga et al. (2017c)
		Murine norovirus	
CuO/PHBV	Electrospinning	Murine norovirus	Castro-Mayorga et al. (2018b)

3-HAs 3-hydroxycarboxylic acids; *P(3HB-co-4HB)* poly(3-hydroxybutyrate-co-4-hydroxybutyrate), *P(3-HB-co-3-HV)* poly(3-hydroxybutyrate-co-3-hydroxybutyrate, *P(3-HB-co-3-PHBHHx)* poly(3-hydroxybutyrate-co-3-hydroxyhexanoate), *PHBV* poly(3-hydroxybutyrate-co-3-hydroxyvalerate), *NPs* nanoparticles quorum sensing regulated the production of pyocyanin (Radivojevic et al. 2016). Hema et al. suggested that co-polymers P(3HB-co-4HB) produced by *Cupriavidus* sp. USMAA1020 showed significant improvement in physiological properties such as mechanical, optical and thermal over P(3HB) polymers through the modification using montmorillonite organoclays (Cloisite 30B and Claytone APA) with the solvent-casting process (Hema et al. 2013). These composite exhibited anti-microbial activity against *S. aureus* as a Gram-positive bacteria only. Whereas these were not founded effective towards *Escherichia coli* as Gram-negative bacteria, this phenomenon might be associated with different cell structure. Co-polymers PHBV modified using cinnamaldehyde and corn zein protein through electrospinning process as a multi-layer system showed high anti-microbial activity against *Listeria monocytogenes* (Cerqueira et al., 2016). Similarly, chitosan grafted polycaprolactone and P(3HB-co-3-HB) blends synthesized using 70:30 (chitosan/polycaprolactone: PHBHHx) by electrospinning process exhibited very high anti-microbial activity against both Gram-positive (*Bacillus subtilis* ATCC 6633 and *Staphylococcus aureus* ATCC 12600) and Gram-negative (*Escherichia coli* ATCC 11303 and *Pseudomonas aeruginosa* USM-AR2) (Diez-Pascual and Diez-Vicente 2016).Also, chemically modified chitosan exhibited significant higher anti-microbial activity than those to control (Patel et al. 2018f).

Nanoparticles (NPs) and their composite materials has been widely used for the biotechnological applications such as biocatalyst and anti-microbial agents due to their useful properties including high surface area, synthesized as controllable sizes and shapes and porous-nature or toxic nature (Patel et al. 2014b, 2018e; Kim et al. 2016; Otari et al. 2016, 2017a, b, 2018; Anwar et al. 2017; Kumar et al. 2018a, b, c). Nanoparticles, including silver (Ag), zinc (Zn), copper (Cu) and titanium (Ti) and PHAs-based composite materials preparation through biological synthesis, electrospinning, dip-coating, solvent-casting and physical mixing have showed their potential application as an anti-microbial and anti-viral agents (Castro-Mayorga et al. 2014, 2017a, 2018a; Diez-Pascual and Diez-Vicente 2014; Rennukka et al. 2014; Rodriguez-Contreras et al. 2017). The anti-microbial potential of NPs or their composite material is primarily associated with their high volume ratio to surface area. Rennukka et al. have demonstrated the fabrication of P(3HB-co-4HB)/chitosan-based Ag nanocomposite material through solvent casting and chemical cross-linking by glutaraldehyde (GA) to improve the hybrid material properties such as mechanical and tensile strength (Rennukka et al. 2014). Here, the anti-bacterial potential of synthesized nano-composite was significantly influenced by Ag NPs content in the hybrid. The hybrid nano-composite materials containing 7% of Ag NPs showed complete inhibition of *Escherichia coli* ATCC 11303 after 24 h of incubation. Whereas, a higher content of 9% of AgNPs are required for the complete inhibition of *Bacillus subtilis* ATCC 6633, *Pseudomonas aeruginosa* USM-AR2 and *Staphylococcus aureus* ATCC 12600 (Rennukka et al. 2014). Similarly, Castro-Mayorga et al. have briefly evaluated the influence of AgNPs and PHA-based composite materials prepared by biological synthesis, physical mixing and electrospinning process as a anti-bacterial agents against food-borne Gram-positive (*Listeria monocytogenes* CECT 7467) and Gram-negative (*Salmonella enterica* CECT 4300) pathogens (Castro-Mayorga et al.

2014, 2017a, 2018a). Biologically produced Ag/PHB hybrid with an average diameter of 76–95 nm by *Cupriavidus necator* strain H16 showed good potential antimicrobial activity towards these pathogens (Castro-Mayorga et al. 2018a). Whereas, a significantly lower concentration of 0.040 wt% Ag NPs in the composite materials prepared using mixed culture-based poly(3-hydroxybutyrate-co-3-hydroxyvalerate) [P(3HB-co-3 HV)] by physical mixing process has demonstrated the bactericide effect against *L. monocytogenes* CECT 7467 and *S. enterica* CECT 4300) pathogens pathogens (Castro-Mayorga et al. 2014). In contrast, AgNPs-PHA composite synthesized by electrospinning coating technique exhibited very high anti-bacterial effects at 20-folds lower concentration 0.002 wt% Ag NPs against (Castro-Mayorga et al. 2017a).

Cellulose, ZnO, and P(3HB-co-4HB)-based composite hybrid exhibited almost 100% anti-bacterial activity against both Gram-negative (*Escherichia coli*) and Gram-positive (*Staphylococcus aureus*) organisms (Abdalkarim et al. 2017). Here, the extent of inhibition was highly dependent on the ZnO NPs content in the composite material with the maximum inhibition at 5.0 wt% cellulose/ZnO/P(3HB-co-4HB). The lower inhibition at 10 and 15 wt% cellulose/ZnO- P(3HB-co-4HB) was associated with the poorer ZnO NPs dispersion in the composite material. Similarly, the ZnO NPs incorporated PHBV18 containing 18 mol% of valerate synthesized by three methods (i) direct melt-mixing, (ii) electrospinning and (iii) coating showed better thermal stability, barrier and optical properties over PHBV18 film and also exhibited effective and prolonged anti-microbial activity against *L. monocytogenes* CECT 7467 pathogen (Castro-Mayorga et al. 2017b). ZnO/PHBV composite synthesized using 4.0 wt% ZnO NPs by solvent-casting process showed anti-bacterial activity against the human pathogen bacteria *E. coli* ATCC 25922 and *S. aureus* ATCC 12600 (Diez-Pascual and Diez-Vicente 2014). Rodriguez-Contreras et al. demonstrated effective application of PHAs and Ti disc composite material as antimicrobial agents towards both Gram-positive (*S. aureus*) and Gram-negative (*E. coli*) bacteria (Rodriguez-Contreras et al. 2017). In addition, Ti/PHBH hybrid was also very effective for drug delivery application to improve the implant surface to avoid pathogenic microorganism inhibition. On the other hand, PHBV (18 mol% valerate) derived from microbial mixed culture and 0.05 wt% CuO NPs based CuO-PHBV composite exhibited significant anti-bacterial and anti-viral against food borne pathogens (*L. monocytogenes* CECT 7467 and *S. enterica* CECT 4300) and murine norovirus with a biodegradation properties within 35 days (Castro-Mayorga et al. 2018b). Similarly, Ag NP coated PHBV as an Ag/PHBV3/PHBV18 composite showed a suitable for anti-viruses including feline calicivirus and murine norovirus (MNV-1 strain) applications (Castro-Mayorga et al. 2017c).

8.4 Conclusions

The microbial production of PHA using biowaste as a feed or through integrative approaches has suggested as an economical process over the costly feed as carbohydrate sugars. PHA has been wide studied for the biomedical applications, including anti-microbial agents due to their biodegradable and biocompatible properties. Since the directly synthesized microbial PHA did not exhibit good potential or partial inhibition for their anti-microbial applications against. Therefore, the PHA modifications through the chemical or biological synthesis by incorporation of materials such as nanoparticles showed their potential application as anti-microbial agents including bacteria as Gram-negative [*Escherichia coli, Pseudomonas aeruginosa Salmonella* spp. (*S. enterica* and *S. typhimurium*), *Shigella flexneri* and *Vibrio* spp. (*V. alginolyticus, V. fischeri, V. harveyi, V. parahaemolyticus* and *V. vulnificus*)], and Gram-positive [*Bacillus subtis, Listeria monocytogenes, Micrococcus gypseum, Rhodococcus opacus, Staphylococcus* spp. [*S. aureus* and *S. epidermidis*], and fungi such as *Aspergillus* spp. (*A. flavus, A. fumigatus, A. parasiticus, A. niger, A. ochraceus*), *Candida albicans, Penicillium* spp. (*P. clavigerum, P. viridicatum* and *Penicillium sp.*] and *Rhizopus* stolonifera. Overall, these studies suggested that biologically synthesized PHA properties such as mechanical, optical, thermal can be significantly improved by various approaches for their broad biotechnological applications, including anti-microbial agents.

8.5 Opinion

A numbers of reports has been suggested that the complementation of microbial producers or different feeds can improve the properties of the synthesized PHA over largely produced PHB as a homopolymer with the high crystalline nature and very low strength (Singh et al. 2009, 2015; Kumar et al. 2013, 2016; Ray et al. 2018). Also, the screening and bioinformatics strategies have been recommended to identify the suitable of novel PHA producers (Kalia et al. 2003; Porwal et al. 2008). Further, genomic analysis, metabolic and enzyme engineering has demonstrated to improve the catalytic properties of biocatalyst for the effective biotransformation processes, including PHA production (Singh et al. 2009; Rahm et al. 2010; Ramachandran et al. 2016; Selvaraj et al. 2016; Bhatia et al. 2015, 2018; Gao et al. 2019; Patel et al. 2018g). The use of biowaste materials or their mixed feed has suggested to enhancing the biopolymer contents as well as their composition to overcome the expensive used of pure sugars (Singh et al. 2009; Kumar et al. 2016). In addition, the integrative approaches of biofuels and PHA has been effective using both pure sugars and biowaste materials (Patel et al. 2012b, 2015a, b, 2016f). The biotechnological application of PHA are highly dependent on their properties such as molecular weight, bioactive groups on their surface, size etc. (Defoirdt et al. 2009; Zhang et al. 2018). Further, the several approaches to modifying properties, including chemical and biological synthesis, solvent-casting, electrospinning, chemical cross-linking, physical mixing, dip-coating, melting and mixing processes for the suitable biotechnological applications have been considered as suitable for anti-microbial agents against bacteria, fungi and virus (Jendrossek and Handrick 2002; Defoirdt et al. 2009; Rennukka et al. 2014; Radivojevic et al. 2016; Castro-Mayorga et al. 2018b; Zhang et al. 2018). The NPs such as Ag, Cu, Ti and Zn has been well known for their anti-microbial properties (Otari et al. 2016;

Castro-Mayorga et al. 2017a, 2018a, b; Rodriguez-Contreras et al. 2017). The feasibility eco-friendly of Ag NPs-based synthesis of PHA hybrid for the anti-microbial applications using *Cupriavidus necator* strain H16 suggested that biologically synthesized hybrid has great potential for broad biotechnological applications due to their biocompatible nature over the toxic NPs. Therefore, more research is recommended to improve the properties of PHA via: (i) potential microbial strain, co-culture or mixed culture to synthesized bioactive PHA molecules and (ii) biosynthesis of different NPs-based PHA hybrids to improve the biotechnological application of modified PHA or hybrids, including anti-microbial agents.

Achnowledgements This research was supported by Basic Science Research Program through the National Research Foundation of Korea (NRF) funded by the Ministry of Education, Science and Technology (NRF-2013R1A1A2012159 and NRF-2013R1A1A2007561). This research was supported by the KU Research Professor program of Konkuk University.

References

Abdalkarim SYH, Yu HY, Wang D, Yao J (2017) Electrospun poly(3-hydroxybutyrate-co-3-hydroxyvalerate)/cellulose reinforced nanofibrous membranes with ZnO nanocrystals for antibacterial wound dressings. Cellulose 24:2925–2938. https://doi.org/10.1007/s10570-017-1303-0

Allen AD, Daley P, Ayorinde FO, Gugssa A, Anderson WA, Eribo BE (2012) Characterization of medium chain length (R)-3-hydroxycarboxylic acids produced by *Streptomyces* sp. JM3 and the evaluation of their antimicrobial properties. World J Microbiol Biotechnol 28:2791–2800. https://doi.org/10.1007/s11274-012-1089-z

Anwar MZ, Kim DJ, Kumar A, Patel SKS, Otari S, Mardina P, Jeong JH, Sohn JH, Kim JH, Park JT, Lee JK (2017) SnO_2 hollow nanotubes: a novel and efficient support matrix for enzyme immobilization. Sci Rep 7:15333. https://doi.org/10.1038/s41598-017-15550-y

Bhatia SK, Shim Y-H, Jeon J-M, Brigham CJ, Kim Y-H, Kim H-J, Seo H-M, Lee J-H, Kim J-H, Yi D-H, Lee YK, Yang Y-H (2015) Starch based polyhydroxybutyrate production in engineered *Escherichia coli*. Bioprocess Biosyst Eng 38:1479–1484. https://doi.org/10.1007/s00449-015-1390-y

Bhatia SK, Bhatia RK, Yang Y-H (2016) Biosynthesis of polyesters and polyamide building blocks using microbial fermentation and biotransformation. Rev Environ Sci Biotechnol 15:639–663. https://doi.org/10.1007/s11157-016-9415-9

Bhatia SK, Kim J-H, Kim M-S, Kim J, Hong JW, Hong YG, Kim H-J, Jeon J-M, Kim S-H, Ahn J, Lee H, Yang Y-H (2018) Production of (3-hydroxybutyrate-co-3-hydroxyhexanoate) copolymer from coffee waste oil using engineered Ralstonia eutropha. Bioprocess Biosyst Eng 41:229–235. https://doi.org/10.1007/s00449-017-1861-4

Castro-Mayorga JL, Martinez-Abad A, Fabra MJ, Olivera C, Reis M, Lagaron JM (2014) Stabilization of antimicrobial silver nanoparticles by apolyhydroxyalkanoate obtained from mixed bacterial culture. Int J Biol Macromol 71:103–110. https://doi.org/10.1016/j.ijbiomac.2014.06.059

Castro-Mayorga JL, Fabra MJ, Pourrahimi AM, Olsson RT, Lagaron JM (2017a) The impact of zinc oxide particle morphology as an antimicrobial and when incorporated in poly(3-hydroxybutyrate-co-3-hydroxyvalerate) film for food packaging and food contact surfaces applications. Food Bioprod Process 101:32–44. https://doi.org/10.1016/j.fbp.2016.10.007

Castro-Mayorga JL, Fabra MJ, Cabedo L, Lagaron JM (2017b) On the use of the electrospinning coating technique to produce antimicrobial polyhydroxyalkanoate materials containing in situ-stabilized silver nanoparticles. Nanomaterials 7(4). https://doi.org/10.3390/nano701000

Castro-Mayorga JL, Randazzo W, Fabra MJ, Lagaron JM, Aznar R, Sanchez G (2017c) Antiviral properties of silver nanoparticles against norovirus surrogates and their efficacy in coated polyhydroxyalkanoates systems. LWT-Food Sci Technol 79:503–510. https://doi.org/10.1016/j.lwt.2017.01.065

Castro-Mayorga JL, Freitas F, Reis MAM, Prieto MA, Lagaron JM (2018a) Biosynthesis of silver nanoparticles and polyhydroxybutyrate nanocomposites of interest in antimicrobial applications. Int J Biol Macromol 108:426–435. https://doi.org/10.1016/j.ijbiomac.2017.12.007

Castro-Mayorga JL, Rovira MJF, Mas LC, Moragas GS, Cabello JML (2018b) Antimicrobial nanocomposites and electrospun coatings based on poly(3-hydroxybutyrate-co-3-hydroxyvalerate) and copper oxide nanoparticles for active packaging and coating applications. J Appl Polym Sci 135:45673. https://doi.org/10.1002/APP.45673

Cerqueira MA, Fabra MJ, Castro-Mayorga JL, Bourbon AI, Pastrana LM, Vicente AA, Lagaron JM (2016) Use of electrospinning to develop antimicrobial biodegradable multilayer systems: encapsulation of cinnamaldehyde and their physicochemical characterization. Food Bioprocess Technol 9:1874–1884. https://doi.org/10.1007/s11947-016-1772-4

Chung MG, Kim HW, Kim BR, Kim YB, Rhee YH (2012) Biocompatibility and antimicrobial activity of poly(3-hydroxyoctanoate) grafted with vinylimidazole. Int J Biol Macromol 50:310–316. https://doi.org/10.1016/j.ijbiomac.2011.12.007

Defoirdt T, Boon N, Sorgeloos P, Verstraete W, Bossier P (2009) Short-chain fatty acids and poly-β-hydroxyalkanoates: (new) biocontrol agents for a sustainable animal production. Biotechnol Adv 27:680–685. https://doi.org/10.1016/j.biotechadv.2009.04.026

Diez-Pascual AM, Diez-Vicente AL (2014) ZnO-reinforced poly(3-hydroxybutyrate-co-3-hydroxyvalerate) bionanocomposites with antimicrobial function for food packaging. ACS Appl Mater Interfaces 6:9822–9834. https://doi.org/10.1021/am502261e

Diez-Pascual AM, Diez-Vicente AL (2016) Electrospun fibers of chitosan-grafted polycaprolactone/poly(3-hydroxybutyrate-co-3-hydroxyhexanoate) blends. J Mater Chem B 4:600–612. https://doi.org/10.1039/c5tb01861g

Gao H, Li J, Sivakumar D, Kim T-S, Patel SKS, Kalia VC, Kim I-W, Zhang Y-W, Lee J-K (2019) NADH oxidase from lactobacillus reuteri: a versatile enzyme for oxidized cofactor regeneration. Int J Biol Macromol 123:629–636. https://doi.org/10.1016/j.ijbiomac.2018.11.096

Hema R, Ng PN, Amirul AA (2013) Green nanobiocomposite: reinforcement effect of montmorillonite clays on physical and biological advancement of various polyhydroxyalkanoates. Polym Bull 70:755–771. https://doi.org/10.1007/s00289-012-0822-y

Jendrossek D, Handrick R (2002) Microbial degradation of polyhydroxyalkanoates. Annu Rev Microbiol 56:403–432. https://doi.org/10.1146/aanurev.micro.56.012302.160838

Kalia VC (2013) Quorum sensing inhibitors: an overview. Biotechnol Adv 31:224–245. https://doi.org/10.1016/j.biotechadv.2012.10.004

Kalia VC (2014) Microbes, antimicrobials and resistance: the battle goes on. Indian J Microbiol 54:1–2. https://doi.org/10.1007/s12088-013-0443-7

Kalia VC, Purohit HJ (2008) Microbial diversity and genomics in aid of bioenergy. J Ind Microbiol Biotechnol 35:403–419. https://doi.org/10.1007/s10295-007-0300-y

Kalia VC, Purohit HJ (2011) Quenching the quorum sensing system: potential antibacterial drug targets. Crit Rev Microbiol 37:121–140. https://doi.org/10.3109/1040841X.2010.532479

Kalia VC, Chauhan A, Bhattacharyya G, Rashmi (2003) Genomic databases yield novel bioplastic producers. Nat Biotechnol 21:845–846. https://doi.org/10.1038/nbt0803-845

Kalia VC, Lal S, Cheema S (2007) Insight in to the phylogeny of polyhydroxyalkanoate biosynthesis: horizontal gene transfer. Gene 389:19–26. https://doi.org/10.1016/j.gene.2006.09.010

Kalia VC, Prakash J, Koul S (2016) Biorefinery for glycerol rich biodiesel industry waste. Indian J Microbiol 54:262–267. https://doi.org/10.1007/s12088-016-0583-7

Kalia VC, Prakash J, Koul S, Ray S (2017) Simple and rapid method for detecting biofilm forming bacteria. Indian J Microbiol 57:109–111. https://doi.org/10.1007/s12088-016-0616-2

Kalia VC, Patel SKS, Kang YC, Lee JK (2019) Quorum sensing inhibitors as antipathogens: biotechnological applications. Biotechnol. Adv. 37:68–90. https://doi.org/10.1016/j.biotechadv.2018.11.006

Kehail AA, Brigham CJ (2018) Anti-biofilm activity of solvent-cast and electrospun polyhydroxyalkanoate membranes treated with lysozyme. J Polym Environ 26:66–72. https://doi.org/10.1007/s10924-016-0921-1

Kim T-S, Patel SKS, Selvaraj C, Jung W-S, Pan C-H, Kang YC, Lee J-K (2016) A highly efficient sorbitol dehydrogenase from *Gluconobacter oxydans* G624 and improvement of its stability through immobilization. Sci Rep 6:33438. https://doi.org/10.1038/srep33438

Kiran GS, Lipton AN, Priyadharshini S, Anitha K, Suárez LEC, Arasu MV, Choi KC, Selvin J, Al-Dhabi NA (2014) Antiadhesive activity of poly-hydroxy butyrate biopolymer from a marine *Brevibacterium casei* MSI04 against shrimp pathogenic vibrios. Microb Cell Factories 13:114. https://doi.org/10.1186/s12934-014-0114-3

Kiran GS, Jackson SA, Priyadharsini S, Dobson ADW, Selvin J (2017) Synthesis of Nm-PHB (nanomelanin-polyhydroxybutyrate) nanocomposite film and its protective effect against biofilmforming multi drug resistant *Staphylococcus aureus*. Sci Rep 7:9167. https://doi.org/10.1038/s41598-017-08816-y

Koul S, Prakash J, Mishra A, Kalia VC (2016) Potential emergence of multi-quorum sensing inhibitor resistant (MQSIR) bacteria. Indian J Microbiol 56:1–18. https://doi.org/10.1007/s12088-015-0558-0

Kumar T, Singh M, Purohit HJ, Kalia VC (2009) Potential of *Bacillus* sp. to produce polyhydroxybutyrate from biowaste. J Appl Microbiol 106:2017–2023. https://doi.org/10.1111/j.13652672.2009.04160.x

Kumar P, Patel SKS, Lee JK, Kalia VC (2013) Extending the limits of *Bacillus* for novel biotechnological applications. Biotechnol Adv 31:1543–1561. https://doi.org/10.1016/j.biotechadv.2013.08.007

Kumar P, Singh M, Mehariya S, Patel SKS, Lee JK, Kalia VC (2014) Ecobiotechnological approach for exploiting the abilities of *Bacillus* to produce co-polymer of polyhydroxyalkanoate. Indian J Microbiol 54:151–157. https://doi.org/10.1007/s12088-014-0457-9

Kumar P, Mehariya S, Ray S, Mishra A, Kalia VC (2015a) Biotechnology in aid of biodiesel industry effluent (glycerol): biofuels and bioplastics. In: Kalia VC (ed) Microbial factories. Springer, New Delhi, pp 105–119. https://doi.org/10.1007/978-81-322-2598-0

Kumar P, Ray S, Patel SKS, Lee JK, Kalia VC (2015b) Bioconversion of crude glycerol to polyhydroxyalkanoate by *Bacillus thuringiensis* under non-limiting nitrogen conditions. Int J Biol Macromol 78:9–16. https://doi.org/10.1016/j.ijbiomac.2015.03.046

Kumar P, Sharma R, Ray S, Mehariya S, Patel SKS, Lee J-K, Kalia VC (2015c) Dark fermentative bioconversion of glycerol to hydrogen by *Bacillus thuringiensis*. Bioresour Technol 182:383–388. https://doi.org/10.1016/j.biortech.2015.01.138

Kumar P, Koul S, Patel SKS et al (2015d) Heterologous expression of quorum sensing inhibitory genes in diverse organisms. In: Kalia VC (ed) Quorum sensing vs quorum quenching: a battle with no end in sight. Springer India, New Delhi, pp 343–356. https://doi.org/10.1007/978-81-322-1982-8_28

Kumar P, Ray S, Kalia VC (2016) Production of co-polymers of polyhydroxyalkanoates by regulating the hydrolysis of biowastes. Bioresour Technol 200:413–419. https://doi.org/10.1016/j.biortech.2015.10.045

Kumar A, Kim I-W, Patel SKS, Lee J-K (2018a) Synthesis of protein-inorganic nanohybrids with improved catalytic properties using $Co_3(PO_4)_2$. Indian J Microbiol 58:100–104. https://doi.org/10.1007/s12088-017-0700-2

Kumar A, Patel SKS, Mardan B, Pagolu R, Lestari R, Jeong S-H, Kim T, Haw JR, Lim S-Y, Kim I-W, Lee J-K (2018b) Immobilization of xylanase using a protein-inorganic hybrid system. J Microbiol Biotechnol 28:638–644. https://doi.org/10.4014/jmb.1710.10037

Kumar A, Park GD, Patel SKS, Kondaveeti S, Otari S, Anwar MZ, Kalia VC, Singh Y, Kim SC, Cho B-K, Sohn J-H, Kim D-R, Kang YC, Lee J-K (2018c) SiO microparticles with carbon nanotube-derived mesopores as an efficient support for enzyme immobilization. Chem Eng J. https://doi.org/10.1016/j.cej.2018.11.052

Leja K, Lewandowicz G (2010) Polymer biodegradation and biodegradable polymers – a review. Pol J Environ Stud 19:255–266. www.pjoes.com/pdf/19.2/255-266.pdf

Mardina P, Li J, Patel SKS, Kim I-W, Lee J-K, Selvaraj C (2016) Potential of immobilized whole-cell *Methylocella tundrae* as a biocatalyst for methanol production from methane. J Microbiol Biotechnol 26:1234–1241. https://doi.org/10.4014/jmb.1602.02074

Narayanan A, Neera, Mallesha, Ramana KV (2013) Synergized antimicrobial activity of eugenol incorporated polyhydroxybutyrate films against food spoilage microorganisms in conjunction with pediocin. Appl Biochem Biotechnol 170:1379–1388. https://doi.org/10.1007/s12010-013-0267-2

Otari SV, Patel SKS, Jeong JH, Lee JH, Lee J-K (2016) A green chemistry approach for synthesizing thermostable antimicrobial peptide-coated gold nanoparticles immobilized in an alginate biohydrogel. RSC Adv 6:86808–86816. https://doi.org/10.1039/c6ra1488k

Otari SV, Kumar M, Anwar MZ, Thorat ND, Patel SKS, Lee D, Lee JH, Lee J-K, Kang YC, Zhang L (2017a) Rapid synthesis and decoration of reduced graphene oxide with gold nanoparticles by thermostable peptides for memory device and photothermal applications. Sci Rep 7:10980. https://doi.org/10.1038/srep41598-017-10777-1

Otari SV, Pawar SH, Patel SKS, Singh RK, Kim S-Y, Lee J-H, Zhang L, Lee J-K (2017b) *Canna edulis* leaf extract-mediated preparation of stabilized silver nanoparticles: characterization, antimicrobial activity, and toxicity studies. J Microbiol Biotechnol 27:731–738. https://doi.org/10.4014/jmb.1610.10019

Otari SV, Patel SKS, Kim S-Y, Haw JR, Kalia VC, Kim I-W, Lee J-K (2018) Copper ferrite magnetic nanoparticles for the immobilization of enzyme. Indian J Microbiol. https://doi.org/10.1007/s12088-018-0768-3

Patel SKS, Kalia VC (2013) Integrative biological hydrogen production: an overview. Indian J Microbiol 53:3–10. https://doi.org/10.1007/s12088-012-0287-6

Patel SKS, Purohit HJ, Kalia VC (2010) Dark fermentative hydrogen production by defined mixed microbial cultures immobilized on ligno-cellulosic waste materials. Int J Hydrog Energy 35:10674–10681. https://doi.org/10.1016/j.ijhydene.2010.03.025

Patel SKS, Singh M, Kalia VC (2011) Hydrogen and polyhydroxybutyrate producing abilities of *Bacillus* spp. from glucose in two stage system. Indian J Microbiol 51:418–423. https://doi.org/10.1007/s12088-011-0236-9

Patel SKS, Kumar P, Kalia VC (2012a) Enhancing biological hydrogen production through complementary microbial metabolisms. Int J Hydrog Energy 37:10590–10603. https://doi.org/10.1016/j.ijhydene.2012.04.045

Patel SKS, Singh M, Kumar P, Purohit HJ, Kalia VC (2012b) Exploitation of defined bacterial cultures for production of hydrogen and polyhydroxybutyrate from pea-shells. Biomass Bioenergy 36:218–225. https://doi.org/10.1016/j.biombioe.2011.10.027

Patel SKS, Kumar P, Mehariya S, Purohit HJ, Lee JK, Kalia VC (2014a) Enhancement in hydrogen production by co-cultures of *Bacillus* and *Enterobacter*. Int J Hydrog Energy 39:14663–14668. https://doi.org/10.1016/j.ijhydene.2014.07.084

Patel SKS, Kalia VC, Choi JH, Haw JR, Kim IW, Lee JK (2014b) Immobilization of laccase on SiO_2 nanocarriers improves its stability and reusability. J Microbiol Biotechnol 24:639–647. https://doi.org/10.4014/jmb.1401.01025

Patel SKS, Kumar P, Singh S, Lee JK, Kalia VC (2015a) Integrative approach for hydrogen and polyhydroxybutyrate production. In: Kalia VC (ed) Microbial factories: waste treatment. Springer, New Delhi, pp 73–85. https://doi.org/10.1007/978-81-322-2598-0_5

Patel SKS, Kumar P, Singh S, Lee JK, Kalia VC (2015b) Integrative approach to produce hydrogen and polyhydroxybutyrate from biowaste using defined bacterial cultures. Bioresour Technol 176:136–141. https://doi.org/10.1016/j.biortech.2014.11.029

Patel SKS, Choi SH, Kang YC, Lee J-K (2016a) Large-scale aerosol-assisted synthesis of biofriendly Fe_2O_3 yolk-shell particles: a promising support for enzyme immobilization. Nanoscale 8:6728–6738. https://doi.org/10.1039/C6NR00346J

Patel SKS, Jeong J-H, Mehariya S, Otari SV, Madan B, Haw JR, Lee J-K, Zhang L, Kim I-W (2016b) Production of methanol from methane by encapsulated *Methylosinus sporium*. J Microbiol Biotechnol 26:2098–2105. https://doi.org/10.4014/jmb.1608.08053

Patel SKS, Mardina P, Kim D, Kim S-Y, Kalia VC, Kim I-W, Lee J-K (2016c) Improvement in methanol production by regulating the composition of synthetic gas mixture and raw biogas. Bioresour Technol 218:202–208. https://doi.org/10.1016/j.biortech.2016.06.065

Patel SKS, Mardina P, Kim S-Y, Lee J-K, Kim I-W (2016d) Biological methanol production by a type II methanotroph *Methylocystis bryophila*. J Microbiol Biotechnol 26:717–724. https://doi.org/10.4014/jmb.1601.01013

Patel SKS, Selvaraj C, Mardina P, Jeong J-H, Kalia VC, Kang Y-C, Lee J-K (2016e) Enhancement of methanol production from synthetic gas mixture by *Methylosinus sporium* through covalent immobilization. Appl Energy 171:383–391. https://doi.org/10.1016/j.apenergy.2016.03.022

Patel SKS, Lee JK, Kalia VC (2016f) Integrative approach for producing hydrogen and polyhydroxyalkanoate from mixed wastes of biological origin. Indian J Microbiol 56:293–300. https://doi.org/10.1007/s12088-016-0595-3

Patel SKS, Otari SV, Kang YC, Lee JK (2017a) Protein-inorganic hybrid system for efficient his-tagged enzymes immobilization and its application in L-xylulose production. RSC Adv 7:3488–3494. https://doi.org/10.1039/c6ra24404a

Patel SKS, Singh R, Kumar A, Jeong JH, Jeong SH, Kalia VC, Kim I-W, Lee J-K (2017b) Biological methanol production by immobilized *Methylocella tundrae* using simulated biohythane as a feed. Bioresour Technol 241:922–927. https://doi.org/10.1016/j.biortech.2017.05.160

Patel SKS, Lee JK, Kalia VC (2017c) Dark-fermentative biological hydrogen production from mixed biowastes using defined mixed cultures. Indian J Microbiol 57:171–176. https://doi.org/10.1007/s12088-017-0643-7

Patel SKS, Choi SH, Kang YC, Lee J-K (2017d) Eco-friendly composite of Fe_3O_4-reduced graphene oxide particles for efficient enzyme immobilization. ACS Appl Mater Interfaces 9:2213–2222. https://doi.org/10.1021/acsami.6b05165

Patel SKS, Lee JK, Kalia VC (2018a) Nanoparticles in biological hydrogen production: an overview. Indian J Microbiol 58:8–18. https://doi.org/10.1007/s12088-017-0678-9

Patel SKS, Anwar MZ, Kumar A, Otari SV, Pagolu RT, Kim S-Y, Kim I-W, Lee J-K (2018b) Fe_2O_3 yolk-shel particle-based laccase biosensor for efficient detection of 2,6-dimethoxyphenol. Biochem Eng J 132:1–8. https://doi.org/10.1016/j.bej.2017.12.013

Patel SKS, Kondaveeti S, Otari SV, Pagolu RT, Jeong SH, Kim SC, Cho BK, Kang YC, Lee JK (2018c) Repeated batch methanol production from a simulated biogas mixture using immobilized *Methylocystis bryophila*. Energy 145:477–485. https://doi.org/10.1016/j.energy.2017.12.142

Patel SKS, Kumar V, Mardina P, Li J, Lestari R, Kalia VC, Lee J-K (2018d) Methanol peoduction from simulated biogas mixtures by co-immobilized *Methylomonas methanica* and *Methylocella tundrae*. Bioresour Technol 263:25–32. https://doi.org/10.1016/j.biortech.2018.04.096

Patel SKS, Otari SV, Li J, Kim DR, Kim SC, Cho B-K, Kalia VC, Kang YC, Lee J-K (2018e) Synthesis of cross-linked protein-metal hybrid nanoflowers and its application in repeated batch decolorization of synthetic dyes. J Harad Mater 347:442–450. https://doi.org/10.1016/j.jhazmat.2018.01.003

Patel SKS, Kim JH, Kalia VC, Lee JK (2018f) Antimicrobial ativity of amino-derivatized cationic polysaccharides. Indian J Microbiol. https://doi.org/10.1007/s12088-018-0764-7

Patel SKS, Lee JK, Kalia VC (2018g) Beyond the theoretical yields of dark-fermentative biohydrogen. Indian J Microbiol 58:529–530. https://doi.org/10.1007/s12088-018-0759-4

Porwal S, Kumar T, Lal S, Rani A, Kumar S, Cheema S, Purohit HJ, Sharma R, Patel SKS, Kalia VC (2008) Hydrogen and polyhydroxybutyrate producing abilities of microbes from diverse habitats by dark fermentative process. Bioresour Technol 99:5444–5451. https://doi.org/10.1016/j.biortech.2007.11.011

Prakash J, Gupta RK, Priyanka XX, Kalia VC (2018a) Bioprocessing of biodiesel industry effluent by immobilized bacteria to produce value-added products. Appl Biochem Biotechnol 185:179–190. https://doi.org/10.1007/s12010-017-2637-7

Prakash J, Sharma R, Patel SKS, Kim IW, Kalia VC (2018b) Bio-hydrogen production by co-digestion of domestic wastewater and biodiesel industry effluent. PLoS One 13:e0199059. https://doi.org/10.1371/journal.pone.0199059

Radivojevic J, Skaro S, Senerovic L, Vasiljevic B, Guzik M, Kenny ST, Maslak V, Nikodinovic-Runic J, O'Connor KE (2016) Polyhydroxyalkanoate-based 3-hydroxyoctanoic acid and its derivatives as a platform of bioactive compounds. Appl Microbiol Biotechnol 100:161–172. https://doi.org/10.1007/s00253-015-6984-4

Ramachandran P, Jagtap SS, Patel SKS, Li J, Kang YC, Lee J-K (2016) Role of the non-conserved amino acid asparagine 285 in the glycone-binding pocket of *Neosartorya fischeri* β-glucosidase. RSC Adv 6:48137–48144. https://doi.org/10.1039/c5ra28017f

Ray S, Kalia VC (2016) Microbial co-metabolism and polyhydroxyalkanoate co-polymers. Indian J Microbiol 57:39–47. https://doi.org/10.1007/s12088-016-0622-4

Ray S, Kalia VC (2017a) Biomedical applications of polyhydroxyalkanoates. Indian J Microbiol 57:261–269. https://doi.org/10.1007/s12088-017-0651-7

Ray S, Kalia VC (2017b) Co-metabolism of substrates by *Bacillus thuringiensis* regulates polyhydroxyalkanoate co-polymer composition. Bioresour Technol 224:743–747. https://doi.org/10.1016/j.biortech.2016.11.089

Ray S, Kalia VC (2017c) Polyhydroxyalkanoate production and degradation patterns in *Bacillus* species. Indian J Microbiol 57:387–392. https://doi.org/10.1007/s12088-017-0676-y

Ray S, Sharma R, Kalia VC (2018) Co-utilization of crude glycerol and biowaste for producing polyhydroxyalkanoates. Indian J Microbiol 58:33–38. https://doi.org/10.1007/s12088-017-0702-0

Reddy CSK, Ghai R, Rashmi KVC (2003) Polyhydroxyalkanoates: an overview. Bioresour Technol 87:137–146. https://doi.org/10.1016/S09608524(02)002122

Rehm BHA (2010) Bacterial polymers: biosynthesis, modifications and applications. Nat Rev Microbiol 8:578–592. https://doi.org/10.1038/nrmicro2354

Rennukka M, Sipaut CS, Amirul AA (2014) Synthesis of poly(3-hydroxybutyrate-*co*-4-hydroxybutyrate)/chitosan/silver nanocomposite material with enhanced antimicrobial activity. Biotechnol Prog 30:1469–1479. https://doi.org/10.1002/btpr.1986

Rodriguez-Contreras A, Garcia Y, Manero JM, Ruperez E (2017) Antibacterial PHAs coating for titanium implants. Eur Polym J 90:66–78. https://doi.org/10.1016/j.eurpolymj.2017.03.004

Selvaraj C, Krishnasamy G, Jagtap SS, Patel SKS, Dhiman SS, Kim T-S, Singh SK, Lee J-K (2016) Structural insights into the binding mode of D-sorbitol with sorbitol dehydrogenase using QM-polarized ligand docking and molecular dynamics simulations. Biochem Eng J 114:244–256. https://doi.org/10.1016/j.bej.2016.07.008

Singh M, Patel SKS, Kalia VC (2009) *Bacillus subtilis* as potential producer for polyhydroxyalkanoates. Microb Cell Factories 8:38. https://doi.org/10.1186/1475-2859-8-38

Singh M, Kumar P, Patel SKS, Kalia VC (2013) Production of polyhydroxyalkanoate co-polymer by *Bacillus thuringiensis*. Indian J Microbiol 53:77–83. https://doi.org/10.1007/s12088-012-0294-7

Singh M, Kumar P, Ray S, Kalia VC (2015) Challenges and opportunities for the customizing polyhydroxyalkanoates. Indian J Microbiol 55:235–249. https://doi.org/10.1007/s12088-015-0528-6

Van Immerseel F, De Buck J, Pasmans F, Velge P, Bottrew E, Fievez V, Haesebrouck F, Ducatelle R (2003) Invasion of Salmonella enteritidis in avian intestinal epithelial cells in vitro is influenced by short-chain fatty acids. Int J Food Microbiol 85:237–248. https://doi.org/10.1016/S0168-1605(02)00542-1

Van Immerseel F, Russell JB, Flythe MD, Gantois I, Timbermont L, Pasmans F, Haesebrouck F, Ducatelle R (2006) The use of organic acids to combat Salmonella in poultry: a mechanistic explanation of the efficacy. Avian Pathol 35:182–188. https://doi.org/10.1080/03079450600711045

Xavier JR, Babusha ST, George J, Ramana KV (2015) Material properties and antimicrobial activity of polyhydroxybutyrate (PHB) films incorporated with vanillin. Appl Biochem Biotechnol 176:1498–1510. https://doi.org/10.1007/s12010-015-1660-9

Zhang J, Shishatskaya EI, Volova TG, da Silva LF, Chen GQ (2018) Polyhydroxyalkanoates (PHA) for therapeutic applications. Mater Sci Eng C 86:144–150. https://doi.org/10.1016/j.msec.2017.12.035

Polyhydroxyalkanoates (PHA) – Applications in Wound Treatment and as Precursors for Oral Drugs

9

Larissa de Souza and Srividya Shivakumar

Abstract

Polyhydroxyalkanoates (PHA) are a family of thermoplastic linear polymers that are naturally synthesized by microbes which is an alternate nutrient (carbon) source, utilized during unfavourable conditions. PHA possess numerous important qualities like bio-compatibility, biodegradability, etc. which allows its safe usage, especially in the medical and environmental sectors. This chapter addresses the recent advancements accomplished in the last decade for the potential utilization of PHA in two specific avenues in the medical field: (1) for skin and wound healing treatment and (2) as precursors for oral drugs production. The advantages of utilizing PHA as biomaterials in medical devices are innumerable. Biomaterials are substances of either natural or artificial origin which are designed for its utilization in medically treating a biological system. In addition to biocompatibility, PHA have mechanical stability, strength, structure like extracellular matrix (ECM), non-toxic degradation products, etc. which make it highly desirable for medical applications. In their natural form, PHA lack antimicrobial properties and are non-antigenic. However, studies have shown that certain degradation products of PHA hydrolysis can evoke an inhibitory effect on microbial growth. In addition, numerous reports of chemically incorporating antibiotics, metal nanoparticles, etc. onto a PHA film can generate a composite blend that has antimicrobial properties. PHA are composed of monomers made up of 3-hydroyalkanoic acids (3-HA) and as they are enantiomerically pure, PHA constitute an untapped source of chiral building blocks which can be used for the manufacture of oral drugs. The ease with which these chiral molecules can be obtained from PHA through the action of dehydrogenases and hydrolases in addition to direct biosynthesis by microbes, makes this a viable and eco-friendly

L. de Souza · S. Shivakumar (✉)
Department of Microbiology, School of Sciences (SoS) Centre for PG Studies, Jain University, Bangalore, Karnataka, India
e-mail: sk.srividya@jainuniversity.ac.in

V. C. Kalia (ed.), *Biotechnological Applications of Polyhydroxyalkanoates*,
https://doi.org/10.1007/978-981-13-3759-8_9

method as compared to chemical synthesis which involves the usage of harsh reaction conditions and chemicals. It is evident from the quantity of studies being conducted in these avenues, that PHA application as antimicrobial wound dressings and as oral drug precursors has great potential and scope.

Keywords
Polyhydroxyalkanoates · Biomaterials · 3-hydroxy acids · Drug precursors · Wound dressings · Skin healing

9.1 Introduction

Polyhydroxyalkanoates (PHA) are naturally biosynthesized thermoplastic polymers, of microbial origin formed during conditions of nutrients stress, often in the presence of a limiting nutrient and an abundant source of carbon. PHA are utilized by microbes as a reserve for carbon and energy to help tide over unfavourable conditions like fluctuating environmental and growth parameters – temperature, hydrogen peroxide, UV irradiation, high osmotic pressure, nutrient deficiency, etc. PHA accumulates within cells as discrete granules (Subin and Bhat 2014) as much as 97% of total weight of the cells, with a molecular mass of approximately 50,000–1,000,000 Da. These biological polymers are composed of monomers made up of 3-hydroyalkanoic acids (3-HA) and over 100 different monomers have been identified. PHA monomers have a general structure that consists of an R (pendant) group attached to the main alkyl chain in the C-3 or β position. The R group signifies various alkyl groups. The stereospecificity of 3-HA are found to be of a single kind (R configuration) because of the PHA polymerase enzyme. An ester bond is formed between the hydroxyl and carboxyl group of two adjacent monomers (Anderson and Dawes 1990; Doi et al. 1992). The combination of monomers with different alkyl groups gives rise to diverse polymers with unique properties. For example, Poly-β-hydroxyvalerate (PHV) has an ethyl (CH_3CH_2) group while Poly-β-hydroxyoctanoate (PHO) has a pentyl ($CH_3(CH_2)_4$) group. Among the numerous PHA polymers, the best understood polymer is a four carbon compound called Poly-β-hydroxybutyrate (PHB), composed of (*R*)-3-hydroxybutyric acid (3-HB) monomers with a methyl (CH_3) group.

PHA are classified into three categories on the basis of the carbon quantity in the monomers: short chain length (scl) with 3–5 carbon atoms, medium chain length (mcl) with 6–14 carbon atoms and long chain length (lcl) with above 14 carbon atoms (Zinn et al. 2001). PHA production occurs widely across innumerable genera. Noteworthy examples are *Alcaligenes latus, Bacillus megaterium, Bacillus thuringiensis, Bacillus subtilis, Cupriavidus necator, Pseudomonas putida,* etc. Within the last two decades, it has come to light that certain archaeal species are capable of PHA production. *Haloferax mediterranei,* belonging to the Halobacteriaceae family, is one of the best studied archael PHA producers. Other PHA producing archaea include *Haloarcula, Halobacterium, Halococcus,*

Haloquadratum, Halorubrum, Haloterrigena, Natrialba, Natronobacterium, Natronococcus, etc. (Chen et al. 2006; Koller et al. 2007; Poli et al. 2011).

The PHA synthases system is responsible for the biosynthesis of PHA in all producers. It is composed of genes for β-ketothiolase (*pha*A), NADP-dependent acetoacetyl-CoA reductase (*pha*B) and PHA synthase (*pha*C). These three genes constitute, what is known as, the *pha*CAB operon. The versatility of PHA synthases can be attributed to their capacity to act on various HA. However, the PHA synthases enzyme is highly specific towards *R* stereoisomeric substrates and thus PHA biosynthesis in micro-organisms only involves (*R*)-isomers as substrates for the polymerization. Consequently, all PHA that has been characterized so far from micro-organisms have *R*-configuration (Anderson and Dawes 1990; Rehm and Steinbuchel 2002; Subin and Bhat 2014). The PHA synthases system is divided into four classes based on the composition of the enzyme subunits, the substrate preference of the enzymes and the length of the polymer (Rehm and Steinbuchel 2002; Subin and Bhat 2014). Generally, PHA synthesis occurs in three steps. During normal bacterial growth, β-ketothiolase (first enzyme) is inhibited by unbound coenzyme-A (CoA) generated through functioning of the Krebs cycle. However, under nutrient limiting conditions, utilization of acetyl-CoA in Krebs cycle is curtailed, creating a drop in the CoA levels, in turn activating β-ketothiolase (phaA). Consequently, the accumulating acetyl-CoA is directed to the PHA cycle. β-ketothiolase condenses it to acetoacetyl-CoA, which in turn is converted to 3-hydroxyalkanoic acid-CoA by phaB (for example, 3-hydroxybutyryl-CoA in the case of PHB biosynthesis). In the final step, PHA polymerase (phaC) joins together the 3-hydroxyalkanoic acid-CoA through esterification, with the release of free CoA, to generate the polymer (Zinn et al. 2001; Rehm and Steinbuchel 2002).

PHA producers are categorized into two groups based on the phase of growth and the condition which favour PHA production. Group I organisms require a limiting nutrient with excess Carbon. Examples include *Bacillus thuringiensis, Cupriavidus necator, Methylobacterium extorquens,* etc. Group II organisms can grow and biosynthesize PHA during its exponential phase of growth. Examples include *Bacillus megaterium, Alcaligenes latus, Azotobacter vinelandii, Bacillus mucoides,* etc. (Thakur et al. 2001; Zinn et al. 2001; Rehm 2009; Gowda and Shivakumar 2014; Orita et al. 2014; Subin and Bhat 2014).

The degradation of PHA polymers (intracellularly and extracellularly) is carried out through the action of PHA hydrolases and depolymerases (which are carboxyesterases) (Anderson and Dawes 1990; Jendrossek and Handrick 2002). PHA depolymerases are enzymes, composed of catalytic and substrate-binding domains connected to each other by a 'linker' domain. After the bonding of the substrate-binding domain to PHA, the catalytic domain initiates cleavage of the polymer chain (Numata et al. 2009; Ong et al. 2017). Based on the location of enzyme activity, depolymerases have been classified into two groups: (1) Intracellular depolymerases and (2) Extracellular depolymerases. The biodegradation of PHA can be clearly understood by examining the hydrolysis cycle of PHB. Native granules of PHB have amorphous polymer within them and the granules are covered by a thick layer comprising of phasins, PHB depolymerases, etc. PHB is depolymerized by

PHB depolymerase (phaZ) and 3-hydroxybutyrate-CoA is liberated. This is then subjected to oxidation by a dehydrogenase to form acetoacetyl-CoA and finally, is converted into acetyl-CoA by β-ketothiolase (Uchino et al. 2007; Eggers and Steinbuchel 2013; Ong et al. 2017). Examples of PHA degraders include *Comamonas testosteroni, C. acidovorans, Ralstonia pickettii* T1 (formerly known as *Alcaligenes faecalis*), *Pseudomonas lemoignei, P. stutzeri, P. fluorescens* GK13, *B. megaterium, Xanthomonas* sp. JS02, *Streptomyces* sp. KJ-72, *P. alcaligenes, Rhodospirillum rubrum,* etc. (Anderson and Dawes 1990; Jendrossek and Handrick 2002; Subin and Bhat 2014; Vigneswari et al. 2015).

9.1.1 Current Scenario of Biomaterials and Biopolymers in Medicine

Biomaterials are substances of either natural or artificial origin which are designed for its utilization in medically treating a biological system. It is necessary for biomaterials to have biocompatibility, that is, they should function efficiently without an immunological response (Ulery et al. 2011). For a biodegradable biomaterial to be utilized in a medical environment, it must fulfil the necessary criteria to ensure a safe and efficient product. Some of these criteria include:

1. Being non-antigenic, non-immunogenic and non-inflammatory to the host system
2. If antimicrobial capacities are present, they should be clearly demonstrated through in vitro and in vivo studies and durability of antimicrobial action should be for atleast a year or 2
3. Should possess desired mechanical properties (appropriate permeability, processability, resistance to surgical and post-surgical stress, etc.)
4. Should biodegrade within the biological system at the correct time
5. The products of its biodegradation should be non-toxic and be naturally metabolized by the biological system (Ulery et al. 2011; Gallo et al. 2014; Francis et al. 2016).

Antimicrobial biomaterials have been categorized into two categories based on their mode of action: (1) through 'contact' killing and (2) through elution of the antimicrobial agent. In addition, biomaterials may be also categorized based on their functionality: (1) mono-functional (involves performing only a single function like providing structural support, prevention of bacterial infections and biofilm formation, initiation and enhancement of cell growth, etc.) and (2) multi-functional (involves carrying out more than one function simultaneously; for example, providing a scaffold for skin cells proliferation while preventing microbial colonization) (Gallo et al. 2014).

Biopolymers are polymeric substances obtained from natural sources and have innate traits like biocompatibility, low antigenicity, biodegradability, non-toxic degradation products, high bio-activity, malleability, appropriate porosity, promote cell growth and proliferation, desired mechanical strength, etc. (Yadav et al. 2015). Thus, biopolymers can be safely utilized as biomaterials for medical devices. Biopolymers can be classified depending on their basal chemical composition and origin:

1. Polyesters: Polyhydroxyalkanoates, poly(lactic acid) (PLA)
2. Proteins: Collagen, gelatin, elastin, etc.
3. Polysaccharides
   Bacterial: Xanthan gum, dextran, cellulose etc.
   Fungal: Pullulan, yeast glucans, etc.
   Plant/Algal: Starch, cellulose, agar, alginate, carrageenan, etc.
   Animal: Chitin, hyaluronic acid
4. Lipids/Surfactants: Waxes, etc.
5. Polyphenols: Tannin, lignin, etc. (Mano et al. 2007; Yadav et al. 2015)

Numerous biopolymers are currently in use in the medical field in different subsectors depending on the characteristics of the biopolymer in question. Most commonly used is collagen as it is the primary structural component of vertebrates. Collagen, synthesized by fibroblasts, is easily absorbed by the body, has low antigenicity, non-toxic, biocompatible, hydrophilic and biodegradable. It finds application in delivery of drugs to treat infected tissues. In addition, collagen sponges are used to manage burn wounds while collagen implants have been used as bone substitutes, etc. (Chattopadhyay and Raines 2014; Yadav et al. 2015).

Chitosan is derived from chitin, a natural polysaccharide, and is biocompatible, biodegradable, antimicrobial, selectively permeable, etc. These properties enable chitosan to be used in controlled release of drugs, anti-bacterials, promotion of osteogenesis, anti-oxidants, etc. Cellulose is the most abundant naturally occurring polymer. Cellulose fibers in use in medicine are natural cotton fibers, man-made fibers like viscose, modal, etc. In addition, cellulose ether is used in controlled drug release and regenerative medicine (Hoenich 2007; Bacakova et al. 2014). Hyaluronic acid is found in the ECM and the joint liquid of mammals. It is a biocompatible mucoadhesive polysaccharide which is biodegradable. Hyaluronan was used as a vitreous replacement during eye surgery in the late 1950s. Gum Arabic, xantham gum, carrageenan, etc., are other biomaterials obtained for natural sources (Yadav et al. 2015).

9.1.2 The Need for PHA in Medicine

Biopolymers, as a family, are very desirable for construction of medical devices and treatments. Despite sharing similar properties (biocompatibility, non-antigenic, non-immunogenic traits etc.) with other biopolymers, PHA have certain unique qualities that render an edge over the rest of the other biopolymers.

An important trait that makes PHA such a superior commodity is its capacity for complete biodegradation, without the release of any toxic by-products. In terms of biomedical applications, non-toxic degradation products which can be easily metabolized by the biological system and removed from the body are desirable. The degradation products of PHA fit this criterion. Studies have shown that as most PHA polymers degrade, their degradation products are far less acidic than those released on hydrolysis of many currently used synthetic absorbable polymers (PGA and PLA) (Williams and Martin 2002). In terms of the environment, biodegradability is necessary especially for plastic products. Plastics, with their countless applications, have irreversibly integrated themselves across industries and avenues. Consequently, the amount of waste generated is staggering. Reports have revealed that the per capita plastics products consumption by the average Indian is around 9.7 Kg, rivalled by 109 Kg/person and 65 Kg/person for USA and Europe, respectively (PlastIndia 2013). Eriksen et al. (2014) published a study on the amount of plastic particles (5.25 trillion particles) present in oceans across the globe, weighing an estimated 268,940 tons. As a result, plastic wastes (including medical wastes) need to be minimized and can be done so using PHA.

The discovery of over hundred different monomers constituting PHA has enabled the prospect of creating and designing numerous types of biodegradable polymers with extensive properties (Anderson and Dawes 1990; Zinn et al. 2001). Thus, depending on the desired physical and mechanical traits for a medical device, various monomers can be combined to form co-polymers exhibiting these traits. PHA provides an avenue for the creation of tailor-made natural biopolymers. Additionally, PHA can easily combine with other components and form a composite blend with antibiotics, metals, non-metals, etc.

PHA have been found to be non-harmful to human and animal tissues, that is, they are bio-compatible. Bio-compatibility is a desirable property when dealing with medical products like sutures, implants, dressings, grafts etc. The criteria for a material to be considered bio-compatible include having non-immunogenic and non-toxic reactions. Additionally, it is preferential that an inflammatory response is either absent or negligible. In cases where degradation is expected of a biodegradable medical device, the products of degradation must not be toxic and should be easily metabolized, and naturally cleared from the biological system (Mano et al. 2007; Puppi et al. 2011; Brigham and Sinskey 2012).

The advantages of utilizing PHA as biomaterials in medical devices are countless. Mechanical stability, strength, similar in structure to ECM, non-toxic degradation products, etc. contribute to the benefits of PHA.

In addition to the medical aspect, PHA also greatly contributes to the environmental health, which in turn, boosts human health. As they originate from a biological source, PHA can be produced using different substrates (pure sugars, waste materials like agro-wastes, molasses, bagasse, glycerol liquid phase, etc.). As a result, this greatly contributes to lowering the cost of production and additionally, pressure from fossil fuel reserves is averted (from which conventional plastics are manufactured). Moreover, PHA are unique polymers that possess no carbon fingerprint as they do not contribute to global warming. The reason for this is the ability of these polymers to be produced from renewable carbon resources, and thus be unaffected by any depletion in fossil fuels and inflation in crude oil prices. Thus, their neutrality with respect to carbon dioxide emission, released from final mineralization, is attributed to the biosynthesis from renewable sources of carbon and in turn, fixation of released carbon dioxide by plants through photosynthesis regenerates renewable carbon sources. Consequently, the synthesis and degradation of carbon in biopolymers is balanced and is completely compatible with the carbon cycle (Koller et al. 2010; Subin and Bhat 2014).

9.1.3 Current Applications of PHA in Medicine

Through decades of research, PHA has entered the medical sector for biomedical devices. Numerous companies in developed countries have begun manufacturing PHA based biodegradable and biocompatible commodities.

TephaFLEX® absorbable sutures were the first product made from PHA that gained FDA approval for application in clinical settings. This product is prepared from Poly(4-hydroxybutyrate) [P(4HB)] and is superior in its highly elastic qualities that allow it to stretch ten times its original length before any breaking is observed (Fernandez Bunster 2015). In addition, this US-based company, Tepha, manufactures medical devices like artery augments, vascular grafts, pericardial patches, cardiological stents, heart valves, dressings, prodrugs and microparticulate carriers using PHA, dusting powders, etc. (Williams and Martin 2002; Subin and Bhat 2014).

The last decade has seen numerous improvements with respect to the utilization of PHA and, PHB. Available to health professionals are PHA based devices including articular cartilages, sutures, repair devices, nerve guide, slings, cardiovascular patches, vascular grafts, pericardial patches, cardiovascular stents, heart valves, orthopaedic pins, stents, repair patches, bone marrow scaffold, wound dressings, etc. (Chen and Wu 2005a, b; Hazer and Steinbuchel 2007; Subin and Bhat 2014; Fernandez Bunster 2015). In addition, through chemical modification, amphiphilic graft copolymers have potential usage in blood-contacting devices (Hazer and Steinbuchel 2007).

9.1.4 Antimicrobial Properties of PHA

In their natural form, PHA do not have any antimicrobial properties. PHA polymers are non-antigenic and are usually enantiomerically pure (Fernandez Bunster 2015). However, studies have shown that certain degradation products of PHA hydrolysis can evoke an inhibitory effect on microbial growth. In addition, numerous reports of chemically incorporating antibiotics, metal nanoparticles, etc. onto a PHA film can generate a composite blend that has antimicrobial properties (Hu et al. 2003; Pramanik et al. 2015; Francis et al. 2016). The reason for using antibiotics is self-explanatory. The use of metals as antimicrobial agents is somewhat unconventional. There have been numerous studies conducted on the antimicrobial effects of metals (Petrini et al. 2006; Panacek et al. 2009; Grass et al. 2011) although their usage in medicine is done so extremely carefully due to the cytotoxic effects to human cells. One study demonstrated the effective inhibition of *Candida albicans* by a concentration of silver nanoparticles below human fibroblasts cytotoxic levels, that is, 30 mg/L (Panacek et al. 2009).

Certain non-metals have also demonstrated a capability to inhibit microbial growth (Tran and Webster 2011; Wang and Webster 2012; Holinka et al. 2013). A study using titanium implants covalently bound with Selenium to prevent *Staphylococcus aureus* and *Staphylococcus epidermidis* attachment without affecting osteoblast viability was conducted. Selenium acts through the catalysis of superoxide radicals formation thereby inhibiting microbial biofilm formation (Tran and Webster 2011; Holinka et al. 2013). Scientists have experimented with coating antimicrobial agents onto the surface of various implant devices (Gaoet al. 2011; Forbes et al. 2013; Kazemzadeh-Narbat et al. 2013). However, the scope of this chapter is to put forth the progress of these techniques using PHA as the material of choice instead of a metal matrix.

In the subsequent sections, numerous studies conducted within the last decade have been discussed. There have been novel and innovative ideas about PHA and its modification that have been proposed and tested for application in the medical field. One such extremely promising approach hypothesizes the inhibition or interference quorum signalling or sensing which could potentially affect the invading bacteria from moving away from the antimicrobial agent thereby allowing a window of opportunity for the agents to act (Gallo et al. 2014). Another approach puts forth the utilization of biomaterials with antimicrobial surfaces that could prevent the colonization and formation of biofilms. These biomaterials are designed to have surfaces that are either repellent so that microbes can't adhere to them or contain antimicrobial agents that destroy microbes near the biomaterial (Dinjaski et al. 2014).

This chapter addresses the recent advancements accomplished in the past decade for the potential utilization of PHA in two specific avenues of medicine: (1) for skin and wound healing treatment and (2) as precursors for oral drugs production.

9.2 Recent Advancements in PHA Applications – Skin and Wound Healing

The largest organ of vertebrate, the skin, is composed of the epidermis, dermis and hypodermis with an array of interconnections with nerve and blood vessels. The epidermis is the topmost layer and provides a barrier against microbes and antigens. The next layer is the dermis and consists of collagen, elastin and glycosaminoglycans (GAGs). The hypodermis is the final layer having well vascularized adipose tissue. This layer imparts thermoregulation and mechanical vigour to the skin (Metcalfe and Ferguson 2006). Skin comprises of numerous different cell types: the most commonly found are keratinocytes, followed by the cells from the lower epidermis, melanocytes and the lower dermal layer cells, fibroblasts. The keratinocytes are the barrier surface layer, the melanocytes provide colour of the skin and the fibroblasts produce proteases and collagenases which are responsible for wound healing and strength (MacNeil 2007; Metcalfe and Ferguson 2007).

9.2.1 Process of Wound Healing

The healing process of injured tissues, skin and wounds is very complex and involves a number of cellular, molecular and biochemical activities. Wound healing therapy proceeds on the basic principle of debridement (medical removal of injured/damaged tissues to enhance healing of remaining healthy tissues) and optimal wound healing conditions (Gumel et al. 2015; Shishatskaya et al. 2016).

As part of the repair complex of a biological system, wound healing occurs through four precise phases: (1) Hemostasis, (2) Inflammation, (3) Proliferation and (4) Remodelling.

Hemostasis, the first phase of wound healing, commences as soon as an injury occurs with initiation of vascular constriction and formation of a fibrin clot. This prompts the release of cytokines, growth factors like platelet-derived growth factor (PDGF), fibroblast growth factor (FGF), epidermal growth factor (EGF), etc.

After the control of bleeding is achieved, the inflammatory cells migrate to the site of the wound through chemotaxis and the second phase, inflammation, begins. This phase is characterized by an infiltration of lymphocytes, neutrophils and macrophages. The role played by T-lymphocytes is still a matter of debate but they are known to enter the site of injury along with macrophages. In addition, their optimum quantity is reached between the third and fourth phases of wound healing. Neutrophils deal with invading microbes and cell debris near the wound through production of reactive oxygen species (ROS) and proteases. Macrophages have numerous functions. Immediately after injury, macrophages release cytokines which induce an inflammatory response and aid in attracting and activating leukocytes. Secondly, macrophages induce and dispose of apoptotic cells (including neutrophils) and this subsides the inflammation because the disposed-off cells transform into a reparative state. This state is responsible for stimulating cells for tissue regeneration (fibroblasts, keratinocytes, etc.) and thus begins the proliferation phase.

The proliferation phase involves the proliferation of epithelial cells along with migration (re-epithelialization). The formation of collagen, capillary growth and tissue is achieved by fibroblasts and endothelial cells. The former produces collagen, proteoglycans and glycosaminoglycans, leading to formation of the ECM.

The final remodelling phase observes the remodelling of ECM to a structure that is characteristic of a normal tissue and the regression of newly synthesized capillaries. The wound undergoes physical contraction during this final phase, brought about by the action of contractile fibroblasts (Metcalfe and Ferguson 2007; Guo and DiPietro 2010).

In order to achieve successful healing, it is extremely necessary that all the phases activate and complete their function in the correct order and for the proper duration of time (Guo and DiPietro 2010; Gumel et al. 2015). Any disturbances in even one of the phases could lead to impaired healing that in turn develops into chronic non-healing wounds or delayed wound healing.

9.2.2 Factors Affecting Wound Healing

Numerous factors are responsible for influencing efficient wound healing like wound location, depth and severity of the injury, microbial invasion, conditions and health history of the patient, etc. (Guo and DiPietro 2010; Shishatskaya et al. 2016). These factors can be categorized into two broad groups namely: (1) Local and (2) Systemic. In general, factors which have a direct influence on the wound features are local factors while those factors which determine the health state of an individual and the patient's ability to heal are systemic factors (Guo and DiPietro 2010). Some local factors are infection, oxygenation, etc. while stress, obesity, age, gender, substance abuse, etc. are systemic factors. The factors from both these groups, at times, tend to be related and work through each other to interfere with optimal healing.

With regard to skin and wound healing, the dressing materials and the epidermal and dermal skin cells play a vital role in efficient and seamless repair of damaged skin and tissues. An ideal dressing material should fulfil the abilities of normal and healthy skin. Some abilities include providing mechanical protection, barrier against microbial colonization, facilitating moist environment for healing, permitting ease of exchange of gases, and desired elasticity (Shishatskaya et al. 2016). Scientists and medical professionals throughout the world are investing time and resources to design and test novel skin and wound dressing materials which are biocompatible (ensuring that an immunological response will be absent), biodegradable (ensuring non-pollution of the environment), and biological (ensuring efficient utilization of microbial resources for production of such superior biomaterials).

Through 35 years' worth of extensive research, tissue engineering using cells (keratinocytes) have been successful in treating patients who have sustained burn injuries, a technique that had started back in the 1980s. Through research conducted in the following decade, a clearer picture of the co-action of other skin cells like fibroblasts with keratinocytes was obtained (MacNeil 2007). However, after the millennium, research progressed further into understanding the wound healing process. Consequently, new ideas and techniques were hypothesized targeting specific phases and events within the broad wound healing event. Desire for solutions to problems like infections during healing, dressings adhering to healing tissues, improper wound healing due to extensive tissue damage, delayed wound healing and chronic wounds, etc. initiated research into unexplored avenues and fields. Some of the new ideas and avenues involved utilizing biocompatible biodegradable biomaterials with varying features designed to target the inadequacies of conventionally used dressings and methods. As a result, the last decade has seen significant work in this respect, especially in the utilization of PHA for biomedical applications.

Literature on the theories, challenges and various other aspects of tissue engineering with respect to skin engineering and wound healing is readily available (MacNeil 2007; Metcalfe and Ferguson 2007; Farrar 2011; Rahmani Del Bakhshayesh et al. 2018). However, the scope of this chapter is to elucidate the progress that has been made in the utilization of PHA, specifically, in the treatment of skin and wound injuries within the last decade.

9.2.3 Skin Cell Lines – Types and Significance

Before addressing the progress made in the field of medicine with regard to PHA applications in skin and wound healing, it is necessary to note the importance of cell lines. In vitro cytotoxicity tests are used to establish whether a test agent may/may not be toxic to a biological system through the utilization of cell lines. A cell line is defined as one that is brought forth through subculture from a primary culture. A primary culture is one which originates from biological samples (cells/tissues/organs) harvested directly from organisms (Ekwall et al. 1990). Depending on the intended location in the human body of the test agent to be analysed, numerous cell lines are available for cytotoxicity studies. And in the study of novel materials to be used in the medical field, a demonstration of in vitro and in vivo cytotoxicity is necessary.

Irritation of the skin involves inflammatory reactions of the cells that form the human dermis and epidermis. As a result, various cell lines have been derived from skin cells in order to test skin related agents, medicines, dressings, etc. The cell lines most commonly used to test immunogenic reactions of the skin include human keratinocytes, HaCaT cell line, 3T3 and L929 mouse fibroblasts, Schwann cells, etc. (Hoh and Maier 1993; MacNeil 2007; Suwantong et al. 2007; Keshel et al. 2014).

Keratinocytes are important cell lines used for cytotoxicity tests because they are considered to exhibit the metabolism of the skin more precisely than other cell types do. HaCaT cell line originated through spontaneously transformation of the human keratinocytes cell line. They possess numerous characteristics of keratinocytes and are known to stratify to a lesser degree than human keratinocytes (Hoh and Maier 1993).

9.2.4 Recent Advances in PHA Application for Skin and Wound Healing

The recent advances made in the utilization of PHA for skin and wound healing are numerous. The advancements discussed in this chapter have been categorized and arranged chronologically into two groups based on the biomaterial utilized: (1) PHA polymers (2) Modified PHA polymers (Fig. 9.1).

9.2.4.1 PHA Polymers

This group addresses the progress made in the utilization of different PHA polymers for biomedical applications. The polymers have been constructed into films and are without any chemical modification, that is, the entire film is composed of only PHA.

Numerous studies within the past two decades have employed various techniques of obtaining or manufacturing the biodegradable PHA polymers utilized in their experimentations. Some of these include: solvent casting method, electrospinning process, etc.

Electrospinning is a process which utilizes an electrostatic field to distort the natural shape of liquid polymer (pendant) into another form (conical). Once a threshold point is reached, a finely charged stream of liquid polymer is ejected onto a collecting sheet or vessel. This is called a charged jet and it moves in the direction

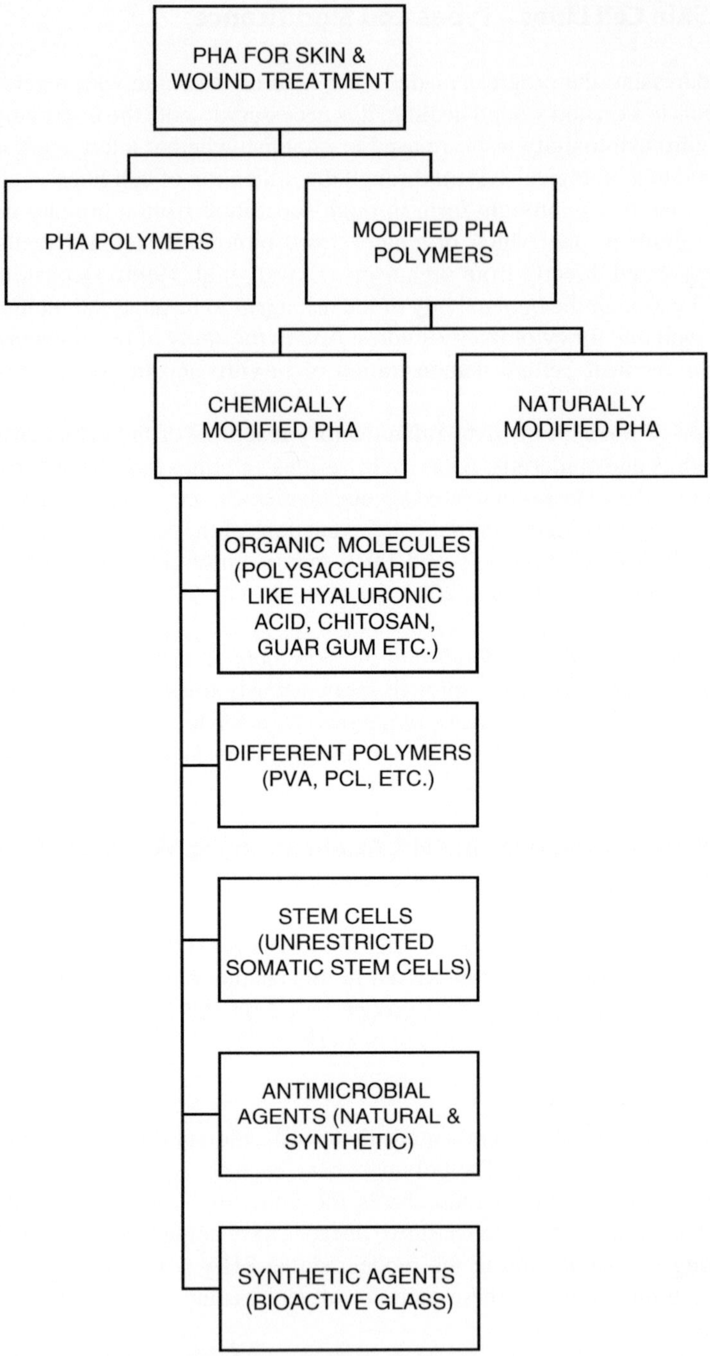

Fig. 9.1 Representation of the different forms of PHA studied for application in skin and wound treatment

of polarity opposite to its own. Due to the highly viscous liquid polymer, the jet retains its stream, resulting in a non-woven polymer membrane with ultra-fine fibers (Deitzel et al. 2001; Suwantong et al. 2007). The advantage of using electrospun non-woven membranes is their structural similarity to the collagen fibers that are present in living and functional tissues (Shishatskaya et al. 2016).

Electrospun PHA mats have started gaining recognition for potential application in regeneration of skin and nerve tissues, skin engineering, etc. Studies on these potential applications have been explored. One study describes the analysis of fiber mats of PHB and poly(3-hydroxybutyrate-co-3-hydroxyvalerate) (PHBV) to treat skin injuries and to regenerate damaged nerve tissues (i.e. attachment and proliferation capacity). These fiber mats were electrospun to an ultrafine quality and were tested on Schwann cells (RT4-D6P2T) as well as mouse fibroblasts (L929). Indirect cytotoxicity tests revealed acceptance of the mats by both cell lines. Regeneration tests were conducted and a comparison was made between films and tissue-culture polystyrene plate (TCPS). It was found that L929 attachment on fibrous scaffolds was significantly better and superior in cell proliferation, than on the TCPS and films. However, the opposite was observed in the case of RT4-D6P2T. Scanning electron microscopy (SEM) images showed active L929 on all of the surfaces investigated. RT4-D6P2T exhibited growth on PHB films (Suwantong et al. 2007). Another study explored the application of a terpolymer composed of 3-hydroxybutyrate (3-HB or HB), 3-hydroxyvalerate (3-HV or HV) and 3-hydroxyhexanoate (3-HHx or HHx) for in skin engineering. The study was conducted using human keratinocyte cell line HaCaT, revealing that these cells demonstrated the highest viability and proliferation on PHBVHHx films as compared to other biodegradable biomaterials like PLA, PHB, PHBV, PHBHHx and poly(3-hydroxybutyrate-co-4-hydroxybutyrate) [P(3HB-co-4HB)]. The superiority of this novel polymer was attributed to its surface roughness and its capacity to stimulate a rapid increase of calcium influx within the cytosol, in turn contributing to an enhancement in cell growth. In addition, the degradation product HB, aided in promoting cell proliferation (Ji et al. 2008). These results support the hypothesis that ultrafine electrospun PHB, PHBV fiber mats and PHBVHHx polymer is suitable for skin engineering material development.

Reconstructive surgery is a challenging procedure under normal circumstances and even more so especially if materials used in the surgical process are lacking in terms of mechanical and biocompatible traits. However, within the last decade there have been a few studies on the potential utilization of PHA in surgical reconstruction. A study to assess the potential usage of PHA in the repair of the tarsi of the eyelid was conducted. Tarsal plates (tarsi) are thick and elongated connective tissue plates, present directly above the eyelid margins. They aid in eyelid support and form. Each eyelid has two tarsi (2.5 cm) (Gray et al. 2016). The PHA polymer studied was poly(3-hydroxybutyrate-co-3-hydroxyhexanoate) (PHBHHx). Scaffolds composed of PHBHHx were produced and implanted into Sprague-Dawley rats with tarsal defects. A comparison was made using blank defect controls and commercial acellular dermal matrices (ADM). At weekly intervals, samples of the eyelids containing the polymer, ADM and controls were collected and subjected to histological examination. The results of this study suggested that both the ADM and PHBHHx scaffolds provided fairly satisfactory repair in comparison to the controls. This was observed despite the presence of inflammation caused by the polymer scaffolds. In addition, fibrous encapsulation along with scaffold degradation was observed with regard to the PHBHHx implants (Zhou et al. 2010). Thus, the biocompatibility coupled with strong mechanical properties ensures that PHBHHx can be used as a suitable biomaterial for tarsal repair.

Many novel approaches have been hypothesized for efficient and enhanced skin and wound healing. One such approach is the utilization of a bilayer skin substitute to promote healing. A study was conducted on this approach and involved a biodegradable PHA polymer, PHBV. The liquid polymer was subjected to solvent casting to produce a fine nano-porous membrane to be used for epidermal cell growth and a freeze-drying method which produced a 3D porous scaffold for usage in dermal cell proliferation. Cytotoxicity tests were assessed using human keratinocytes (hKC) and dermal fibroblasts (hDFb).

The resulting biomaterial had high water retention capacity, appropriate mechanical traits and was susceptible to enzymatic degradation. Cytotoxicity tests confirmed that PHBV was biocompatible with a biological system. In addition, the polymer was discovered as being satisfactory to support cell adhesion and proliferation. An interesting observation was made that during co-culture in different predefined conditions, the hKC cells had the ability to grow and arrange themselves in multilayers. The proliferating cells were in the basal layers while the cells which expressed terminal differentiation marker were located in the upper layer (Zonari et al. 2014). Thus, the bilayer structures composed of PHBV clearly have properties which enhance proliferation and in turn, improve wound healing.

Recently, a group of scientists investigated the production and utilization of biocompatible and biodegradable PHA-based wound dressing materials. A copolymer of P(3HB-co-4HB) was synthesized by *Cupriavidus eutrophus* B10646 and once purified, were electrospun into nonwoven membranes made up of ultrafine fibers. The MTT assay was used to test cell viability. Fibroblast cells were layered onto the electrospun fiber mats and their efficiency of healing skin and wound defects were tested on Wistar rats induced with model skin damages. The experiment consisted of animals divided into three sets, wherein two were treatment/test groups and third group served as a control group. Of the test groups, one was treated with non-woven membranes and the second was treated with non-woven membranes layered with fibroblast cells from adipose tissues. The control group was treated with plain and sterile gauze.

The results of this investigation revealed the presence of 3-HB and 4-HB monomer units in a ratio of 90/10 mol% having average molecular weight (Mw) 473 KDa and water contact angle between 75.7 and 76.9°. The solvent cast P(3HB/4HB) films had round pores (2 μm diameter) while the electrospun membranes had distinct fibers (1.7 μm diameter and 1.3–1.67 μm interspaces). Stability studies were carried out through in vitro methods using phosphate buffer solutions at different pH and stabilized blood for 100 days. The films were found to be completely stable in the buffer solutions while the enzymes in the blood medium degraded the films by 25% in 100 days. Cell viability tests showed a 25% and 30% higher number on nonwoven membranes as compared to that on polymer and the control (polystyrene), respectively. Hence, the non-woven membranes were chosen for in vivo studies as potential wound dressings.

The fibroblasts on the dressing material clearly enhanced the healing of wounds with regeneration of cells. In addition, the cells were observed to have secreted ECM which facilitated migration and flow of cells of the epidermis from surrounding healthy tissues. The decrease in wound surface area measurements of non-woven membranes layered with epidermal cells was found to be the highest at 0.4–0.65 cm^2/day with 57% of total wound area at day 7 and 9% at day 14. The second group with non-woven membranes without fibroblast cells was lower at 0.3–0.4 cm^2/day with 78% of total wound area at day 7 and 9.1% at day 14.The decrease in wound surface area of the control group was found to be the least at 0.16–0.23 cm^2/day with 23% of total wound area at day 7 and the rate of healing decreased at day 14 with 30%. Thus, the polymer with the layered fibroblasts healed the injured skin 1.4 and 3.5 times than the membranes without the cells and the control, respectively. Complete healing was noted at 14 days (Shishatskaya et al. 2016).

This study clearly indicates the advantages of using electrospun non-woven biodegradable PHA-based polymers [P(3HB/4HB)] layered with fibroblast cells for potentially successful application in wound dressings. The dressing was found to initiate phase three of wound healing which is the cell proliferation stage by effectively reducing inflammation and promoting and enhancing the angiogenesis.

9.2.4.2 Modified PHA Polymers

9.2.4.2.1 Chemically Modified PHA Polymers

Numerous studies over the last decade have involved the modification of the PHA polymer to generate and enhance the innate qualities of the nascent polymer. Certain antimicrobial groups, hydrophilicity enhancing entities, other natural polymers, etc. are some of the add-ons that have been incorporated into the PHA polymer.

(a) PHA Polymers blended with Organic Components

Multiple studies have been conducted which involve the blending of PHA polymers with organic components like acids, polysaccharides, etc. An evaluation of the protein adsorption, antibacterial activity and fibroblast activity was performed on a PHBV membrane. The membrane was subjected to ozone treatment and was grafted with acrylic acid. The grafted membrane was esterified with either chitooligosaccharide (COS) or chitosan (CS). Post esterification, the CS or COS grafted membranes were immobilized with hyaluronic acid (HA). The antibacterial activity of these grafts was studied and compared. Cytotoxicity tests were conducted on L929 fibroblasts. The results revealed that antimicrobial activity of CS and COS was preserved after HA immobilization. Antimicrbial activity was tested against *Staphylococus aureus*, *Escherichia coli* and *Pseudomonas aeruginosa*. Antibacterial activity tests showed that CS/PHBV membranes demonstrated higher activity than

COS/PHBV membranes. In vitro tests demonstrated that the grafting with CS/COS enhanced L929 fibroblasts attachment and protein adsorption. However, it was observed that the cell numbers decreased. Post HA immobilization greatly improved cell proliferation while decreased the level of protein adsorption and cell attachment (Hu et al. 2003). Thus, one can conclude that the individual components had limitations which were overcome effectively through combination with each other. The biocompatibility, enhanced cell proliferation, protein adhesion and antibacterial activity against known nosocomial pathogens of the composite films suggests that they could potentially be utilized in clinical settings for the treatment and management of skin and wound healing.

Targeting the regeneration of damaged skin, a group of scientists experimented on a novel absorbable film. The aim was to assess the potential application of this new biomaterial as a scaffold for keratinocyte growth and proliferation. The polymers used were poly(3-hydroxybutyrate) [P(3HB)] and poly(4-hydroxybutyrate) [P(4HB)]. Blends of these polymers were combined with minute quantities of alginic acid, chitosan (CH), hyaluronic acid (HA) and pectin. Through solvent casting, these different polymers and polysaccharides were moulded into porous films. Mechanical properties (tensile strength, flexibility etc.) of the films were analysed. Comparison was made with two commercially available films PLA copolymer and HA benzyl ester. In vitro cytotoxicity tests were carried out using HaCaT keratinocytes. In addition, cell viability and proliferation studies were conducted for all the films and blends. Results of the mechanical studies showed that the porous films had favourable mechanical properties, that it, the tensile strength tests demonstrated flexibility of P(4HB) to a high degree. This was preserved in the P(4HB)-polysaccharide composite films while the P(3HB) film (being more brittle in nature than 4HB) and its composites, were not as flexible. Cytotoxicity tests revealed that the blends were not only harmless but significantly enhanced the growth and proliferation of the keratinocytes. This was concluded to be due to the porous nature of the surface of the PHA polymers/HA/CH blends. The quantity of HaCaT cells which grew on P(4HB) films was large. Cell numbers were determined and found to be in the descending order of: P(4HB)/HA > P(4HB)/CH > P(3HB)/HA > P(3HB)/CH > P(3HB)/pectin > P(3HB)/alginic acid. Therefore, the peak cell count was found on the P(4HB)/HA composite films. On blends of P(3HB) and P(4HB), the films were non-porous in nature and the cells proliferation rate increased with decreasing hydrophobicity: P(4HB) > P(3HB)/P(4HB) blend > P(3HB). SEM and confocal laser scanning microscopy revealed the formation of cell clusters on P(3HB) and its composites, while confluent layers of cells grew on P(4HB) and its composites (Peschel et al. 2008). Thus, P(4HB)/HA composite films are promising biocompatible biomaterial capable of enhancing keratinocyte growth and proliferation.

A study conducted by scientists focused on constructing a nano-fiber composed of collagen peptides which were incorporated in P(3HB-co-4HB). The objective of the study was to enhance cell growth and proliferation while increasing surface wettability. These traits are desirable for efficient biodegradable and biocompatible wound dressings.

The nano-fibers used in the experiments were produced through an electrospinning technique which involved simultaneous construction using a dual syringe system. Wettability studies were carried out and compared using different concentrations of the 4-HB polymer. Cytotoxicity tests of the nano-fibers were conducted on mouse fibroblast cells (L929). In vivo studies used rats (Sprague Dawley rats) to understand the immunological and biological responses to the nano-fibers.

The results revealed that the wettability of the nano-fiber constructs increased with the concentration rise of 4-HB, that is, from 20 mol% 4HB [53.2°] to 35 mol% 4-HB [48.9°] to 50 mol%4-HB [44.5°] to finally 82 mol%4-HB [37.7°]. In vitro cytotoxicity studies displayed a peak in cell growth and proliferation. In vivo experiments demonstrated that the nano-fibrous construct had a significant effect on contractions of wound healing with 79% as the highest percentage of wound closure reported (Vigneswari et al. 2016a).

A study, by the same group of scientists, on increasing the hydrophilicity of PHA polymers for enhanced wound healing, was conducted recently demonstrated the effectiveness of incorporating fish-scale collagen peptides (FSCPs) onto P(3HB-co-4HB) films. This polymer is known to be hydrophobic with a small number of cell attachment recognition sites. In addition, P(3HB-co-4HB) scaffolds had FSCPs incorporated into it. Results revealed that the water contact angle decreased with the increase in concentration of FSCPs. An In vitro study elucidated that L929 mouse fibroblast cells were much better able to attach and grow onto these scaffolds while an in vivo study demonstrated a significant effect on wound healing where the highest percentage of closing of the wound was 61% in 7 days by P(3HB-co-4HB)/1.5 wt% FSCPs (Vigneswari et al. 2016b).

(b) PHA Polymers blended with Different Polymers

Another avenue for exploration was revealed when scientists combined PHA with different polymers, both natural and synthetic in nature.

An evaluation on the potential utilization of a blend of two polymers, polyvinyl alcohol [(PVA), synthetic polymer] and PHB, scaffolds for tissue engineering was conducted. Nano-fibrous blends were prepared from the two polymers by the electrospinning method. Physical characterization of the polymers and their blend was studied using differential scanning calorimetry (DSC), Fourier transform infrared spectroscopy (FTIR) and electron microscopy. Degradation behaviour of the blends was studied. In vitro cytotoxicity, cell viability and adhesion studies were conducted on two cell lines: human keratinocytes (HaCaT) and dermal fibroblast cells. Observations of the characterization revealed that the PVA and PHB polymers were miscible with good compatibility properties. A noticeable depression in the levels of crystallinity was observed in the PHB component of the PHB/PVA blends correlating to a rise PVA quantity in the blend. PHB degradation was increasingly rapid with the increment of the PVA fraction. Cell culture experiments showed that peak adhesion and proliferation occurred on pure PHB nano-fibers. Interestingly, 5% PVA in the PHB/PVA blend was inhibitory to HaCaT cells but not to the fibroblast cells while a 50/50 concentration of PHB and PVA greatly promoted the adhesion and proliferation of HaCaT cells and inhibited fibroblastic growth (Asran et al. 2010). Thus, depending on the desired area and the type of cell growth, these blends could possibly be used as biomaterials for constructing strong and stable scaffolds for regeneration of damaged tissues.

Research on the potential healing wounds using poly(3-hydroxyalkanoates)-co-(6-hydroxyhexanoate)-ε-caprolactone [(PHA-PCL), PCL is a synthetic polyester] composite hydrogel was conducted using rats. The advantage of using a hydrogel matrix as a wound dressing lies with its high water absorption power which in turn, affords the hydrogel the capacity to retain hydration in the wound while restoring the skin and facilitating the replacement of dressing material (Dong et al. 2014; Gumel et al. 2015). The scientists enzymatically synthesized PHA-PCL using mcl-PHA from *Pseudomonas putida* Bet001, ε-caprolactone and immobilized enzyme (Novozym 435). The bacterium was grown in production medium with undecanoic acid which served as the only source of C. 77.3 KDa molecular weight with a monomeric composition (mole%) of C7 (20.1%), C9 (20.3%), C10 (1.9%) and C11 (57.7%) was established. Cross-linking of biosynthesized PHA-PCL macromere and polyethyleneglycol methacrylate (PEGMA) was undertaken to produce the hydrogel. Twenty-four rats were randomly grouped into four groups of six where group I and group II were the negative and secondary controls, respectively. Group I rats were treated with sterile gum acacia paste and group II with only PEGMA hydrogel (PH). The test group III rats were subjected to PHA-PCL cross-linked PEGMA hydrogel (PPH) and group IV rats (positive control) received Intrasite® gel treatment.

The MTT test for cytotoxicity using human liver cells (WRL68) was conducted and showed no difference in proliferation and viability despite providing longer reaction time. The researchers reported that wound closure was higher in PPH or Intrasite® gel groups when compared to groups I and II. In addition, it was observed that treatment with PPH or Intrasite® gel displayed an improvement in the deposition of collagen, fibrosis, etc. on the 14th day when compared to group I. Although there was an improvement in healing of group II, there was no significant difference between it and group I in the tests (Gumel et al. 2015). Thus, these inferences suggest a topical application of PPH accelerated the rats' wound healing process. This promising hydrogel composite could potentially find application in accelerated wound healing in humans and animals.

(c) PHA Polymers blended with Stem Cells

In a study that joined together two separate fields, stem cell research and polymer research, unrestricted somatic stem cells (USSCs) were loaded onto a biodegradable biopolymer, PHBV. The work involved designing and preparing a nano-fibrous PHBV scaffold using electrospinning. The scaffold was then subjected to modification through the immobilization of collagen, achieved through the plasma radiation method. Physical and mechanical characterization was conducted. In vivo studies were performed on 50 Wistar white rats (male) grouped into five sets of ten rats each. The first group was treated with PHBV nano-fibrous mat, the second with PHBV loaded with USSCs, the third with PHBV/collagen blends, the fourth with

PHBV/collagen blends loaded with USSCs and finally the fifth served as a control. A wet compress (standard issue) was used to cover the wounds in order to combat the detachment and possible desiccation of the scaffold. Observations were recorded after 21 days. The results revealed a contact angle of 105° for PHBV films and 56° for collagen-coated nanofibers. This is due to the ability of collagen to enhance the hydrophilicity of a material and is evident in this study along with numerous other studies mentioned in this text. The tensile modulus and tensile strength of the scaffolds was found to be equal and comparable, respectively, to that of human skin. Good resilience and compliance to movement was noted. In vivo studies revealed negligible to no inflammation with a statistically significant pronounced enhancement in wound closure as seen on day 21. The collagen-coated nano-fibrous mat and collagen-coated nano-fibrous mat loaded with USSCs demonstrated more than 82% and 89% closing of the wound area, respectively. Wounds in group B examined through histological studies exhibited the growth of the epidermal and dermal layers. Immunostaining revealed the benefit of using scaffolds implanted with cells because of which a thicker layer of epidermis was formed. USSC-collagen-PHBV scaffold yielded better results during skin healing in rat model (Keshel et al. 2014). Thus, the study successfully demonstrated the application of PHBV/collagen/USSCs blended nano-fibrous mats for full-skin healing in rats and perhaps, potentially for human application.

(d) PHA Polymers blended with Antimicrobial Agents

Research was conducted on the possibility of utilization of PHA with guar gum and curcumin for an effective antimicrobial wound dressing material. Guar gum, a natural polysaccharide, has a high-molecular weight and water soluble properties. It is composed of mannose backbone (β-1,4-linked) and galactose residues joined (1,6-linked) to the backbone at regular intervals. Traits like biocompatibility, biodegradability, hydrophilicity etc. make guar gum a sought after material that could potentially be utilized as a biomaterial for wound dressings. Curcumin (diferuloylmethane,) is a bioactive polyphenol agent obtained from *Curcuma longa* (commonly known as turmeric), a spice most often used in the Indian cuisine. It has been reported that it plays an essential role in combatting cancer, infections and in treatment of wounds (Pramanik et al. 2015). The polymer PHBV was produced by *Alkaliphilus oremlandii* ohILAs strain. After extraction and purification, the polymer was blended with guar gum and curcumin using solvent casting to give different composite films (GPC) with varying concentrations of the components. Physical and mechanical characterization of the different blends was carried out, including ATR-FTIR spectra and TGA analyses. Protein adsorption studies were performed using Bovine Serum Albumin (BSA). As curcumin is a known antimicrobial agent, the antibacterial activity of pure curcumin and curcumin incorporated films was undertaken. Testing was carried out through well diffusion and disc diffusion method, respectively. Test pathogens were *Vibrio vulnificus, Bacillus subtilis, Staphylococcus aureus, E. coli* XL1B, *Pseudomonas aeruginosa* and *Enterobacter aerogenes.* In vitro drug release studies were conducted using guar gum/PHBV/curcumin films. Cell cytotoxicity and viability studies were performed on fibroblast

NIH 3T3 cells and assessed using the MTT assay. In vivo experimentation involved usage of male mice with a thigh wound. The wound was dressed uniformly with the guar gum/PHBV/curcumin films and protected by commercial gauze. Two groups of control mice were maintained: one treated with only the test film and the other with only commercially available gauze. The wound healing and contraction was measured for days 0, 2, 4 and 7.

Thirty percent PHBV content in guar gum/PHBV film showed maximum rigidity and tensile strength of 19.8 MPa, attributed to non-covalent interactions. Additionally, rough surface topography (viewed through electron microscopy) of guar gum/PHBV/curcumin facilitated the adherence and proliferation of cells. Protein adsorption, indicated by the increasing order of BSA content, was found to be 64% and 70% on guargum/PHBV and guar gum/PHBV/curcumin blends, respectively. Antibacterial activity was determined through the inhibition of several microorganisms in the presence of pure curcumin and curcumin loaded film. Significant antimicrobial activity was demonstrated against gram-positive strains. In vitro drug release studies indicated that curcumin release from guargum/PHBV/curcumin blend films occurred in a sustained process of 62.86% in 36 h. Notably, the rate of the drug release decreased with increasing PHBV content. The cytotoxic effects of curcumin depend on the dose and the type of cell line. It was found that a similar number of cells were adhering to both 7:3 guargum/PHBV and curcumin loaded guargum/PHBV blend films. Hence, it was concluded that guargum/PHBV/curcumin composite is suitable for fibroblast cell adhesion as well as cell proliferation because of lack of cytotoxic responses. The in vivo studies revealed that the guar gum/PHBV/curcumin composite film enhanced healing with nearly 90% of closing of wound area (Pramanik et al. 2015). Thus, attributing to the success of this novel blend was the addition of curcumin and guar gum which aided in enhancing the innate properties of the biocompatible PHBV. Potential application of guargum/PHBV/curcumin films in wound dressings has been proven possible.

One of the major local factors inhibiting natural wound-healing processes is bacterial biofilm and infections (Guo and DiPietro 2010; Marcano et al. 2017). Especially troublesome is *Staphylococcus aureus* which is notorious for biofilm formation on surgical implants and its mutant Methicillin-resistant *S. aureus* (MRSA) has developed a resistance to almost every β-lactam antibiotics. The challenges faced as a consequence of infections and biofilm formation are twofold. With regard to biofilms, a slime layer is produced which serves as a protective shield against any antimicrobial agents and host defences. As a result, the biofilm continues to grow and spread. With regard to infections, both, the bacteria and the endotoxins produced, cause a prolonged influx of cytokines which extend the inflammation phase. The wound can likely become chronic and fail to heal if there is continuous extension of inflammation. Consequently, an increase in the matrix metalloproteases (MMPs) level (brought about by prolonged inflammation), destroys the ECM, thus severely affecting the proliferation phase (Guo and DiPietro 2010; Dinjaski et al. 2014; Gallo et al. 2014).

In order to combat the formation of biofilms on surgical implants and in open wounds, a recent study was conducted to demonstrate the effectiveness of an asymmetric PHA – polyvinylpyrrolidone (PVP) blend, incorporated with Dispersin B

(DB), an antibiofilm protein, prepared using wet induced phase separation (WIPS). The results showed that whatever was the molecular weight, the porogen content up to 30% increased membrane surface porosity and consequently protein uptake, above which pore size and the physical integrity/mechanical robustness both decreased. Antibiofilm activity was evaluated against *Staphylococcus epidermidis* biofilms which clearly showed up to 33% halting of biofilm formation and 26% destruction of pre-formed biofilm (Marcano et al. 2017).

A recent study demonstrated the superior traits of a PHA polymer PHBV which has graphene oxide (GO) and collagen incorporated into it. The resulting composite was a nano-fibrous scaffold obtained through electrospinning the polymer. The nano-fibrous films were characterized using Field Emission Scanning Electron Microscopy (FESEM) and FTIR Spectroscopy. Antimicrobial activity of the composite nano-fibrous mats was tested against *E. coli* and *S. aureus*. The results of the FESEM showed that the diameter of the nano-fibrous mats decreased with an increase in porosity attributed to the addition of GO and collagen. No chemical interaction was observed between the three components. PHBV offered a sturdy structure which resembles ECM, in turn contributing to fibroblast cell migration and proliferation. Antimicrobial activity against *E. coli* and *S. aureus* was confirmed to be due to GO. In addition, the incorporation of GO was found to increase the strength of the scaffold accompanied by a decrease in hydrophilicity and pore size. However, the decrease in hydrophilicity was compensated by the addition of collagen without significantly altering the strength and pore size. Further, the collagen was responsible for enhancement of cell growth and proliferation as compared to combinations: PHBV/GO and just PHBV alone (Zine and Sinha 2015). Thus it is evident that GO and collagen together form a favourable combination with PHBV that can be potentially utilized as a biomaterial for wound healing.

(e) PHA Polymers blended with Synthetic Agents

A major factor determining successful regeneration of skin is a satisfactory scaffold (Shishatskaya et al. 2016). Targeting the first phase of wound healing, hemostasis, a study based on the incorporation of bioactive glass particles of nanoscale (n-BGs) onto a biodegradable polymer PHB or P(3HB) material was undertaken. Bioactive glass is a material that is developed using silica that is incorporated with calcium and phosphorous. The function of this material was for repair of broken bones. This is possible as bioactive glass mimics bone thereby stimulating regrowth (Krishnan and Lakshmi 2013).

Hemostatic bioactive glass (BG) releases Ca^{2+} ions on hydration, a phenomenon that is necessary to facilitate thrombosis. In addition to which the polar surfaces of BG activate blood clotting because on exposure to foreign polar surfaces, various blood-clotting factors (prekallilrein, kininogen etc.) are triggered, which in turn activate thrombin and subsequent fibrin polymerization. Thus, BG is considered to be the ideal material for hemostatis (Francis et al. 2016). Previous studies have shown that BG provides an efficient structure for thrombosis and in an optimum ratio with Ca^{2+} ions commences the blood clotting cascade (Ostomel et al. 2006; Francis et al. 2016). Thus, n-BG was designed as such to serve as a multifunctional structural material. The chemical composition of the n-BG used in the study was similar to 45S5Bioglass® microscale bioactive glass (m-BGs) that are commercially available. The difference, however, is the spherical shape of the n-BG along with a uniform surface as compared to the irregular size of m-BGs.

The study was divided into two main parts. The first involved the production of P(3HB)/n-BG microspheres, followed by compression into films and characterization of the same. The second involved the immersion of P(3HB)/(n-BG) microsphere films in simulated body fluid (SBF) for a period of 1, 3 and 7 days to induce HA formation on its surface, thereby giving rise to a functionalized film material. Analysis of the P(3HB)/(n-BG) microsphere composite films and the functionalized composite film was conducted to assess its influence on different structural and cellular factors including keratinocyte adhesion and viability (Francis et al. 2016).

The results of this study revealed a plethora of interesting findings. The characteristics of the P(3HB)/n-BG microspheres were determined to be as follows: diameter of 1.7 μm and a specific surface area of 9.4 $m^2 g^{-1}$; a smooth surface with agglomerations of n-BG particles; an rms roughness value of 4.05 μm; static contact angle of 56.89° and 323.5 μg of protein adsorption, etc. Cytotoxicity tests showed 100% cell viability with the composite film. In the thrombo elastograph tests it was concluded that clot detection time decreased linearly with the increase in n-BG concentration.

The second phase of the work elucidated the following characteristics: surface morphology was rough on days 3 and 7 as compared to day 1 (smooth); the rms roughness value of P(3HB)/n-BG composite microsphere films on day 1, 3 and 7 was 4.3 μm, 5.99 μm and 7.2 μm, respectively and a decreased water contact angle was observed on days 1, 3 and 7 was 12.60%, 19.29%, and 22.62%, respectively. Day 1 displayed the highest cell viability (198%) while 116% and 147% were observed for days 3 and 7, respectively (Francis et al. 2016).

Thus, the conclusions of this research enforced certain important attributes that a potential wound healing material should possess. Some of these are an optimal surface roughness, particles on the surface topography which enhance cell adhesion, proliferation and increase hydrophilicity of the dressing. As proved by this study, in wound healing preparations, the presence of n-BG dressing surfaces would be beneficial for increased hydrophilicity, biomineralization, improved attachment of cells, blood clotting and an increase in protein adsorption.

9.2.4.2.2 Naturally Modified PHA Polymers

A study was conducted using a novel PHA (PHACOS) biosynthesized from *Pseudomonas putida* KT2442 (wild-type strain) and *P. putida* KT42FadB. The latter is mutated in the fadB gene belonging to the β-oxidation pathway that plays a role in mcl-PHA biosynthesis. The novelty of this polymer is its monomer composition which contains thioester groups in the side chains. The groups confer new properties to the polymer, in addition to antibacterial properties (Escapa et al. 2011; Dinjaski et al. 2014).

The experimentation involved the assessment of the antibacterial properties of PHACOS and comparing it with a non-reactive natural polymer poly(3-hydroxyoctanoate-*co*-hydroxyhexanoate) [P(HO-co-HHx)] and a synthetic polymer poly(ethylene terephthalate) (PET). PHACOS and P(HO-co-HHx) were coated onto sterile disks of PET through solvent casting, allowed to dry at room temperature for 72 h and subsequently, were sterilized using ethylene oxide. The capacity of the polymers to prevent bacterial biofilm formation was tested using *Staphylococcus aureus* subsp. *aureus* and *Pseudomonas aeruginosa* CECT 4122. The disks were placed in 24 well plates and inoculated with the cultures. The biofilms formation was analyzed by environmental scanning electron microscopy (ESEM), crystal violet staining and CFU counting. Anti-adherence tests were carried out in a similar way and LIVE/DEAD *Bac*Light bacterial viability kitL-13152 staining technique was used for detection of adherents. The antibacterial activity of the disks was conducted according to the ISO 22196:2011. In vitro cytotoxicity was evaluated using either murine RAW 264.7 macrophages/BALB 3T3 fibroblasts while inflammatory activity was monitored only on RAW 264.7 cells and cell proliferation studies only on fibroblasts BALB 3T3. In vivo studies were carried out using sterile disks, subcutaneously implanted in BALB/c mice. Control mice were treated the same but received no in order to account for surgery-associated inflammation. Implants were then surgically removed after euthanasia and analysed for inflammatory cells.

The monomer content of PHACOS was determined to be 29.7% of non-functionalized monomers [17.5% of 3-hydroxyoctanoate (OH-C8), 10.3% of 3-hydroxydecanoate, and 1.9% of 3-hydroxyhexanoate (OH-C6) monomers] and 70.3% of functionalized monomers (46.4% of 3-hydroxy-6-acetylthiohexanoate and 23.9% of 3-hydroxy-4-acetylthiobutanoate monomers) while P(HO-co-HHx) consisted of 8.5% OH-C6 and 91.5% OH-C8. Electron microscopy revealed significantly lower (3.2-fold) formation of biofilm by *S. aureus* than *P. aeruginosa* on PHACOS while biofilms formed on P(HO-co-HHx) by both were close. Adhered cells were always observed in both polymers. PHACOS displayed significantly less *S. aureus* counts than P(HO-co-HHx). The percentage of viable *S. aureus* on PHACOS was much lower (20%) than those on P(HO-co-HHx) (83%), suggesting that PHACOS possesses anti-staphylococcal activity. Fatty acids with structures resembling PHACOS monomers (hexanoic, octanoic and 6-acetylthiohexanoicacids) were assessed for antimicrobial activity. The results suggested that the antimicrobial ability is due to the functionalized thioester side chains. PHACOS caused negligible immunological responses from murine macrophages while supporting fibroblast attachment. The demonstration of a similar percentage of inflammatory cells, found in tissues surrounding sterile PHACOS and *S. aureus* pre-colonized PHACOS implants and significantly lower numbers than *S. aureus* pre-colonized control polymers suggest that PHACOS has a contact active surface mode of antibacterial action (Dinjaski et al. 2014). Thus, this study successfully establishes the potential usage of functionalized polyhydroxyalkanoates (PHACOS) as an infection-resistant biomaterial against MRSA.

Table 9.1 summarizes the steady progress of work in bringing us closer to using PHA polymers and modified PHA polymers for improved treatment to skin damages and to healing wounds. The studies discussed here have laid a solid foundation and have opened up an interesting avenue for exploration.

Table 9.1 Summary of PHA applications in skin and wound healing

Polymer	Production process	Cell line(s)/ animal models	Application	References
PHA polymers				
PHB and PHBV	Electrospinning	L929 and Schwann cells (RT4-D6P2T)	Skin and nerve regeneration	Suwantong et al. (2007)
PHBVHHx		HaCaT	Skin engineering	Ji et al. (2008)
PHBHHx		Sprague-Dawley rats	Bio-substitute for tarsal repair	Zhou et al. (2010)
PHBV	Solvent casting and freeze-drying	Human keratinocytes (hKC) and dermal fibroblasts (hDFB)	Wound healing	Zonari et al. (2014)
P(3HB-co-4HB)	*Cupriavidus eutrophus* B10646 Electrospinning	Fibroblast cells from adipose tissues of rats	Wound dressing	Shishatskaya et al. (2016)
Modified PHA polymers – chemically modified PHA polymers				
(a) PHA polymers blended with organic components				
PHBV/CS, PHBV/COS, PHBV/CS/HA and PHBV/COS/HA	Esterification and immobilization	Fibroblast cells (L929)	Antibacterial wound dressing	Hu et al. (2003)
PHA polymers P(3HB)/P(4HB)/HA/CH blends	Solvent casting	HaCaT keratinocytes	Scaffold for keratinocyte growth and proliferation	Peschel et al. (2008)
P(3HB-co-4HB)/collagen peptides	Dual-syringe electrospinning	Mouse fibroblast cells (L929)	Wound dressing for enhanced cell growth	Vigneswari et al. (2016a)
P(3HB-co-4HB)/FSCPs	Aminolysis	Mouse fibroblast cells (L929)	Wound healing	Vigneswari et al. (2016b)
(b) PHA polymers blended with different polymers				
PHB/PVA	Electrospinning	Human keratinocytes (HaCaT) and dermal fibroblast	Scaffold for tissue engineering	Asran et al. (2010)
PHA-PCL	Enzyme	Human liver cells WRL68	Accelerated wound healing	Gumel et al. (2015)
(c) PHA polymers blended with stem cells				
PHBV, PHBV/USSC, PHBV/collagen, PHBV/collagen/USSC	Electrospinning	Wistar rats	Scaffold for full-skin regeneration	Keshel et al. (2014)

(continued)

Table 9.1 (continued)

Polymer	Production process	Cell line(s)/ animal models	Application	References
(d) PHA polymers blended with antimicrobial agents				
Guar gum/ PHBV/ cur-cumin films	Solvent casting	Fibroblast cells NIH 3T3	Antimicrobial wound dressing	Pramanik et al. (2015)
PHA/PVP/ disper-sin B	Wet induced phase separation (WIPS)	*S. epidermidis* biofilm	Antimicrobial wound dressing	Marcano et al. (2017)
PHBV/GO/ collag-en	Electrospinning		Antimicrobial scaffold	Zine and Sinha (2015)
(e) PHA polymers blended with synthetic agents				
P(3HB)/n-BG composite microsphere films		HaCaT CELLS	Healing of wounds	Francis et al. (2016)
Naturally modified PHA polymers				
PHACOS	Biosynthesis and solvent casting onto pet	Murine RAW 264.7 macrophages/ BALB 3T3 fibroblasts	Antimicrobial wound dressing against MRSA	Dinjaski et al. (2014)

9.3 Recent Advancements in PHA Applications – As Precursors for Oral Drugs

PHA are made up of (*R*)-3-hydroyalkanoic acids (3-HA or R3-HA) and serve as an alternate nutrition source. The monomers of PHA polymers, 3-HA, have great importance in the biomedical field, specifically in the pharmaceutical sector. 3-HA are enantiomerically pure, chiral by nature and have the presence of two important functional groups, that is, the hydroxyl and the carboxyl groups. In addition, ease of introducing a second chiral centre and modifying both of the functional groups adds to the benefits of using 3-HA as building blocks (Chen and Wu 2005a, b; Ren et al. 2010). Consequently, they have immense importance in the manufacture of drugs, vitamins, antibiotics, pharmaceuticals, perfumes, pheromones and flavour compounds.

The main reason for the preference of chiral (or enantiomerically pure) drugs is due of its efficient activity at low dosages. In addition, these drugs are significantly safer. An undesirable enantiomer, giving rise to an enantiomerically impure drug, causes innumerable side effects with a great reduction in drug activity (Chen and Wu 2005a, b; Holscher et al. 2010; Ren et al. 2010). An important example is the drug thalidomide. Marketed in the 1950s by a German company, it was prescribed as a sedative and as treatment for morning sickness in pregnant women. However, countless reports of babies born with birth defects led to the ban of thalidomide. Through extensive investigation that continues till date, it was determined that this drug induces a range of birth defects, the most common one being defects in developing limbs. Recently, it was discovered that thalidomide prevents the angiogenesis of limbs during the early limb formation phase. What is now referred to as the 'Thalidomide Tragedy' marked the beginning of intensive toxicity testing (using animal models and clinical trials) for any novel drug before its commercialization and public use (Smithells and Newman 1992; Therapontos et al. 2009; Kim and Scialli 2011).

Studies report the utilization of 3-HA, produced from (*R*)-3-hydroxyalkanoates, as chiral building blocks for macrolide antibiotics (the macrocyclic part of Elaiophylin), peptide antibiotics (the hydroxyacyl hydrazine part of Visconsin), fungicides (Vermicullin), pharmaceuticals (β-lactams), etc. Certain studies reported that 3-HB was used as a precursor of 4-acetoxyazetidinone for the synthesis of carbapenem antibiotics (Lee and Lee 2003; Chen and Wu 2005a, b; Holscher et al. 2010; Ren et al. 2010; Tseng et al. 2010; Subin and Bhat 2014).

9.3.1 Synthesis of 3-HA

The production of enantiomerically pure chemicals from resources which are renewable has begun to attract attention and this has been observed especially in the pharmaceutical industry.

Conventional chemical processes for the production of selective stereo-chemicals, are known to utilize expensive raw materials and chemicals like chiral catalysts, solvents, etc. In addition, harsh physical conditions are required and an excess of by-products are formed. This decreases the overall output of the desired chemical, resulting in a less-than efficient process. Hence, there is a need for an improved and efficient process to combat the drawbacks of conventional chemical production processes. Hypothesized processes like enzyme catalysis, PHA depolymerisation, etc. have been proposed and tested for the production of 3-HA.

Enzyme catalyzed reactions result in the production of only one of the two enantiomers due to the highly specific nature of the enzymes. The fact that the reaction can be accomplished under mild conditions in an aqueous system greatly benefits the entire process. Thus, enzyme catalysed reactions are considered to be an efficient manufacturing method. It has been observed that chiral compounds are normally obtained in enzyme-catalyzed reactions. As a result, replacing chemical processes by biological ones from renewable sources has been explored (Chen and Wu 2005a, b; Ren et al. 2010; Tseng et al. 2010).

The synthesis of 3-HA (Fig. 9.2) can be broadly categorized into two on the basis of the nature of the production process: (1) Chemical Synthesis and (2) Biological Synthesis.

9.3.1.1 Chemical Synthesis of 3-HA

The chemical synthesis of 3-HA involves different approaches. One approach utilizes the enantio-selective reduction of 3-keto esters while another applies the Sharpless' epoxidation and hydroxylation of allylic alcohols. These approaches are categorized as 'de novo' synthetic methods. In addition, an approach where PHA

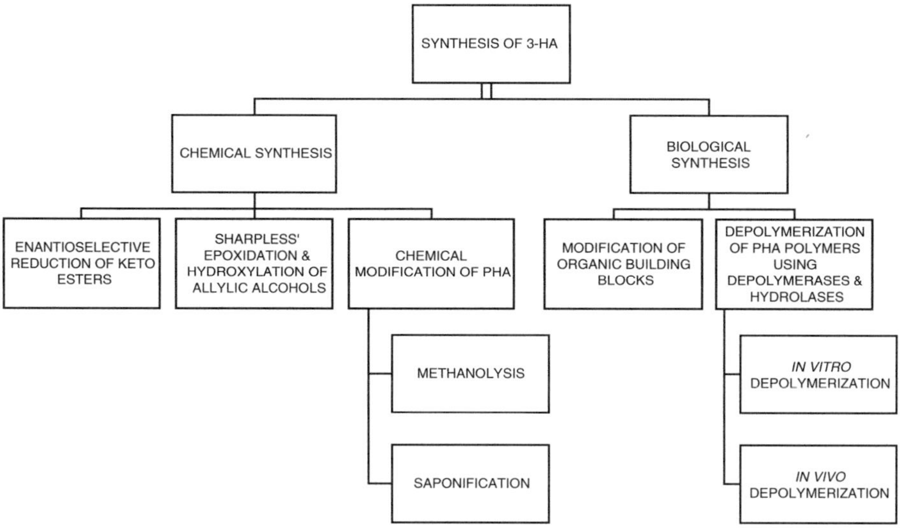

Fig. 9.2 Representation of 3-HA synthesis

are subjected to different chemical reactions (methanolysis, saponification, etc.) to synthesize 3-HA has been in use (Ren et al. 2010).

(a) *De novo* Chemical Synthesis

(*R*)-3-hydroxyalkanoic acids, that are enantiomerically pure, are difficult to synthesize chemically. However, a method for this synthesis which involved microwave-assisted catalytic transfer hydrogenation and a facile microwave-assisted hydrolysis of an N-methoxy-N-methyl (Weinreb) amide was proposed (Chen and Wu 2005a, b). The incorporation of a chiral centre during production of enantiomerically pure 3-HA is fraught with challenges. A number of methods have been proposed and tested. Through the Sharpless' asymmetric epoxidation and hydroxylation or through asymmetric allylboration developed by Brown, stereo-selective oxidation can be achieved (Brown and Ramachandran 1991; Ren et al. 2010). The enantio-selective reduction process has given rise to 3-hydroxy esters which are enantiomerically pure. The raw materials or the precursors used were 3-keto esters (Noyori et al. 2004; Ren et al. 2010). However, numerous problems are encountered while applying such processes and include the requirement of expensive chemical building blocks and catalysts, contamination of end products, harsh reaction environments (high pressure, special media, low/high temperatures), etc. In addition, precursor synthesis is a necessity in most manufacturing processes and this complicates production and could contribute in lowering the yields of the desired products (Brown and Ramachandran 1991; Ren et al. 2010). Another major drawback could be a lowering in the enantiomeric excesses (ee) as compared to biochemical processes (Ren et al. 2010).

(b) Chemical Modification of PHA

PHA represents a potential source of chiral hydroxyl acid feed stock for the fine chemical industry. All constituent monomeric units are enantiomerically pure and exist in (*R*)-configuration. As a result, it was deduced that different enantiomerically pure 3-HA could be easily prepared by PHA depolymerization (Chen and Wu 2005a, b; Ren et al. 2010).

One of the first reports of chemically modifying or degrading PHA was used to produce R-3HB and R-3HV from PHB and PHBV, respectively (Seebach et al. 1992). A decade later, a group of scientists used hydrolytic degradation to produce mcl-3-HA. The parent polymer was produced by *Pseudomonas putida*. The polymer was degraded using acid methanolysis and the resulting 3-HA methyl esters were subjected to distillation. Saponification of the methyl esters yielded the desired 3-HA (de Roo et al. 2002; Ren et al. 2010).

9.3.1.2 Biological Synthesis of 3-HA

Biological synthesis (biotransformation) of 3-HA involves the utilization of enzymes: to produce 3-HA and to degrade PHA to release 3-HA. Numerous strategies have been hypothesized.

(a) *De novo* Biological Synthesis

Numerous studies have been conducted on the production of 3-HA using microbes and their enzymes to modify and introduce chiral centres into chemicals (which are precursors of 3-HA). These studies involve the utilization of other organic building blocks (fatty acids, keto acids, carbohydrates, etc.) and not PHA for 3-HA synthesis. Through fermentative carbohydrate conversion, 3-hydroxypropionic acid was produced. A different study involving the concerted action of nitrile hydratase and amidase from *Comamonas testosteroni* 5MGAM4D was performed. The study involved utilization of these enzymes for the hydrolysis of 3-hydroxyalkanenitriles to 3-HA. However, it must be noted that the chirality of resulting 3-HA wasn't determined. If these 3-HA are to be utilized for pharmaceuticals, a separation of the R and S enantiomers is a necessity (Hann et al. 2003; Ren et al. 2010).

(b) Enzymatic Depolymerization of PHA polymers

Various enantiomerically pure 3-HA can be prepared from the depolymerisation of PHA polymers (Lee and Lee 2003). There are various techniques that can be applied to achieve depolymerisation which involve either of two methods: (1) In vitro depolymerization and (2) In vivo depolymerization. These are discussed as follows:

(i) *In vitro* depolymerization

In vitro depolymerization is achieved by secretion of extracellular enzymes like PHA depolymerases, lipases etc. The rate of degradation was strongly dependent upon polymer composition, environmental conditions, etc. The action of extracellular PHA depolymerases results in monomer and/or dimer 3-HA. A study conducted on 3-HB production revealed that a thermophilic strain of *Streptomyces* sp. MG had the ability to hydrolyze PHB. Being thermophilic (and stable at 50 °C) ensured less contamination tendencies and greatly facilitated downstream processing as the cells aggregated post fermentation (Calabia and Tokiwa 2006; Ren et al. 2010).

Through years of research, it was observed that most lipases have the ability to breakdown polyesters comprising of Ω-hydroxyalkanoic acid such as PHHx or P(4HB). PHA were partially hydrolysed by these lipases. Extracellular lipases produced by *P. aeruginosa* yielded a dimeric ester of HHx on PCL hydrolysis (Ren et al. 2010).

(ii) *In vivo* depolymerization

In vivo depolymerization can be achieved in two ways. The first method involves the depolymerization within wild type cells, using the cell's own depolymerization machinery. The second method involves the genetic modification of the cell and includes the introduction of genes into non-PHA producers, switching off existing genes within producers, alteration of PHA metabolic pathways within wild type and recombinant PHA producers, etc.

In vivo depolymerisation results in enantiomerically pure (*R*)-3-hydroxyalkanoic acids with 4–12 carbon atoms. The principle of this method is based on providing conditions wherein high intracellular PHA depolymerase activity coupled with low dehydrogenase activity is present. Numerous studies have put this theory to the test and have successfully achieved production of 3-HA (Chen and Wu 2005a, b). *P. putida* cells were suspended in phosphate buffer of varying pH. It was observed that alkaline pH supported the best degradation and production of 3-HA. Efficient degradation [over 90% (w/w)] in 9 h with corresponding monomer yields over 90% was noted in PHA with monomer composition of 3-HO and 3-HHx (Ren et al. 2005, 2010).

E. coli was genetically engineered by introducing *Cupriavidus necator* genes (*pha* CAB operon and depolymerase genes), resulting in the production of enantiomerically pure 3-HB. The study revealed that 3-HA production could be accomplished through the removal of CoA from (*R*)-3-hydroxyacyl-CoA via in vivo depolymerisation. 3-HB dimers were produced by this organism grown on a glucose production medium. The resulting 3-HB dimers were converted to monomers through mild alkaline heat treatment (Lee and Lee 2003; Park et al. 2004; Chen and Wu 2005a, b). Other studies on recombinant technology using *E. coli* were conducted with success (Gao et al. 2002; Wu et al. 2003; Park et al. 2004). In addition, 3-hydroxydecanoyl was produced through recombination of *E. coli* using *P. putida* genes (Zhao et al. 2003; Zheng et al. 2004). Thus, it was evident that depolymerases (in vivo and in vitro) could be employed to degrade scl- and mcl-PHA, thereby providing a renewable source of 3-HA.

9.3.2 Physiological Role of 3-HA

3-HB and 4-HB have been investigated for their physiological role in a biological system. 3-HB, being the simplest monomer, is one of the three major ketone bodies produced in the body, especially during disease or starvation. Some of the properties of 3-HB is its ability to transcend the blood-brain barrier, tolerance at low levels by the body, etc. (Tieu et al. 2003; Chen and Wu 2005a, b). Under normal conditions, 3-HBis produced by the liver, released and is absorbed by primarily the heart and skeletal muscles, where 3-HB is transformed into acetyl-CoA and participates in the TCA (Chen and Wu 2005a, b). Studies have shown that partial protection is provided against dopamine neuro-degeneration and motor defects in mice (possible application for Parkinson's disease treatment) (Tieu et al. 2003). In addition, it confers stability to neurons during conditions of glucose deprivation, increases cardiac output and prevents brain damage (Ugwu et al. 2008). High levels of 3-HB have been observed in diabetic ketoacidosis. In the human system, complex oligo (*R*)-3-hydroxybutyrate (O3HB) have been found in atherosclerotic plaques. O3HB levels were compared in kidney, plasma, eye, aorta, brain, etc., of healthy control and streptozotocin-diabetic Sprague-Dawley rats. O3HB levels were high in tissues of diabetic rats. This indicated that O3HB is connected to diabetes-affected organs (Reusch et al. 2003; Chen and Wu 2005a, b).

4-HB has neuro-modulatory properties. Succinic semialdehyde dehydrogenase (SSADH) deficiency is a rare innate error in human metabolism that interferes with the normal metabolism of gamma-aminobutyric acid (GABA), an inhibitory neurotransmitter. The biological system responds to this disorder through the accumulation of 4-HB in physiologic fluids. The sodium salt of 4-HB has been analysed for treatment of withdrawal symptoms of patients with substance-abuse disorders (alcoholism). The results revealed that 78% of the patients were successful in abstaining from alcoholism due to treatment with 4-HB without any systemic or single-organ consequences. In addition, significant reduction in cravings was observed (Chen and Wu 2005a, b).

9.3.3 Potential Applications of 3-HA

3-HA comprise of a chiral centre and two modifiable functional groups. Thus, they are high-value starting materials (synthons) for pharmaceuticals like antibiotics, vitamins, etc. (Chen and Wu 2005a, b; Ren et al. 2010).

The potential applications of 3-HA are numerous. An important application of 3-HA is its utilization as chiral synthons for the production of organic compounds. For example, the synthesis of 3,5-dihydroxydecanoic acid (a δ-lactone) can be undertaken using (*R*)-3-hydroxyoctanoate. The drawback of the conventional chemical synthesis is the necessity of expensive complex catalysts (metals) to incorporate a chiral centre. Hence, use of 3-HA would greatly simplify the synthesis of saturated and unsaturated aliphatic β-lactones (Ren et al. 2010).

Chiral 3-HA find special application in the creation of tailor-made polymers. The challenge faced in polymer synthesis is the goal to achieve high molecular weights. The important factor, for condensation of 3-HA for polyester manufacture, is carboxylic group activation and the removal of H_2O to shift the equilibrium to the side of the polyester product. Seebach and Fritz (1999) produced PHB from its monomers through the addition of $COCl_2$ and pyridine at beyond freezing temperatures (−78 °C) carboxylic group activation (Ren et al. 2010).

3-HA have been reported to aid in the manufacture of β-amino acids. Peptides containing β-amino acids are known to be resistant to enzymatic degradation because proteases and peptidases are unable to break amide bonds close to the β-amino acid. Thus, their utilization as scaffolds for peptide mimics is justified (Park et al. 2001; Ren et al. 2010). 3-HA have also been reported to find use in nutrition and as supplements for injured patients. This aspect has been discussed in detail in the next section.

Certain 3-HA have demonstrated antimicrobial activities against selective pathogens. A study performed tested the antimicrobial activities of R-3-HO, (*R*)-3-hydroxy-8-nonenoic acid, and (*R*)-3-hydroxy-10-undecenoic acid. It was observed that these hydroxy acids exhibited much higher activities against *Listeria* species and *S. aureus* than their racemic mixtures or their non-hydroxylated free fatty acid counterparts (Ruth et al. 2007; Ren et al. 2010).

9.3.4 Recent Advancements in PHA Application for Synthesis of 3-HA

9.3.4.1 *In Vitro* Depolymerization

An experiment on the production of 3-HO through the depolymerase activity of *Pseudomonas fluorescens* GK13 was undertaken. The extracellular enzyme was purified and immobilized onto a polypropylene support. This enzyme catalyses the hydrolysis of poly(3-hydroxyoctanoic acid) [P(3HO)]. The immobilized enzyme, after 24 h of hydrolysis, yielded the complete degradation of P(3HO) polymer to (*R*)-3-HO monomers while the soluble enzyme yielded dimeric esters of 3-HO in the same reaction conditions (Gangoiti et al. 2010).

9.3.4.2 *In Vivo* Depolymerization

The formation of enantiomerically pure (*R*)-3-hydroxycarboxylic acids from PHA biosynthesized by *Pseudomonas putida* GPo1 was evaluated in a recent study. Conditions were provided to the organisms to induce and favour in vivo depolymerase activity. Once depolymerization was complete, the monomers were ejected into the extracellular environment and were subjected to separation and purification by acidic precipitation, preparative reversed-phase column chromatography, and finally, solvent extraction. Eight(*R*)-3-hydroxycarboxylic acids were isolated, having a purity of over 95 wt%: (*R*)-3-hydroxyoctanoic acid, (*R*)-3-hydroxyhexanoicacid, (*R*)-3-hydroxy-10-undecenoic acid, (*R*)-3-hydroxy-8-nonenoic acid, (*R*)-3-hydroxy-6-heptenoic acid, (*R*)-3-hydroxyundecanoic acid, (*R*)-3-hydroxynonanoic acid, and (*R*)-3-hydroxyheptanoic acid. The overall yield based on released monomers was around 78 wt% for (*R*)-3-hydroxyoctanoic acid (Ruth et al. 2007).

Cupriavidus necator VCM 11282 was subjected to UV mutagenesis in a study to obtain (R)-3HB in culture supernatant. The mutants generated as a result of such treatment was able to produce (R)-3HB. In addition, acetoacetate esters were added to the production medium. Through UV mutagenesis, a disruption in the phbB gene (phbB knock-out) occurred and this directed the production of (R)-3HB in mutant strains. The addition of the esters significantly increased the (R)-3HB production. Thus, the conversion of acetoacetyl-CoA (an intermediate of the PHB pathway) to acetoacetate, through the action of (R)-3HB dehydrogenase resulted in extracellular production of (R)-3HB (Ugwu et al. 2008). (R)-3HB dehydrogenase catalyzes the reversible oxidation of hydroxybutyrate to acetoacetate, utilizing NAD+ as coenzyme. This enzyme is important as the transformation it catalyzes is necessary for determination of ketone bodies (acetoacetate and 3-HB), used in diagnosis of insulin-dependent diabetes mellitus (type I) (Kruger et al. 1999). Evidence suggests that biosynthetic pathway exists in *C. necator* for formation of (R)-3HB. This is the first report via acetoacetate for (R)-3HB formation (Ugwu et al. 2008).

In a study conducted, a methylotrophic bacterium was genetically modified to produce 3-HB from methanol. The dehydrogenase gene in *Methylobacterium rhodesianum* MB 126 was inactivated. This resulted in a mutant which was able to grow (low rate) on 3-HB medium. This indicated that an alternative pathway was present by which 3-HB could be utilized. A second mutation was induced within the mutant and a double mutant devoid of lipoic acid synthase (LipA) was obtained. This mutant was unable to grow on 3-HB and thus, could utilize methanol for PHB synthesis and degradation to yield 3-HB (Holscher et al. 2010).

A spike in the number of studies on recombinant *E. coli* has been observed over the last 15 years. In recombinant *E. coli,* the biosynthetic pathway for (*S*)-3-hydroxybutyric acid (S-3-HB) production was established. The medium used was a glucose based-medium. The success of this study was attributed to the introduction of phaA from *R. eutropha*, the (*S*)-3-hydroxybutyryl-CoA dehydrogenase gene from *R. eutropha* or *Clostridium acetobutylicum* ATCC824 and the 3-hydroxyisobutyryl-CoA hydrolase gene from *Bacillus cereus* ATCC14579. This recombinant *E. coli* was capable of producing enantiomerically pure S-3-HB. A similar study revealed that the chirality of 3-HB could be controlled by metabolic pathway engineering (Lee et al. 2008; Tseng et al. 2009; Ren et al. 2010). Recently, a study was undertaken to optimize strategies for 3-HB production in a recombinant strain *E.coli* strain AF1000pJBGT3RX. This strain is known to produce extracellular 3-HB due to the expression of PHB degradation pathway genes from *Halomonas boliviensis*. This halophile accumulates PHB under conditions of excess carbon in the presence of a limiting nutrient like P or N. An ammonium/phosphate limited process was used as a comparison for 3-HB productivity at different nutrient limitations, in order to control the flux from acetyl-CoAto3-HB. The highest 3-HB volumetric productivity of 1.5 g/L/h was observed under phosphate-limiting conditions (Guevara-Martínez et al. 2015).

In addition to genetically modified *E. coli,* pseudomonads have been genetically modified and studied for their potential as producers of 3-HA. An economical and high yielding 3-HA method was developed for 4-HV and 3-HV production using renewable resources (levulinic acid). The study involved fermentation of levulinic acid by *P. putida* KT2440 expressing the tesB gene from *E. coli* (Ren et al. 2010). Subsequently, a similar study was conducted on the production of mcl-3-HA, R-3-HHx and R-3-HO, using *P. putida* KT2442. The organism was genetically modified to overexpress the depolymerase gene along with a lcl fatty acid transport gene (fadL) of *P. putida* KT2442 and acyl-CoA synthetase gene (fadD) of *E. coli* MG1655. After 48 h fermentation of 3-L sodium octanoate mineral medium, 5.8 g/L of extracellular R-3-HHx and R-3-HO was obtained (Yuan et al. 2008; Ren et al. 2010). Using a different genetically modified pseudomonad, *P. putida* KTOY01, was capable of direct mcl-3-HA synthesis. This was due to the inactivation (knock-out mutant) of the PHA synthesis operon. The organism was able to produce R-3HHx, R-3HO, R-3HD and (*R*)-3-hydroxydodecanoate (R-3HDD) using dodecanoate as the sole carbon source (Chung et al. 2009; Ren et al. 2010). In a related study, a metabolic pathway was engineered to express the *tesB* gene (encoding thioesterase). This is able to catalyze acyl-CoA to free fatty acids. The *tesB* of *E. coli* (thioesterase II) was incorporated into *P. entomophila* LAC31 (derived from wild type *P. entomophila* L48) which has PHA synthase and the β-oxidation pathway deleted. The results revealed that on addition of tetradecanoic acid or dodecanoic acid as related carbon sources in the medium, 6.65 g/l 3-hydroxytetradecanoic acid (3-HTD) and 4.6 g/l 3-hydroxydodecanoic acid (3-HDD) were obtained, respectively. In addition, *P. entomophila* LAC31 also harboured the *PTE1* gene from *Saccharomyces cerevisiae.*1.8 g/l of 3-hydroxydecanoic (3-HD) acid was produced by using the corresponding fatty acid decanoic acid. In an interesting turn of events, it was observed that *P. entomophila* LAC31 with *PTE1* favoured 3HDD and 3HD production while *P. entomophila* LAC31 with *tesB* favoured 3HTD production. The significance of these studies provide evidence that fine chemical precursors (3-HA) can be efficiently produced through genetic manipulation of the β-oxidation in *Pseudomonas* species (Chung et al. 2013).

In a recent study, a novel pathway for the synthesis of different 3-HA was designed. The enzymes involved in this pathway were capable of condensing acyl-CoA molecules, reducing the ketone formed in a stereo-selective manner and hydrolysing CoA thioester to release the free acid. In this method, the standard substrate, acetyl-CoA, is one substrate while the second substrate varies depending on the desired final product. The second substrate was obtained through intracellular production from the supply of a starting raw material. Feeding of butyrate, isobutyrate and glycolate resulted in the production of 3-hydroxyhexanoate,3-hydroxy-4-methylvalerate and 3,4-dihydroxybutyric acid3-hydroxy-γ-butyrolactone, respectively. Thus, the final product desired in this study, 3,4-dihydroxybutyricacid, was obtained. This product has immense application as a precursor for the production of 3-hydroxy-γ-butyrolactone, which in turn is used in the manufacture of fine chemicals (Martin et al. 2013).

Thus, it is evident that genetic recombination has played a vital role in propelling research on the utilization of PHA as renewable resources for 3-HA building blocks.

Table 9.2 Summary of 3-HA synthesis as building blocks for oral drugs using PHA

Monomer	Source	Process	References
In vitro depolymerization			
3-HO	*Pseudomonas fluorescens* GK13	Extracellular PHA depolymerase	Gangoiti et al. (2010)
In vivo depolymerization			
8 3-HA with 78 wt % for 3-HO	*Pseudomonas putida* GPo1	PHA depolymerases	Ruth et al. (2007)
(R)-3-HB	*Cupriavidus necator* VCM 11282	UV mutagenesis, (R)-3-HBdehydrogenase	Ugwu et al. (2008)
3-HB	*Methylobacterium rhodesianum* MB 126	Double mutant: Inactivation of dehydrogenase gene and devoid of lipoic acid synthase (LipA)	Holscher et al. (2010)
Genetically modified *E. coli*			
S-3-HB	Recombinant *E. coli*	Recombinant: phaA from *R. eutropha* Dehydrogenase gene from *R. eutropha* or *Clostridium acetobutylicum*ATCC824 Hydrolase gene from *Bacillus cereus* ATCC14579	Lee et al. (2008), Tseng et al. (2009), and Ren et al. (2010)
Extracellular 3-HB	*E.coli* strain AF1000pJBGT3RX	Recombinant: PHB degradation genes from *Halomonas boliviensis*	Guevara-Martínez et al. (2015)
Genetically modified pseudomonads			
mcl-3-HA: R-3-HHx and R-3-HO	*P. putida* KT2442	Recombinant: Depolymerase gene from *P. putida* KT2442 Acyl-CoA synthetase gene (fadD) from *E. coli* MG1655	Yuan et al. (2008) and Ren et al. (2010)

(continued)

Table 9.2 (continued)

Monomer	Source	Process	References
R-3-HHx, R-3-HO, R-3-HD and R-3-HDD	*P. putida* KTOY01	Knockout mutant of the PHA synthesis operon	Chung et al. (2009)
	P. putida KT2440	Recombinant:	Ren et al. (2010)
		TesB gene from *E. coli*	
3-HTD and 3-HDD	*Pseudomonas entomophila* L48		Chung et al. (2013)
3-HD and 3-HDD	*P. entomophila* LAC31	Mutagenesis of wild type *P. entomophila* L48	Chung et al. (2013)
		tesB of *E. coli*	
		PTE1 gene from *Saccharomyces cerevisiae*	
3,4-dihydroxybutyricacid		Novel pathway for 3,4-dihydroxybutyricacid which is a precursor for 3-hydroxy-γ-butyrolactone (used in fine chemicals production)	Martin et al. (2013)

However, it is noticeable that the amount of research in this avenue is beginning to lose some momentum. There may be various contributing factors as to why this may be. Table 9.2 gives a summary of the advancements made in the synthesis of 3-HA using PHA as raw materials.

9.4 PHA as Nutritional Supplements

Scientists have discovered that intermediates of PHA catabolism, 3-HA, could potentially be used as a source of carbon for human and animal consumption. This is because 3-HA exhibits good penetration and hence, can be rapidly metabolized in a biological system.

3-HB, the product of degradation, is normally present in blood at low concentrations. 3-HB dimers and trimers were tested in rat and human tissues as potential food supplements. The results of the study showed that these were converted rapidly to monomers within the tissues, thus, concluding that 3-HB could be utilized as an energy source for sick and injured patients. In addition, it is also used to combat myocardial damage, hemorrhagic shock, extensive burns and cerebral hypoxia, anoxia and ischemia (Tasaki et al. 1999; Chen and Wu 2005a, b; Ren et al. 2010; Subin and Bhat 2014). Oligomers of R-3-HB were evaluated in vivo for the release of the ketone body over a long time interval. The potential applications resulting from this study could possibly be appetite suppression, reduction of catabolism of proteins, controlling metabolic diseases, treatment of lifestyle diseases like diabetes, parenteral nutrition requirements, etc. (Williams and Martin 2002). Another study conducted using 3-HB, obtained through degradation of PHB, clearly demonstrated that it can be used as an intravenous or oral carbon supply. The subjects in the study were obese patients who were undergoing therapeutic starvation for 14 days and interestingly, none ever complained of hunger (Fernandez Bunster 2015). This supports the claim that PHA has ideal biocompatibility properties.

A study was conducted in rats to assess the utilization of P(4HB) as a prodrug for 4-HB. The rats were treated with the polymer and their serum was tested for presence of the monomer. It was found that 4-HB in the serum increased to ~86 μM in 30 min and remained elevated for 8 h. In contrast, the 4-HB in the serum increased to 182 μM in 30 min after dosage with the monomer and decreases rapidly within 2 h. Hence, it was concluded that the treatment of P(4HB) for prolonged release of 4-HB would be beneficial for alcohol withdrawal, narcolepsy, etc. (Williams and Martin 2002; Subin and Bhat 2014).

In addition to studies with respect to human nutrition, scientists have conducted studies on the potential application of polymers, PHB and PHBV, in animal nutrition. One such study was performed by Peoples et al. (1999) wherein the digestion of PHB and poly(3HO-co-3HHx) were evaluated in broiler chicks. The study revealed that the energy available through these polymers lay between that obtained from carbohydrates and oils (Williams and Martin 2002).

Developed countries have taken the initiative to explore and produce a small range of products made from PHA possibly owing to the fact that they have the resources and the infrastructure to successfully undertake these endeavours. These products have been used in food packaging, disposable items, medical devices etc. A well-known American based company named Metabolix has been successful in developing various PHA based commodities. One such commodity for which they received FDA approval was for a blend of PHB and PHO, known by its commercial name, Metabolix PHA, for utilization in the production of food additives (Babu et al. 2013; Alavi et al. 2014; Subin and Bhat 2014).

The research in this area is less than what it had been a decade ago. However, it has been included in this text in order to spark up new ideas. It is an avenue which still generates great interest. PHA and its monomers have the potential to serve as an alternative food source, especially in areas that are war zones, riddled by diseases, famines, droughts etc. In addition, as proven in the studies conducted, they can be used as a source of energy for recuperating patients. Further studies are required to test the feasibility of PHA utilization in this area.

9.5 Opinion

Scientists have the ability and the responsibility to push the boundaries of science and knowledge, exploring new avenues and asking questions which have not been asked before. As is evident from the extensive research being conducted in skin and tissues engineering, scientists have hypothesized novel approaches to deal with the short-comings of the techniques which are currently in use. For example, a study conducted by a group of scientists demonstrated the advantage of using bioactive glass nanoparticles embedded onto PHA films for antimicrobial wound dressings. Another study involved incorporating bacteriophages onto a PHA film to target a specific pathogen which can cause infections on open wounds. Consequently, as each study is conducted and published, there are newer ideas sparking in the minds of scientists which continue to add momentum to this upcoming area of polymer and biomedical research. With regard to utilizing PHA as raw materials for chiral building blocks, genetic engineering has played a major role. It is through recombinant technology that scientists have engineered micro-organisms to utilize a wide variety of raw materials for 3-HA synthesis either through direct synthesis or through degradation of PHA polymers. PHA as food supplements is an area that needs extensive research in order to become a reality. The potential of utilizing PHA as an alternate food source holds great promise to areas experiencing food shortage. Regardless of the field, PHA research has immense potential and will continue to thrive with each advancement that is made.

9.6 Conclusion

Through decades of research, which is still in full swing, scientists have made giant strides towards the ultimate goal of efficient utilization of PHA without compromising its innate biodegradability and biocompatibility. Specifically, the advancements achieved in the field of medicine with regard to PHA application have been possible through the joint efforts of scientists across the globe. Currently, sutures, tissue repair biomaterials, feminine hygiene products, drug delivery systems, biocontrol agents, food supplements, etc. are a few commodities in which PHA find usage and this is just the tip of the iceberg. It is evident from the quantity of studies being conducted, especially those addressed in this chapter (wound treatment and oral drug manufacture) that PHA has great scope for numerous applications as potential antimicrobial wound dressings and as oral drug precursors. PHA, in unmodified or modified forms, has been successfully proven to enhance wound healing, provide a sturdy scaffold for tissue regeneration, provide antimicrobial action against invading pathogens, attract keratinocytes for wound repair, etc. 3-HA synthesis from PHA continues to improve with each study and this holds the possibility for PHA utilization as the exclusive source of raw materials for oral drugs precursors.

References

Alavi S, Thomas S, Sandeep KP, Kalarikkal N, Varghese J, Yaragalla S (2014) Polymers for packaging applications. CRC Press, Boca Raton, 459 p

Anderson AJ, Dawes EA (1990) Occurrence, metabolism, metabolic role, and industrial uses of bacterial polyhydroxyalkanoates. Microbiol Rev 54:450–472

Asran AS, Razghandi K, Aggarwal N, Michler GH, Groth T (2010) Nanofibers from blends of polyvinyl alcohol and polyhydroxy butyrate as potential scaffold material for tissue engineering of skin. Biomacromolecules 11:3413–3421. https://doi.org/10.1021/bm100912v

Babu RP, O'Connor K, Seeram R (2013) Current progress on bio-based polymers and their future trends. Prog Biomater 2:8. https://doi.org/10.1186/2194-0517-2-8

Bacakova L, Novotná K, Parizek M (2014) Polysaccharides as cell carriers for tissue engineering: the use of cellulose in vascular wall reconstruction. Physiol Res Suppl 1:S29–S47

Brigham CJ, Sinskey AJ (2012) Applications of polyhydroxyalkanoates in the medical industry. Int J Biotechnol Wellness Ind 1:52–60. https://doi.org/10.6000/1927-3037.2012.01.01.03

Brown HC, Ramachandran PV (1991) The boron approach to asymmetric synthesis. Pure Appl Chem 63:307–316. https://doi.org/10.1351/pac199163030307

Bunster GF (2015) Polyhydroxyalkanoates: production and use. In: Mishra M (ed) Encyclopedia of biomedical polymers and polymeric biomaterials. CRC Press, Boca Raton, pp 6412–6421. https://doi.org/10.1081/E-EBPP-120049915. isbn:9781439898796

Calabia BP, Tokiwa Y (2006) A novel PHB depolymerase from a thermophilic *Streptomyces* sp. Biotechnol Lett 28:383–388. https://doi.org/10.1007/s10529-005-6063-5

Chattopadhyay S, Raines RT (2014) Review collagen-based biomaterials for wound healing. Biopolymers 101:821–833. https://doi.org/10.1002/bip.22486

Chen GQ, Wu Q (2005a) Microbial production and applications of chiral hydroxyalkanoates. Appl Microbiol Biotechnol 67:592–599. https://doi.org/10.1007/s00253-005-1917-2

Chen GQ, Wu Q (2005b) The application of polyhydroxyalkanoates as tissue engineering materials. Biomaterials 26:6565–6578. https://doi.org/10.1016/j.biomaterials.2005.04.036

Chen CW, Don TM, Yen HF (2006) Enzymatic extruded starch as a carbon source for the production of poly (3-hydroxybutyrate-co-3-hydroxyvalerate) by *Haloferax mediterranei*. Process Biochem 41:2289–2296. https://doi.org/10.1016/j.procbio.2006.05.026

Chung A, Liu Q, Ouyang SP, Wu Q, Chen GQ (2009) Microbial production of 3-hydroxydodecanoic acid by PHA operon and fadBA knockout mutant of *Pseudomonas putida* KT2442 harboring tesB gene. Appl Microbiol Biotechnol 83:513–519. https://doi.org/10.1007/s00253-009-1919-6

Chung AL, Zeng GD, Jin HL, Wu Q, Chen JC, Chen GQ (2013) Production of medium-chain-length 3-hydroxyalkanoic acids by β-oxidation and phaC operon deleted *Pseudomonas entomophila* harboring thioesterase gene. Metab Eng 17:23–29. https://doi.org/10.1016/j.ymben.2013.02.001

de Roo G, Kellerhals MB, Ren Q, Witholt B, Kessler B (2002) Production of chiral R-3-hydroxyalkanoic acids and R-3-hydroxyalkanoic acid methylesters via hydrolytic degradation of polyhydroxyalkanoate synthesized by pseudomonads. Biotechnol Bioeng 77:717–722. https://doi.org/10.1002/bit.10139

Deitzel JM, Kleinmeyer JD, Hirvonen JK, Tan NB (2001) Controlled deposition of electrospun poly (ethylene oxide) fibers. Polymer 42:8163–8170. https://doi.org/10.1016/S0032-3861(01)00336-6

Dinjaski N, Fernández-Gutiérrez M, Selvam S, Parra-Ruiz FJ, Lehman SM, San Román J, García E, García JL, García AJ, Prieto MA (2014) PHACOS, a functionalized bacterial polyester with bactericidal activity against methicillin-resistant *Staphylococcus aureus*. Biomaterials 35:14–24. https://doi.org/10.1016/j.biomaterials.2013.09.059

Doi Y, Kanesawa Y, Tanahashi N, Kumagai Y (1992) Biodegradation of microbial polyesters in the marine environment. Polym Degrad Stab 36:173–177. https://doi.org/10.1016/0141-3910(92)90154-W

Dong Y, Hassan WU, Kennedy R, Greiser U, Pandit A, Garcia Y, Wang W (2014) Performance of an in situ formed bioactive hydrogel dressing from a PEG-based hyperbranched multifunctional copolymer. Acta Biomater 10:2076–2085. https://doi.org/10.1016/j.actbio.2013.12.045

Eggers J, Steinbüchel A (2013) Poly (3-hydroxybutyrate) degradation in *Ralstonia eutropha* H16 is mediated stereoselectively to (S)-3-hydroxybutyryl coenzyme A (CoA) via crotonyl-CoA. J Bacteriol 195:3213–3223. https://doi.org/10.1128/JB.00358-13

Ekwall B, Silano V, Paganuzzi-Stammati A, Zucco F (1990) Toxicity tests with mammalian cell cultures. In: Bourdeau P, Sommers E, Richardson G, Hickman JR (eds) Short-term toxicity tests for non-genotoxic effects. Wiley, New York, pp 75–99. https://doi.org/10.1002/jat.2550110220. isbn:978-0471925064

Eriksen M, Lebreton LC, Carson HS, Thiel M, Moore CJ, Borerro JC, Galgani F, Ryan PG, Reisser J (2014) Plastic pollution in the world's oceans: more than 5 trillion plastic pieces weighing over 250,000 tons afloat at sea. PLoS One 9:e111913. https://doi.org/10.1371/journal.pone.0111913

Escapa IF, Morales V, Martino VP, Pollet E, Avérous L, García JL, Prieto MA (2011) Disruption of β-oxidation pathway in *Pseudomonas putida* KT2442 to produce new functionalized PHAs with thioester groups. Appl Microbiol Biotechnol 89:1583–1598. https://doi.org/10.1007/s00253-011-3099-4

Farrar D (2011) Advanced wound repair therapies. Elsevier, New York, 672 p

Forbes S, McBain AJ, Felton-Smith S, Jowitt TA, Birchenough HL, Dobson CB (2013) Comparative surface antimicrobial properties of synthetic biocides and novel human apolipoprotein E derived antimicrobial peptides. Biomaterials 34:5453–5464. https://doi.org/10.1016/j.biomaterials.2013.03.087

Francis L, Meng D, Locke IC, Knowles JC, Mordan N, Salih V, Boccaccini AR, Roy I (2016) Novel poly (3-hydroxybutyrate) composite films containing bioactive glass nanoparticles for wound healing applications. Polym Int 65:661–674. https://doi.org/10.1002/pi.5108

Gallo J, Holinka M, Moucha CS (2014) Antibacterial surface treatment for orthopaedic implants. Int J Mol Sci 15:13849–13880. https://doi.org/10.3390/ijms150813849

Gangoiti J, Santos M, Llama MJ, Serra JL (2010) Production of chiral (R)-3-hydroxyoctanoic acid monomers, catalyzed by *Pseudomonas fluorescens* GK13 poly (3-hydroxyoctanoic acid) depolymerase. Appl Environ Microbiol 76:3554–3560. https://doi.org/10.1128/AEM.00337-10

Gao HJ, Wu Q, Chen GQ (2002) Enhanced production of D-(−)-3-hydroxybutyric acid by recombinant *Escherichia coli*. FEMS Microbiol Lett 213:59–65. https://doi.org/10.1111/j.1574-6968.2002.tb11286.x

Gao G, Lange D, Hilpert K, Kindrachuk J, Zou Y, Cheng JT, Kazemzadeh-Narbat M, Yu K, Wang R, Straus SK, Brooks DE (2011) The biocompatibility and biofilm resistance of implant coatings based on hydrophilic polymer brushes conjugated with antimicrobial peptides. Biomaterials 32:3899–3909. https://doi.org/10.1016/j.biomaterials.2011.02.013

Gowda V, Shivakumar S (2014) Agrowaste-based Polyhydroxyalkanoate (PHA) production using hydrolytic potential of *Bacillus thuringiensis* IAM 12077. Braz Arch Biol Technol 57:55–61. https://doi.org/10.1590/S1516-89132014000100009

Grass G, Rensing C, Solioz M (2011) Metallic copper as an antimicrobial surface. Appl Environ Microbiol 77:1541–1547. https://doi.org/10.1128/AEM.02766-10

Gray H, Standring S, Anand N, Birch R, Collins P, Crossman AR, Gleeson M, Jawaheer G, Smith AL, Spratt JD, Stringer MD (2016) Gray's anatomy: the anatomical basis of clinical practice. Elsevier, New York, 1576 p

Guevara-Martínez M, Gällnö KS, Sjöberg G, Jarmander J, Perez-Zabaleta M, Quillaguamán J, Larsson G (2015) Regulating the production of (R)-3-hydroxybutyrate in *Escherichia coli* by N or P limitation. Front Microbiol 6:844. https://doi.org/10.3389/fmicb.2015.00844

Gumel AM, Razaif-Mazinah MR, Anis SN, Annuar MS (2015) Poly (3-hydroxyalkanoates)-co-(6-hydroxyhexanoate) hydrogel promotes angiogenesis and collagen deposition during cutaneous wound healing in rats. Biomed Mater 10:045001. https://doi.org/10.1088/1748-6041/10/4/045001

Guo SA, DiPietro LA (2010) Factors affecting wound healing. J Dent Res 89:219–229. https://doi.org/10.1177/0022034509359125

Hann EC, Sigmund AE, Fager SK, Cooling FB, Gavagan JE, Ben-Bassat A, Chauhan S, Payne MS, Hennessey SM, DiCosimo R (2003) Biocatalytic hydrolysis of 3-hydroxyalkanenitriles to 3-hydroxyalkanoic acids. Adv Synth Catal 345:775–782. https://doi.org/10.1002/adsc.200303007

Hazer B, Steinbüchel A (2007) Increased diversification of polyhydroxyalkanoates by modification reactions for industrial and medical applications. Appl Microbiol Biotechnol 74:1–12. https://doi.org/10.1007/s00253-006-0732-8

Hoenich NA (2007) Cellulose for medical applications: past, present, and future. BioResources 1:270–280

Hoh A, Maier K (1993) Comparative cytotoxicity test with human keratinocytes, HaCaT cells, and skin fibroblasts to investigate skin-irritating substances. In: Bernd A, Bereiter-Hahn J, Hevert F, Holzmann H (eds) Cell and tissue culture models in dermatological research. Springer, Berlin/Heidelberg, pp 341–347. https://doi.org/10.1007/978-3-642-77817-9. isbn:978-3-642-77819-3

Holinka J, Pilz M, Kubista B, Presterl E, Windhager R (2013) Effects of selenium coating of orthopaedic implant surfaces on bacterial adherence and osteoblastic cell growth. Bone Joint J 95:678–682. https://doi.org/10.1302/0301-620X.95B5.31216

Hölscher T, Breuer U, Adrian L, Harms H, Maskow T (2010) Production of the chiral compound (R)-3-hydroxybutyrate by a genetically engineered methylotrophic bacterium. Appl Environ Microbiol 76:5585–5591. https://doi.org/10.1128/AEM.01065-10

Hu SG, Jou CH, Yang MC (2003) Protein adsorption, fibroblast activity and antibacterial properties of poly (3-hydroxybutyric acid-co-3-hydroxyvaleric acid) grafted with chitosan and chitooligosaccharide after immobilized with hyaluronic acid. Biomaterials 24:2685–2693. https://doi.org/10.1016/S0142-9612(03)00079-6

Huang TY, Duan KJ, Huang SY, Chen CW (2006) Production of polyhydroxyalkanoates from inexpensive extruded rice bran and starch by *Haloferax mediterranei*. J Ind Microbiol Biotechnol 33:701–706. https://doi.org/10.1007/s10295-006-0098-z

http://plastindia.org/insights-may2013-pdf.html#book/ (accessed on 10-01-2018)

Jendrossek D, Handrick R (2002) Microbial degradation of polyhydroxyalkanoates. Annu Rev Microbiol 56:403–432. https://doi.org/10.1146/annurev.micro.56.012302.160838

Ji Y, Li XT, Chen GQ (2008) Interactions between a poly (3-hydroxybutyrate-co-3-hydroxyvalerate-co-3-hydroxyhexanoate) terpolyester and human keratinocytes. Biomaterials 29:3807–3814. https://doi.org/10.1016/j.biomaterials.2008.06.008

Kazemzadeh-Narbat M, Lai BF, Ding C, Kizhakkedathu JN, Hancock RE, Wang R (2013) Multilayered coating on titanium for controlled release of antimicrobial peptides for the prevention of implant-associated infections. Biomaterials 34:5969–5977. https://doi.org/10.1016/j.biomaterials.2013.04.036

Keshel SH, Biazar E, Rezaei Tavirani M, Rahmati Roodsari M, Ronaghi A, Ebrahimi M, Rad H, Sahebalzamani A, Rakhshan A, Afsordeh K (2014) The healing effect of unrestricted somatic stem cells loaded in collagen-modified nanofibrous PHBV scaffold on full-thickness skin defects. Artif Cells Nanomed Biotechnol 42:210–216. https://doi.org/10.3109/21691401.2013.800080

Kim JH, Scialli AR (2011) Thalidomide: the tragedy of birth defects and the effective treatment of disease. Toxicol Sci 122:1–6. https://doi.org/10.1093/toxsci/kfr088

Kim HW, Chung MG, Rhee YH (2007) Biosynthesis, modification, and biodegradation of bacterial medium-chain-length polyhydroxyalkanoates. J Microbiol 45:87–97

Koller M, Hesse P, Bona R, Kutschera C, Atlić A, Braunegg G (2007) Biosynthesis of high quality polyhydroxyalkanoate co-and terpolyesters for potential medical application by the archaeon *Haloferax mediterranei*. In: Slomkowski S (ed) Macromolecular symposia, vol 253. Wiley-VCH Verlag, Weinheim, pp 33–39. https://doi.org/10.1002/masy.200790041

Koller M, Atlić A, Dias M, Reiterer A, Braunegg G (2010) Microbial PHA production from waste raw materials. In: Chen GQ (ed) Plastics from bacteria. Springer, Berlin/Heidelberg, pp 85–119. https://doi.org/10.1007/978-3-642-03287-5. isbn:978-3-642-03287-5

Krishnan V, Lakshmi T (2013) Bioglass: a novel biocompatible innovation. J Adv Pharm Technol Res 4:78–83. https://doi.org/10.4103/2231-4040.111523

Krüger K, Lang G, Weidner T, Engel AM (1999) Cloning and functional expression of the D-β-hydroxybutyrate dehydrogenase gene of *Rhodobacter* sp. DSMZ 12077. Appl Microbiol Biotechnol 52:666–669

Lee SY, Lee Y (2003) Metabolic engineering of *Escherichia coli* for production of enantiomerically pure (R)-(−)-hydroxycarboxylic acids. Appl Environ Microbiol 69:3421–3426. https://doi.org/10.1128/AEM.69.6.3421-3426.2003

Lee SH, Park SJ, Lee SY, Hong SH (2008) Biosynthesis of enantiopure (S)-3-hydroxybutyric acid in metabolically engineered *Escherichia coli*. Appl Microbiol Biotechnol 79:633–641. https://doi.org/10.1007/s00253-008-1473-7

Mano JF, Silva GA, Azevedo HS, Malafaya PB, Sousa RA, Silva SS, Boesel LF, Oliveira JM, Santos TC, Marques AP, Neves NM (2007) Natural origin biodegradable systems in tissue engineering and regenerative medicine: present status and some moving trends. J R Soc Interface 4:999–1030. https://doi.org/10.1098/rsif.2007.0220

Marcano AM, Bou Haidar N, Marais S, Valleton JM, Duncan AC (2017) Designing biodegradable PHA-based 3D scaffolds with antibiofilm properties for wound dressings: optimization of the micro/nanostructure. ACS Biomater Sci Eng 3:3654–3661. https://doi.org/10.1021/acsbiomaterials.7b00552

Martin CH, Dhamankar H, Tseng HC, Sheppard MJ, Reisch CR, Prather KL (2013) A platform pathway for production of 3-hydroxyacids provides a biosynthetic route to 3-hydroxy-γ-butyrolactone. Nat Commun 4:1414. https://doi.org/10.1038/ncomms2418

MacNeil S (2007) Progress and opportunities for tissue-engineered skin. Nature 445:874–880. https://doi.org/10.1038/nature05664

Metcalfe AD, Ferguson MW (2006) Tissue engineering of replacement skin: the crossroads of biomaterials, wound healing, embryonic development, stem cells and regeneration. J Royal Soc Interface 4:413–437. https://doi.org/10.1098/rsif.2006.0179

Noyori R, Kitamura M, Ohkuma T (2004) Toward efficient asymmetric hydrogenation: architectural and functional engineering of chiral molecular catalysts. Proc Natl Acad Sci U S A 101:5356–5362. https://doi.org/10.1073/pnas.0307928100

Numata K, Abe H, Iwata T (2009) Biodegradability of poly (hydroxyalkanoate) materials. Materials 2:1104–1126. https://doi.org/10.3390/ma2031104

Ong SY, Chee JY, Sudesh K (2017) Degradation of polyhydroxyalkanoate (PHA): a review. J Sib Fed Univ Biol 10:211–225. https://doi.org/10.17516/1997-1389-0024

Orita I, Nishikawa K, Nakamura S, Fukui T (2014) Biosynthesis of polyhydroxyalkanoate copolymers from methanol by *Methylobacterium extorquens* AM1 and the engineered strains under cobalt-deficient conditions. Appl Microbiol Biotechnol 98:3715–3725. https://doi.org/10.1007/s00253-013-5490-9

Ostomel TA, Shi Q, Tsung CK, Liang H, Stucky GD (2006) Spherical bioactive glass with enhanced rates of hydroxyapatite deposition and hemostatic activity. Small 2:1261–1265. https://doi.org/10.1002/smll.200600177

Panáček A, Kolář M, Večeřová R, Prucek R, Soukupová J, Kryštof V, Hamal P, Zbořil R, Kvítek L (2009) Antifungal activity of silver nanoparticles against *Candida* spp. Biomaterials 30:6333–6340. https://doi.org/10.1016/j.biomaterials.2009.07.065

Park SH, Lee SH, Lee SY (2001) Preparation of optically active β-amino acids from microbial polyester polyhydroxyalkanoates. J Chem Res 2001:498–499. https://doi.org/10.3184/030823401103168640

Park SJ, Lee SY, Lee Y (2004) Biosynthesis of (R)-3-hydroxyalkanoic acids by metabolically engineered *Escherichia coli*. Appl Biochem Biotechnol 114:373–379. https://doi.org/10.1385/ABAB:114:1-3:373

Peschel G, Dahse HM, Konrad A, Wieland GD, Mueller PJ, Martin DP, Roth M (2008) Growth of keratinocytes on porous films of poly (3-hydroxybutyrate) and poly (4-hydroxybutyrate) blended with hyaluronic acid and chitosan. J Biomed Mater Res A 85:1072–1081. https://doi.org/10.1002/jbm.a.31666

Peoples OP, Saunders C, Nichols S, Beach L (1999) Animal nutrition compositions, PCT Patent Applications No. WO. 99:34687

Petrini P, Arciola CR, Pezzali I, Bozzini S, Montanaro L, Tanzi MC, Speziale P, Visai L (2006) Antibacterial activity of zinc modified titanium oxide surface. Int J Artif Organs 29:434–442

Poli A, Di Donato P, Abbamondi GR, Nicolaus B (2011) Synthesis, production, and biotechnological applications of exopolysaccharides and polyhydroxyalkanoates by archaea. Archaea 2011:693253. https://doi.org/10.1155/2011/693253

Pramanik N, Mitra T, Khamrai M, Bhattacharyya A, Mukhopadhyay P, Gnanamani A, Basu RK, Kundu PP (2015) Characterization and evaluation of curcumin loaded guar gum/polyhydroxyalkanoates blend films for wound healing applications. RSC Adv 5:63489–63501. https://doi.org/10.1039/C5RA10114J

Puppi D, Chiellini F, Dash M, Chiellini E (2011) Biodegradable polymers for biomedical applications. In: Felton G (ed) Biodegradable polymers: processing, degradation & applications. Nova Science Publishers, Hauppauge, pp 545–604. isbn:978-1-61209-534-9

Rahmani Del Bakhshayesh A, Annabi N, Khalilov R, Akbarzadeh A, Samiei M, Alizadeh E, Alizadeh-Ghodsi M, Davaran S, Montaseri A (2018) Recent advances on biomedical applications of scaffolds in wound healing and dermal tissue engineering. Artif cells nanomed biotechnol 46:691–705. https://doi.org/10.1080/21691401.2017.1349778

Rehm BH (2009) Microbial production of biopolymers and polymer precursors: applications and perspectives. Horizon Scientific Press, Poole, 289 p

Rehm BH, Steinbüchel A (2002) Polyhydroxyalkanoate (PHA) synthases: the key enzymes of PHA synthesis. Biopolymers Online. https://doi.org/10.1002/3527600035.bpol3a06

Ren Q, Grubelnik A, Hoerler M, Ruth K, Hartmann R, Felber H, Zinn M (2005) Bacterial poly (hydroxyalkanoates) as a source of chiral hydroxyalkanoic acids. Biomacromolecules 6:2290–2298. https://doi.org/10.1021/bm050187s

Ren Q, Ruth K, Thöny-Meyer L, Zinn M (2010) Enatiomerically pure hydroxycarboxylic acids: current approaches and future perspectives. Appl Microbiol Biotechnol 87:41–52. https://doi.org/10.1007/s00253-010-2530-6

Reusch RN, Bryant EM, Henry DN (2003) Increased poly-(R)-3-hydroxybutyrate concentrations in streptozotocin (STZ) diabetic rats. Acta Diabetol 40:91–94. https://doi.org/10.1007/s005920300011

Ruth K, Grubelnik A, Hartmann R, Egli T, Zinn M, Ren Q (2007) Efficient production of (R)-3-hydroxycarboxylic acids by biotechnological conversion of polyhydroxyalkanoates and their purification. Biomacromolecules 8:279–286. https://doi.org/10.1021/bm060585a

Seebach D, Fritz MG (1999) Detection, synthesis, structure, and function of oligo (3-hydroxyalkanoates): contributions by synthetic organic chemists. Int J boil macromol 25:217–236. https://doi.org/10.1016/S0141-8130(99)00037-9

Seebach D, Beck AK, Breitschuh R, Job K (1992) Direct degradation of the biopolymer poly [(R)-3-hydroxybutyric acid] to (R)-3-hydroxybutanoic acid and its methyl ester. Org Synth 1:39

Shishatskaya EI, Nikolaeva ED, Vinogradova ON, Volova TG (2016) Experimental wound dressings of degradable PHA for skin defect repair. J Mater Sci Mater Med 27:165. https://doi.org/10.1007/s10856-016-5776-4

Smithells RW, Newman CG (1992) Recognition of thalidomide defects. J Med Genet 29:716–723

Subin R, Bhat SG (2014) Bacterial Polyhydroxyalkanoates production and its applications. In: Bhat SG, Nambisan P (eds) Microbial Bioproducts. Directorate of Public Relations and Publications for Department of Biotechnology, Cochin University of Science and Technology, pp 70–96. ISBN. isbn:978-93-80095-51-6

Suwantong O, Waleetorncheepsawat S, Sanchavanakit N, Pavasant P, Cheepsunthorn P, Bunaprasert T, Supaphol P (2007) In vitro biocompatibility of electrospun poly (3-hydroxybutyrate) and poly (3-hydroxybutyrate-co-3-hydroxyvalerate) fiber mats. Int J Biol Macromol 40:217–223. https://doi.org/10.1016/j.ijbiomac.2006.07.006

Tasaki O, Hiraide A, Shiozaki T, Yamamura H, Ninomiya N, Sugimoto H (1999) The dimer and trimer of 3-hydroxybutyrate oligomer as a precursor of ketone bodies for nutritional care. J Parenter Enter Nutr 23:321–325. https://doi.org/10.1177/0148607199023006321

Thakur PS, Borah B, Baruah SD, Nigam JN (2001) Growth-associated production of poly-3-hydroxybutyrate by *Bacillus mycoides*. Folia Microbiol 46:488–494. https://doi.org/10.1007/BF02817991

Therapontos C, Erskine L, Gardner ER, Figg WD, Vargesson N (2009) Thalidomide induces limb defects by preventing angiogenic outgrowth during early limb formation. Proc Natl Acad Sci U S A 106:8573–8578. https://doi.org/10.1073/pnas.0901505106

Tieu K, Perier C, Caspersen C, Teismann P, Wu DC, Yan SD, Naini A, Vila M, Jackson-Lewis V, Ramasamy R, Przedborski S (2003) D-β-hydroxybutyrate rescues mitochondrial respiration and mitigates features of Parkinson disease. J Clin Invest 112:892–901. https://doi.org/10.1172/JCI200318797

Tran PA, Webster TJ (2011) Selenium nanoparticles inhibit *Staphylococcus aureus* growth. Int J Nanomed 6:1553–1558. https://doi.org/10.2147/IJN.S21729

Tseng HC, Martin CH, Nielsen DR, Prather KL (2009) Metabolic engineering of *Escherichia coli* for enhanced production of (R)-and (S)-3-hydroxybutyrate. Appl Environ Microbiol 75:3137–3145. https://doi.org/10.1128/AEM.02667-08

Tseng HC, Harwell CL, Martin CH, Prather KL (2010) Biosynthesis of chiral 3-hydroxyvalerate from single propionate-unrelated carbon sources in metabolically engineered *E. coli*. Microb Cell Factories 9:96. https://doi.org/10.1186/1475-2859-9-96

Uchino K, Saito T, Gebauer B, Jendrossek D (2007) Isolated poly (3-hydroxybutyrate) (PHB) granules are complex bacterial organelles catalyzing formation of PHB from acetyl coenzyme A (CoA) and degradation of PHB to acetyl-CoA. J Bacteriol 189:8250–8256. https://doi.org/10.1128/JB.00752-0

Ugwu CU, Tokiwa Y, Aoyagi H, Uchiyama H, Tanaka H (2008) UV mutagenesis of Cupriavidus necator for extracellular production of (R)-3-hydroxybutyric acid. J Appl Microbiol 105:236–242. https://doi.org/10.1111/j.1365-2672.2008.03774.x

Ulery BD, Nair LS, Laurencin CT (2011) Biomedical applications of biodegradable polymers. J Polym Sci B Polym Phys 49:832–864. https://doi.org/10.1002/polb.22259

Vigneswari S, Lee TS, Bhubalan K, Amirul AA (2015) Extracellular polyhydroxyalkanoate depolymerase by *Acidovorax* sp. DP5. Enzym Res 2015:212159. https://doi.org/10.1155/2015/212159

Vigneswari S, Murugaiyah V, Kaur G, Khalil HA, Amirul AA (2016a) Biomacromolecule immobilization: grafting of fish-scale collagen peptides onto aminolyzed P (3HB-co-4HB) scaffolds as a potential wound dressing. Biomed Mater 11:055009. https://doi.org/10.1088/1748-6041/11/5/055009

Vigneswari S, Murugaiyah V, Kaur G, Khalil HA, Amirul AA (2016b) Simultaneous dual syringe electrospinning system using benign solvent to fabricate nanofibrous P(3HB-co-4HB)/collagen peptides construct as potential leave-on wound dressing. Mater Sci Eng C 66:147–155. https://doi.org/10.1016/j.msec.2016.03.102

Wang Q, Webster TJ (2012) Nanostructured selenium for preventing biofilm formation on polycarbonate medical devices. J Biomed Mater Res A 100:3205–3210. https://doi.org/10.1002/jbm.a.34262

Williams SF, Martin DP (2002) Applications of PHAs in medicine and pharmacy. Biopolymers Online 4:91–103. https://doi.org/10.1002/3527600035.bpol4004

Wu Q, Zheng Z, Xi JZ, Gao H, Chen GQ (2003) Production of hydroxyalkanoate monomers by microbial fermentation. J Chem Eng Jpn 36:1170–1173. https://doi.org/10.1252/jcej.36.1170

Yadav P, Yadav H, Shah VG, Shah G, Dhaka G (2015) Biomedical biopolymers, their origin and evolution in biomedical sciences: a systematic review. J Clin Diagn Res 9:ZE21–ZE25. https://doi.org/10.7860/JCDR/2015/13907.6565

Yuan MQ, Shi ZY, Wei XX, Wu Q, Chen SF, Chen GQ (2008) Microbial production of medium-chain-length 3-hydroxyalkanoic acids by recombinant *Pseudomonas putida* KT2442 harboring genes fadL, fadD and phaZ. FEMS Microbiol Lett 283:167–175. https://doi.org/10.1111/j.1574-6968.2008.01164.x

Zhao K, Tian G, Zheng Z, Chen JC, Chen GQ (2003) Production of D-(−)-3-hydroxyalkanoic acid by recombinant *Escherichia coli*. FEMS Microbiol Lett 218:59–64. https://doi.org/10.1111/j.1574-6968.2003.tb11498.x

Zheng Z, Zhang MJ, Zhang G, Chen GQ (2004) Production of 3-hydroxydecanoic acid by recombinant *Escherichia coli* HB101 harboring phaG gene. Antonie Van Leeuwenhoek 85:93–101. https://doi.org/10.1023/B:ANTO.0000020275.23140.ca

Zhou J, Peng SW, Wang YY, Zheng SB, Wang Y, Chen GQ (2010) The use of poly (3-hydroxyb utyrate-co-3-hydroxyhexanoate) scaffolds for tarsal repair in eyelid reconstruction in the rat. Biomaterials 31:7512–7518. https://doi.org/10.1016/j.biomaterials.2010.06.044

Zine R, Sinha M (2015) Nanofibrous poly (3-hydroxybutyrate-co-3-hydroxyvalerate)/collagen/graphene oxide scaffolds for wound coverage. Mater Sci Eng C 80:129–134. https://doi.org/10.1016/j.msec.2017.05.138

Zinn M, Witholt B, Egli T (2001) Occurrence, synthesis and medical application of bacterial polyhydroxyalkanoate. Adv Drug Deliv Rev 53:5–21. https://doi.org/10.1016/S0169-409X(01)00218-6

Zonari A, Cerqueira MT, Novikoff S, Goes AM, Marques AP, Correlo VM, Reis RL (2014) Poly(hydroxybutyrate-co-hydroxyvalerate) bilayer skin tissue engineering constructs with improved epidermal rearrangement. Macromol Biosci 14:977–990. https://doi.org/10.1002/mabi.201400005

Exploiting Polyhydroxyalkanoates for Tissue Engineering

10

Subhasree Ray, Sanjay K. S. Patel, Mamtesh Singh, Gajendra Pratap Singh, and Vipin Chandra Kalia

Abstract

Petroleum based synthetic plastics are an integral part of our daily life. However, their excessive usage has resulted in environmental pollution. The primary reason for this pollution is due to their non-biodegradable nature. On the other hand, polyhydroxyalkanoates (PHAs) are biodegradable polymers, which have been shown to be produced by a wide range of bacteria. The unique feature of this bioplastic production is that they can be produced from renewable substrate materials through a unique metabolic route. These PHAs have the potential to replace petroleum based synthetic plastics. PHAs have high commercial value which make them suitable agent for industrial and medical applications. Although simpler and monomeric forms of PHAs have limited biotechnological applications, however, modified forms of PHA can be used in various medical applications such as, drug delivery, biodegradable implants, anticancer agent, and tissue engineering etc. Among all, tissue engineering has emerged globally to improve the current therapeutic approaches, entailing a revolution in clinical practice. PHAs offer several benefits in tissue engineering. These chemically modified biopolymers can be used in tissue repair, regeneration of tissue, scaffolds preparation etc.

S. Ray
Microbial Biotechnology and Genomics, CSIR – Institute of Genomics and Integrative Biology (IGIB), Delhi, India

Academy of Scientific & Innovative Research (AcSIR), New Delhi, India

S. K. S. Patel · V. C. Kalia (✉)
Department of Chemical Engineering, Konkuk University, Seoul, Republic of Korea

M. Singh
Department of Zoology, Gargi College, University of Delhi, Delhi, India

G. P. Singh
Mathematical Sciences and Interdisciplinary Research Lab (MathSciIntR-Lab), School of Computational and Integrative Sciences, Jawaharlal Nehru University, New Delhi, India

V. C. Kalia (ed.), *Biotechnological Applications of Polyhydroxyalkanoates*,
https://doi.org/10.1007/978-981-13-3759-8_10

Keywords
Polyhydroxyalkanoates · Tissue engineering · Biotechnology · Bacteria · Bacillus · Copolymers

10.1 Introduction

Polyhydroxyalkanoates (PHAs) are biodegradable biopolymers, produced by a diverse range of bacteria under nutrient limiting conditions (Reddy et al. 2003; Porwal et al. 2008; Patel et al. 2011, 2012, 2015a, b, 2016; Kumar et al. 2016; Ray and Kalia 2017a, b). In general, under conditions here high carbon concentration is accompanied by limited quantities of nitrogen, microbes regulate their metabolic pathway in a way that acetyl-CoA gets in to PHAs production pathway rather than going towards TCA cycle (Kumar et al. 2013, 2014, 2015; Singh et al. 2009, 2013, 2015; Ray et al. 2018). These PHAs have gained attention due to the following properties, such as (i) biocompatibility (ii) biodegradability (iii) non-toxicity (iv) cytotoxicity, and (v) non-carcinogenicity as compared to synthetic plastic. Thus, PHAs can be serve as an attractive target for tissue engineering biomaterials (Peppas and Langer 2004; Ray and Kalia 2017c, d; Kalia et al. 2019). Tissue engineering is an emerging field which combines biology, material science and surgical re-construction to help in maintenance and improvement of tissue function through repairing and surgical procedures. Generally, there are three different steps which are being used in engineering of new tissues such as (i) cell substitutes, (ii) materials use to induce tissues, and (iii) use of scaffolds for implantation of cells. Several PHAs, such as poly (3-hydroxybutyrate) P(3HB), poly (3hydroxybutyrate-co-3hydroxyvalerate) P(3HB-co-3HV), poly (4-hydroxybutyrate) P(4HB), poly (3hydroxybutyrate-co-3hydroxyhexanoate) P(3HB-co-3HHx), and poly(3-hydroxyoctanoate) P(3HO) are employed for tissue engineering. The applications involve sutures, wound dressings, scaffolds preparation, bone tissue engineering, subcutaneous tissue engineering, nerve tissue engineering, maxillofacial treatment etc.

10.2 Scaffolds

Tissue engineering involves the scaffold preparation, which helps in the repair and regeneration of defective tissues (Martina and Hutmacher 2007). They provide support for cells to adhere and undergo proliferation process to form an extracellular network (ECM). These scaffolds are composed of bioactive molecules like biodegradable polymers, which play a major role in tissue engineering (Jagur-Grodzinski 2006; Armentano et al. 2010). Scaffolds can be prepared by several methods such as solvent casting, foaming, electro-spinning etc. P (3HB-co-3HV) with pearl powder was prepared for nanofiber scaffold by electrospinning method which promotes cell proliferation (Bai et al. 2015). Curcumin entrapped with polyaniline was conjugated with PHBV for the preparation of scaffold. It was employed in the tissue engineering process. The PHBV scaffolds were characterized by UV−vis and ATR/FT-IR spectrophotometry, thermogravimetry, fluorescence microscopy, and X-ray diffractometric analysis (Pramanik et al. 2016). P (3HB-co-3HV-co-2,3-dHB) produced by recombinant *Ralstonia eutropha* was exploited for scaffold material by utilizing glycolate as a sole carbon source (Insomphun et al. 2016). Scaffold prepared from P(3HB-co-3HHx) can be used as support material for cartilage tissue engineering (Ye et al. 2009). P (3HB-co-3HHx) can be used as scaffold material for fibroblast growth and capsulation. These scaffolds were also found to be favourable for tarsal repair (Zhou et al. 2010). P(3HB-co-3HHX) scaffold blended with hydroxyapatite (HAP) promoted osteoblast growth, chondrocytes proliferation, migration and cartilage repair (Wang et al. 2005, 2008). P(3HB-co-3HHx) enhances smooth muscle cell proliferation and attachment (Qu et al. 2006a, b) (Table 10.1). P(3HB-co-3HV) when grafted with chitosan or chitooligosachharide showed better antibacterial activity against *Staphylococcus aureus, Escherichia coli, Pseudomonas aeruginosa.* The scaffold helped in fibroblast attachment and adsorption of protein and cell proliferation (Hu et al. 2003). The utilization of P(3HB-co-3HHx) in the preparation of scaffolds helped in liver tissue engineering. P(3HB-co-3HHx) fibres and tubes were used to treat Achilles tendon injury in rats (Xu et al. 2010; Webb et al. 2013). These scaffolds proved effective in tissue remodelling. Scaffolds prepared from co-polymerization of PHB with bacterial cellulose improved condition of artificial ligaments and tendon repair with high biocompatibility and biodegradable properties. Co-polymerization of PHB and PHO resulted in scaffold formation, which have been used in cartilage repair (Ching et al. 2016). Similarly, co-polymerization of PHB with poly (ethylene glycol) (PEG) was used in scaffolds preparation which repaired defective bone tissue (Bonartsev et al. 2016).

10.3 Subcutaneous Tissue Engineering (TE)

PHAs can be used in subcutaneous tissue engineering. P(3HB) was used as the suture material in skin tissue, which showed anti-inflammatory properties (Volova et al. 2003). PHO another class of PHA, when used as grafts in soft tissue reduced inflammation (Stock et al. 2000; Hazer et al. 2009). PHBV, poly (L-lactic acid) and poly (glycerol sebacate) were exploited for the preparation of 3D microfibrous material. As a result, a thick myocardial patch was developed to replace myocardial infarctions (Kenar et al. 2010). Scaffold containing PHB/PHBHHx was seeded along with stem cells derived from differentiated human adipose could produce cartilage -like tissue when it was implanted into the subcutaneous layer of the nude mice (Ye et al. 2009). These scaffolds have also been found effective *in vivo* tendon repair model (Webb et al. 2013). PHA copolymers of medium chain length (mcl) size, were biosynthesized from frying oil. Combination of mcl-PHA (at more than 10% by wt) and P(3HB) proved effective in improving the brittle properties of P(3HB). Such blended materials have application in soft TE, which requires a material having desired mixture of tensile strength, stiffness and ductility. The soft and flexible blended biopolymer showed higher biocompatibility as evident from the high viability and proliferation of C2C12 mouse myoblast cells (Lukasiewic et al. 2018).

Table 10.1 Applications of polyhydroxyalkanoates and their composites in tissue engineering

Bio-products	Source	Particles	Method	Applications	References
Gram-positive					
PHA	*Bacillus* sp.	Scaffold	SLP	Enhance growth of new proliferative cells.	Getachew et al. (2016)
P(3HB)	*B. cereus* SPV	3D-scaffold	PLT	Cartilage tissue engineering	Akaraonye et al. (2016)
PHB + BG + Vitamin E	*B. cereus* SPV	Scaffolds preparation	SCM, PLT	Bone tissue engineering	Misra et al. (2010)
P(3HB-co-3HV)	*Alkaliphilus oremlandii* OhiLas	3D Porus scaffolds	SLP	Electroconductive material for tissue engineering	Pramanik et al. (2016)
Gram-negative					
PHB + nHA	*Escherichia coli*	Scaffolds preparation	SCM	Enhances porosity	Meischel et al. (2016)
P (3HB-co-3HHx)	*Aeromonas hydrophila* 4AK4	Scaffold		SMCs-scaffold fabrication	Qu et al. (2006a)
P(3HO)	*Pseudomonas mendocina*	Cardiac patches	SCPL	Replacement of infarcted cardiac tissue	Bagdadi et al. (2016)
P(3HO)	*P. mendocina*	Polymer nano composite	SCM	Keratinocyte regeneration	Rai et al. (2017)
				Skin tissue engineering, wound healing	
Commercial					
P(3HB)-collagen blends	Commercial	Composite fibrous meshes	ES	Anisotropic tissue engineering	Salvatore et al. (2018)
PHBHHx	Commercial	Scaffolds preparation	SLP	Support cell attachment and proliferation of osteoblasts	Wang et al. (2005)
P(3HB-co-3HV)	Commercial	Scaffold	SCM	Tarsal repairing in eyelid	Zhou and Yu (2014)
P(3HB-co-3HHx)	Commercial	3D-scaffold	WPS	Pre-osteoblast proliferation and differentiation	Mota et al. (2017)

P(3HB-co-4HB)	Commercial	Composite fibers	ES	Tissue engineering	Zhijiang et al. (2016)
P(3HB-3HV)::97:3	Commercial	Nerve conduit	SCPL	Regeneration of peripheral nerve	Young et al. (2002)
P(3HB); P(3HB-3HV)::72:28; P(3HB-4HV)::92:8; P(3HB-3HV-4HV)::73:10:17	Commercial	Film preparation	SLP	Enhances cell attachment and proliferation	Ansari and Amirul (2013)
P(3HB-co-3HV)- chitosan; Chitooligosaccharide	Commercial	Grafted membrane	ES	Antibacterial activity	Hu et al. (2003)
P(3HB)- glial growth factor (GGF) in hydrogel	Commercial	Conduit	ES	Repairing of nerve gap	Mohanna et al. (2003)
P(3HB-co-3HHx)	Commercial	Scaffold	SCM	Cartilage repair, chondrocytes proliferation, migration and differentiation	Wang et al. (2008)
PHB + poly (p-dioxanone) (PPD)	Commercial	Blend preparation	SCM	Enhances thermal stability and mechanical strength in bone tissue engineering	Dias et al. (2008)
PHB +PEG	Commercial	Blend preparation	SCM	Support cell spreading and proliferation	Cheng et al. (2003)
PHBHHx/PHB	Commercial	Scaffolds preparation	SLP	Improve cell proliferation in chondrocytes cartilage tissue engineering	Deng et al. (2002)

SCPL Solvent casting particle leaching, *PLT* Particulate leaching technique, *SCM* Solvent casting method, *WPS* Wet-spinning system, *ES* Electrospining, *SLP* Salt leaching process

10.4 Nerve Tissue Engineering (TE)

Nerve injuries result in axonal disruption which cause degenerative changes. Thus, gap formation occurs between nerves where repair is not possible. In this case, nerve grafts act as a bridge to support axonal growth (Arslantunali et al. 2014). Several synthetic nerve conduits have been prepared for the repair of peripheral nerve faults. PHAs are modified to improve neural prosthesis. A porous and fibrous type of polymer prosthesis is favourable for neural regeneration (Mosahebi et al. 2002; Bian et al. 2009). P(3HB), as a neuronal conduit exhibited axonal regeneration which showed low level of inflammatory infiltration (Hazari et al. 1999; Mosahebi et al. 2002). P(3HB-co-3HHx) was found to be helpful in neuronal regeneration (Bian et al. 2009). PHA coated films were found to improve the survival rate of neural stem cells and neural progenitor cells and differentiated in to neurons (Lu et al. 2013).

(PHBV-P(L)-PLGA) with (PHBV-PLGA) was used as a nerve conduit which showed good mechanical properties. PHB conduit was found helpful in peripheral nerve regeneration. PHB conduit was composed of glial growth factor and alginate hydrogel resulted in a progressive and sustainable nerve regeneration. P (3HB-co-4HB) was exploited for the preparation of composite nanofibrous membrane. This membrane was developed by electro-spinning of P(3HB-co-4HB) and cellulose acetate blend solution (Zhijiang et al. 2016). P(3HB-co-3HHx) produced by microbial fermentation was found to be a suitable candidate for artificial nerve conduit due to their proper mechanical strength and biodegradability. This helps to repair nerve damage. These nerve conduits are prepared by particle leaching method (Bian et al. 2009). Neural stem cells which were grown on PHA scaffolds were reported to be useful for repairing injury to the central nervous system (Xu et al. 2010).

10.5 Bone Tissue Engineering (TE)

Bone TE is developed to eliminate the risk associated with the bone graft transplantation process, supply of a limited quantity of bone grafts, and pitfalls associated with transmission of the disease. It is a complex process with the migration of osteoprogenitor cells (Table 10.1). The process composed of proliferation, differentiation, matrix formation, mineralization and finally the remodelling of the bone. Scaffolds prepared for bone TE should be osteoconductive which helps to attract the stem cells. In the presence of suitable growth factors, scaffolds containing stem cells differentiate into pre-osteoblasts, which in turn get transformed to osteoblasts and ECM. As a result, bone remodelling occurs with osteocytes formation. Biopolymers are exploited for bone tissue repairing, through metallic parts and antibiotic carriers to the infected site of bone tissues (Jagur-Grodzinski et al. 2006). P(3HB-co-3HV) as graft was found to be the best biomaterial for osteoblast attachment, proliferation, and differentiation of bone marrow cells. P(3HB-co-3HHx) also showed better attachment, proliferation and differentiation of osteoblasts.

Combination of Hydroxyapatite (HA) and PHA enhances osteoblastic activity and bone integrity. P(3HB-co-3HV) when conjugated with calcium phosphate-reinforcing phases such as HA, submicron-sized calcined hydroxyapatite (cHA) and submicron-sized (β-TCP) showed anti-inflammatory properties and improved osteogenic properties (Cool et al. 2007). PHB composites fabricated with different quantities of zirconium dioxide and herafill (a bone filler loaded with a antibiotics), when implanted in the femora of growing rats proved to have high strain and tensile strength, which were as good as the actual bone (Meischel et al. 2016). Different PHA scaffolds as blends or as composites with hybrid materials have also been shown to be effective in bone tissue engineering (Lim et al. 2017). Electrospun fiber mesh made up of PHB/PHBV in combination with stem cell derived human adipose tissue proved effective in improving vascularization in engineered bone tissues (Goonoo et al. 2017).

10.6 Cartilage Tissue-Tendon and Ligament Tissue Engineering (TE)

PHAs play a vital role in cartilage tissue engineering. When cartilage tissue is damaged, it results in an osteoarthritis and functional loss of joints. PHA implants help in neocartilage formation. It regenerates hyaline cartilage in the defective site (Hazel Fox and Webb 2013). PHBV matrices cause early cartilage formation. Collagen matrices containing calcium phosphate (Cap-Gelfx) and P (3HB-co-3HV) were designed for novel cartilage by tissue engineering which showed better healing properties (Kose et al. 2004, 2005). P(3HB-co-3HHx) was exploited for producing neocartilage (Ye et al. 2009) (Table 10.1).

10.7 Skin Tissue Engineering (TE)

Skin protects the human body from the surrounding environment by protecting the underlying organs from pathogens. Auto-healing property of skin may be damaged during burns, diabetic wounds etc. Several methods to treat burns have been employed, such as autografts, and allografts. However, these two methods face problems due to the limited availability, disease transmission, risk of donor site morbidity and immune rejection. Thus, there is a need to develop substitutes, which can mimic human skin to replace damaged skin. PHAs and its co-polymers when blended with polysaccharides such as P(4HB) and Hyaluronic acid, increased keratinocyte proliferation rate (Groeber et al. 2011). Electrospun nanofibers have been used as polyvinyl alcohol – PHB scaffolds for engineering skin tissue (Sundaramurthi et al. 2014). In vitro study had shown the use of such scaffolds to enhance the proliferation of keratinocytes and fibroblast cells (Asran et al. 2011).

10.8 Conclusion

PHAs have applications in diverse fields. PHAs and their derivatives are employed in medical purposes which have the highest level of application. PHAs with chemical modifications have great potential in tissue engineering. Various PHA-based tissue engineered products have been employed in several clinical use.

Acknowledgements This work was supported by Brain Pool grant (NRF-2018H1D3A2001746) by National Research Foundation of Korea (NRF) to work at Konkuk University.

References

Akaraonye E, Filip J, Safarikova M, Salih V, Keshavarz T, Knowles JC, Roy I (2016) Composite scaffolds for cartilage tissue engineering based on natural polymers of bacterial origin, thermoplastic poly (3-hydroxybutyrate) and micro-fibrillated bacterial cellulose. Polym Int 65:780–791. https://doi.org/10.1002/pi.5103

Ansari NF, Amirul AA (2013) Preparation and characterization of polyhydroxyalkanoates macroporous scaffold through enzyme-mediated modifications. Appl Biochem Biotechnol 170:690–709. https://doi.org/10.1007/s12010-013-0216-0

Armentano I, Dottori M, Fortunati E, Mattioli S, Kenny JM (2010) Biodegradable polymer matrix nanocomposites for tissue engineering: a review. Polym Degrad Stab 95:2126–2146. https://doi.org/10.1016/j.polymdegradstab.2010.06.007

Arslantunali D, Budak G, Hasirci V (2014) Multiwalled CNT-pHEMA composite conduit for peripheral nerve repair. J Biomed Mater Res A 102:828–841. https://doi.org/10.1002/jbm.a.34727

Asran A, Razghandi K, Aggarwal N, Michler GH, Groth T (2011) Nanofibers from blends of polyvinyl alcohol and polyhydroxy butyrate as potential scaffold material for tissue engineering of skin. Biomacromolecules 11:3413–3421. https://doi.org/10.1021/bm100912v

Bagdadi AV, Safari M, Dubey P, Basnett P, Sofokleous P, Humphrey E, Locke I, Edirisinghe M, Terracciano C, Boccaccini AR, Knowles JC, Harding SE, Roy I (2016) Poly (3-hydroxyoctanoate), a promising new material for cardiac tissue engineering. J Tissue Eng Regen Med 12:e495–e512. https://doi.org/10.1002/term.2318

Bai J, Dai J, Li G (2015) Electrospun composites of PHBV/pearl powder for bone repairing. Progress Nat Sci Mater Int 25:327–333. https://doi.org/10.1016/j.pnsc.2015.07.004

Bian YZ, Wang Y, Aibaidoula G, Chen GQ, Wu Q (2009) Evaluation of poly (3-hydroxybutyrate-co-3-hydroxyhexanoate) conduits for peripheral nerve regeneration. Biomaterials 30:217–225. https://doi.org/10.1016/j.biomaterials.2008.09.036

Bonartsev AP, Zharkova II, Yakovlev SG, Myshkina VL, Makhina TK, Zernov AL, Kudryashova KS, Feofanov AV, Akulina EA, Ivanova EV, Zhuikov VA (2016) 3D-scaffolds from poly (3-hydroxybutyrate) poly (ethylene glycol) copolymer for tissue engineering. J Biomater Tissue Eng 6:42–52. https://doi.org/10.1166/jbt.2016.1414

Cheng G, Cai Z, Wang L (2003) Biocompatibility and biodegradation of poly (hydroxybutyrate)/poly (ethylene glycol) blend films. J Mater Sci Mater Med 14:1073–1078. https://doi.org/10.1023/B:JMSM.0000004004.37103.f4

Ching KY, Andriotis OG, Li S, Basnett P, Su B, Roy I, Stolz M (2016) Nanofibrous poly (3-hydroxybutyrate)/poly (3-hydroxyoctanoate) scaffolds provide a functional microenvironment for cartilage repair. J Biomater Appl 31:77–91. https://doi.org/10.1177/0885328216639749

Cool SM, Kenny B, Wu A, Nurcombe V, Trau M, Cassady AI, Grondahl L (2007) Poly (3-hydroxybutyrate-co-3-hydroxyvalerate) composite biomaterials for bone tissue regeneration: in vitro performance assessed by osteoblast proliferation, osteoclast adhesion and resorption, and macrophage proinflammatory response. J Biomed Mater Res 3:599–610. https://doi.org/10.1007/s00253-011-3099-4

Deng Y, Zhao K, Zhang XF, Hu P, Chen GQ (2002) Study on the three-dimensional proliferation of rabbit articular cartilage-derived chondrocytes on polyhydroxyalkanoate scaffolds. Biomaterials 23:4049–4056. https://doi.org/10.1016/S0142-9612(02)00136-9

Dias M, Antunes MCM, Santos AR, Felisberti MI (2008) Blends of poly (3-hydroxybutyrate) and poly (p-dioxanone): miscibility, thermal stability and biocompatibility. J Mater Sci Mater Med 19:3535–3544. https://doi.org/10.1007/s10856-008-3531-1

Getachew A, Berhanu A, Birhane A (2016) Production of sterilized medium chain length polyhydroxyalkanoates (Smcl-PHA) as a biofilm to tissue engineering application. J Tissue Sci Eng 7:2. https://doi.org/10.4172/2157-7552.1000167

Goonoo N, Bhaw-Luximon A, Passanha P, Esteves SR, Jhurry D (2017) Third generation poly(hydroxyacid) composite scaffolds for tissue engineering. J Biomed Mater Res B Appl Biomater 105:1667–1684. https://doi.org/10.1002/jbm.b.33674

Groeber F, Holeiter M, Hampel M, Hinderer S, Schenke-Layland K (2011) Skin tissue engineering-in vivo and in vitro applications. Adv Drug Deliv Rev 63:352–366. https://doi.org/10.1016/j.addr.2011.01.005

Hazari A, Johansson-Ruden G, Junemo-Bostrom K, Ljungberg C, Terenghi G, Green C, Wiberg M (1999) A new resorbable wrap-around implant as an alternative nerve repair technique. J Hand Surg Eur 24:291–295. https://doi.org/10.1054/jhsb.1998.0001

Hazel Fox QC, Webb P (2013) The law of state immunity. Oxford University Press, Oxford

Hazer DB, Hazer B, Kaymaz F (2009) Synthesis of microbial elastomers based on soybean oily acids. Biocompatibility studies. Biomed Mater 4:035011. https://doi.org/10.1088/1748-6041/4/3/035011

Hu SG, Jou CH, Yang MC (2003) Protein adsorption, fibroblast activity and antibacterial properties of poly (3-hydroxybutyric acid-co-3-hydroxyvaleric acid) grafted with chitosan and chitooligosaccharide after immobilized with hyaluronic acid. Biomaterials 24:2685–2693. https://doi.org/10.1016/S0142-9612(03)00079-6

Insomphun C, Chuah JA, Kobayashi S, Fujiki T, Numata K (2016) Influence of hydroxyl groups on the cell viability of polyhydroxyalkanoate (PHA) scaffolds for tissue engineering. ACS Biomater Sci Eng 3:3064–3075. https://doi.org/10.1021/acsbiomaterials.6b00279

Jagur-Grodzinski J (2006) Polymers for tissue engineering, medical devices, and regenerative medicine. Concise general review of recent studies. Polym Adv Technol 17:395–418. https://doi.org/10.1002/pat.729

Kalia VC, Patel SKS, Kang YC, Lee JK (2019) Quorum sensing inhibitors as antipathogens: biotechnological applications. Biotechnol. Adv. 37:68–90. https://doi.org/10.1016/j.biotechad.2018.11.006

Kenar H, Kose GT, Hasirci V (2010) Design of a 3D aligned myocardial tissue construct from biodegradable polyesters. J Mater Sci Mater Med 21:989–997. https://doi.org/10.1007/s10856-009-3917-8

Kose GT, Korkusuz F, Korkusuz P, Hasirci V (2004) In vivo tissue engineering of bone using poly (3-hydroxybutyric acid-co-3-hydroxyvaleric acid) and collagen scaffolds. Tissue Eng 10:1234–1250. https://doi.org/10.1089/ten.2004.10.1234

Kose GT, Korkusuz F, Özkul A, Soysal Y, Özdemir T, Yildiz C, Hasirci V (2005) Tissue engineered cartilage on collagen and PHBV matrices. Biomaterials 26:5187–5197. https://doi.org/10.1016/j.biomaterials.2005.01.037

Kumar P, Patel SKS, Lee JK, Kalia VC (2013) Extending the limits of *Bacillus* for novel biotechnological applications. Biotechnol Adv 31:1543–1561. https://doi.org/10.1016/j.biotechadv.2013.08.007

Kumar P, Singh M, Mehariya S, Patel SKS, Lee JK, Kalia VC (2014) Ecobiotechnological approach for exploiting the abilities of *Bacillus* to produce co-polymer of polyhydroxyalkanoate. Indian J Microbiol 54:151–157. https://doi.org/10.1007/s12088-014-0457-9

Kumar P, Ray S, Patel SK, Lee JK, Kalia VC (2015) Bioconversion of crude glycerol to polyhydroxyalkanoate by *Bacillus thuringiensis* under non-limiting nitrogen conditions. Int J Biol Macromol 78:9–16. https://doi.org/10.1016/j.ijbiomac.2015.03.046

Kumar P, Ray S, Kalia VC (2016) Production of co-polymers of polyhydroxyalkanoates by regulating the hydrolysis of biowastes. Bioresour Technol 200:413–419. https://doi.org/10.1016/j.biortech.2015.10.045

Lim J, You M, Li J, Li Z (2017) Emerging bone tissue engineering via Polyhydroxyalkanoate (PHA)-based scaffolds. Mater Sci Eng C Mater Biol Appl 79:917–929. https://doi.org/10.1016/j.msec.2017.05.132

Lu XY, Wang LL, Yang ZQ, Lu HX (2013) Strategies of polyhydroxyalkanoates modification for the medical application in neural regeneration/nerve tissue engineering. Adv Biosci Biotechnol 4:731–740. https://doi.org/10.4236/abb.2013.46097

Lukasiewicz B, Basnett P, Nigmatullin R, Matharu R, Knowles JC, Roy I (2018) Binary polyhydroxyalkanoate systems for soft tissue engineering. Acta Biomater 71:225–234. https://doi.org/10.1016/j.actbio.2018.02.027

Martina M, Hutmacher DW (2007) Biodegradable polymers applied in tissue engineering research: a review. Polym Int 56:145–157. https://doi.org/10.1002/pi.2108

Meischel M, Eichler J, Martinelli E, Karr U, Weigel J, Schmöller G, Tschegg EK, Fischerauer S, Weinberg AM, Stanzl-Tschegg SE (2016) Adhesive strength of bone-implant interfaces and in-vivo degradation of PHB composites for load-bearing applications. J Mech Behav Biomed Mater 53:104–118. https://doi.org/10.1016/j.jmbbm.2015.08.004

Misra SK, Ansari TI, Valappil SP, Mohn D, Philip SE, Stark WJ, Roy I, Knowles JC, Salih V, Boccaccini AR (2010) Poly (3-hydroxybutyrate) multifunctional composite scaffolds for tissue engineering applications. Biomaterials 31:2806–2815. https://doi.org/10.1016/j.biomaterials.2009.12.045

Mohanna PN, Young RC, Wiberg M, Terenghi G (2003) A composite poly-hydroxybutyrate–glial growth factor conduit for long nerve gap repairs. J Anat 203:553–565. https://doi.org/10.1046/j.1469-7580.2003.00243.x

Mosahebi A, Fuller P, Wiberg M, Terenghi G (2002) Effect of allogeneic Schwann cell transplantation on peripheral nerve regeneration. Exp Neurol 173:213–223. https://doi.org/10.1006/exnr.2001.7846

Mota C, Wang SY, Puppi D, Gazzarri M, Migone C, Chiellini F, Chern GQ, Chiellini E (2017) Additive manufacturing of poly [(R)-3-hydroxybutyrate-co-(R)-3-hydroxyhexanoate] scaffolds for engineered bone development. J Tissue Eng Regen Med 11:175–186. https://doi.org/10.1002/term.1897

Patel SKS, Singh M, Kalia VC (2011) Hydrogen and polyhydroxybutyrate producing abilities of *Bacillus* spp. from glucose in two stage system. Indian J Microbiol 51:418–423. https://doi.org/10.1007/s12088-011-0236-9

Patel SKS, Singh M, Kumar P, Purohit HJ, Kalia VC (2012) Exploitation of defined bacterial cultures for production of hydrogen and polyhydroxybutyrate from pea-shells. Biomass Bioenergy 36:218–225. https://doi.org/10.1016/j.biombioe.2011.10.027

Patel SKS, Kumar P, Singh S, Lee JK, Kalia VC (2015a) Integrative approach for hydrogen and polyhydroxybutyrate production. In: Kalia VC (ed) Microbial factories: waste treatment. Springer, New Delhi, pp 73–85. https://doi.org/10.1007/978-81-322-2598-0_5

Patel SKS, Kumar P, Singh S, Lee JK, Kalia VC (2015b) Integrative approach to produce hydrogen and polyhydroxybutyrate from biowaste using defined bacterial cultures. Bioresour Technol 176:136–141. https://doi.org/10.1016/j.biortech.2014.11.029

Patel SKS, Lee JK, Kalia VC (2016) Integrative approach for producing hydrogen and polyhydroxyalkanoate from mixed wastes of biological origin. Indian J Microbiol 56:293–300. https://doi.org/10.1007/s12088-016-0595-3

Peppas NA, Langer R (2004) Origins and development of biomedical engineering within chemical engineering. AICHE J 50:536–546. https://doi.org/10.1002/aic.10048

Porwal S, Kumar T, Lal S, Rani A, Kumar S, Cheema S, Purohit HJ, Sharma R, Patel SKS, Kalia VC (2008) Hydrogen and polyhydroxybutyrate producing abilities of microbes from diverse habitats by dark fermentative process. Bioresour Technol 99:5444–5451. https://doi.org/10.1016/j.biortech.2007.11.011

Pramanik N, Dutta K, Basu RK, Kundu PP (2016) Aromatic π-conjugated curcumin on surface modified polyaniline/polyhydroxyalkanoate based 3d porous scaffolds for tissue engineering applications. ACS Biomater Sci Eng 2:2365–2377. https://doi.org/10.1021/acsbiomaterials.6b00595

Qu XH, Wu Q, Liang J, Zou B, Chen GQ (2006a) Effect of 3-hydroxyhexanoate content in poly (3-hydroxybutyrate-co-3-hydroxyhexanoate) on in vitro growth and differentiation of smooth muscle cells. Biomaterials 27:2944–2950. https://doi.org/10.1016/j.biomaterials.2006.01.013

Qu XH, Wu Q, Chen GQ (2006b) In vitro study on hemocompatibility and cytocompatibility of poly (3-hydroxybutyrate-co-3-hydroxyhexanoate). J Biomater Sci Polym Ed 17:1107–1121. https://doi.org/10.1163/156856206778530704

Rai R, Roether JA, Knowles JC, Mordan N, Salih V, Locke IC, Michael PG, MnCormick A, Mohn D, Stark WJ, Keshavarz T, Bccaccini AR, Roy I (2017) Highly elastomeric poly (3-hydroxyoctanoate) based natural polymer composite for enhanced keratinocyte regeneration. Int J Polym Mater Polym Biomater 66:326–335. https://doi.org/10.1080/00914037.2016.1217530

Ray S, Kalia VC (2017a) Co-polymers of substrates by *Bacillus thuringiensis* regulates polyhydroxyalkanoates co-polymer compositions. Bioresour Technol 224:743–747. https://doi.org/10.1016/j.biortech.2016.11.089

Ray S, Kalia VC (2017b) Microbial cometabolism and polyhydroxyalkanoate co-polymers. Indian J Microbiol 57:39–47. https://doi.org/10.1007/s12088-016-0622-4

Ray S, Kalia VC (2017c) Biomedical applications of polyhydroxyalkanoates. Indian J Microbiol 57:261–269. https://doi.org/10.1007/s12088-017-0651-7

Ray S, Kalia VC (2017d) Polyhydroxyalkanoate production and degradation patterns in *Bacillus* species. Indian J Microbiol 57:387–392. https://doi.org/10.1007/s12088-017-0676-y

Ray S, Sharma R, Kalia VC (2018) Co-utilization of crude glycerol and biowastes for producing polyhydroxyalkanoates. Indian J Microbiol 58:33–38. https://doi.org/10.1007/s12088-017-0702-0

Reddy CSK, Ghai R, Kalia VC (2003) Polyhydroxyalkanoates: an overview. Bioresour Technol 87:137–146. https://doi.org/10.1016/S0960-8524(02)00212-2

Salvatore L, Carofiglio VE, Stufano P, Bonfrate V, Calò E, Scarlino S, Nitti P, Centrone D, Cascione M, Leporatti S, Sannino A (2018) Potential of electrospun poly (3-hydroxybutyrate)/collagen blends for tissue engineering applications. J Health Eng 2018:6573947. https://doi.org/10.1155/2018/6573947

Singh M, Kumar P, Ray S, Kalia VC (2015) Challenges and opportunities for customizing polyhydroxyalkanoates. Indian J Microbiol 55:235–249. https://doi.org/10.1007/s12088-015-0528-6

Singh M, Patel SKS, Kalia VC (2009) *Bacillus subtilis* as potential producer for polyhydroxyalkanoates. Microb Cell Factories 8:38. https://doi.org/10.1186/1475-2859-8-38

Singh M, Kumar P, Patel SKS, Kalia VC (2013) Production of polyhydroxyalkanoate copolymer by *Bacillus thuringiensis*. Indian J Microbiol 53:77–83. https://doi.org/10.1007/s12088-012-0294-7

Stock UA, Nagashima M, Khalil PN, Nollert GD, Herdena T, Sperling JS, Moran A, Lien J, Martin DP, Schoen FJ, Vacanti JP (2000) Tissue-engineered valved conduits in the pulmonary circulation. J Thorac Cardiovasc Surg 119:732–740. https://doi.org/10.1016/S0022-5223(00)70008-0

Sundaramurthi D, Krishnan UM, SethuKumaran S (2014) Electrospun nanofibers as scaffolds for skin tissue engineering. Polym Rev 54:348–376. https://doi.org/10.1080/15583724.2014.881374

Volova T, Shishatskaya E, Sevastianov V, Efremov S, Mogilnaya O (2003) Results of biomedical investigations of PHB and PHB/PHV fibers. Biochem Eng J 16:125–133. https://doi.org/10.1016/S1369-703X(03)00038-X

Wang YW, Wu Q, Chen J, Chen GQ (2005) Evaluation of three-dimensional scaffolds made of blends of hydroxyapatite and poly (3-hydroxybutyrate-co-3-hydroxyhexanoate) for bone reconstruction. Biomaterials 26:899–904. https://doi.org/10.1016/j.biomaterials.2004.03.035

Wang Y, Bian YZ, Wu Q, Chen GQ (2008) Evaluation of three-dimensional scaffolds prepared from poly (3-hydroxybutyrate-co-3-hydroxyhexanoate) for growth of allogeneic chondrocytes for cartilage repair in rabbits. Biomaterials 29:2858–2868. https://doi.org/10.1016/j.biomaterials.2008.03.021

Webb WR, Dale TP, Lomas AJ, Zeng G, Wimpenny I, El Haj AJ, Forsyth NR, Chen GQ (2013) The application of poly (3-hydroxybutyrate-co-3-hydroxyhexanoate) scaffolds for tendon repair in the rat model. Biomaterials 34:6683–6694. https://doi.org/10.1016/j.biomaterials.2013.05.041

Xu XY, Li XT, Peng SW, Xiao JF, Liu C, Fang G, Chen GQ (2010) The behaviour of neural stem cells on polyhydroxyalkanoate nanofiber scaffolds. Biomaterials 31:3967–3975. https://doi.org/10.1016/j.biomaterials.2010.01.132

Ye C, Hu P, Ma MX, Xiang Y, Liu RG, Shang XW (2009) PHB/PHBHHx scaffolds and human adipose-derived stem cells for cartilage tissue engineering. Biomaterials 30:4401–4406. https://doi.org/10.1016/j.biomaterials.2009.05.001

Young PM, Cocconi D, Colombo P, Bettini R, Price R, Steele DF, Tobyn MJ (2002) Characterization of a surface modified dry powder inhalation carrier prepared by "particle smoothing". J Pharm Pharmacol 54:1339–1344. https://doi.org/10.1211/002235702760345400

Zhijiang C, Jianru J, Qing Z, Shiying Z, Jie G (2016) Preparation and characterization of novel poly (3-hydroxybutyrate-co-4-hydroxybutyrate)/cellulose acetate composite fibers. Mater Lett 173:119–122. https://doi.org/10.1016/j.matlet.2016.03.019

Zhou M, Yu D (2014) Cartilage tissue engineering using PHBV and PHBV/bioglass scaffolds. Mol Med Rep 10:508–514. https://doi.org/10.3892/mmr.2014.2145

Zhou J, Peng SW, Wang YY, Zheng SB, Wang Y, Chen GQ (2010) The use of poly (3-hydroxyb utyrate-co-3-hydroxyhexanoate) scaffolds for tarsal repair in eyelid reconstruction in the rat. Biomaterials 31:7512–7518. https://doi.org/10.1016/j.biomaterials.2010.06.044

CH_4-Based Polyhydroxyalkanoate Production: A Step Further Towards a Sustainable Bioeconomy

11

Juan C. López, Yadira Rodríguez, Víctor Pérez, Raquel Lebrero, and Raúl Muñoz

Abstract

Methane, a near-term climate forcer exhibiting a global warming capacity 90-fold higher compared to carbon dioxide on a 10-year horizon, is the second most prevalent greenhouse gas (GHG) within the global GHG inventory. Both physical/chemical and biological technologies are available for the treatment of CH_4 emissions, the later also allowing for the production of high-added value products such as polyhydroxyalkanoates (PHAs) at significantly lower raw material costs. Among CH_4 biotechnologies, turbulent contactors such as stirred tank reactors or bubble column bioreactors have been the most employed for the methanotrophic production of PHAs under nutrient-limited conditions. However, the moderate biomass productivities achieved at the expense of high energy consumption in these systems has headed the most recent research towards the development of new bioreactor configurations and the implementation of strategies to enhance CH_4 mass transfer, such as the internal gas recycling or the use of non-aqueous phases. Also, the specificity of type II methanotrophs to accumulate these bioproducts, mainly as poly-3-hydroxybutyrate (PHB), has stimulated the development of novel enrichment procedures based on modifications in the nitrogen source and concentration, the oxygen content, the pH and the composition of micronutrients (e.g. Cu^{2+}) to effectively select this type of methanotrophs. Finally, current research is focused on the use of co-substrates during methanotrophic cultivation to increase PHA yields and modify the composition of the biocomposite, thus enhancing the thermal and mechanical properties of the product and boosting its widespread production at industrial-scale.

J. C. López · Y. Rodríguez · V. Pérez · R. Lebrero · R. Muñoz (✉)
Department of Chemical Engineering and Environmental Technology, School of Industrial Engineerings, University of Valladolid, Valladolid, Spain
e-mail: mutora@iq.uva.es

V. C. Kalia (ed.), *Biotechnological Applications of Polyhydroxyalkanoates*,
https://doi.org/10.1007/978-981-13-3759-8_11

Keywords
Bioconversion · Biopolymer · Bioreactor · Methane · Methanotroph · PHA · PHB · PHV

11.1 Introduction

11.1.1 Key Figures and Implications of GHG Emissions on Climate Change

Over the past two centuries, the combustion of fossil fuels such as oil, natural gas and coal, deforestation, land-use changes and other anthropogenic activities have supported a significant increase in the atmospheric concentration of heat-trapping greenhouse gases (GHGs) such as methane (CH_4), carbon dioxide (CO_2), and nitrous oxide (N_2O) (IPCC 2014). The emissions of these GHGs represent nowadays around 98% of the total inventory worldwide, and this component is likely to enhance in this century based on their industrial and organic-based nature and the forthcoming scenario of increasing human population (EEA 2017). In this context, the total anthropogenic GHGs emissions increased at 2.2% year^{-1} in the period 2000–2010 compared to the rates recorded in the period 1970–2000 (1.3% year^{-1}). Thus, despite the increasing number of climate change mitigation policies implemented worldwide, the total GHG emissions peaked at 49 $GtCO_2$-eq year^{-1} in 2010 (IPCC 2014). As a result of this increase in GHG emissions, the average temperature of the Earth has gradually increased at rates never reported in the past 50 years. Indeed, the average global temperature across ocean and land surface areas for 2016 was 0.94 °C more than the average of that recorded in twentieth century (NOAA 2017). GHG emissions are promoting a disruption in the Earth's energy budget, which has resulted in recent changes in the rainfall patterns, snow and ice cover area or in the level and acidification of oceans (IPCC 2014; EPA 2017). For instance, satellite inspections have recently revealed that the Greenland and West Antarctic ice covers are shedding about 125 billion tons of ice year^{-1} (equivalent to 0.35 mm year^{-1}), which will directly mediate a rise of the sea level between 0.27 and 0.6 m by the end of 2099 as warming sea water expands (IPCC 2014). More importantly, global warming is boosting the migration of hundreds of both marine and land-based plants and animals to cope with the extreme temperatures encountered in the past decades in many parts of the globe (EPA 2017).

Climate change represents nowadays one of the greatest environmental concerns all over the world, and therefore governments are gradually implementing policies in order to limit the impacts of their GHG emissions and comply with the Paris Climate Agreement without compromising industrial growth. In this sense, most members of the United Nations have committed to reduce their GHG emissions by at least 18% lower than that of 1990 levels during a span of period from 2013 to 2020 (UNFCCC 2013). Some of the measures taken include adaptation policies implemented across all the categories of European governance, to integrate coastal and water management, into land planning, environmental protection, and managing disaster risk. Likewise, governments of US and Canada are also engaged in incremental adaptation assessment and planning, together with proactive adaptation devoted to protect longer-term investments made in to energy and public infrastructure (IPCC 2014).

11.1.2 Methane Emissions – Nature, Future Trends and Available Abatement Technologies

CH_4 is the second most prevalent GHG within the global GHG inventory. It contributes to approximately 14% of the worldwide annual emissions (on a 100-year horizon) and its atmospheric concentration has increased by ~160% in the past 250 years (IPCC 2014; EPA 2017). CH_4 possesses a global warming potential (GWP) 25 times higher than that of CO_2 (on a 100-year horizon) due to its high chemical stability, which results in a lifespan in the atmosphere of 12.4 years (EPA 2017). However, the higher GWP values for CH_4 estimated on 10- and 20-years horizons (90 and 72, respectively) demonstrate that near-term climate forcers such as CH_4 might have contributions to climate change comparable to those of CO_2 for short-time horizons (Dessus et al. 2009; IPCC 2013).

In nature, most CH_4 emissions to the atmosphere originate from the anaerobic decomposition of organic matter in ecosystems such as wetlands or oceans, though more than 60% of the CH_4 emissions worldwide are anthropogenic (EEA 2017). Anthropogenic CH_4 emissions by 2015 accounted for 464 and 650 Mt CO_2-eq in the EU-28 and US, respectively, with livestock farming (45%), anaerobic waste treatment (18%), coal mining (6%), natural gas handling (5%) and wastewater treatment and discharge (4%) as the main contributors (EEA 2017; EPA 2017). The concentration of CH_4 in these emissions greatly varies from 0 to 0.2 g m^{-3} (compost piles, animal farming) to 20–100 g m^{-3} (old landfills) or even up to 250–450 g m^{-3} (biogas from wastewater treatment plants (WWTPs)). Diluted CH_4 emissions (<30% (*v/v*) CH_4 content), which are not suitable for energy recovery or incineration, are commonly released to the atmosphere without prior treatment (Ménard et al. 2012; López et al. 2013). In this context, more than 55% of the anthropogenic CH_4 emissions possess concentrations below the lower explosive limit of CH_4-air mixtures (5% *v/v*) (Avalos-Ramirez et al. 2012). In addition, despite the potential for heat and electricity generation of the biogas (40–70% (*v/v*) CH_4) produced from anaerobic sludge digestion at WWTPs, the share of biogas used for this purpose has been estimated to be less than 1% in the US, being typically burned or vented to the atmosphere without any treatment (Muñoz et al. 2013; López et al. 2017). This fact has been attributed to (i) the presence of undesirable compounds (e.g., siloxanes, H_2S, halocarbons) within biogas, (ii) the high investment costs needed for energy recovery on-site or (iii) the high costs associated to biomethane production (1.08 € m^{-3} in the EU market compared to 0.30–0.67 € m^{-3} for natural gas) (Muñoz et al. 2013; Vestman et al. 2014; ESE 2016).

To this end, end-of-the-pipe physical/chemical or biological technologies can be implemented for CH_4 mitigation. Both are based on the initial mass transfer of CH_4 from the target gas emission to a solid or a liquid phase – via adsorption or absorption, respectively – and a further chemical/biological oxidation (Estrada et al. 2012a). However, the physical/chemical treatment methods reported to date – i.e. thermal and catalytic oxidation processes – may be inefficient or costly at the flowrates and CH_4 concentrations usually found on-site, besides exhibiting a high CO_2 footprint due to their intensive energy use (Muñoz et al. 2015). In contrast, off-gas treatment biotechnologies can become, once they are properly tailored and operated, low-cost and environmentally-friendly alternatives to their physical/chemical counterparts (Estrada et al. 2012b). Particularly CH_4 abatement biotechnologies, which are based on the catalytic oxidation carried out by methane-oxidizing bacteria (namely methanotrophs), are gaining even more attraction in the past 10 years due to the possibility to create additional value out of CH_4 mitigation when coupled to the production of high-added value products (Strong et al. 2016a; Pieja et al. 2017) or to other pollution treatment processes such as denitrification (López et al. 2017). Thus, the use of this massive resource by methanotrophs at industrial facilities such as WWTPs or landfill installations would imply not only the sequestration of GHG emissions, but also potential raw material costs savings and the development of a sustainable GHG biorefinery (Cantera et al. 2017).

11.1.3 Methane to Bioplastics: The Significance of the Carbon Source for a Sustainable GHG Biorefinery

More than a dozen of different bioproducts have been listed to be produced by methanotrophs, polyhydroxyalkanoates (PHAs) representing one of the most profitable in terms of productivity, production-derived costs and proximity to market (Strong et al. 2015, 2016b). PHAs, such as poly-3-hydroxybutyrate (PHB) or poly-3-hydroxyvalerate (PHV), are linear thermoplastic polymers with the potential to substitute fossil fuel-based plastics (e. g. polyethylene, polypropylene) due to their biodegradability, biocompatibility and mechanical properties (Cantera et al. 2017). They are usually accumulated as intracellular storage granules under nutrient-limiting conditions and excess of carbon source, and serve as reservoir of energy, carbon and reducing-power during starvation periods. Both PHA yield and monomeric composition of the biocomposite may be influenced by culture conditions – e.g. pH, temperature, carbon source, presence/absence of co-substrates, oxygen concentration, macro- and micronutrient concentrations (Karthikeyan et al. 2015a; Sundstrom and Criddle 2015). Despite the extensive research carried out on the optimization of microbial PHA accumulation over the past 40 years, their higher production costs compared to those of conventional plastics represent to date the major issue to be addressed prior large-scale production and further application. In this regard, the PHA price has considerably decreased in the past 5 years and currently varies from 2 to 20 € kg^{-1}, which greatly depends on the carbon source, the product purity, the microbial strain used and the monomer composition (Bugnicourt et al. 2016).

In addition to biopolymer downstreaming, carbon source acquisition still represents one of the major constraints concerning industrial PHA production, since it accounts up to 40–50% of the total production costs (Khosravi-Darani et al. 2013). Plant-derived feedstocks (corn starch, sugar cane, and vegetable oil) and waste streams (dairy whey waste, syngas or glycerol) have been postulated as feasible feedstocks for PHA production, though they still suffer from competition with food supply, and a limited abundance and distribution, respectively (Nikodinovic-Runic et al. 2013; Levett et al. 2016). In contrast, the abundance and wide availability of CH_4, either diluted in residual streams or concentrated within biogas, offer an opportunity to save more than 50% of the raw material costs during PHA production (Levett et al. 2016). This fact has promoted the creation to date of at least two companies – Newlight Technologies and Mango Materials – devoted to the production of PHA from these CH_4 sources (Pieja et al. 2017). Also, recent life cycle assessments indicated that CH_4 derived from biogas is an energy-richer feedstock with higher CO_2 sequestration potential during the production of bioplastics compared to alternative renewable substrates. Thus, energy requirements of 37 MJ kg^{-1} PHB and carbon fixation yields of 2 kg CO_2-eq fixed kg^{-1} CH_4 have been reported for this gas substrate compared to those values obtained for corn-derived sugar (42 MJ kg^{-1} PHB and 0.1 kg CO_2-eq fixed kg^{-1} sugar) (Rostkowski et al. 2012). Finally, mixed methanotrophic cultures have proved to be robust and self-regulating under long-term operation, which enables the production of PHAs at full-scale under non-sterile conditions and, therefore, the reduction of operating costs (Strong et al. 2016b).

This chapter aims at reviewing the state-of-the-art of the available biotechnologies reported to date and their pros and cons for the abatement of CH_4 and the co-production of PHAs. Moreover, the microbiology and the metabolic pathways involved in the production of these commodities are detailed. Finally, the most recent strategies carried out on CH_4-driven PHA production studies to increase both yield and quality of the biopolymer considering further industrial implementation are discussed.

11.2 CH_4 Abatement Biotechnologies

11.2.1 Basis, Limitations and Enhancement Strategies

Biotechnologies for CH_4 mitigation have been intensively implemented in the past 20 years and are based on the enzymatic oxidation of this GHG catalyzed by the methane monooxygenase (MMO) enzyme present in all methanotrophs. This oxidation occurs at ambient temperature and pressure, does not require an external supply of chemicals (apart from water and nutrients) and takes place in the aqueous phase (or inside the biofilm) following CH_4 mass transport from the gas emission (Fig. 11.1a) (Ménard et al. 2012). The absence of extreme operating conditions and hazardous chemicals constitutes an additional advantage from a safety point of view for on-site staff and operators. The use of biological methods for CH_4 abatement has gradually increased based on their proven high robustness and efficiency (in a wide

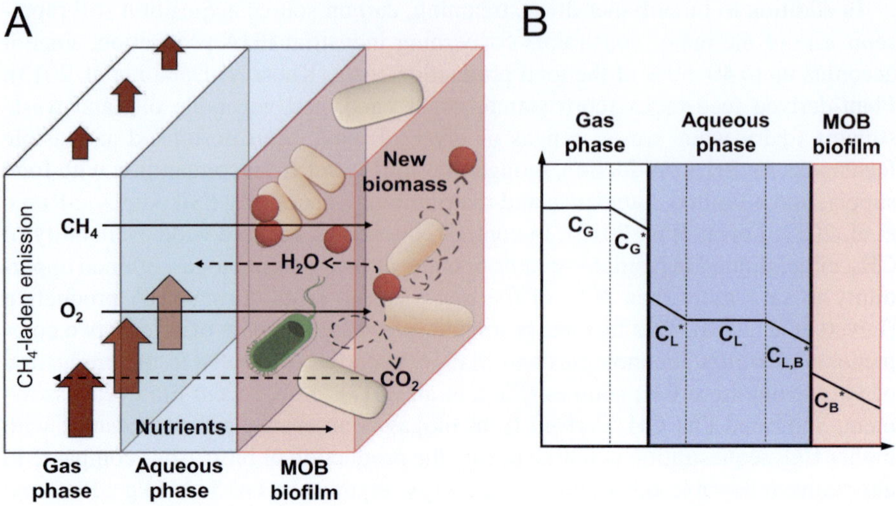

Fig. 11.1 (**a**) Schematic of CH_4 and O_2 mass transport and biodegradation in bioreactors, and (**b**) CH_4 (or O_2) concentration profile representing the mass transfer occurring within a gas treatment bioreactor. C_G = Bulk concentration of CH_4 in the gas phase, C_G* = Gas phase concentration of CH_4 at the gas–liquid interphase, C_L* = Liquid phase concentration of CH_4 in equilibrium with the gas phase, C_L = Bulk concentration of CH_4 in the liquid phase, $C_{L,B}$* = Liquid phase concentration of CH_4 at the liquid–biofilm interphase, C_B* = Biofilm concentration of CH_4 in equilibrium with the liquid phase

range of flowrates and at CH_4 concentrations below its lower explosive limit of 5% *v/v*), and on their lower operating costs compared to their physical/chemical counterparts (lower energy demands and no chemicals requirements). In addition, CH_4 abatement biotechniques typically show lower environmental impacts as a result of their lower CO_2 footprint and generation of innocuous or less-contaminating biodegradation products (CO_2, H_2O and new biomass) (Estrada et al. 2012a). In this context, biotechnologies only require: (i) the presence of an aqueous phase to support the metabolic reactions underlying CH_4 biodegradation and (ii) the availability of macronutrients (N, P, K, S) and micronutrients (usually heavy metals such as Cu, Fe, Zn or Mn needed for the synthesis and functioning of enzymes) to support microbial activity; which are both usually given in the form of a mineral salt medium (MSM) (Muñoz et al. 2015).

In the particular case of CH_4 abatement biotechnologies, their cost-efficient application is often constrained by the poor mass transport of this GHG and of the O_2 required for its oxidation from the gas stream to the methane-oxidizing bacteria (MOB) biofilm due to their low aqueous solubility (dimensionless Henry's law constant, $H = C_G/C_L$, of 29.0 and 31.3 respectively, at 25 °C and 1 atm) (Sander 2015). These low partitioning coefficients result in difficulties for CH_4 and O_2 solubilization in the liquid phase, and thus, in a reduced CH_4 biodegradation performance

(Fig. 11.1b) (López et al. 2013; Estrada et al. 2014). A common metric used to evaluate the performance of different bioreactor configurations and mass transfer enhancement approaches is the volumetric mass-transfer coefficient (k_La). Indeed, considering that k_La values for CH_4 and O_2 are quite similar under identical operational conditions and can be inferred between them, those strategies devoted to improve O_2 mass transfer might be applied to enhance CH_4 removals to the same extent (Yu et al. 2006; Estrada et al. 2013a). However, since this parameter is highly sensitive to cell density and broth properties, more direct metrics such as biomass productivity, CH_4 elimination capacities (ECs) or removal efficiencies (REs) are employed to assess process performance (Stone et al. 2017).

The poor mass transfer rates given for both CH_4 and O_2 usually entails the implementation of higher contact times between the liquid and gas phases to achieve reasonable CH_4 REs, thus forcing the design of bioreactors with excessive volume and increasing the fixed costs of these technologies. Also, vigorous mixing and agitation systems are usually implemented to ensure turbulence and promote mass transfer, though at the expense of a higher power consumption, which increases operating costs. Indeed, recent techno-economic assessments pointed out that the higher capital investment and operating costs of CH_4 pressurized bioreactors at full-scale might be balanced with the consequent increase in CH_4 solubility (Levett et al. 2016). Thus, the development of innovative compact bioreactor configurations with a high mass transfer potential at reduced energy requirements constitutes a key research niche to boost the widespread implementation of GHG abatement biotechnologies. In this regard, two different approaches have gained interest in the past years due to their superior performance during the abatement of poorly soluble compounds such as CH_4: (i) the implementation of an internal gas recycling, and (ii) the use of a secondary non-aqueous phase (NAP) (Cantera et al. 2018). On the one hand, internal gas recycling systems allow for the decoupling of the actual gas residence time and turbulence in the microbial broth from the overall empty bed residence time (EBRT) within the bioreactor, thus boosting the mass transfer without compromising the contact time between the gas and liquid phases (Estrada et al. 2014; García-Pérez et al. 2018). On the other hand, NAPs are non-toxic, non-biodegradable and non-volatile compounds with higher affinity for CH_4 and O_2 compared to that of the liquid phase, where these substances are dispersed. Once the amount of NAP within the system is optimized, this pollutant reservoir allows for both a reduced gas bubble size (which increases the interfacial area) and a reduced gas-liquid interfacial tension, hence enhancing pollutant mass transfer (Cantera et al. 2016a; Stone et al. 2017).

Finally, the economic sustainability of the biological oxidation of CH_4 has demonstrated to be positively impacted through the co-production of high-added value products such as PHAs (Cantera et al. 2017). For this purpose, the configuration selected must facilitate not only a continuous or quasi-continuous biomass production, but also biomass harvesting without compromising CH_4 abatement performance. The bioreactor configurations available for CH_4 mitigation together with their feasibility to co-produce PHAs will be hereunder discussed.

11.2.2 Bioreactor Configurations

Traditionally, gas treatment bioreactors have been divided into packed bed and suspended growth systems as a function of the presence or absence of a support media for biomass attachment, respectively. In addition, these systems can be also classified into laminar and turbulent systems based on the hydrodynamics given within the reactors, which will be the classification used from now onwards for clarity purposes (Kraakman et al. 2011).

- **Laminar bioreactors**. In laminar contactors, the gas-aqueous contact area is usually provided by a structure (packing media) which increases contact time and allows for biomass retention. Despite these systems do not require agitation and, thus, usually exhibit lower power consumptions, they suffer from lower mass transfer rates and higher reactor sizes compared to turbulent bioreactors (Kraakman et al. 2011; Stone et al. 2017). In addition, laminar bioreactors often hinder biomass recovery and, by extension, the extraction of high-added value compounds produced within the cells. Biofilters and biotrickling filters are well-stablished laminar technologies for CH_4 mitigation, though reactor designs at an early stage of development enabling biomass harvesting, such as horizontal flow biofilm reactors, membrane biofilm reactors and capillary biofilm reactors, have been also recently employed for CH_4 abatement.
  (i) Biofilters (BFs). Methanotrophs-based BFs are likely the most widespread bioreactor configuration employed worldwide for the treatment of CH_4 emissions from landfills, coal mines or livestock farms (Nikiema et al. 2007). The off-gas emission in these laminar contactors gets humidified before passing through a packed bed supporting the biofilm responsible for pollutant biodegradation, while a nutrient aqueous solution or water is irrigated intermittently (the latter if nutrients are uptaken from the support) (Fig. 11.2a). Two types of BFs are distinguished – actively and passively vented – depending on whether the flow of the CH_4-laden emission occurs forcibly or naturally, respectively (López et al. 2013). The intensive research conducted in the field of biofiltration and material science in the past years has resulted in important advances in packing materials. Packing material is a key design parameter determining both water activity (and therefore microbial activity) and the specific surface available for biofilm formation and gas-biofilm mass transfer (Cantera et al. 2017). Thus, inorganic and organic beds from simple materials such as compost, gravel, coal, stone, clay or pine bark to more sophisticated structured packings such as autoclaved aerated concrete, polymeric rings and foams have been used (López et al. 2013). Fortunately, BFs exhibit large gas-biofilm interfacial areas, a rapid start-up and low operating costs (annual cost of ~2 € m^{-3} h^{-1}). On the contrary, they present a limited height in order to maintain low pressures drops (<10 cm H_2O) and a large footprint due to the high EBRTs needed for an efficient CH_4 removal (20–70 min and 1–14 h for removal efficiencies of 60–90% in actively and passively vented BFs, respectively) (Estrada et al. 2012b; López et al. 2013; Muñoz et al. 2015). In this

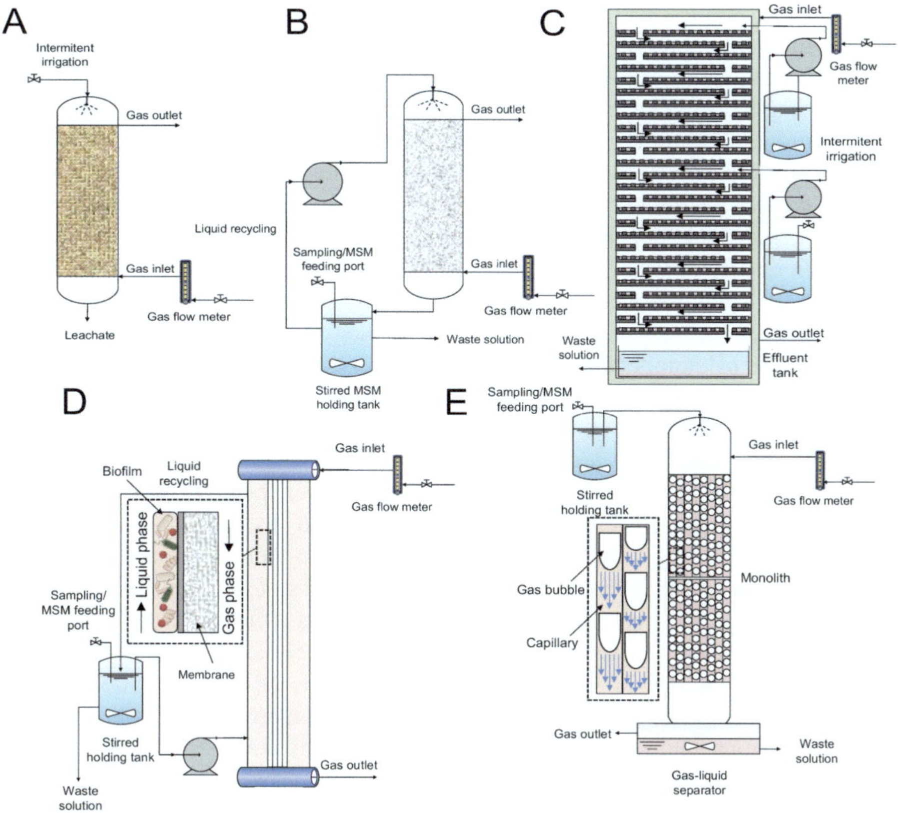

Fig. 11.2 Schematic representation of laminar bioreactors available to date for CH_4 mitigation. (**a**) Biofilter (BF), (**b**) Biotrickling filter (BTF), (**c**) Horizontal flow biofilm reactor (HFBR), (**d**) Membrane biofilm reactor (MBfR) and (**e**) Capillary biofilm reactor (CBfR)

regard, the use of fungi in biofiltration, either standalone or in combination with bacteria, has been recently proposed as a microbiological platform to achieve higher REs at lower EBRTs than those supported for bacterial BFs due to the large surface areas and the higher affinity for hydrophobic compounds of these microbes (Lebrero et al. 2016; Oliver and Schilling 2016). Another potential drawback that must be addressed during CH_4 biofiltration is biomass overgrowth, which can compromise the abatement performance and increase the pressure drop in biofilters under long-term operation, thus leading to higher operating costs. In this context, several low-cost biomass control strategies such as the addition of chemicals, protozoa and mite predation, continuous ozone addition, step feed biofiltration or famine periods have been recently proposed (Estrada et al. 2013b; López et al. 2018a; Covarrubias-García et al. 2017).

(ii) Biotrickling filters (BTFs). BTFs are also laminar contactors where the polluted gas emission passes through a biofilm immobilised onto a packed bed continuously irrigated with an aqueous nutrient solution at rates typically ranging from 5 to 30 L m^{-3}_{bed} min^{-1} (Fig. 11.2b). The CH_4 initially absorbed in the trickling aqueous solution is subsequently degraded in the fixed biofilm using the nutrients contained in the trickling solution to support biomass growth. The investment costs of this bioreactor configuration (10–41 € m^{-3} h^{-1}) are scale-dependent and can be up to two-times higher than those of BFs. In contrast, the operating costs of BTFs are lower (annual costs of 1.2 € m^{-3} h^{-1}) than those of BFs as a result of their lower pressure drops (Muñoz et al. 2015; Lebrero et al. 2015). This biotechnology supports a more efficient control of pH, temperature, water activity and nutrient supply compared to BFs. Despite the improved biomass growth control compared to traditional BFs, strategies for biomass harvesting and high-added value products extraction in both BFs and BTFs are not feasible given the low accessibility to the biomass and its adherence to the packing media. Moreover, the presence of a continuous water layer over the biofilm entails a limited abatement performance for hydrophobic gas pollutants such as CH_4 due to their poor gas-liquid mass transport, thus resulting in REs ≤30% at EBRTs of 4–5 min (Estrada et al. 2014). In this context, recent studies have been focused on the addition of NAPs such as silicone oil or Brij 35 with a higher affinity for CH_4 to boost absorption, and subsequently, achieve higher CH_4 REs up to 40–50% (Rocha-Ríos et al. 2009; Ávalos-Ramírez et al. 2012; Lebrero et al. 2015).

(iii) Horizontal flow biofilm reactors (HFBRs). This novel trickled technology comprises a stack of multiple horizontal plastic sheets, positioned vertically one above the other, with integrated frustums to sustain biofilm growth (Fig. 11.2c). Compared to other packed bed technologies, this configuration enables easier access to the different reactor parts, hence allowing for (i) the examination of the pollutant biodegradation profile and microbial structure, and (ii) biomass harvesting during the co-production of high added-value products. Both the liquid and gas phases flow horizontally across the sheets before moving to the sheet below, which ensures a good contact with the biofilm attached to the sheets and alleviates conventional drawbacks of biofiltration such as clogging, channeling or compaction (Kenelly et al. 2014). Even though k_La values have not been reported in literature, the structure of this bioreactor configuration and the gas/liquid flow patterns indicate that unfeasible reactor volumes are required to treat CH_4-laden streams at the flowrates commonly found at industrial-scale. Average CH_4 REs of 50% at EBRTs of 45–50 min and negligible pressure drops (<10 mm H_2O) have been recorded in this bioreactor configuration under long-term operation (>1 year) and even at temperatures as low as 10 °C (Kennelly et al. 2012, 2014). The addition of one or two NAPs to the trickling aqueous phase can also significantly enhance CH_4 REs in HFBRs up to 65–78% at the aforementioned EBRTs (Kennelly et al. 2014). Notwithstanding, further research on the scalability of HFBR is required, especially to ensure a sustained biomass production/harvesting without compromising the CH_4 abatement performance.

(iv) Membrane biofilm reactors (MBfRs). MBfRs are based on the passive or active diffusion of the gas pollutant from one side of a membrane into the liquid-biofilm located on the other side, driven by a concentration or a pressure gradient (Fig. 11.2d). This bioreactor configuration can overcome the CH_4 mass transfer limitations aforementioned due to the high permeability and affinity of some membrane materials for hydrophobic pollutants such as CH_4. Thus, both the membrane material and membrane configuration must be properly selected to ensure an optimum pollutant abatement performance (Kumar et al. 2008). The few studies reported on CH_4 abatement with MBfRs have employed micro- (polyvinylidene difluoride, 0.40 μm pore size) and ultrafiltration (Zenon and acrylonitrile, <0.1 μm pore size) hollow fiber membranes, the later supporting the highest gas-liquid interfacial areas for mass transfer. These systems have been studied for the simultaneous removal of NH_4^+, NO_3^-, NO_2^- and CH_4 with a minimized carbon footprint (Sun et al. 2013; Shi et al. 2013). However, little attention was paid to the CH_4 REs in these works, which were focused on the use of CH_4 as electron donor for denitrification, either coupling aerobic methane oxidation to denitrification or denitrifying anaerobic methane oxidation (DAMO) to anammox (Shi et al. 2013; Cai et al. 2015; Sánchez et al. 2016). In this context, MBfRs represent a promising technology to support the growth of DAMO microorganisms due to their higher capacity for biomass retention. However, the production of PHAs through an intensive accumulation of methanotrophic biomass may lead to membrane fooling issues and, consequently, low biodegradation performance (Cai et al. 2015). Despite small changes in the pressure and/or hydraulic retention time can be applied as strategies for the detachment of the excessive biomass, the REs might be jeopardized during several days (Cai et al. 2015; Chen et al. 2016.

(v) Capillary biofilm reactors (CBfRs). CBfRs are composed of parallel straight microcapillaries (Ø of 1–5 mm) separated by a thin wall (Fig. 11.2e), where capillary forces are dominant over gravity or viscosity forces (Kraakman et al. 2011). This novel configuration, often referred as monolith packing or monolith catalyst, pushes liquid and gas to separately pass through the channels, ensuring a plug flow behavior. The preferred flow pattern in monolith reactors is known as Taylor flow and it is a bubble train of alternate liquid and small gas bubbles flowing upwards or downwards in the same direction. Recent studies have demonstrated that the gas-liquid mass transfer under this Taylor flow patterns is almost independent of the channel size, which suggests the use of larger channels in order to reduce the pressure drop in this configuration at industrial-scale (Ebrahimi et al. 2006). The research conducted to date have revealed that the pollutant RE is not compromised by biomass overgrowth within the monolith channels, even when trickling liquid flushing strategies are applied, which suggests the feasibility of CBfRs to continuously produce high-added value products such as PHAs (Jin et al. 2006). Also, since each channel acts independently, the scale-up concerns are drastically minimized with this technology. CBfRs present mass transfer coefficients equivalent to those of turbulent bioreactors at around one order of magnitude lower power consumptions (Ebrahimi et al. 2006; Kraakman et al. 2011; Rocha-Ríos et al. 2013). In contrast, the economic viability of monolith reactors nowadays is limited by the expensive materials and the complex designs required for their construction (Ebrahimi et al. 2006). Moreover, this CH_4 abatement technology has been barely validated at lab-scale and its use for CH_4 abatement remain in an embryonic stage and still requires further research. In this regard, the only work published to date on CH_4 abatement capillary bioreactors reported 50% higher CH_4 REs at significant lower reactor sizes and pressure drops compared to turbulent contactors supplemented with a NAP (Rocha-Ríos et al. 2013).

- **Turbulent bioreactors**. Turbulent contactors are characterized by higher mass transfer rates (by one or two orders of magnitude) and energy requirements due to an intensive mixing and pumping compared to laminar bioreactors (e. g. BFs and BTFs). Among them, stirred tank and bubble column bioreactors have been the most applied, especially in industrial fermentative processes, due to their manageable monitoring and easier control of culture broth conditions (Kraakman et al. 2011; Stone et al. 2017). In turbulent bioreactors, biomass is generally suspended, easily harvested and can reach higher concentrations in the culture broth without generating operational failures. Due to the intensive power consumptions given in these systems, novel reactor designs such as the forced loop bioreactors, and strategies such as the internal gas recycling and NAP addition have been recently employed to enhance gas mass transfer and, therefore, boost their implementation at full-scale (Cantera et al. 2017, 2018).

  (i) Stirred tank reactors (STRs). STRs are turbulent contactors where the continuous agitation, often mechanical, ensures perfect mixing and biomass suspension (Fig. 11.3a) (Nienow et al. 2014). The energy input for mixing is used to break down the bubbles of the polluted gas supplied at the bottom of the reactor (which increases the gas-liquid interfacial area) and to reduce the liquid film width (thus increasing the overall volumetric mass transfer coefficient of the pollutant from the gas to the liquid phase) (Kraakman et al. 2011). Hence, REs of 50–60% have been achieved in STRs devoted to CH_4 abatement at relatively low EBRTs (4–10 min) (Rocha-Ríos et al. 2009; Cantera et al. 2016a). The addition of a NAP in this type of bioreactor has been also recently assessed, which consequently enhanced CH_4 REs up to 70% at an EBRT of 4.8 min (Rocha-Ríos et al. 2010). However, despite the promising results obtained in STRs, their higher power consumption compared to packed bed systems has resulted in the preferential implementation of both BFs and BTFs in full-scale applications (Kraakman et al. 2011). In this sense, STRs are especially recommended for applications where the volumetric gas flows treated are relatively modest, while the treatment of higher gas stream volumes is preferentially carried out in bubble column or loop bioreactors (Petersen et al. 2017).

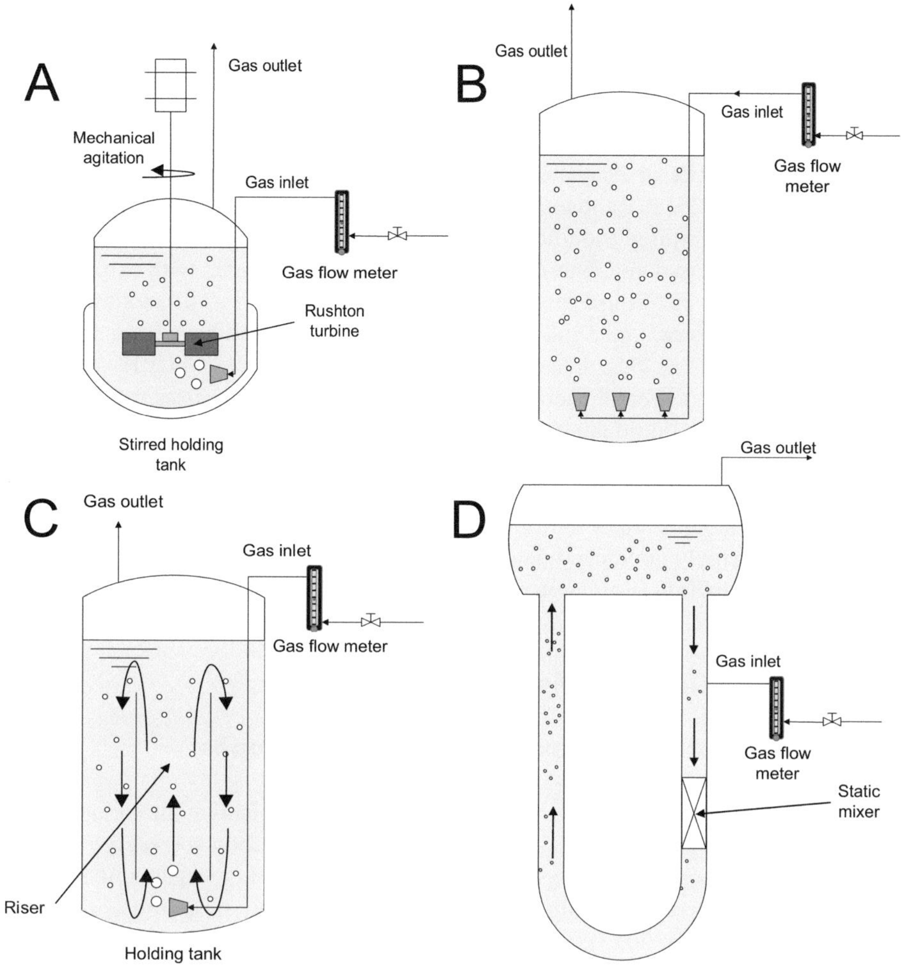

Fig. 11.3 Schematic diagram of turbulent bioreactors reported to date for CH_4 mitigation. (**a**) Stirred tank bioreactor (STR), (**b**) Bubble column bioreactor (BCB), (**c**) Airlift reactor (ALR) and (**d**) Forced circulation loop bioreactor (FCLB)

(ii) Bubble column bioreactors (BCBs), airlift reactors (ALRs) and forced circulation loop bioreactors (FCLBs). These three suspended growth configurations differ from the STRs on the fact that no mechanical agitation is provided. Thus, adequate mixing and effective pollutant mass transfer to the microorganisms is typically achieved either using micropore diffusers in BCBs, or installing a concentric draft-tube (riser) to create a density gradient within the culture broth in order to enhance the turbulence in ALRs (Fig. 11.3b, c) (López et al. 2013). Similar to the STRs, these suspended growth bioreactors are ideal for both community structure monitoring and biomass harvesting during the co-production of bioproducts from CH_4. Also, their lower power requirements, as a result of the lack of mechanical agitation, guarantee a higher cost-effectiveness of BCBs and ALRs compared to STRs. However, either the addition of NAPs or the implementation of gas-recycling strategies is hitherto mandatory to boost CH_4 biodegradation over the basal REs of 20–30% typically recorded in ALRs and BCBs (Rocha-Rios et al. 2011; Rahnama et al. 2012). In recent years, several patents and studies regarding modifications of ALRs have emerged. Most of them include internal and external recirculation pipes equipped with extra fittings to enhance gas-liquid mass transfer such as static mixers, nozzles for gas side streams or additional liquid pumping systems, and have been sub-classified as FCLBs (Yazdian et al. 2012; Petersen et al. 2017). FCLBs generally include two different sections, one devoted to maximize mass transfer from the gas to the liquid phase and another one to separate gas and liquid streams, thus allowing for an easy biomass harvesting (Fig. 11.3d). It is noteworthy that this unique structure simultaneously enables the gas stream monitoring and the monitoring of the culture broth conditions. Hence, ALRs and FCLBs are extremely good candidates for GHG abatement combined with bioproducts production (Rahnama et al. 2012).

11.3 Microbiology of Methane Oxidation

11.3.1 Distribution and Taxonomy of Methanotrophs

Aerobic methane oxidizers are methylotrophic microorganisms with the ability to oxidize CH_4 in the presence of O_2 and use it as their main carbon and energy source. Despite most aerobic methanotrophs are bacteria, other microorganisms such as yeasts, fungi and green microalgae have been also tentatively proposed as methane oxidizers (López et al. 2013). Additionally, anaerobic oxidation of methane (AOM) has been also observed by anaerobic archaea and bacteria in marine sediments, where the process occurs coupled to sulfate, nitrate, nitrite, manganese and iron reduction (Cui et al. 2015). Aerobic methane oxidizers, herein after named methanotrophs, are widely distributed in environments such as wetlands, peat bogs, forests, rice paddies, groundwater, landfill cover soils, sewage sludge or marine sediments, and represent the major terrestrial sink for CH_4 (Rostkowski et al. 2013). Their growth preferentially occurs in the interphase between aerobic and anaerobic zones, where significant CH_4 fluxes are given (Semrau et al. 2010).

Methanotrophs are usually classified into three different groups according to their physiological and morphological characteristics: type I, type II and type X methanotrophs (Karthikeyan et al. 2015a). Type I includes methanotrophs that (i) present intracytoplasmatic membranes as bundles of vesicular discs, (ii) use the ribulose monophosphate (RuMP) pathway for carbon assimilation, and (iii) contain phospholipid fatty acids of 14/16 carbons length. Type II methanotrophs are characterized by (i) an intracytoplasmic membrane aligned along the peripheral part of the cell, (ii) the use of the serine pathway for carbon assimilation, and (iii) phospholipid fatty acids of 18 carbons length. On the other hand, type X (considered as a subset of type I) share characteristics of both type I and II methanotrophs, thus including the RuMP pathway to assimilate formaldehyde and exhibiting the enzyme ribulose biphosphate carboxylase from the serine pathway to fix CO_2 (López et al. 2013; Knief 2015). Current classification of known aerobic methanotrophs based on 16S rRNA encloses a wide phylogenetic distribution within two different phyla, *Proteobacteria* (subdivisions *α*- and *γ-Proteobacteria*) and *Verrucomicrobia.* The most representative genera within the *α-Proteobacteria* phylum are *Methylocystis*, *Methylosinus*, *Methylocella* and *Methylocapsa*, which are type II methanotrophs and cluster within *Methylocystaceae* and *Beijerinckiaceae* families. Conversely, those methanotrophs within *γ-Proteobacteria* (e. g. *Methylomonas*, *Methylobacter*, *Methylosarcina* or *Methylococcus* genera, among others) belong to the *Methylococcaceae* family and are known as type I methanotrophs. Finally, methanotrophs from *Verrucomicrobia* phylum are essentially affiliated within the *Methylacidiphilacecae* family, and more specifically to the *Methylacidiphilum* and *Methylacidimicrobium* genera (Semrau et al. 2010; Pawlowska 2014). Based on their different ecophysiology and taxonomical affiliation, *Verrucromicrobia* methane oxidizers have been recently referred as type III methanotrophs (Sazinsky and Lippard 2015). The phylogenetic relationships between methanotrophs are usually examined as a function of the enzyme MMO, which catalyses the conversion of methane to methanol. Thus, the particulate methane monooxygenase (pMMO) is found in most methanotrophs, which is cytoplasmic membrane bound, where as the soluble methane monooxygenase (sMMO) is cytoplasmic in nature. It can be expressed either as the sole form of MMO or together with pMMO. The enzyme sMMO has been traditionally associated with type II methanotrophs, though few genera of type I methanotrophs with the capability to synthesize sMMO have been recently identified (Knief 2015).

11.3.2 Methane Biodegradation Pathways in Methanotrophs

Methane biodegradation comprises a multistage process, where the monooxygenase-mediated CH_4 oxidation to methanol represents the first step-reaction. Formaldehyde and formic acid are the successive metabolic intermediates formed, thus eventually ending up with the production of CO_2 (López et al. 2013). It must be stressed that formaldehyde may be either converted to formic acid or used as building blocks for the synthesis of new cell material through the RuMP pathway in type I/X methanotrophs, or through the serine pathway in type II methanotrophs (Pawlowska 2014). Notwithstanding, methanotrophs cannot achieve a complete conversion of methane into microbial biomass since assimilatory pathways result in a net production of at least 12% CO_2 according to previous studies (Bédard and Knowles 1989).

Methanotrophic growth stoichiometry may be expressed (i) based on the oxygen and nitrogen demands (Eqs. (11.1) and (11.2), for the RuMP and the serine pathways, respectively) (Karthikeyan et al. 2015a):

$$CH_4 + 1.50O_2 + 0.12NH_4^+ \rightarrow 0.12(C_4H_8O_2N) + 0.53CO_2 + 1.71H_2O + 0.12H^+ \tag{11.1}$$

$$CH_4 + 1.57O_2 + 0.10NH_4^+ \rightarrow 0.10(C_4H_8O_2N) + 0.59CO_2 + 1.75H_2O + 0.10H^+ \tag{11.2}$$

or (ii) as a function of the fraction of electrons required by the electron donor for growth. According to Rostkowski et al. (2013), methanotrophic cell synthesis implies the use of a fraction of electrons (f_e) from CH_4 oxidation to reduce O_2 to H_2O (thus producing energy), while the remaining fraction (f_s) is devoted to cell synthesis. Thus, it can be assumed that CH_4 metabolism requires the use of O_2 as (a) terminal electron acceptor to produce energy (Eq. 11.3), and (b) reactant in the initial attack on CH_4 within the electron donor reaction (Eq. 11.4):

$$f_e\left(1/4\left[O_2 + 4H^+ + 4e^- \rightarrow 2H_2O\right]\right) \tag{11.3}$$

$$1/4\left[CH_4 + O_2 \rightarrow CO_2 + 4H^+ + 4e^-\right] \tag{11.4}$$

The nature of the micro- and macronutrients greatly influences the stoichiometry of the cell synthesis half reaction, which can be given as in Eq. (11.5) when nitrate is the sole nitrogen source and the empirical formula $C_5H_7O_2N$ for biomass is assumed:

$$f_s\left(1/28\left[5CO_2 + NO_3^- + 29H^+ + 28e^- \rightarrow C_5H_7O_2N + 11H_2O\right]\right) \tag{11.5}$$

Thus, the overall stoichiometric reaction for the microbial CH_4 biodegradation (Eq. 11.6) can be obtained by summing up Eqs. (11.3), (11.4), and (11.5):

$$\begin{aligned} &1/4CH_4 + (1/4 + f_e/4)O_2 + (f_s/28)NO_3^- + (29/28f_s + f_e - 1)H^+ \\ &\rightarrow (1/4 - 5f_s/28)CO_2 + (f_e/2 + 11f_s/28)H_2O + (f_s/28)C_5H_7O_2N \end{aligned} \tag{11.6}$$

It can be then concluded that the number of moles of O_2 consumed per mole of CH_4 oxidized is $1 + f_e$, where $f_e + f_s = 1$. Also, the biomass yield (Y_X, expressed as gTSS (total suspended solids) per gCH_4) depends on the nitrogen source and can be calculated as $(113f_s/28)/4$ when nitrate is employed.

Both Monod and Michaelis-Menten models are commonly used to describe microbial CH_4 oxidation kinetics. However, the different experimental setups and models used for the estimation of the kinetic parameters usually hamper direct comparison of the data available in literature. Among the kinetic parameters, the Monod constant (K_S) characterizes the affinity of microorganisms for the substrates, and typically ranges from 1.0×10^{-6} to 4.7×10^{-4} M in the particular case of CH_4. pMMO-bearing bacteria, usually type I methanotrophs, have been reported to exhibit a higher affinity for CH_4 (lower K_S values) than those expressing sMMO (Amaral and Knowles 1995; López et al. 2014). Moreover, the maximum specific biodegradation rates (q_{max}) for CH_4 typically range from 4.16×10^{-5} to 1.3×10^{-4} gCH_4 $gTSS^{-1}$ h^{-1} and are greatly influenced by culture conditions (Cantera et al. 2016b; AlSayed et al. 2018a). In this context, microorganisms with a high q_{max} and a high affinity for CH_4 (low K_S) are desirable to guarantee an efficient biocatalytic activity and to reduce the start-up period of bioreactors devoted to CH_4 mitigation. However, it must be noted that most CH_4 biodegradation kinetic studies reported to date were carried out at high biomass concentrations, which did not ensure the absence of mass transport limitations and, therefore, the validity of the kinetic parameters obtained (López et al. 2013).

Regarding the CH_4 biodegradation efficiency achieved through the two aforementioned metabolic pathways, the RuMP pathway (from type I methanotrophs) provides a higher CH_4 conversion efficiency since only 1 ATP molecule is required to assimilate 3 formaldehyde molecules during CH_4 oxidation. Conversely, despite type II methanotrophs are superior in terms of CH_4 storage as phospholipid fatty acids and polyhydroxyalkanoates (PHAs), they require 3 ATP and 2 NADH molecules to fix 2 formaldehyde and 1 CO_2 molecules through the serine pathway (Karthikeyan et al. 2015a). Hence, type I methanotrophs tend to prevail over type II strains, unless cultivation conditions provide a competitive situation for the latter (AlSayed et al. 2018b).

11.3.3 Ecophysiology of Methanotrophs

Main driving forces enabling the selection of either type I or type II methanotrophs from mixed cultures are O_2 and CH_4 concentrations, temperature, pH, nitrogen source type and concentration, and the concentration of micronutrients such as copper. In this sense, it must be highlighted that an appropriate selection strategy to enrich type II over type I methanotrophs in mixed methanotrophic cultures is crucial prior to proceeding with PHA accumulation stages.

Hence, according to the CH_4 biodegradation stoichiometry aforementioned, molar O_2:CH_4 ratios should be maintained $\geq$1.5:1 to enable CH_4 oxidation (Bédard and Knowles 1989). However, previous studies demonstrated that low O_2:CH_4 ratios may favour the growth of type II methanotrophs, while high O_2:CH_4 ratios can promote the growth of type I methanotrophs (Henckel et al. 2000). These findings support the hypothesis that the enzymes sMMO (predominantly found in type II methanotrophs) and pMMO are respectively expressed at high and low CH_4 concentrations (Van Teeseling et al. 2014). Also, microbial kinetics of the type II methanotrophs *Methylosinus trichosporium* OB3b and *Methylocystis parvus* OBBP were recently studied under different oxygen partial pressure conditions and nitrogen sources (0.10–0.40 atm O_2 in the headspace; N_2, NH_4^+ and NO_3^- as nitrogen sources) (Rostkowski et al. 2013). These authors found that both strains clearly showed sensitivity to oxygen when molecular nitrogen was used as nitrogen source, which was attributed to the inactivation of the nitrogenase enzyme of these strains.

To date, studies assessing the influence of temperature as a selective pressure for either type I or type II methanotrophs have not been reported. In this regard, methanotrophs usually exhibit maximum biodegradation rates under mesophilic conditions (25–30 °C), though thermotolerant and thermophilic type I methanotrophs such as *Methylotermus*, *Methylocaldum*, *Methylococcus* genera or several *Verrucomicrobia* strains have been isolated from active silts, hot springs and aquifers and volcanic soils, with optimal growths in the range of 42–65 °C (Pol et al. 2007; Knief 2015). Similarly, psychrophilic strains of *Methylobacter*, *Methylosphaera* and *Methylomonas* have been isolated from tundra soils, Antarctic meromictic lakes and deep igneous groundwater, exhibiting an optimal growth range of 3.5–15 °C (Pol et al. 2007; Kalyuzhnaya et al. 2013).

Regarding the pH, most methanotrophs preferentially grow at neutral values (6.8–7.5), although recent studies have demonstrated that acidic environments promote the selection of type II methanotrophs over type I methanotrophs in mixed cultures (Table 11.1) (Pieja et al. 2011a). Not only species belonging to the *Mehylocapsa* genera (obligate acidophilic), but also species from the *Methylocella* and *Methylocystis* genera tolerate low pH and have been found to grow optimally at pH values of 5–6 and 3.6–4.5, respectively (Pieja et al. 2011a; López et al. 2013). The fact that type II exhibit more tolerance than type I methanotrophs to low pH has been attributed to their CO_2 requirements as an input of the serine cycle (Karthikeyan et al. 2015a). In contrast, the CH_4 biodegradation ability of the halophilic type I *Methylomicrobium alcaliphilum* strain 20Z is enhanced at pH values as high as 9.0 (Kalyuzhnaya et al. 2013).

Solely type II methanotrophs (and some type I genera belonging to *Methylosoma, Methylosphaera* and *Methylococcus* genera) have demonstrated the capability to fix N_2, since they possess the *nifH* genes for the synthesis of the nitrogenase enzyme (Rostkowski et al. 2013). Nonetheless, nitrate and ammonium are preferred over nitrite as nitrogen sources for both type I and II methanotrophs (Dunfield and Knowles 1995). Interestingly, a shift in a methanotrophic consortium towards a type II-dominated culture was feasibly achieved in a fluidized bed reactor (FBR) by using N_2 (instead of nitrate or ammonium) as nitrogen source and decreasing the dissolved oxygen levels (Pfluger et al. 2011). However, several studies have successfully addressed the enrichment of type II methanotrophs (i. e. *Methylocystis* species, >60%) from fresh activated sludge even in less than 1 month using ammonium as nitrogen source (Myung et al. 2015a; Fergala et al. 2017). In this regard, it is noteworthy that the maximum specific biodegradations rates reported in these studies were obtained when concentrations of ammonium as low as 10 mM were employed for cultivation. These results are consistent with those findings reported by our research group during the enrichment of methanotrophs from landfill samples under different ammonium concentration (0–200 mM) (López et al. 2018b). The enriched consortia exhibited the highest specific biodegradation rates and the higher type II abundances at the lowest ammonium concentrations (q_{max} of

Table 11.1 Enrichment culture conditions for the preferential selection of type II methanotrophs

References	Configuration	Inoculum	O_2:CH_4	N source (mM)	Other enrichment conditions[a]	Microbial selection (abundance)
Graham et al. (1993)	CSTR[b]	*M. trichosporium* OB3b/*M. albus* BG8 co-culture	(i) 4:1	(i) NO_3^- (1)	(i) No Cu	(i) *M. trichosporium* OB3b (>85%)
			(ii) 1:10	(ii) NO_3^- (0.5)	(ii) –	(ii) *M. trichosporium* OB3b (~100%)
Nyerges et al. (2010)	Batch	*Methylocystis* sp./*Methylomicrobium album* co-culture	1:5	NH_4^+ (50)	–	*Methylocystis* sp.
Pfluger et al. (2011)	FBR[c]	Hot spring sediments	1:4	N_2	–	*Methylocystis*/*Methylosinus*
Pieja et al. (2011a)	Batch	Activated sludge	1:1	NO_3^- (10)	(i) No Cu, diluted medium (10%)	*Methylocystis*/*Methylosinus*
					(ii) pH 5, carbonate at 10 mM	
Myung et al. (2015a)	Fed-batch	Activated sludge	1.5:1	NH_4^+ (nd)	30 °C	*Methylocystis* (>70%)
López et al. (2018b)	Fed-batch	Mixed landfill sample	1:5	NH_4^+ (4)	–	*Methylocystis* (>80%)

[a]Unless specified otherwise, temperature and pH were set at standard values of 25 °C and 6.8–7.5, respectively
[b]CSTR: Continuous stirred tank reactor
[c]FBR: Fluidized bed reactor

$7.8 \times 10^{-4} \pm 10^{-5}$ gCH_4 $gTSS^{-1}$ h^{-1} and >80% *Methylocystis* at 4 mM) (Table 11.1). Ammonium inhibition in methanotrophs has been reported to occur due to either nitrite or hydroxylamine accumulation from ammonium oxidation, which explains the importance of adjusting the concentration of this nitrogen source (Nyerges et al. 2010). Despite the use of nitrate has been reported to lead to higher specific growth rates, the absence of specificity towards type II methanotrophs makes this nitrogen source not suitable for the enrichment processes. In general terms, it can be stated that type I methanotrophs are dominant in environments under low CH_4/N conditions, whereas type II prevail under high CH_4/N conditions and outcompete type I methanotrophs under nutrient-limiting conditions (e. g. N-limitation) (Amaral and Knowles 1995; Pieja et al. 2011a).

Regarding micronutrient culture conditions, copper seems to control the expression of the genes encoding for the MMO enzyme and positively regulate the activity of the enzymes pMMO and sMMO (Semrau et al. 2010). However, copper concentration in the cultivation broth must be adjusted in order to maintain copper homeostasis and prevent metal toxicity. Most methanotrophs grow optimally at copper concentrations lower than 4.3 mM, though previous enzymatic assays have demonstrated that sMMO in type II methanotrophs is properly synthesized at low Cu^{2+} concentrations (below 0.8 mM) (Graham et al. 1993; Bender and Conrad 1995; Semrau et al. 2010). In this regard, the expression of the *pmoA* gene (encoding a subunit of pMMO) has been observed at significant levels regardless of the Cu^{2+} concentration, though the transcript numbers increased concomitantly with the concentration of copper (Murrell et al. 2000). However, the absence of copper as a selection criterion for type II methanotrophs seems to be insufficient. In their study, Pieja et al. (2011a) demonstrated that a combined lack of copper and use of a diluted mineral salts medium (10% *v/v*) may positively select type II methanotrophs from mixed inocula, as well as the combined use of low pH values and 10 mM carbonate (Table 11.1).

11.4 Methanotrophic PHA Biosynthesis

11.4.1 Metabolic PHA Pathways

Controversial information on the methanotrophic species exhibiting capacity to produce PHAs can be found in literature. However, most studies asserting PHA accumulation by type I methanotrophs often suffer from a lack of evidence or are based on qualitative evidences (Heyer et al. 2005). In this regard, a recent study screened seven different methanotrophic genera in terms of PHB production and confirmed the exclusivity of type II methanotrophs, such as *Methylocystis*, *Methylosinus* and *Methylocapsa*, to accumulate PHAs through the serine pathway (Pieja et al. 2011a). Once unbalanced nutrient conditions (usually in the form of a N, P, S, O_2 or Mg limitation) are given in the culture broth, type II methanotrophs deviate part of the carbon from CH_4 oxidation towards the production of these commodities. For this purpose, the formaldehyde formed within the CH_4 oxidation pathway acts as link

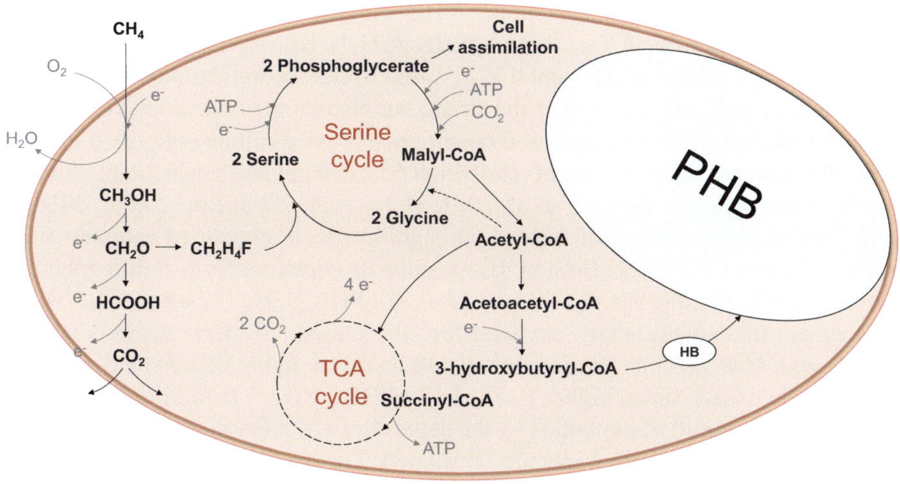

Fig. 11.4 PHB production pathway in obligate type II methanotrophs. Dotted arrows indicate the existence of not-mentioned intermediates

between anabolism and catabolism in type II methanotrophs. The activation of the anabolic pathway is driven by the tetrahydromethanopterin (H_4MPT)-dependent enzymes, which catalyze the reaction between formaldehyde and glycine to constitute serine (Fig. 11.4). Serine is later converted within the serine cycle into acetyl-CoA, which in turn may enter either the Tricarboxylic Acid Cycle (TCA) under nutrient-sufficient conditions, or the PHA cycle under nutrient-limiting conditions (Anthony 2011; Karthikeyan et al. 2015a). The limitation of any of the aforementioned nutrients hinders the production of high amounts of coenzyme A from Krebs cycle, which enables the conversion of acetyl-CoA into acetoacetyl-CoA mediated by the β-ketothiolase (βKT) enzyme encoded by the *phaA* gene (Tan et al. 2014). This substrate is later converted into a hydroxyacyl (HA) coenzyme A thioester by an acetoacetyl-CoA reductase (*phaB* gene), and finally PHA is formed through HA monomers polymerization via ester bond linking mediated by the PHA synthase (*phaC* gene) (Pieja et al. 2011a; Tan et al. 2014).

It must be stressed that PHB of high molecular weight is the PHA synthesized after culture exposure to the nutrient-limiting conditions when CH_4 is used as the sole C and energy source (Wendlandt et al. 2001). The overall equation for PHB accumulation in methanotrophs using the serine pathway is given below:

$$8CH_4 + 12O_2 + FP \rightarrow C_4H_6O_2 + 4CO_2 + 12ATP + FPH_2 \qquad (11.7)$$

where FP represents the oxidised succinate dehydrogenase (involved in the TCA), FPH_2 the reduced succinate dehydrogenase and $C_4H_6O_2$ the empirical formula of the PHB monomer. According to Eq. 11.7, the theoretical yield for bioconversion of CH_4 into PHA (Y_{PHA}) can be estimated at 0.67 gPHA gCH_4^{-1}. However, the fact that part of the CH_4 and O_2 consumed has to be converted to CO_2 to generate $NADP^+$ for the acetoacetyl-CoA reductase in the PHB biosynthetic pathway, results in a decrease of the theoretical Y_{PHA} to 0.54 gPHB gCH_4^{-1} (Khosravi-Darani et al. 2013). The adjustment of the experimental Y_{PHA} value to the theoretical one depends not only on the purity and the type of the strain, but also on the cultivation conditions. In this regard, previous studies based on enriched or pure cultures of type II methanotrophs reported empirical yields of 0.46–0.55 gPHB gCH_4^{-1}, which are close to the theoretical one (Wendlandt et al. 2001; Helm et al. 2008; López et al. 2018c). Moreover, a recent study highlighted the significance of copper to maintain such values (Y_{PHA} of 0.44–0.49 gPHB gCH_4^{-1}), since its depletion from culture medium led to a decrease in the PHA yield to 0.23–0.32 gPHB gCH_4^{-1} (Zhang et al. 2017). In contrast, the stoichiometry obtained for pure cultures of *Methylocystis parvus* OBBP and *Methylosinus trichosporium* OB3b under optimal growth conditions revealed yields two times higher (0.88–1.13 gPHB gCH_4^{-1}) than those previously estimated. This could be explained by the dependence of this value on the parameter f_s, which is highly organism-specific (Rostkowski et al. 2013). Thus, according to previous studies, the overall stoichiometry for PHA accumulation in type II methanotrophs may be also expressed considering f_e and f_s in the equations and assuming that nitrogen is not required for its formation (Eq. 11.8) (Rostkowski et al. 2013):

$$1/4CH_4 + (1/4 + f_e/4)O_2 \rightarrow (1/4 - 2f_s/9)CO_2 + (f_e/2 + f_s/3)H_2O + (f_s/18)C_4H_6O_2 \quad (11.8)$$

Y_{PHA} can be then inferred from Eq. 11.8 ($Y_{PHA} = 86f_s/18/4$) when f_s and f_e values are calculated during the PHA accumulation phase for a given methanotrophic strain under specific culture conditions.

The stored PHB may be exceptionally consumed as energy source (source of intracellular reducing equivalents, e.g. $NADH^+$) by type II methanotrophs, though the number of studies reported to date on this issue are scarce. In this context, the catabolism of PHB into acetyl-CoA has been suggested to occur by the sequential activity of a PHB depolymerase, β-hydroxybutyrate dehydrogenase and an acetoacetate succinyl-CoA transferase as in other PHB-producer microorganisms (Karthikeyan et al. 2015a). A recent work has demonstrated the ability of PHB-rich type II methanotrophs to couple PHB oxidation to the reduction of nitrite to nitrous oxide (step 2 of the coupled aerobic-anoxic nitrous decomposition operation, CANDO), which constitutes a novel approach to be implemented in WWTPs for nitrogen removal (Myung et al. 2015b). More interestingly, recent studies found that PHB-rich cells of *M. parvus* OBBP are able to grow at higher specific growth rates than those of cells without PHB under nutrient-sufficient conditions, while short-term CH_4 starvations do not induce either PHB consumption or cellular replication by this type II methanotroph (Pieja et al. 2011b). Since PHB-rich type II methanotrophs may exhibit an advantage with excess of nutrient supply, feast-famine strategies based on the availability/deprivation of nitrogen have been successfully employed under continuous operation of a bubble column bioreactor to maintain a self-regulated culture of *Methylocystis hirsuta* producing PHB under non-sterile conditions (García-Pérez et al. 2018). Indeed, feast-famine strategies based on the alternation of different nutrient limitations have been assessed as viable criteria to enrich PHB-rich type II methanotrophs within a mixed community and, therefore, enhance the PHA content under long-term operation (Pieja et al. 2012). These authors subjected CH_4-fed sequencing batch reactors to three cyclically stress conditions: (i) repeated nitrogen limitations, (ii) repeated nitrogen and oxygen limitations, and (iii) repeated nitrogen and methane limitations, and concluded that the sequential nitrogen and methane limitations favored PHB accumulation in the strain *M. parvus* OBBP.

11.4.2 Influence of the Microbial Community and Culture Conditions on PHB Accumulation

Most studies on CH_4-driven PHA production are based on the optimization of the culture conditions in order to achieve higher PHA contents, yields and molecular weights for the resulting biocomposite. In this sense, *Methylocystis* and, to a lesser extent *Methylosinus*, are the type II methanotrophic genera commonly used for this purpose. The dominance of *Methylocystis* genus in mixed consortia from different environments and bioreactor configurations and its usually higher PHB contents (up to 68% wt) compared to *Methylosinus* (up to 50% wt) when CH_4 is used as sole carbon and energy source justify the greater number of studies for this genus (Table 11.2). It must be also highlighted that, despite the proven PHB production capacity of the *Methylocapsa* and *Methylocella* genera, there are no studies to date evaluating PHB accumulation by these type II methanotrophs (Pieja et al. 2011a). This fact may be attributed to the low growth rates and difficulties to enrich/isolate these genera from bioreactors devoted to CH_4 mitigation (Crombie and Murrell 2014; López et al. 2018a). Also, few studies have explored the potential of mixed cultures over pure cultures for the production of PHAs, since this may result in a lower accumulation of inhibitory CH_4-derived metabolites (such as methanol or formaldehyde) and allows for operation under non-esterile conditions (Helm et al. 2006). This might be of relevance when the PHB-producer methanotroph is the dominant specie, thus leading under N-limiting conditions to relatively high PHB contents (39–46% wt) compared to those for pure cultures (Helm et al. 2006; Myung et al. 2015a, b). Also, a poorly enriched methanotrophic community dominated by PHB-producer heterotrophs has been recently proposed as an alternative to accumulate PHB at ~49% wt using CH_4 as the unique carbon source, which is comparable to those results obtained for methanotrophic communities (Zhang et al. 2017). However, inadequate enrichments in CH_4-supplemented batch tests led to PHB accumulations as low as 1.5–8.5% wt, possibly due to the predominance of heterotrophs without the ability to accumulate PHB (Karthikeyan et al. 2015b; Chidambarampadmavathy et al. 2017a, b). Also, experiments conducted in different bioreactor configurations enriched methanotrophic communities with low potential to produce PHAs, due to the predominance of type I methanotrophs. In this regard, a recent study achieved maximum PHB accumulations of 5–10% wt in a fluidized bed reactor mainly dominated by *Methylobacter*-like type I methanotrophs (Pfluger

Table 11.2 Comparison of the most significant batch PHA production studies reported to date using CH_4 as the unique carbon and energy source

Reference[a]	Methanotrophic strain	Culture conditions[b]	Maximum PHB content (% wt)	Maximum PHB yield (gPHB gCH_4^{-1}) and molecular weight (Da)	Specific PHB production rate (gPHB g_X^{-1} h^{-1})	Volumetric PHB productivity (gPHB L^{-1} h^{-1})
Asenjo and Suk (1986)	*M. parvus* OBBP	N limitation	68 (1.0 gPHB L^{-1})	0.67 (theoretical)	–	–
Wendlandt et al. (2001)*	*Methylocystis* sp. GB 25	N/P/Mg limitation (growth phase on both NO_3^- and NH_4^+)	51/47/28	0.52/0.55/0.37; $M_w^c = 2.5 \times 10^6$	0.12/0.06/0.05	–
Wendlandt et al. (2005)*	*Methylocystis* sp. GB 25	P limitation	51 (20 gPHB L^{-1})	$M_w = 2.5 \times 10^6$	0.07	–
Helm et al. (2006)*	*Methylocystis* sp. GB 25 (>86%)	P limitation	48	$M_w = 2.4 \times 10^6$	–	1.13
Helm et al. (2008)*	*Methylocystis* sp. GB 25 (>86%)	K/S/Fe limitation	33.6/32.6/10.4	0.45/0.40/0.22; $M_w = (3.1/2.5/1.8) \times 10^6$	0.08/0.05/0.02	–
Pieja et al. (2011a)	*M. trichosporium* OB3b/*M. parvus* OBBP/*Methylocystis* 42/22	N limitation	38/36/25	–	–	–
Pfluger et al. (2011)*	Mixed methanotrophic communities including *Methylosinus* and *Methylocystis* species	N limitation	20–40	–	–	–
Van der Ha et al. (2012)	*M. parvus*	N limitation	30	–	–	–
Rostkowski et al. (2013)	*M. trichosporium* OB3b/*M. parvus* OBBP	N limitation (growth phase on N_2 and NH_4^+, respectively)	45/60	1.13/0.88	–	–

Myung et al. (2015a)	*Methylocystis* sp. (>70–80%)	N limitation (growth phase on both NO_2^- and NH_4^+)	40	–	–	–
Sundstrom and Criddle (2015)	*M. parvus* OBBP	N limitation; Cu and Ca concentrations optimized	49.4 (3.43 gPHB L^{-1})	–	–	–
Karthikeyan et al. (2015b)	Methanotrophic–heterotrophic communities including *Methylocystis* and *Methylosinus* species	No nutrient limitation, CH_4/O_2 ratio varied	2.5–8.5	–	–	–
Myung et al. (2016a)	*Methylocystis* sp. (>70%)	Natural gas (CH_4 + ethane + propane + butane)	42	$M_w = 1.3 \times 10^6$	0.012	–
Myung et al. (2016b)	*M. parvus* OBBP	N limitation, fluorinated oil added (growth phase on NH_4^+)	35	–	–	–
Zhang et al. (2016)	*M. trichosporium* OB3b	N limitation (growth phase on $NO_3^-/NH_4^+/N_2$)	51/45/32	–	–	–
Chidambarampadmavathy et al. (2017a)	Methanotrophic–heterotrophic communities	No nutrient limitation, CH_4/O_2 ratio varied	3	–	–	–
Chidambarampadmavathy et al. (2017b)	Methanotrophic–heterotrophic communities including *Methylosinus* species	No nutrient limitation, optimized Cu^{2+}:Fe^{2+} ratio	5	–	–	–

(continued)

Table 11.2 (continued)

Reference[a]	Methanotrophic strain	Culture conditions[b]	Maximum PHB content (% wt)	Maximum PHB yield (gPHB gCH_4^{-1}) and molecular weight (Da)	Specific PHB production rate (gPHB g_X^{-1} h^{-1})	Volumetric PHB productivity (gPHB L^{-1} h^{-1})
Zhang et al. (2017)	Methanotrophic–heterotrophic communities including *Methylocella*, *Methylosinus*, *Methylocystis* and *Methylocapsa* species	N limitation, Cu and O_2 contents optimized	49	0.49	–	–
López et al. (2018a)	*M. hirsuta*	N limitation, headspace composed of O_2 + CH_4/H_2S-free biogas/Biogas	45–50 regardless of the condition	0.42–0.46 regardless of the condition	–	–
García-Pérez et al. (2018)	*M. hirsuta*	Mn/K/N limitation/N limitation + Fe excess	8/13/28/19	–	–	–

[a]The studies highlighted (*) comprised a growth phase carried out under continuous operation in a STR and a subsequent batchwise accumulation phase
[b]Unless specified otherwise, the nutrient limitation was induced after a growth phase where NO_3^- was used as N source
[c]M_w: molecular weight

et al. 2011). Similarly, a recent study conducted in STRs at different CH_4 concentrations and subjected to sequential nitrogen limitations found that the methanotrophic communities enriched did not exceed 13% wt of PHB content regardless of the CH_4 concentration applied, due to the predominance of type I methanotrophs such as *Methylomicrobium* and *Methylosarcina* (López et al. 2014).

CH_4-driven PHA assays are usually conducted in two-stages, so as the methanotrophic biomass is grown under nutrient-sufficient conditions in a first stage to generate a high-cell density culture, being subsequently harvested by centrifugation and subjected to nutrient-limiting conditions in a second stage in order to promote the accumulation of PHAs within the first 48–72 h. In this regard, the type of nutrient limitation applied during the accumulation phase influences the maximum PHA content achieved and its molecular weight. Though N-limitation usually results in the highest PHA accumulations, the possibility to apply P-, K-, S-, O_2- or Mg-limitations to increase final PHB contents in methanotrophic bacteria should not be dismissed. Several studies demonstrated that *Methylocystis* sp. GB25 was able to accumulate PHB up to 51% wt under P- and N-limitation, whereas contents of 33% wt were not exceeded under K- or S-limitations. Interestingly, the highest molecular weights in the produced biopolymer (3.1×10^6 Da) were obtained for the condition under K-deficiency (Wendlandt et al. 2001, 2005; Helm et al. 2008). Similarly, *Methylosinus trichosporium* IMV 3011 achieved higher PHB accumulations under P- and N-limitation than those under Mg- or S-limitation (20–26 *vs.* 15% wt, respectively) (Zhang et al. 2008). Moreover, other authors found that N-limitation was the most promoting conditions to accumulate PHB up to 30% in *M. hirsuta* compared to other conditions (K- or Mn-limitations), which suggests that the optimum limiting-condition is species-dependent (García-Pérez et al. 2018).

Macro- and micronutrient concentrations in the culture broth during both the growth and the accumulation phases are also key parameters to be optimized to achieve a proper PHA accumulation by methanotrophs. In this regard, it was recently demonstrated that the oxygen concentration and the nitrogen source employed for methanotrophic growth influenced PHB synthesis, nitrogen gas mediating a PHA content of 45% wt and ammonium of 50–60% wt in *M. trichosporium* OB3b and *M. parvus* OBBP, respectively, at O_2 partial pressures of 0.10–0.30 atm (Rostkowski et al. 2013). Furthermore, an exhaustive high-throughput screening was recently carried out to optimize the concentration of N (in the form of NO_3^- or NH_4^+), Ca, Cu, P, K, Mg, and S during the production of PHB by *M. parvus* OBBP (Sundstrom and Criddle 2015). Despite the PHB contents were higher when NH_4^+ concentrations as low as 2.25 mM were applied (compared to those used in the NO_3^- assays), the highest PHB volumetric productions were obtained at 30 mM NO_3^-. The authors also found that PHB accumulation in *M. parvus* OBBP do not depend on the concentration of nutrients such as Mg or S, while phosphate and K limiting concentrations of 120 and 90 μM, respectively, guaranteed the highest PHB contents. Interestingly, Cu and Ca represented the most critical elements to be simultaneously adjusted, so as concentrations of 5 and 7.2 μM, respectively, allowed higher PHB contents and volumetric productions than the control condition (49% wt of PHB and 3.43 g L^{-1} *vs.* 18% wt and 0.65 g L^{-1}). Moreover, a batch study based on the optimization of culture conditions for PHB production in *M. trichosporium* IMV 3011 revealed that Fe^{3+} limitations during the accumulation phase leads to PHB contents two times higher than those under control conditions (Zhang et al. 2008). This fact may be attributed to the preferential intervention of the pMMO enzyme under the deficiency of this metal ion, which seems to be crucial for the correct activity of the counterpart sMMO enzyme (Glass and Orphan 2012). This effect of metal ions was later investigated in *M. trichosporium* OBBP batchwise, exhibiting the highest PHB accumulation (~50% wt) under predominant expression of the pMMO enzyme, which was promoted by cultivation of the strain at 5 μM Cu^{2+} during both the growth and accumulation phases (Zhang et al. 2016). However, further research evaluating the influence of other trace metal ions and micronutrients, pH and temperature on the capability of type II methanotrophs to produce PHAs must be still considered.

11.4.3 The Use of Methanotrophic Bioreactors and Alternative CH_4 Sources: Towards the Full-Scale Implementation

It is noteworthy that most works on PHA production by obligate type II methanotrophs have been carried out batchwise and mainly focused on the use of pure CH_4, which may challenge its implementation under continuous operation in real-case scenarios. Thus, several efforts have been made in order to design adequate bioreactors for the simultaneous CH_4 mitigation and PHB production. In this context, the genus *Methylocystis* has been the only one cultivated for this purpose, reaching accumulations of 20–25% wt and 40–50% wt in sequencing batch reactors (SBRs) and BCBs, respectively (Table 11.3) (Pieja et al. 2012; Rahnama et al. 2012; García-Pérez et al. 2018). The higher PHB contents achieved in the BCBs may be attributed to the higher turbulence guaranteed by the internal gas-recycling system there installed, which allowed for a higher CH_4 mass transfer compared to that of SBRs, rather than to the different strains employed. It must be stressed that several studies have claimed the production of PHB by *Methylocystis* sp. GB25 with a two-stage process in STRs, though the accumulation phases were conducted batchwise (Wendlandt et al. 2001, 2005; Helm et al. 2006, 2008). On the other hand, real CH_4 feedstocks such as biogas or natural gas have been poorly explored as sources for PHA production and further research in this respect is still required. For instance, the sequential use of the microalgae *Scenedesmus* sp. and the methanotroph *M. parvus* OBBP to bioconvert the CH_4 and CO_2 from synthetic biogas into PHB and lipids, respectively, was recently assessed in a 2-stage process, maximum PHB contents of ~30% wt being obtained (Van der Ha et al. 2012). The feasibility of biogas as a CH_4 source for PHA accumulation was further evaluated in *M. hirsuta* cultures, which exhibited PHB contents of ~45% wt regardless of the headspace composition assayed (i.e. H_2S and CO_2 content) (López et al. 2018c). Also, a recent study confirmed the feasibility of pipeline natural gas containing trace amounts of ethane, propane and, to a lesser extent butane, to produce PHB up to 42% wt with *Methylocystis*-enriched cultures (Myung et al. 2016a).

Table 11.3 Comparison of the most significant CH_4-based PHB production studies reported to date in continuous bioreactors

Reference	System (PHA production period)[a]	Methanotrophic strain	Culture conditions[b]	Maximum PHB accumulation (% wt)	Volumetric PHB productivity (gPHB L^{-1} h^{-1})
Shah et al. (1996)	STR[c] (120 h)	*M. trichosporium* OB3b	P limitation	51	–
Pieja et al. (2012)	SBR[d] (11 days)	Mixed type II methanotrophic consortia including *M. parvus* OBBP, *Methylocystis* 42/22, *Methylocystis* KS30 and *Methylosinus* sp. LW4	Sequential N/N+O_2/N+CH_4 limitations (feast-famine strategies)	20/23/25	–
Rahnama et al. (2012)	BCB[e]/VTLB[f] (120/8 h)	*M. hirsuta*	N limitation	43/52	<0.01[g]
López et al. (2014)	STR (~300 days)	Mixed methanotrophic culture including *Methylocystis* sp. and *M. parvus*	Sequential N limitations (feast-famine strategy)	13	–
García-Pérez et al. (2018)	BCB (~30 days)	*M. hirsuta*	Sequential N limitations (feast-famine strategy)	40	0.06

[a]PHA production period was referred to the elapsed time where the methanotrophic bacteria was continuously producing PHB
[b]Unless specified otherwise, the nutrient limitation was induced after a growth phase where NO_3^- was used as N source
[c]STR: stirred tank reactor
[d]SBR: sequencing batch reactor
[e]BCB: bubble column bioreactor
[f]VTLB: vertical tubular loop bioreactor
[g]The volumetric PHB productivity was estimated based on the PHB content and the growth rates reported

11.4.4 Tailor-Made Biopolymers in Methanotrophs Through Co-substrate Addition

Although PHB is by far the most produced PHA, both the contents and/or productivities reported for methanotrophs and the quality of the biopolymer still jeopardise the widespread production at industrial-scale and its further use over conventional plastics. In this regard, a current trend based on co-substrate addition in methanotrophic cultures has been implemented in the past decade to overcome these particular issues. Thus, the addition of C_1 compounds such as formate has been proposed to provide reducing power to effectively slow down PHB consumption in PHB-rich cells of *M. parvus* OBBP (Pieja et al. 2011b). Other authors have proposed the addition of trace amounts of methanol, another C_1 compound included within the CH_4 oxidation pathway, to slightly enhance CH_4 degradation, biomass growth and final PHB contents over the basal content obtained using only CH_4 (by 10–35% wt) (Zhang et al. 2008). Similarly, these authors proposed the addition of citric acid as TCA inhibitor, which enables PHA accumulation even under nutrient-sufficient conditions and allows for two times higher volumetric PHB productions compared to those without acid citric supplementation (Table 11.4). Also, a recent study evaluated the use of volatile fatty acids such as acetic, propionic and butyric acids as co-substrates for *M. hirsuta*, which resulted in a 10–30% increase of Y_{PHA} and PHA accumulation over the basal values obtained when only CH_4 was used as carbon and energy source (López et al. 2018c). More innovative approaches have currently included the addition of alcohols or ω-hydroxyalkanoates in order to generate copolymers such as poly(3-hydroxybutyrate-*co*-3-hydroxyvalerate) (PHBV) or poly(3-hydroxybutyrate-*co*-4-hydroxybutyrate) (P3HB4HB) with lower melting temperature and crystallinity. Thus, valeric acid (or valerate) supplementation in the culture broth has led to the production of PHBV by the strains *M. hirsuta*, *M. parvus* OBBP and *M. trichosporium* OB3b, 3-hydroxyvalerate (HV) fractions of 20–40 mol % within the biocomposite being obtained depending on the co-substrate concentration (Myung et al. 2015a, 2016b; López et al. 2018c). Propionate and *n*-pentanol have been also successfully employed for PHBV production in *Methylocystis* species, though the HV fractions reported were considerably lower than those for valerate co-feeding (Myung et al. 2016b; Cal et al. 2016). It has been also found that PHA and specific HV contents in *Methylocystis* sp. WRRC1 may be increased by both adjusting valerate concentration and limiting copper in the culture broth, thus leading to maximum PHA and HV contents of 78% and >50 mol %, respectively (Cal et al. 2016). Finally, the recent use of alternative ω-hydroxyalkanoates has resulted in the production of new tailor-made copolymers by the strain *M. parvus* OBBP, such as P3HB4HB, poly(3-hydroxybutyrate-*co*-5-hydroxyvalerate-*co*-3-hydroxyvalerate) (P(3HB-*co*-5HV-*co*-3HV)), or poly(3-hydroxybutyrate-*co*-6-hydroxyhexanoate-*co*-4-hydroxybutyrate) (P(3HB-*co*-6HHx*co*-4HB)), which exhibited significantly enhanced mechanical properties and molecular weights ranging from 4.5×10^5 to 1.5×10^6 Da (Myung et al. 2017). However, despite these promising results, future research should focus on the evaluation of the influence of new co-substrates on methanotrophic growth and PHA production, aiming at co-integrating this innovative approach within a biorefinery concept.

Table 11.4 Comparison of the most significant batch PHA production studies reported to date using both CH_4 and cosubstrates as carbon and/or energy sources

Reference	Methanotrophic strain	Cosubstrate(s)	Culture conditions[a]	PHA yield (gPHA $gSubstrate^{-1}$) and molecular weight (Da)	Maximum PHA content[b] (% wt) and concentration	Biocomposite composition
Zhang et al. (2008)	*M. trichosporium* IMV3011	Methanol/citric acid	N limitation; Fe, Cu, Mg and P concentrations optimized (growth phase on both NO_3^- and NH_4^+)	$M_w{}^b = 1.5 \times 10^6$ Da	Without/with methanol: 12/22	100% mol HB
					Without/with citric acid: 26/38 (PHB concentrations of 0.16 and 0.26 g L^{-1})	
Pieja et al. (2011b)	*M. parvus* OBBP	Formate	N limitation	0.24	Without/with formate: 50 (formate delayed only PHB consumption)	100% mol HB
Myung et al. (2015b)	*Methylocystis* sp. (>75%)	Valerate (low/high concentrations)	N limitation (growth phase on NH_4^+)	Low valerate: 0.35 1.2×10^6	Low valerate: 43;	Low valerate: 20% mol HV + 80% mol HB;
				High valerate: 0.36 and 9.3×10^5	High valerate: 30	High valerate: 40% mol HV + 60% mol HB
Cal et al. (2016)	*Methylocystis* sp. WRRC1	*n*-pentanol/valerate	N limitation/N or N+Cu limitation	0.46/0.67	*n*-pentanol: 41; valerate: 60 (+Cu) or 78 (−Cu)	Pentanol: 54% mol HV + 46% mol HB; valerate: 50% mol HV + 50% mol HB (+Cu) or 60% mol HV + 40% mol HB (−Cu)
Myung et al. (2016c)	*M. parvus* OBBP	3-hydroxybutyrate/propionate/valerate	N limitation (growth phase on NH_4^+)	$M_w = 1.8 \times 10^6$ Da (valerate conditions)	3-hydroxybutyrate: 60; propionate: 32; valerate: 60	3-hydroxybutyrate: 100% mol HB; propionate: 10% mol HV + 90% mol HB; valerate: 60% mol HV + 40% mol HB

(continued)

Table 11.4 (continued)

Reference	Methanotrophic strain	Cosubstrate(s)	Culture conditions[a]	PHA yield (gPHA $gSubstrate^{-1}$) and molecular weight (Da)	Maximum PHA content[b] (% wt) and concentration	Biocomposite composition
Myung et al. (2017)	*M. parvus* OBBP	4-hydroxybutyrate/5-hydroxyvalerate/6-hydroxyhexanoate/3-hydroxybutyrate/butyrate/valerate/hexanoate/octanoate	N limitation (growth phase on NH_4^+)	$M_w = 1.22–1.48 \times 10^6$ Da	3-hydroxybutyrate, valerate, butyrate hexanoate and octanoate: 54–59; 4-hydroxybutyrate, 5-hydroxyvalerate and 6-hydroxyhexanoate: 48–50	Butyrate, 3-hydroxybutyrate, hexanoate, octanoate: 100% mol 3HB; 4-hydroxybutyrate: 91.5% mol 3HB, 9.5% mol 4HB; valerate: 75% mol 3HB, 25% mol 3HV; 5-hydroxyvalerate: 95% mol 3HB, 1.4% mol 3HV, 3.5% mol 5HV; 6-hydroxyhexanoate: 97.6% mol 3HB, 1% mol 4HB, 1.4% mol 6HHx
López et al. (2018a)	*M. hirsuta*	Acetic/propionic/butyric/valeric acids	N limitation, biogas headspace (including H_2S and CO_2)	Acetic acid: 0.50; propionic acid: 0.45; butyric acid: 0.54; valeric acid: 0.63	Acetic acid: 52; propionic acid: 48; butyric acid: 52; valeric acid: 54	Acetic and butyric acids: 100% mol HB; propionic acid: 98% mol HB, 2% mol HV; valeric acid: 25% mol HV, 75% mol HB

[a]Unless specified otherwise, the nutrient limitation was induced after a growth phase where NO_3^- was used as N source
[b]M_w molecular weight

11.5 Conclusion

The production of PHAs through CH_4 oxidation might represent a potential cost-efficient alternative for the sequestration of this potent GHG. PHB is the most widely produced biopolymer, though its price in market and its thermal and mechanical properties, compared to those of conventional plastics, still constrain its widespread production at full-scale. Thus, despite the promising results obtained during CH_4-based PHA production, important challenges to be overcome include (i) the simultaneous increase in CH_4 removals and biomass productivities through the design of novel bioreactor configurations devoted to enhance CH_4 mass transfer within the culture broth, and (ii) the optimization of PHB production under nutrient-limited conditions together with the modification of the biocomposite through the addition of co-substrates at low energy inputs. Though additional research is still required, the recent trends in research and development suggest an increasing interest on this area, which might promote the development of a GHG bioeconomy in the near future.

11.6 Opinion

This work was supported by the Spanish Ministry of Economy and Competitiveness and the European Union through the project CTM2015-70.442-R (Retos and FEDER Programs). The authors acknowledge Professor V. C. Kalia from CSIR-Institute of Genomics and Integrative Biology (CSIR-IGIB), Delhi, India, the opportunity of contributing with their expertise on this chapter.

References

AlSayed A, Fergala A, Khattab S, Eldyasti A (2018a) Kinetics of type I methanotrophs mixed culture enriched from waste activated sludge. Biochem Eng J 132:60–67. https://doi.org/10.1016/j.bej.2018.01.003

AlSayed A, Fergala A, Khattab S, ElSharkawy A, Eldyasti A (2018b) Optimization of methane bio-hydroxylation using waste activated sludge mixed culture of type I methanotrophs as biocatalyst. Appl Energ 211:755–763. https://doi.org/10.1016/j.apenergy.2017.11.090

Amaral JA, Knowles R (1995) Growth of methanotrophs in methane and oxygen counter gradients. FEMS Microbiol Lett 126:215–220. https://doi.org/10.1111/j.1574-6968.1995.tb07421.x

Anthony C (2011) How half a century of research was required to understand bacterial growth on C1 and C2 compounds; the story of the serine cycle and the ethylmalonyl-CoA pathway. Sci Prog 94:109–137. https://doi.org/10.3184/003685011X13044430633960

Asenjo JA, Suk JS (1986) Microbial conversion of methane into poly-β-hydroxybutyrate (PHB): growth and intracellular product accumulation in a type II methanotroph. J Ferment Technol 64:271–278. https://doi.org/10.1016/0385-6380(86)90118-4

Avalos-Ramirez A, Jones JP, Heitz M (2012) Methane treatment in biotrickling filters packed with inert materials in presence of a non-ionic surfactant. J Chem Technol Biotechnol 87:848–853. https://doi.org/10.1002/jctb.3811

Bédard C, Knowles R (1989) Physiology, biochemistry and specific inhibitors of CH_4, NH_4 and CO oxidation by methanotrophs and nitrifiers. Microbiol Rev 53:68–84

Bender M, Conrad R (1995) Effect of CH_4 concentrations and soil conditions on the induction of CH_4 oxidation activity. Soil Biol Biochem 27:1517–1527. https://doi.org/10.1016/0038-0717(95)00104-M

Bugnicourt E, Cinelli P, Lazzeri A, Alvarez V (2016) The main characteristics, properties, improvements, and market data of polyhydroxyalkanoates. In: Thakur VK, Thakur MK (eds) Handbook of sustainable polymers: processing and applications, 1st edn. Pan Stanford Publishing Pte. Ltd., Boca Ratón, pp 899–928. https://doi.org/10.1201/b19600-25. isbn:978-9-814-61353-8

Cai C, Hu S, Guo J, Shi Y, Xie G-J, Yuan Z (2015) Nitrate reduction by denitrifying anaerobic methane oxidizing microorganisms can reach a practically useful rate. Water Res 87:211–217. https://doi.org/10.1016/j.watres.2015.09.026

Cal AJ, Sikkema WD, Ponce MI, Franqui-Villanueva D, Riiff TJ, Orts WJ, Pieja AJ, Lee CC (2016) Methanotrophic production of polyhydroxybutyrate-*co*-hydroxyvalerate with high hydroxyvalerate content. Int J Biol Macromol 87:302–307. https://doi.org/10.1016/j.ijbiomac.2016.02.056

Cantera S, Estrada JM, Lebrero R, García-Encina PA, Muñoz R (2016a) Comparative performance evaluation of conventional and two-phase hydrophobic stirred tank reactors for methane abatement: mass transfer and biological considerations. Biotechnol Bioeng 113:1203–1212. https://doi.org/10.1002/bit.25897

Cantera S, Lebrero R, García-Encina PA, Muñoz R (2016b) Evaluation of the influence of methane and copper concentration and methane mass transport on the community structure and biodegradation kinetics of methanotrophic cultures. J Environ Manag 171:11–20. https://doi.org/10.1016/j.jenvman.2016.02.002

Cantera S, Frutos OD, López JC, Lebrero R, Muñoz R (2017) Technologies for the bio-conversion of GHGs into high added value products: current state and future prospects. In: Álvarez-Fernández R, Zubelzu S, Martínez R (eds) Carbon footprint and the industrial life cycle, from urban planning to recycling. Springer International Publishing, Cham, pp 359–388. https://doi.org/10.1007/978-3-319-54984-2. isbn:978-3-319-54983-5

Cantera S, Muñoz R, Lebrero R, López JC, Rodríguez Y, García-Encina PA (2018) Technologies for the bioconversion of methane into more valuable products. Curr Opin Biotechnol 50:128–135. https://doi.org/10.1016/j.copbio.2017.12.021

Chen X, Liu Y, Peng L, Yuan Z, Ni BJ (2016) Model-based feasibility assessment of membrane biofilm reactor to achieve simultaneous ammonium, dissolved methane, and sulfide removal from anaerobic digestion liquor. Sci Rep 6:25114. https://doi.org/10.1038/srep25114

Chidambarampadmavathy K, Karthikeyan OP, Huerlimann R, Maes GE, Heimann K (2017a) Response of mixed methanotrophic consortia to different methane to oxygen ratios. Waste Manag 61:220–228. https://doi.org/10.1016/j.wasman.2016.11.007

Chidambarampadmavathy K, Karthikeyan OP, Huerlimann R, Maes GE, Heimann K (2017b) Responses of mixed methanotrophic consortia to variable Cu^{2+}/Fe^{2+} ratios. J Environ Manag 197:159–166. https://doi.org/10.1016/j.jenvman.2017.03.063

Covarrubias-García I, Aizpuru A, Arriaga S (2017) Effect of the continuous addition of ozone on biomass clogging control in a biofilter treating ethyl acetate vapors. Sci Total Environ 584–585:469–475. https://doi.org/10.1016/j.scitotenv.2017.01.031

Crombie AT, Murrell JC (2014) Trace-gas metabolic versatility of the facultative methanotroph *Methylocella silvestris*. Nature 510:148–151. https://doi.org/10.1038/nature13192

Cui M, Ma A, Qi H, Zhuang X, Zhuang G (2015) Anaerobic oxidation of methane: an "active" microbial process. MicrobiologyOpen 4:1–11. https://doi.org/10.1002/mbo3.232

Dessus B, Laponche B, Le Treut H (2009) The importance of a methane reduction policy for the 21st century. http://www.global-chance.org/IMG/pdf/BDBLHT5Methane.pdf. Accessed 23 Aug 2017

Dunfield P, Knowles R (1995) Kinetics of inhibition of methane oxidation by nitrate, nitrite, and ammonium in a humisol. Appl Environ Microbiol 61:3129–3135

Ebrahimi S, Kleerebezem R, Kreutzer MT, Kapteijn F, Moulijn JA, Heijnen JJ, van Loosdrecht MCM (2006) Potential application of monolith packed columns as bioreactors, control of biofilm formation. Biotechnol Bioeng 93:238–245. https://doi.org/10.1002/bit.20674

Environmental Protection Agency (2017) Inventory of US greenhouse gas emissions and sinks: 1990–2015. http://epa.gov/climatechange/ghgemissions/usinventoryreport.html. Accessed 7 June 2017

Estrada JM, Lebrero R, Quijano G, Kraakman NJR, Muñoz R (2012a) Strategies for odour control. In: Belgiorno V, Naddeo V, Zarra T (eds) Odour impact assessment handbook. Wiley, Hoboken, pp 85–124. https://doi.org/10.1002/9781118481264.ch4. isbn:978-1-119-96928-0

Estrada JM, Kraakman NJRB, Lebrero R, Muñoz R (2012b) A sensitivity analysis of process design parameters, commodity prices and robustness on the economics of odour abatement technologies. Biotechnol Adv 30(6):1354–1363. https://doi.org/10.1016/j.biotechadv.2012.02.010

Estrada JM, Dudek A, Muñoz R, Quijano G (2013a) Fundamental study on gas–liquid mass transfer in a biotrickling filter packed with polyurethane foam. J Chem Technol Biotechnol 89:1419–1424. https://doi.org/10.1002/jctb.4226

Estrada JM, Quijano G, Lebrero R, Muñoz R (2013b) Step-feed biofiltration: a low-cost alternative configuration for off-gas treatment. Water Res 47:4312–4321. https://doi.org/10.1016/j.watres.2013.05.007

Estrada JM, Lebrero R, Quijano G, Pérez R, Figueroa-González I, García-Encina PA, Muñoz R (2014) Methane abatement in a gas-recycling biotrickling filter: evaluating innovative operational strategies to overcome mass transfer limitations. Chem Eng J 253:385–393. https://doi.org/10.1016/j.cej.2014.05.053

European Environment Agency (2017) Annual European union greenhouse gas inventory 1990–2015 and inventory report 2017. http://www.eea.europa.eu//publications/european-union-greenhouse-gas-inventory-2017. Accessed 7 June 2017

Eurostat Statistics Explained (2016) Natural gas price statistics. http://ec.europa.eu/eurostat/statistics-explained/index.php/Natural_gas_price_statistics. Accessed 8 June 2017

Fergala A, AlSayed A, Eldyasti A (2017) Factors affecting the selection of PHB accumulating methanotrophs from waste activated sludge while utilizing ammonium as their nitrogen source. J Chem Technol Biot (In Press). https://doi.org/10.1002/jctb.5502

García-Pérez T, López JC, Passos F, Lebrero R, Revah S, Muñoz R (2018) Simultaneous methane abatement and PHB production by *Methylocystis hirsuta* in a novel gas-recycling bubble column bioreactor. Chem Eng J 334:691–697. https://doi.org/10.1016/j.cej.2017.10.106

Glass JB, Orphan VJ (2012) Trace metal requirements for microbial enzymes involved in the production and consumption of methane and nitrous oxide. Front Microbiol 3:61. https://doi.org/10.3389/fmicb.2012.00061

Graham DW, Chaudhary JA, Hanson RS, Arnold RG (1993) Factors affecting competition between type I and type II methanotrophs in two-organism, continuous-flow reactors. Microb Ecol 25:1–17. https://doi.org/10.1007/BF00182126

Helm J, Wendlandt KD, Rogge G, Kappelmeyer U (2006) Characterizing a stable methane-utilizing mixed culture used in the synthesis of a high-quality biopolymer in an open system. J Appl Microbiol 101:387–395. https://doi.org/10.1111/j.1365-2672.2006.02960.x

Helm J, Wendlandt KD, Jechorek M, Stottmeister U (2008) Potassium deficiency results in accumulation of ultra-high molecular weight poly-β-hydroxybutyrate in a methane-utilizing mixed culture. J Appl Microbiol 105:1054–1061. https://doi.org/10.1111/j.1365-2672.2008.03831.x

Henckel T, Roslev P, Conrad R (2000) Effects of O_2 and CH_4 on presence and activity of the indigenous methanotrophic community in rice field soil. Environ Microbiol 2:666–679. https://doi.org/10.1046/j.1462-2920.2000.00149.x

Heyer J, Berger U, Hardt M, Dunfield PF (2005) *Methylohalobius crimeensis* gen. nov., sp. nov., a moderately halophilic, methanotrophic bacterium isolated from hyper-saline lakes of Crimea. Int J Syst Evol Microbiol 55:1817–1826. https://doi.org/10.1099/ijs.0.63213-0

Intergovernmental Panel on Climate Change (2013) Climate change 2013, the physical science basis. Working group I contribution to the fifth assessment report of the intergovernmental panel on climate change. http://www.ipcc.ch/report/ar5/wg1/. Accessed 23 Aug 2017

Intergovernmental Panel on Climate Change (2014) Fifth assessment report: climate change 2014. Synthesis report. https://www.ipcc.ch/report/ar5/syr/. Accessed 7 June 2017

Jin Y, Veiga MC, Kennes C (2006) Development of a novel monolith-bioreactor for the treatment of VOC-polluted air. Environ Tech 27:1271–1277. https://doi.org/10.1080/09593332708618744
Kalyuzhnaya MG, Yang S, Rozova ON, Smalley NE, Clubb J, Lamb A, Nagana-Gowda GA, Raftery D, Fu Y, Bringel F, Vuilleumier S, Beck DAC, Trotsenko YA, Khmelenina VN, Lidstrom ME (2013) Highly efficient methane biocatalysis revealed in a methanotrophic bacterium. Nat Commun 4:2785. https://doi.org/10.1038/ncomms3785
Karthikeyan OP, Chidambarampadmavathy K, Cirés S, Heimann K (2015a) Review of sustainable methane mitigation and biopolymer production. Crit Rev Environ Sci Technol 45:1579–1610. https://doi.org/10.1080/10643389.2014.966422
Karthikeyan OP, Chidambarampadmavathy K, Nadarajan S, Lee PKH, Heimann K (2015b) Effect of CH_4/O_2 ratio on fatty acid profile and polyhydroxybutyrate content in a heterotrophic-methanotrophic consortium. Chemosphere 141:235–242. https://doi.org/10.1016/j.chemosphere.2015.07.054
Kennelly C, Clifford E, Gerrity S, Walsh R, Rodgers M, Collins G (2012) A horizontal flow biofilm reactor (HFBR) technology for the removal of methane and hydrogen sulphide at low temperatures. Water Sci Technol 66:1997–2006. https://doi.org/10.2166/Wst.2012.411
Kennelly C, Gerrity S, Collins G, Clifford E (2014) Liquid phase optimisation in a horizontal flow biofilm reactor (HFBR) technology for the removal of methane at low temperatures. Chem Eng J 242:144–154. https://doi.org/10.1016/j.cej.2013.12.071
Khosravi-Darani K, Mokhtari ZB, Amai T, Tanaka K (2013) Microbial production of poly(hydroxybutyrate) from C_1 carbon sources. Appl Microbiol Biotechnol 97:1407–1424. https://doi.org/10.1007/s00253-012-4649-0
Knief C (2015) Diversity and habitat preferences of cultivated and uncultivated aerobic methanotrophic bacteria evaluated based on *pmoA* as molecular marker. Front Microbiol 6:1346. https://doi.org/10.3389/fmicb.2015.01346
Kraakman NJR, Rocha-Rios J, van Loosdrecht MCM (2011) Review of mass transfer aspects for biological gas treatment. Appl Microbiol Biotechnol 91:873–886. https://doi.org/10.1007/s00253-011-3365-5
Kumar A, Dewulf J, van Langenhove H (2008) Membrane-based biological waste gas treatment. Chem Eng J 136:82–91. https://doi.org/10.1016/j.cej.2007.06.006
Lebrero R, Hernández L, Pérez R, Estrada JM, Muñoz R (2015) Two-liquid phase partitioning biotrickling filters for methane abatement: exploring the potential of hydrophobic methanotrophs. J Environ Manag 151:124–131. https://doi.org/10.1016/j.jenvman.2014.12.016
Lebrero R, López JC, Lehtinen I, Pérez R, Quijano G, Muñoz R (2016) Exploring the potential of fungi for methane abatement: performance evaluation of a fungal-bacterial biofilter. Chemosphere 144:97–106. https://doi.org/10.1016/j.chemosphere.2015.08.017
Levett I, Birkett G, Davies N, Bell A, Langford A, Laycock B, Lant P, Pratt S (2016) Techno-economic assessment of poly-3-hydroxybutyrate (PHB) production from methane – the case of thermophilic bioprocessing. J Environ Chem Eng 4:3724–3733. https://doi.org/10.1016/j.jece.2016.07.033
López JC, Quijano G, Souza TSO, Estrada JM, Lebrero R, Muñoz R (2013) Biotechnologies for greenhouse gases (CH_4, N_2O, and CO_2) abatement: state of the art and challenges. Appl Microbiol Biotechnol 97:2277–2303. https://doi.org/10.1007/s00253-013-4734-z
López JC, Quijano G, Pérez R, Muñoz R (2014) Assessing the influence of CH_4 concentration during culture enrichment on the biodegradation kinetics and population structure. J Environ Manag 146:116–123. https://doi.org/10.1016/j.jenvman.2014.06.026
López JC, Porca E, Collins G, Pérez R, Rodríguez-Alija A, Muñoz R, Quijano G (2017) Biogas-based denitrification in a biotrickling filter: influence of nitrate concentration and hydrogen sulfide. Biotechnol Bioeng 114:665–673. https://doi.org/10.1002/bit.26092
López JC, Merchán L, Lebrero R, Muñoz R (2018a) Feast-famine biofilter operation for methane mitigation. J Clean Prod 170:108–118. https://doi.org/10.1016/j.jclepro.2017.09.157
López JC, Porca E, Collins G, Clifford E, Quijano G, Muñoz R (2018b) Ammonium influences kinetics and structure of methanotrophic bacteria (submitted for publication)

López JC, Arnáiz E, Merchán L, Lebrero R, Muñoz R (2018c) Biogas-based polyhydroxyalkanoates production by *Methylocystis hirsuta*: a step further in anaerobic digestion biorefineries. Chem Eng J 333:529–536. https://doi.org/10.1016/j.cej.2017.09.185

Ménard C, Avalos-Ramirez A, Nikiema J, Heitz M (2012) Biofiltration of methane and trace gases from landfills: a review. Environ Rev 20:40–53. https://doi.org/10.1139/A11-022

Muñoz R, Souza TSO, Glittmann L, Pérez R, Quijano G (2013) Biological anoxic treatment of O2-free VOC emissions from the petrochemical industry: a proof of concept study. J Hazard Mater 260:442–450. https://doi.org/10.1016/j.jhazmat.2013.05.051

Muñoz R, Malhautier L, Fanlo JL, Quijano G (2015) Biological technologies for the treatment of atmospheric pollutants. Int J Environ Anal Chem 95:950–967. https://doi.org/10.1080/03067319.2015.1055471

Murrell JC, Gilbert B, McDonald IR (2000) Molecular biology and regulation of methane monooxygenase. Arch Microbiol 173:325–332. https://doi.org/10.1007/s002030000158

Myung J (2016) Microbial enrichment enabling stable production of poly(3-hydroxybutyrate) using pipeline natural gas. In: Recovery of resources and energy using methane-utilizing bacteria: synthesis and regeneration of biodegradable, tailorable bioplastics and production of nitrous oxide. Dissertation, Stanford University (California, U.S.)

Myung J, Galega WM, van Nostrand JD, Yuan T, Zhou J, Criddle CS (2015a) Long-term cultivation of a stable *Methylocystis*-dominated methanotrophic enrichment enabling tailored production of poly(3-hydroxybutyrate-*co*-3-hydroxyvalerate). Bioresour Technol 198:811–818. https://doi.org/10.1016/j.biortech.2015.09.094

Myung J, Wang Z, Yuan T, Zhang P, Van Nostrand JD, Zhou J, Criddle CS (2015b) Production of nitrous oxide from nitrite in stable type II methanotrophic enrichments. Environ Sci Technol 49:10969–10975. https://doi.org/10.1021/acs.est.5b03385

Myung J, Kim M, Pan M, Criddle CS, Tang SKY (2016a) Low energy emulsion-based fermentation enabling accelerated methane mass transfer and growth of poly(3-hydroxybutyrate)-accumulating methanotrophs. Bioresour Technol 207:302–307. https://doi.org/10.1016/j.biortech.2016.02.029

Myung J, Flanagan JCA, Waymouth RM, Criddle CS (2016b) Methane or methanol-oxidation dependent synthesis of poly(3-hydroxybutyrate-*co*-3-hydroxyvalerate) by obligate type II methanotrophs. Process Biochem 51:561–567. https://doi.org/10.1016/j.procbio.2016.02.005

Myung J, Flanagan JCA, Waymouth RM, Criddle CS (2017) Expanding the range of polyhydroxyalkanoates synthesized by methanotrophic bacteria through the utilization of omega-hydroxyalkanoate co-substrates. AMB Express 7:118. https://doi.org/10.1186/s13568-017-0417-y

National Oceanic and Atmospheric Administration (2017) National centers for environmental information, state of the climate: global climate report for annual 2016. https://www.ncdc.noaa.gov/sotc/global/201613. Accessed 8 June 2017

Nienow AW (2014) Stirring and stirred-tank reactors. Chem Ing Tech 86:2063–2074. https://doi.org/10.1002/cite.201400087

Nikiema J, Brzezinski R, Heitz M (2007) Elimination of methane generated from landfills by biofiltration: a review. Rev Environ Sci Biotechnol 6:261–284. https://doi.org/10.1007/s11157-006-9114-z

Nikodinovic-Runic J, Guzik M, Kenny ST, Babu R, Werker A, O Connor KE (2013) Carbon-rich wastes as feedstocks for biodegradable polymer (polyhydroxyalkanoate) production using bacteria. In: Sariaslani S, Gadd GM (eds) Advances in applied microbiology, vol 84. Academic, Burlington, pp 139–200. https://doi.org/10.1016/B978-0-12-407673-0.00004-7. isbn:978-0-12-407673-0

Nyerges G, Han SK, Stein LY (2010) Effects of ammonium and nitrite on growth and competitive fitness of cultivated methanotrophic bacteria. Appl Environ Microbiol 76:5648–5651. https://doi.org/10.1128/AEM.00747-10

Oliver JP, Schilling JS (2016) Capture of methane by fungi: evidence from laboratory-scale biofilter and chromatographic isotherm studies. T ASABE 59:1791–1801. https://doi.org/10.13031/trans.59.11595

Pawlowska M (2014) Biological oxidation as a method for mitigation of LFG emission. In: Pawlowska M (ed) Mitigation of landfill gas emissions. CRC Press, London, pp 59–101. isbn:9780415630771

Petersen LAH, Villadsen J, Jørgensen SB, Gernaey K (2017) Mixing and mass transfer in a pilot scale U-loop bioreactor. Biotechnol Bioeng 114:344–354. https://doi.org/10.1002/bit.26084

Pieja AJ, Rostkowski KH, Criddle CS (2011a) Distribution and selection of poly-3-hydroxybutyrate production capacity in methanotrophic proteobacteria. Microb Ecol 62:564–573. https://doi.org/10.1007/s00248-011-9873-0

Pieja AJ, Sundstrom ER, Criddle CS (2011b) Poly-3-hydroxybutyrate metabolism in the type II methanotroph *Methylocystis parvus* OBBP. Appl Environ Microbiol 77:6012–6019. https://doi.org/10.1128/AEM.00509-11

Pieja AJ, Sundstrom ER, Criddle CS (2012) Cyclic, alternating methane and nitrogen limitation increases PHB production in a methanotrophic community. Bioresour Technol 107:385–392. https://doi.org/10.1016/j.biortech.2011.12.044

Pieja AJ, Morse MC, Cal AJ (2017) Methane to bioproducts: the future of the bioeconomy? Curr Opin Chem Biol 41:123–131. https://doi.org/10.1016/j.cbpa.2017.10.024

Pol A, Heijmans K, Harhangi HR, Tedesco D, Jetten MSM, Op den Camp HJ (2007) Methanotrophy below pH 1 by a new Verrucomicrobia species. Nature 450:874–878. https://doi.org/10.1038/nature06222

Rocha-Rios J, Bordel S, Hernández S, Revah S (2009) Methane degradation in two-phase partition bioreactors. Chem Eng J 152:289–292. https://doi.org/10.1016/j.cej.2009.04.028

Rocha-Rios J, Muñoz R, Revah S (2010) Effect of silicone oil fraction and stirring rate on methane degradation in a stirred tank reactor. J Chem Technol Biotechnol 85:314–319. https://doi.org/10.1002/jctb.2339

Rocha-Rios J, Quijano G, Thalasso F, Revah S, Muñoz R (2011) Methane biodegradation in a two-phase partition internal loop airlift reactor with gas recirculation. J Chem Technol Biotechnol 86:353–360. https://doi.org/10.1002/jctb.2523

Rocha-Rios J, Kraakman NJR, Kleerebezem R, Revah S, Kreutzer MT, van Loosdrecht MCM (2013) A capillary bioreactor to increase methane transfer and oxidation through Taylor flow formation and transfer vector addition. Chem Eng J 217:91–98. https://doi.org/10.1016/j.cej.2012.11.065

Rostkowski KH, Criddle CS, Lepech MD (2012) Cradle-to-gate life cycle assessment for a cradle-to-cradle cycle: biogas-to-bioplastic (and back). Environ Sci Technol 46:9822–9829. https://doi.org/10.1021/es204541w

Rostkowski KH, Pfluger AR, Criddle CS (2013) Stoichiometry and kinetics of the PHB-producing type II methanotrophs *Methylosinus trichosporium* OB3b and *Methylocystis parvus* OBBP. Bioresour Technol 132:71–77. https://doi.org/10.1016/j.biortech.2012.12.129

Sánchez A, Rodríguez-Hernández L, Buntner D, Esteban-García AL, Tejero I, Garrido JM (2016) Denitrification coupled with methane oxidation in a membrane bioreactor after methanogenic pre-treatment of wastewater. J Chem Technol Biotechnol 91:2950–2958. https://doi.org/10.1002/jctb.4913

Sander R (2015) Compilation of Henry's law constants (version 4.0) for water as solvent. Atmos Chem Phys 15:4399–4981. https://doi.org/10.5194/acp-15-4399-2015

Sazinsky MH, Lippard SJ (2015) Methane monooxygenase: functionalizing methane at iron and copper. Met Ions Life Sci 15:205–256. https://doi.org/10.1007/978-3-319-12415-5_6

Semrau JD, DiSpirito AA, Yoon S (2010) Methanotrophs and copper. FEMS Microbiol Rev 34:496–531. https://doi.org/10.1111/j.1574-6976.2010.00212.x

Shah NN, Hanna ML, Taylor RT (1996) Batch cultivation of *Methylosinus trichosporium* OB3b: V. Characterization of poly-beta-hydroxybutyrate production under methane-dependent growth conditions. Bioeng Biotechnol 49:161–171. https://doi.org/10.1002/(SICI)1097-0290(19960120)49:2<161::AID-BIT5>3.0.CO;2-O

Shi Y, Hu S, Lou J, Lu P, Keller J, Yuan Z (2013) Nitrogen removal from wastewater by coupling anammox and methane-dependent denitrification in a membrane biofilm reactor. Environ Sci Technol 47:11577–11583. https://doi.org/10.1021/es402775z

Stone KA, Hilliard MV, He QP, Wang J (2017) A mini review on bioreactor configurations and gas transfer enhancements for biochemical methane conversion. Biochem Eng J 128:83–92. https://doi.org/10.1016/j.bej.2017.09.003

Strong PJ, Xie S, Clarke WP (2015) Methane as a resource: can the methanotrophs add value? Environ Sci Technol 49:4001–4018. https://doi.org/10.1021/es504242n

Strong PJ, Kalyuzhnaya M, Silverman J, Clarke WP (2016a) A methanotroph-based biorefinery: potential scenarios for generating multiple products from a single fermentation. Bioresour Technol 215:314–323. https://doi.org/10.1016/j.biortech.2016.04.099

Strong PJ, Laycock B, Mahamud SNS, Jensen PD, Lant PA, Tyson G, Pratt S (2016b) The opportunity for high-performance biomaterials from methane. Microorganisms 4:11. https://doi.org/10.3390/microorganisms4010011

Sun FY, Dong WY, Shao MF, Lv XM, Li J, Peng LY, Wang HJ (2013) Aerobic methane oxidation coupled to denitrification in a membrane biofilm reactor: treatment performance and the effect of oxygen ventilation. Bioresour Technol 145:2–9. https://doi.org/10.1016/j.biortech.2013.03.115

Sundstrom E, Criddle CS (2015) Optimization of methanotrophic growth and production of poly(3-hydroxybutyrate) in a high-throughput microbioreactor system. Appl Environ Microbiol 81:4767–4773. https://doi.org/10.1128/AEM.00025-15

Tan GYA, Chen CL, Li L, Ge L, Wang L, Razaad IMN, Li Y, Zhao L, Mo Y, Wang JY (2014) Start a research on biopolymer polyhydroxyalkanoate (PHA): a review. Polym Basel 6:706–754. https://doi.org/10.3390/polym6030706

United Nations Framework Convention on Climate Change (2013) Report of the conference of the parties serving as the meeting of the parties to the Kyoto protocol on its eighth session, held in Doha from 26 November to 8 December 2012. http://unfccc.int/bodies/body/6409/php/view. Accessed 10 June 2017

Van der Ha D, Nachtergaele L, Kerckhof FM, Rameiyanti D, Bossier P, Verstraete W, Boon N (2012) Conversion of biogas to bioproducts by algae and methane oxidizing bacteria. Environ Sci Technol 46:13425–13431. https://doi.org/10.1021/es303929s

Van Teeseling MC, Pol A, Harhangi HR, van der Zwart S, Jetten MSM, Op den Camp HJ, van Niftrik L (2014) Expanding the verrucomicrobial methanotrophic world: description of three novel species of *Methylacidimicrobium* gen. nov. Appl Environ Microbiol 80:6782–6791. https://doi.org/10.1128/AEM.01838-14

Vestman J, Liljemark S, Svensson M (2014) Cost benchmarking of the production and distribution of biomethane/CNG in Sweden. http://www.sgc.se/ckfinder/userfiles/files/SGC296_v2.pdf. Accessed 8 June 2017

Wendlandt KD, Jechorek M, Helm J, Stottmeister U (2001) Producing poly-3-hydroxybutyrate with a high molecular mass from methane. J Biotechnol 86:127–133. https://doi.org/10.1016/S0168-1656(00)00408-9

Yazdian F, Shojaosadati S, Nosrati M, Pesaran M, Ebrahim V-F (2012) Mixing studies in loop bioreactors for production of biomass from natural gas. Iran J Chem Chem Eng 31:91–101

Yu Y, Ramsay JA, Ramsay BA (2006) On-line estimation of dissolved methane concentration during methanotrophic fermentations. Biotechnol Bioeng 95:788–793. https://doi.org/10.1002/bit.21050

Zhang Y, Xin J, Chen L, Song H, Xia C (2008) Biosynthesis of poly-3-hydroxybutyrate with a high molecular weight by methanotroph from methane and methanol. J Nat Gas Chem 17:103–109. https://doi.org/10.1016/S1003-9953(08)60034-1

Zhang T, Zhou J, Wang X, Zhang Y (2016) Coupled effects of methane monooxygenase and nitrogen source on growth and poly-β-hydroxybutyrate (PHB) production of *Methylosinus trichosporium* OB3b. J Environ Sci 52:49–57. https://doi.org/10.1016/j.jes.2016.03.001

Zhang T, Wang X, Zhou J, Zhang Y (2017) Enrichments of methanotrophic-heterotrophic cultures with high poly-β-hydroxybutyrate (PHB) accumulation capacities. J Environ Sci (In Press). https://doi.org/10.1016/j.jes.2017.03.016

Novel Biocontrol Agents: Short Chain Fatty Acids and More Recently, Polyhydroxyalkanoates

12

Vaishnavi Gowda and Srividya Shivakumar

Abstract

Antibiotic resistance is a risk which has gradually become common knowledge. Steps are being actively taken to limit the use of antibiotics. The meat industry (poultry, piggery and sea food) are prone to infections. Containing these infections becomes cardinal for the meat which is ultimately meant for human consumption. Alternatives are therefore sought. Two alternatives which show promising pathogen containment are short chain fatty acids (SCFA) and Polyhydroxyalkanoates (PHA). SCFA are known to impart bactericidal effect apart from other benefits to the host. PHAs which have been viewed as alternatives to conventional plastic are rich sources of SCFA monomer pools. The most widely studied PHA is Polyhydroxybutyrate (PHB), a homopolymer made of 3-Hydroxybutrate (3HB) monomers. Many studies have been conducted on both SCFA and PHA on their use as biocontrol agents to understand the physiological mechanism of action for both.

Keywords

Biocontrol · Short chain fatty acid · SCFA · Polyhydroxyalkanoate · PHB · 3HB

12.1 Introduction

Supplementation of antibiotics in animal feed was a common practice for terrestrial animal production owing to growth promoting effects. Reports suggest that animals which were fed antibiotic in feed showed 4–5% faster growth than the control animals (Ferber 2003). In marine culture, antibiotics have been used traditionally as

V. Gowda · S. Shivakumar (✉)
Department of Microbiology, School of Sciences (SoS) Centre for PG Studies, Jain University, Bangalore, Karnataka, India
e-mail: sk.srividya@jainuniversity.ac.in

V. C. Kalia (ed.), *Biotechnological Applications of Polyhydroxyalkanoates*,
https://doi.org/10.1007/978-981-13-3759-8_12

primary prophylactics (Cabello 2006). Indiscriminate use of antibiotics has facilitated increase in antibiotic resistance. Use of antibiotics as growth promoter is now shunned in Europe with effect from 2006 (European Parliament and Council Regulation No 1831/2003).

Complete cessation of antibiotic use poses a bigger problem: increased prevalence of pathogenic bacteria like *Salmonella* which would translate as rise in the number of human and animal infections (Phillips et al. 2004). The need of the hour is therefore an alternative which is as effective as antibiotics as a biocontrol agent but without the associated risk of developing resistance. Currently, there are two promising biocontrol agents: Short Chain Fatty Acids (SCFA) and Polyhydroxyalkanoates.

12.2 SCFA as Biocontrol Agent

SFCA like butyric, propionic, valeric, formic, lactic and acetic acid, have been reported to possess antimicrobial activity. Growth of certain enterobacteria such as *Escherichia coli,* yeast, *Salmonella typhimurium* and *Shigella flexneri,* is actively inhibited by SCFA. These are in use as biocontrol agents to control *Salmonella* in commercial mixtures for poultry feed. The bactericidal activity of short chain fatty acid is attributed to its ability to pass through the cell membrane. In an alkaline environment, like the cytoplasm, SCFA disintegrates to increase the cytoplasmic concentration of protons. As a result, the cellular processes are diverted towards balancing proton gradient to equilibrate optimum pH conditions. Consequently, energy diverted to this process limits other metabolic processes thereby inhibiting cell proliferation.

12.2.1 Mechanism of SCFA Dependent Toxicity

Primary targets for biocidial compounds are cytoplasmic membrane, cell wall or specific metabolic functions in the cytoplasm linked to function, protein synthesis or replication. Even though, antibacterial effects of SCFA are not clearly elucidated, they exhibit notable bactericidal and bacteriostatic property (Ricke 2003). Such activity is dependent on physiological state of the organism and the physiochemical nature of the environment. All SCFA are weak acids, thereby the most likely route of action is through variation in pH (Davidson 2001). Undissociated forms of SCFA readily penetrate the lipid membrane of bacterial cell wall.

Toxicity of SCFA is attributed to the formation of non-ionized acids within the bacterial cytoplasm. These non-ionized acids are small uncharged molecules which can freely permeate through the cell membrane. Cytoplasm, in general has a circumneutral pH. Equilibrium is established within the cell between undisassociated and disassociated fatty acids. Relative levels of each are dependent on pH and pK_a of a given fatty acid (in accordance to Henderson-Hasselbalch equation). Inside the cytoplasm, SCFA disassociate leading to accumulation of respective anions and protons. Proton motive force is disrupted by the influx of protons lowering cytoplasmic pH (Axe and Bailey 1995). This leads to compromised metabolic reactions and energy conservation.

Osmotic balance is also disrupted by an increase in SCFA anions in the cytoplasm. Toxicity of diffused SCFA is dependent on pH, less the pH greater the disassociation. Transmembrane pH gradient is significantly affected by internal pH. As a result, influx of acids is disrupted. Cytoplasm of a bacterial cell is resistant to pH fluctuations owing to the impermeability of protons and buffering capacity of amino acids in the cytoplasm. In the presence of low external pH, the organism is more stringent towards maintaining a neutral pH and faces a greater transmembrane pH gradient. Thus, when acid influx is enhanced, susceptibility to SCFA induced toxicity is greater as compared to cells that can with stand acidic internal environment (Diez-Gonzalez and Russell 1997). Response to SCFA induced toxicity is dependent on the nature of the organism (Kirkpatrick et al. 2001).

12.2.2 SCFA and Colonization Resistance

Gut of multicellular organisms is populated by a specifically diverse microorganism population. Composition of each strain determines host health and susceptibility to diseases. The presence of a complex interaction between the microorganism and host makes elucidation of the microbiotas contribution to health ambiguous. Scientific evidence supports the role of gut microbiota in protecting the host from infection and colonization by enteric pathogens. This phenomenon is termed as 'colonization resistance' (Lawley and Walker 2013).

Of the many likely mechanisms to explain colonization resistance, the most credible mechanism is through the nature of interaction between the gut microflora and host immune system to regulate antagonistic response towards pathogen colonization. In essence, the presence of a well flourished microbiota in the gut presents itself as a physical barrier to foreign pathogens. There are two mechanisms of resistance: nutrition competition and production of bacteriocin. Indigenous microorganisms of the gut are well acclimatized to the physical and nutritional constraints of the host. In the event of an invasion, the residing microbiota competes with pathogens as demonstrated in *C. difficile* or *Escherichia coli*. Here, colonization of non-pathogenic strains prevents subsequent challenge of pathogens (Leatham et al. 2009). Additionally, many microorganisms residing in the gut secrete peptides with anti-microbial functions (bacteriocins) which possess the ability to target and kill invading pathogens. Antimicrobial activity of purified bacteriocins in-vitro has been reported (Millette et al. 2008). Therefore, live bacteriocin producing organisms can be potentially used as probiotics to promote overall intestinal health and protect the consumer from enteric pathogen infections (Dobson et al. 2012)

Chemical environment of the gut is defined by the residing microbiota. SCFA are released through breakdown of non-digestible carbohydrates as by-products of fermentation. Two major constituents of intestinal SCFA are propionate and butyrate. There are two different pathways proposed for production of butyrate in the gut post the formation of butyryl-CoA from condensation of two units of acetyl-CoA. The first case is explained by the process prevalent in *Clostridium acetobutylicum*, where butyryl CoA is converted to butyrate with the production of an intermediate butyryl-phosphate by two separate enzymes phosphotransbutyrlase and butyrate kinase.

An alternative pathway involves butyryl-CoA: acetate-CoA transferase enzyme to catalyse the transfer of CoA moiety from acetate to butyrate (Duncan et al. 2002). To identify the predominant process of butyrate production, 38 intestinal butyrate producing isolates were subjected to degenerate PCR. Enzyme assays suggested that the second pathway is mostly the route of intestinal butyrate production (Louis et al. 2004).

Propionate is formed through carbon fixation of succinyl CoA as demonstrated through in vitro analysis of a pure culture of *Bacteroides fragilis.* It is important to understand the metabolic pathways of production of propionate and butyrate to device specific molecular markers based on genes coding for metabolic enzymes in order to study the functionality of gut microflora (Hosseini et al. 2011).

Nature of released fermentation products is dependent on the structure of available carbohydrates. Composition and nature of intestinal SCFA is therefore influenced by the microbial community and diet of the host (Roy et al. 2006). Distribution of intestinal SCFA is dictated by a specific spatial organization of intestinal microbiota (Nava et al. 2011; Pedron et al. 2012). Therefore, no two regions in the large and small intestine exhibit the same SCFA profile creating microenvironments with varying pH (Cherrington et al. 1991). Large intestine withstands a higher load of microflora as compared to small intestine (Walter and Ley 2011). Hence, differences in spatial distribution lead to variation of relative proportion of propionate, butyrate and acetate in the intestines.

12.2.3 Bacterial Susceptibility to SCFA

Apart from modulating host functions, SCFA are carbon sources for endogenous microbiota (Fischbach and Sonnenburg 2011). At high dosage, SCFA induce toxicity towards proliferation of bacteria. Toxic effect of SCFA is linked to nonionizable variants of related acids which are abundant at acidic pH. Early studies on SCFA toxicity have established pleiotropic effects of weak organic acid as an inhibitor for oxidative reactions (Weiner and Draskoczy 1961) and as a trigger for chemotactic responses (Repaske and Adler 1981).

12.2.4 Role of SCFA in Host Organism

SCFA has a well-defined activity in the host organism and against bacteria. In the host organism, as mentioned above, the chemical environment is established by the residing microbiota. Butyrate produced affects energy homeostasis. It is consumed as a primary source of energy (Wong et al. 2006). Donohoe et al. (2011) observed that in germ free mice lacking SCFA, intermediary metabolism was hampered. As a result energy and nutrient sensor AMPK was activated which eventually ended with autophagy. When butyrate is supplied externally, GF mice were shielded from AMPK activation. Thus, the necessity of butyrate for normal colonocyte metabolism in mice was established.

Apart from serving as a metabolic substrate, SCFA regulate host immune functioning. Development of dendritic cells is blocked when propionate or butyrate is taken up by immune cells *via* SLC5A8 transporter through the activation of histone deacetylase (HDAC). Presence of butyrate triggers selective apoptosis of wild type T cell. However, FasL- deficient or Fas-deficient T cells are unaffected. This shows that mechanism of butyrate action is through induction of Fas dependent apoptosis of T cells (Zimmerman et al. 2012). Activation of dendritic cells by butyrate from SCFA mobilized in the gut is through increase in IL-23 concentration with concomitant decrease IL-12 (Berndt et al. 2012).

A specific family of G-protein-coupled receptors (FFAR) recognizes SCFA. Recognition triggers signalling at both systemic and gut epithelium sites. Binding of propionate to FFAR3 (GPR41) or FFAR2 (GPR43)regulates the production of gut hormone, inflammation and obesity (Xiong et al. 2004; Lin et al. 2012; Layden et al. 2013). Mice which lacked FFAR3 or FFAR2 displayed reduced glucagon like peptides- 1- levels and weakened glucose tolerance. Thus, the role of SCFA in diabetes is implicated (Tolhurst et al. 2012). Adipogenesis was stimulated in mice treated with SCFA in FFAR-independent and FFAR-dependent (Hong et al. 2005) mechanisms.

Gut anchoring of invading pathogens is restricted through the stimulation of antimicrobial peptides (AMPs) (Gallo and Hooper 2012). Termén et al. (2008) elucidated the mechanism of production of AMPs in the gut microbiota. It was found that AMP genesis was dependent on SCFA-mediated induction of LL-37 production. A classic example of this mechanism is demonstrated by colonic epithelial cells in human gut.

A similar case was noticed in poultry fed with SCFA. There was a significant increase in AMP production which led to diminished colonization of *Salmonella* (Sunkara et al. 2011, 2012). Therefore, direct supplementation of SCFA or prebiotics which stimulate the production of SCFA in the gut is a viable alternative to indiscriminate use of antibiotics for control of livestock colonization.

12.2.5 Specific Inhibition of Pathogenicity by SCFA

Salmonella is pathogenic after invading intestinal epithelial cells. This process of anchorage is facilitated by specific genes encoded by *Salmonella* Pathogenicity Island 1. Effect of butyrate on the gene was determined through comparative transcriptome analysis. From the study, it was concluded that the presence of butyrate downregulated the expression of *Salmonella* pathogenicity island 1 genes. The primary target of fatty acid is yet to be understood (Gantois et al. 2006).

Virulence of *Vibrios* in-vivo is regulated by quorum sensing (mechanism of inter bacterial communication through the use of specific chemical molecules). Production of quorum sensing chemicals is energy taxing. When such bacteria are exposed to SCFA, energy is diverted towards regulation of internal pH making it inaccessible to trigger virulence (Ricke 2003).

A significant improvement of bone shrimp survival was noticed by Defoirdt et al. (2007) when challenged with *V. campbelli*. SCFA was added to the culture water in concentration of 20 mM at pH 7. Under in-vitro conditions, there was no noticeable effect on growth of the bacteria. Contradiction in behaviour between in-vivo and in-vitro behaviour of the SCFA could be due to the difference in sub inhibitory concentrations which in turn affects expression of virulence in *Vibrios*.

12.2.6 Commercial Use of SCFA as Biocontrol Agents

SCFAs are extensively in use as replacement biocontrol agents to control pathogenic bacteria like *E. coli* and *Salmonella* ssp. in piggeries and poultry farms (Van Immerseel et al. 2003, 2006). Currently, there are two variations of SCFA feed which are commercially available: coated and uncoated acid products (Van Immerseel et al. 2005).

Uncoated formulations are liquids or powders which are mixed directly with feed or water and fed to the animal. Naked SCFA are readily absorbed by the host and rarely reach the gastrointestinal tract to exude bacteriocidal activity. Coated acid products are therefore more effective as biocontrol agents. In Coated acid products, lipid and mineral carriers are used to transport SCFA to specific sites like colonic epithelial cells.

A proposed alternative approach was to directly stimulate bacteria which produce butyric-acid. Butyric acid producers in the human gut belong to the phylogenetic Clostridium clusters IV and XIVa and species related to *Coprococcus, Faecalibacterium, Eubacterium*, and *Roseburia,* can also produce butyrate. Most butyrate-producing microbiota that is identified predominantly consumes acetic acid (Duncan et al. 2004). The possible existence of such a mechanism in poultry gut microbiota is not very clear.

Humblot et al. (2005) concluded that increase in lactic acid bacterial count in the gut led to increase in butyric acid concentration. This, in turn decreased colonization of *Salmonella. Bifidobacteria* are non-producers of butyric acid and others such as Lactic acid bacteria -lactobacilli and *bifidobacteria*, help others to do so. This system is termed as cross-feeding. Lactic acid produced by *Bifidobacterium adolescentis* is consumed by *Anaerostipes caccae* and *Eubacterium hallii* for producing butyric acid (Duncan et al. 2004).

Van Immerseel et al. (2005) studied the effects of SCFA additive on poultry to control *Salmonella* infection. Preparation used was coated butyric acid and was tested for the ability to reduce *Salmonella* colonization of internal organs and ceca post infection. This study was conducted to bridge the prevalent gaps in understanding the effect of butyric acid on poultry. Effectiveness of powdered butyric acid and coated butyric acid was compared in terms of reduced colonization of *Salmonella* in internal organs and ceca of infected young chickens. During the 1st of two trials, 4 groups of 25 chickens free from pathogenic infections were given either powdered butyric acid, coated butyric acid or a combination of both (dosage received by each group was 0.63 g of butyric acid/kg). Control was fed non-supplemented feed. Specific pathogen free layer chicken was orally infected with10^6 cfu of *S. enteritidis*. Significant decrease in cecal colonization was observed in chicken fed with coated butyric acid after 72 h. Chicken fed with powdered feed, however showed no effect. Long term effects of *Salmonella* colonization in chicken fed with coated butyric acid was studied (0.63 g of active product butyric acid/kg). Ten Ross broiler chicken were infected after 5 days with 10^5 cfu *Salmonella* and housed with 40 healthy chickens. Chicken treated with coated butyric acid had a significantly lower number of broilers shedding Salmonella bacteria, but cecal colonization at slaughter age was equal for both groups. The study concluded that butyric acid actively decreased colonization of cecal shortly after infection leading to a lowered faecal shedding. Complete elimination, however could be achieved through a combined practice of good hygiene and other protective measures since the broilers still contained *S. enteritidis* in the ceca at slaughter age, but at enrichment level.

12.2.7 Limitation of SCFA Use as Biocontrol Agent

SCFAs possess an immense potential for use as biocontrol agents. Primary limitation surfaces when it comes to their use in aquaculture environment. Since SCFAs are highly water soluble they are absorbed as a nutritional source and biocontrol potential is lost. By turning the feed as coated supplements, this problem can be overcome. But, the foul volatile odour of SCFA makes feed preparation unpalatable. Therefore, other biocontrol agents which offer a neutral pallet are sought. A candidate best suited as alternative to SCFA are Polyhydroxyalkonoates (PHA). PHA as a radical alternative to SCFA is swiftly gaining attention.

12.3 Polyhydroxyalkanoates

12.3.1 Polyhydroxyalkonte (PHA) the Reserve Carbon Sources

PHAs are polyesters of 3–6 hydroxyalkanoic acids produced by a wide range of bacterial species under specific nutrient conditions. During non-carbon related starvation periods, the producer channelizes metabolic pathways towards the production of inclusions bodies. PHAs are classified into two groups: short chain length (scl PHA) containing C3–5 monomers of hydroxyacids, and Medium Chain Length (mcl PHA) possessing C6–16 monomers of hydroxyacids. Most of the isolated and identified PHAs are linear and the well-studied amongst these are Poly(3)hydroxybutyrate (PHB). Industrial application of PHB is limited given its high brittleness and low thermal stability. Copolymers such as P(3HB-co-3 HV) which is

3-Hydroxybutyrate and 3-Hydroxyvalerate, offer greater flexibility and toughness making it ideal for applications in products like compost bags, films, moulded products and disposable food service ware. PHB finds alternative use as a rich, enantiomerically pure pool of biologically important 3-hydroxybuyrate monomers.

Polyhydroxybutyrate (PHB) is a homopolymer of β-hydroxybutyrate produced by a wide array of bacteria, archaea and fungi. They are produced as reserve carbon sources in the form of inclusion bodies. Native PHB is in an amorphous state and post cell death, lysis releases degraded partially crystallized form. Since PHB are accumulated as inclusion bodies, every producer possesses an inherent mechanism for PHB mobilization. Alternatively, there are scavenger organisms in nature which hold the potential to digest PHB as an energy source. Specific enzymes called PHB depolymerase are widely distributed among fungi and bacteria. These enzymes are known to degrade PHB to its monomeric units of β-hydroxybutyrate which is a short chain fatty acid naturally occurring in living systems including poultry.

12.3.2 Production and Accumulation of PHA

Presently, PHAs are used as potential replacement for conventional plastics because they exhibit a range of crystallinity from 30% to 70%. Melting temperature of PHA is 50–180 °C. Mechanical properties of PHA are comparable to commercially available isotactic polypropylene with 3.5 GPa Young's modulus and tensile strength of 43 MPa. They exhibit a high degree of polymerisation with molecular mass of several million Daltons. PHA accumulation is regulated by an exclusive set of biosynthesis gene (Fig. 12.1). PHA is deposited as spherical intracellular inclusion bodies containing a hydrophobic amorphous PHA core surrounded by proteins which regulate PHA synthesis and metabolism.

There are three enzymes which are involved in the biosynthesis of scl- PHAs like PHB:

(i) β-ketothiolase (PhaA): Catalyses the condensation of two monomeric units of acetyl-CoA to yield acetoacetyl CoA.
(ii) Acetoacetyly CoA reductase (PhaB): Acetoacetyl CoA produced in the first step is reduced to (R) -3-Hydroxybutyryl-CoA through a stereospecific reduction process catalysed by acetoacetyl CoA reductase.
(iii) PHA synthase (PhaC): PHA synthase catalyses the stereo selective conversion of (R)- 3 Hydroxyacyl-CoA to polyoxoesters, here polyhydroxybutyrate, with concomitant release of CoA. PhaC belongs to the α/β hydrolase family. In the active site, PhaC contains a cysteine and histidine group. The thiol group contained in the conserved cysteine group is activated by histidine contained at the same site. This enables a nucleophilic attack on the thioester bond of (R)-3-Hydroxyacyl-CoA enabling the release of Co-A and simultaneously forming a covalent enzyme-substrate intermediate. It is presumed that another conserved aspartic acid residue in the PhaC active site activates the hydroxyl group of the 3-hydroxy fatty acid followed by an attack on the thioester bond. The second

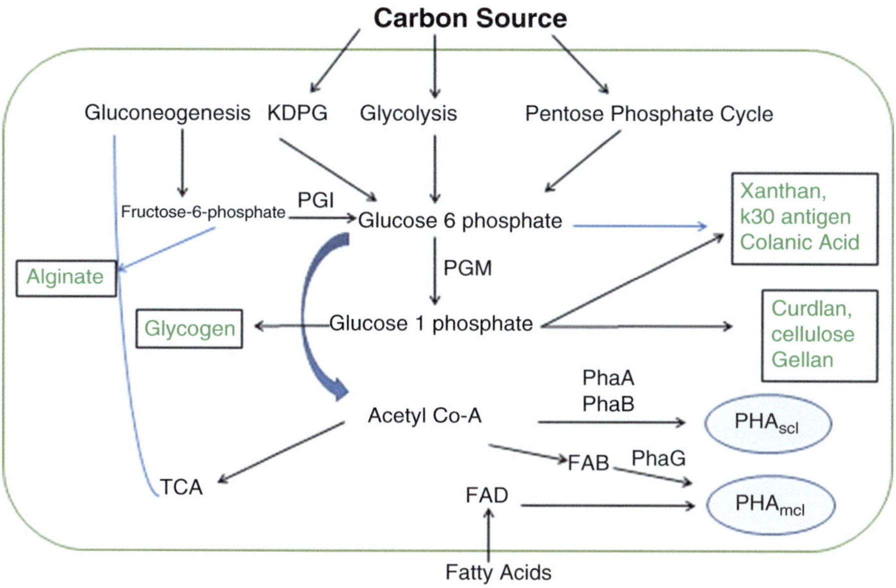

Fig. 12.1 Schematic representation of biosynthesis mechanism of PHA through intermediates of central metabolism. The results represented are metabolic routes which are most predominant in the production of various precursors. Black lines are indicative of direct enzyme-catalysed conversion for the production of intermediates or a link between primary metabolic pathways with intermediates of polymer biosynthesis. Multiple enzymatic steps are indicated by blue lines. Colour of text and its depiction: green-polysaccharides; blue- polyamides. *FAD* fatty acid β oxidation, *FAB* fatty acid de novo biosynthesis, *KDPG* 2-keto, 3-deoxy-6-phosphogluconoate pathway, *PGI* phosphoglucoisomerase, *Pha* polyhydroxyalkanoate synthesis enzyme, *TCA cycle* tricarboxylic acid cycle, PHA_{scl} short chain length PHA, PHA_{mcl} Medium Chain length PHA

thioester bond is between the second hydroxyl fatty acid and cysteines active site of second PhaC subunits. This links both the fatty acids. Chain extension continues with the entry of each unit of substrate by covalently biding to the free cysteine unit, followed by activation of the hydroxyl group and progressive nucleophilic attack. Chain termination occurs when most of the polyester chain is transferred to a second amino-acid exposed on the surface. This hydrolyses the second chain. The primed PhaC which is left behind starts a new cycle of polymerisation. Experimental evidence showed that when the copy number of PhaC was increased, the chain length of the polymer decreased. This suggests that chain length is determined by the amount of PhaC contained within the producer cell.

12.3.3 Fermentation Strategies of PHB

In order to establish an appropriate fermentation strategy, the producer must be capable of storing high concentration of PHA from growth over inexpensive substrates.

Production cost of PHA is five to ten times the cost of plastic synthesised from petrochemicals. Increased cost of production and additional steps of purification is the major hindrance for commercialization of PHAs. Production cost can be brought down if the process cost and/or cost of raw materials were brought down. Experimental evidence show various strains capable of accumulating high levels of polymer using relatively cheaper carbon sources (Charijamrus and Udpuay 2008; Srividya 2011).

Bacteria which synthesise PHA are divided into two groups. In the first group, the PHA synthesizing microorganism accumulates the polymer during stationary phase of growth under nutrient limiting condition and excess carbon. The second group produced PHA during the growth phase (Israni and Shivakumar 2013). Industrially important producer *Ralstonia euthropha* and *Bacillus thuringiensis* IAM12077 (Pal et al. 2009; Gowda and Shivakumar 2013) belongs to the first group.

Currently, either continuous or fed batch method of cultivation of PHA producing bacteria is in use. Fed batch method is ideal for bacteria which belong to the first group. They require two stages of production: the first being the accumulation of high cell density and the second for polymer synthesis by inducing nutrient limiting conditions. If fed batch method is used for producers belonging to the second group, nutrient feeding strategies play a pivotal role in deciding the success of fermentation. A balance between cell growth and PHA accumulation ensures that partial accumulation of PHA or untimely fermentation termination at low concentrations is avoided. In semi-continuous model, another strategy which can be applied is 'feast and famine' (Lemos et al. 2006). Here, the bacteria are subjected to alternative conditions of feed with essential nutrients for growth, immediately followed by imposing a nutritional limitation to promote PHA accumulation. As a result, competitive growth is established wherein organisms which can adapt and maintain balanced growth are favoured (Salehizadeh and Van Loosdrecht 2004). Under these conditions, 66–100% of excess carbon supplied in the feed is diverted towards polymer production. The remaining carbon is channelized towards growth and maintenance.

12.3.4 Degradation of PHA

PHAs are hailed as the suitable alternative for commercially available petroleum based polymers for their biodegradability. PHAs degrade completely to water and carbon dioxide. PHAs are present in two different states within the cell and outside the cell when released into the external environment. PHAs are produced as intracellular granules within the cell. Native PHA (nPHB, granules present within the cell) is in an amorphous state. On exposure to the external environment, they become partially crystalline and denature. Degradation of PHB is therefore categorized into two mechanisms of PHA degradation- intracellular and extracellular degradation. Intracellular degradation is the active mobilization of the nPHB and extracellular degradation is the utilization of denatured PHB by extracellular enzymes.

A classical pathway of PHB degradation exists in most producers. Here, PHA is first depolymerized to constitutive monomeric units catalysed by enzyme PHA depolymerase (PhaZ). PhaZ are specific for intracellular and extracellular degraders

(Fig. 12.2). In most cases, intracellular PhaZ are not capable of degrading crystalline, denatured PHA. Released PHB is degraded by scavenger organisms which possess specific extracellular PhaZ enzymes.

In nature, there is abundant distribution of bacteria and fungi which are capable of degrading PHAs. PHA degrading microorganisms have been isolated successfully from aerobic and anaerobic conditions. Characteristic of all the isolated PhaZ enzymes have been descried. Degradation rates are greatly affected by polymer characteristics such as structure, chemical composition, molecular weight and crystallinity (Jendrossek and Handrick 2002). Hydrolysis product depends on the nature of enzyme; it can be either monomers (Shivakumar 2013; Gowda and Shivakumar 2013) or oligomers (mono-trimers). When oligomers are produced, there is a secondary step of hydrolysis to break down to monomeric units (Jendrossek 1998). Production of the in-vivo PHA degradation enzymes is inhibited when carbon sources like glucose or organic acids are readily available. There are a few untypical reports wherein extracellular PHA depolymerase was detected in the absence of PHA. It can thus be concluded that presence of PHA does not act as an induction mechanism for the expression of *phaZ* genes (Jendrossek and Handrick 2002).

12.3.5 PHA Metabolism in Animal Gut

Since PHAs are homo or hetero-polymers of SCFAs, degradation of PHAs in the gut could result in similar beneficial effects as SCFA laced animal feed. Degradation of PHAs in abiotic conditions is a relatively slow process. Freier et al. (2002) extensively studied the degradation of PHB films in order to produce a reabsorbable gastrointestinal patch. PHB films were used to close bowel defect in mice. For the in-vitro degradation studies, the effect of modification of PHB films was studied. Also, degradation of poly (L-lactide) (PLLA) was examined. Molecular weight of pure PHB films was reduced by half after 1 year of incubation in a buffer solution at pH 7.4 and 37 °C. Blending of the PHB films with atactic PHB (polymer films with arbitrary stereospecific repetitive units) accelerated the process of molecular weight decrease. A similar effect was not noticed with films of lower molecular

Fig. 12.2 Degradation flow chart of PHB

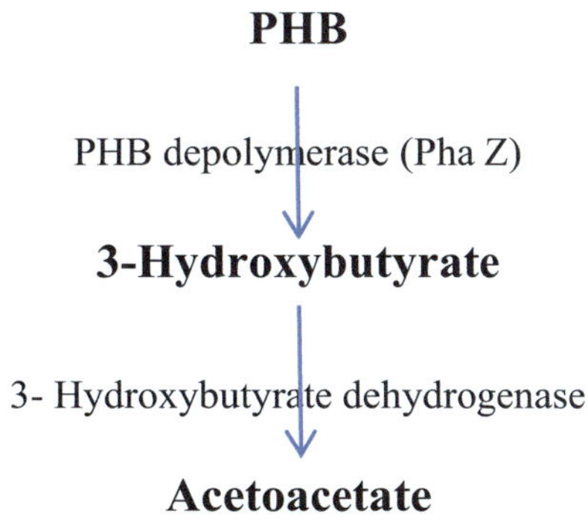

weight. PHB biodegradability was also increased through leaching of water soluble additives added during film preparation. The increase in the rate of degradation, although slightly elevated, was not significant. In contrast, when homophobic additives were incorporated in the films, PHB degradation was retarded.

A significant increase in the rate of PHB degradation in- vitro was noticed when pancreatin was added to the films. PLLA, on the other hand remained unaffected by physical and enzymatic treatments. By comparison to physical PHB degradation methods, addition of pancreatin increased rate of degradation by threefold. From the in-vitro studies, a blend of PHB/atactic and PHB was chosen for in-vivo model of degradation to assess the repair of bowel defect in Wistar rats. Patch films were developed using leaching/dipping method. After 26 weeks of implantation, remnants of the film were detected only in one of the four mice. In all cases, bowel defects were completely closed. The study concluded that materials could resist intestinal secretions for a long time before they were degraded completely. Also, PHB could potentially degrade completely in gastrointestinal tracts of mammals.

Degradation of films in the case of implants was improved when they were pretreated with NaOH followed by neutralization using HCL. On an average, pretreatment increased the rate of degradation by 37%.

Similar results were reported by Forni et al. (1999) wherein implants of PHB treated with NaOH improved digestibility of P (HB-HV) by 40% in sheep. Effect of NaOH pretreatment was also understood from this study. Molecular weight of the resultant polymeric fraction was substantially decreased owing to the formation of monomers. In fact, pretreatment enhanced digestibility by 85% depending on the incubation time and strength of the alkali.

Studies have reported chemical decomposition as alternative mechanisms of PHB degradation. Chemical hydrolysis of PHB can be achieved in both alkali and acidic environment. Under alkaline conditions (0.1–4 M of OH^-), Yu et al. (2005) reported the release of monomeric units from PHB. Similar action was found when PHB was incubated with concentrated acid solutions (80–90% H_2SO_4). An important conclusion drawn from chemical degradation was that amorphous forms of PHB degraded readily when compared to crystalline forms.

An alternative method to improve digestion in the mammalian gut is proposed. Supplementation of feed with PHB degrading microorganisms could potentially increase digestion of the polymer. All the strategies in use will degrade PHAs to their respective SCFA monomers. Given the similarity of PHA activity and SCFA-microbe interaction in the gut, biocontrol capacity of PHA can be explored.

12.3.6 Case Studies of PHAs as Biocontrol Agents

Assessing the Potential of PHB as Biocontrol Agent Through Polymer Direct Feed

- Defoirdt et al. (2007) assessed the protection offered by PHB against *Vibrio campbelli* in *Artemia franciscana*. To determine the efficiency of PHB, degradation of the polymer was first tested. The authors found that PHB particles were partially degraded in the gut of *Artemia franciscana* during initial *in vivo* starvation test. Further improvement in PHB degradation could enhance protection against the *Vibrios*. Effect of amplifying the degradation capacity of the polymer was tested by supplementing the feed with extracellular PHB depolymerase producing bacteria. Different strains of *C. testosterone* were checked for highest PHB degradation activity. *C. testosterone* LMG19554 exhibited the highest PHB deploymerase activity giving maximum zone of clearance on plates containing PHB as the sole carbon source. However, concomitant addition *C. testosterone* LMG19554 without PHB did not affect the survival of the infected *Artemia nauplii.*

- Impact of PHB on the survival of *nauplii* infected with pathogenic *V.campbellii* was also tested. Rate of survival of the infected *A. nauplii* was directly proportional to the concentration of PHB supplemented in the culture water. Increase in concentration of PHB from 100 to 1000 mg/L, showed complete protection at highest concentration. Protective function of PHB was also noticed when PHB was added as preventive treatment. Interestingly, addition of PHB to the culture water 24 h after the addition of the pathogen, showed curative effect. Protection offered by whole PHB was 100 times more efficient than β-hydroxybutyrate supplementation.
- De Schryver et al. (2010) attempted to understand the role of PHB in growth performance of juvenile European sea bass and impact on indigenous gut bacterial composition. Fish which were fed a diet supplemented with 2% and 5% (w/w) showed a positive impact on average weight gain of juvenile sea bass. Over the period of 6 weeks, increase in weight for 2% feed was 243% and 5%, 271% over the control of no-feed. In case of 0% feed, growth increased only by 216%. Higher concentrations of PHB in the feed had a negative effect mainly due to the lack of essential nutrients. Or, the authors postulate that 2 and 5% mark the optimal ecological conditions for fish gut. Growth promoting activity of PHB is similar to that of prebiotics because they doubled as a food source for fish. Since PHB is a homopolymer of fatty acids, they can be viewed as a food source.
- The authors also noted a decrease in gut pH in groups which were fed PHB. Increase in survival of partially fed PHB fish against the non-fed group, and drop in pH indicates that the polymer is partially metabolised in the fish gut. Presence of PHB in the diet also led to changes in the composition of bacterial community in fish gut.
- Nhan et al. (2010) investigated the effect of PHB on performance levels of giant fresh water prawn *Macrobrachium rosenbergii* and the changes in gut microflora. Two feeds were compared: *Artemia nauplii*fed with PHB and enriched with highly unsaturated fatty acids. Fresh water prawn larvae fed with PHB containing *Artemia nauplii* showed improved survival and development. Also, total count of *Vibrio* sp. was lower in this group. This indicated that addition of PHB imparts an inhibitory effect towards pathogenic microorganisms. Best survival and development was noticed when PHB and lipid enrichment was used in combination.

- Sui et al. (2012) assessed the protective effects of PHB on Chinese mitten crab from pathogenic *Vibrio anguillarum. E. sinensis* were subjected to different treatments. There were three groups each for control and test. Test group contained PHB enriched feed with live feed of *Artemia* and rotifers. In the test group, three variables were tested: protective effect of PHB when added 24 h before challenge inoculation (T1), behaviour of PHB when added with challenge inoculum (T2) and curing effect of PHB when added 24 h post challenge (T3). In the control group, C1 was uninfected and non-enriched, C2 was uninfected but enriched and C3 infected and non-enriched. Best results were seen in C2 which was supplemented with live PHB enriched feed. Survival of challenged test groups was significantly higher than that of challenged control C3. Delivery of PHB 24 h before challenge improved survival. Survival of T1, T2 and T3 varied significantly from each other. Development of larva was highest in C2 group. This study concluded that PHB enriched live feed was effective in protecting *E. sinensis* from *Vibrio anguillarum* infection which is highly pathogenic.
- Najdegerami et al. (2012) demonstrated the effect of PHB on changes in gastrointestinal tract (GI) microbiology in Siberian sturgeon (*Acipenser baerii*). Three commercially used diet concentrations of PHB (0, 2 and 5%) were evaluated based on the changes brought about in the structure and activity of microbial community of GI tract. Measures used to determine the changes were Community-level physiological profile (CLPP) and 16sRNA denaturing gradient gel electrophoresis (DGGE). Highest weight gain was noticed in fish fed with 2% PHB as compared to other treatments. Higher feed concentration did not reflect as increased weight gain due to replacement of essential nutrients with PHB. Survival rate was maximum with 2% feed followed by 5% and lowest in 0% feed (96.6%, 94.5% and 89.1%, respectively). pH of the mid gut region was slightly lower in PHB fed sturgeons. PHB feed of 2% showed an improvement in activity of intestinal microbial community in terms of diversity and action. Lorenz curve analysis based on DGGE pattern reflected that 2% PHB feed resulted in higher range-weighed richness and evenness as compared to control and 5% feed. Also, high growth rate and low mortality was noticed in 2% feed. This study established the relationship between PHB and its effect on gut microflora as the reason for improved survival. Sturgeon fingerlings which were fed with a diet containing 2% PHB also showed an increase in uptake of carboxylic acid, amino acids and other carbon sources. Highest aerobic metabolism was noticed in carboxylic acid group with 2% PHB. Health of the sturgeon fingerlings was improved through a balanced diet enriched with 2% PHB and the same was reflected as increase in GI microflora.
- Situmoranga et al. (2016) demonstrated the beneficiary effects of PHB in disease resistance and growth of Nile tilapia (*Oreochromis niloticus*) juveniles. Effects of dietary PHB on growth, body composition, digestive enzyme activities and disease resistance of low trophic level were investigated. An increase in weight was noticed in PHB fed juveniles but was not significantly different from the control group. There was however, a significant increase in the level of lipase activity by 20–40% when PHB was supplemented in the feed. Challenged test on

gnotobiotic larvae using pathogen *Edwardsiella ictaluri*GLY09R showed that the larvae fed with enriched *Artemia* resulted in 20% higher survival than the challenged control larvae. The authors concluded that PHB increased immunity in Nile tilapia juveniles.

- Kiran et al. (2016) demonstrated the disruptive nature of intermediates of PHB degradation on formation of biofilms. Bacteria which possess the capacity to form biofilms are resistant to antibiotics. When the ability to form biofilms is disrupted, pathogenesis and resistance towards antibiotics is reduced in *Vibrio* population. The authors concluded from the study that PHB affects pathogenicity of *Vibrio* by inhibiting or interfering with molecules involved in signalling. Validating this was the absence of peaks in gas chromatography for quorum-quenching activity with respect to AHL molecules. Quorum-quenching ability of PHB was linked to the hydrolysis of PHB by PHB degrading *Vibrio* PUGSK8 to β-hydroxy butyric acid. The authors reported identification of the first ever luminescent *Vibrio* secreting PHB depolymerase evident from the clear zone of clearance visible on PHB agar plates. Disruption of the quorum-sensing systems in *Vibrio* would be an effective and environmentally friendlier alternative to antimicrobial agents for pathogen containment. Thus, PHB treatments could be an eco-friendly non-invasive effectively strategy to contain *Vibrio* infection in shrimp aquaculture.
- Franke et al. (2017) attempted to understand the mechanism of immunostimulatory effects of PHB. PHB was administered directly through the mouth opening of European sea bass (*Dicentrarchus labrax*). Survival of the fish larvae was erratic between tanks independent of the respective treatment. Best larval survival was noticed in the test tank which was fed with PHB. Growth and performance of larvae was independent of PHB administration. However, larvae which were fed PHB showed the highest growth compared to the control group. Through this study it was established that ingestion of PHB prior to the first feed through drinking increased the performance of sea bass larvae. Such an effect was postulated to be the result of either PHB doubling as a secondary source of energy or through the changes induced in intestinal microflora. Gene expressions of the larval immune profile varied between the PHB treatments on 11 and 22 dph, respectively. Analysing the individual genes revealed that the difference was driven by differential expression of two antimicrobial peptides dicentracin (*dic*) and ferritin (*fer*). This explains the lack of identification of the variation in gene expression because the observed changes were moderate. Changes induced at the gene level may be beyond the scope of analyses, for, in the present study only two genes were analysed simultaneously. Immune response in fish larvae is a complex mechanism with synergistic effect of multiple genes. PHB may have induced changes at expression of other immune genes which were not assessed in the present study. Results from this study indicate that mechanism of action of PHB varies between different life cycle and stages. This in turn depends on the maturity of GI microbiota.

Assessing the Potential of PHB as Biocontrol Agent by Feeding Polymer Accumulating Bacteria

- Halet et al. (2007) conducted a study to assess the protective efficacy of PHB accumulating bacteria on *Artemia franciscana* from pathogenic *Vibrio campbellii*. In this study, PHB accumulating bacteria was enriched with activated sludge in a batch reactor using acetate as carbon source under N-limiting conditions. A primary challenge test was conducted to investigate whether the supplemented PHB accumulating bacteria could shield *A. franciscana* from pathogenic *V. campbellii* from infection. Enrichment cultures were grown as aggregates and subjected to different treatment to make the accumulated PHB available to *Artemia*. Naupullii was protected from infection when available PHB content was more than 15% and when the producer cells were subjected to thawing and freezing (to release accumulated amorphous PHB). There was no effect on survival of the naupullii cells when the culture was added untreated or after pasteurization. Pre-treatment of freezing and thawing resulted in smaller, consumable molecules of 50 μm. Size of the molecule plays an important role in uptake by *Artemia* cells. Protective effect exhibited by the addition of PHB accumulating bacteria is credited to the released PHB. Survival rate of infected naupullii cells were directly proportional to the amount of PHB available (similar to the result reported by Defoirdt et al. 2007).
- Of the two PHB accumulating isolates, for the secondary challenge infection, isolate PHB2 (closely related to *Brachymonas denitrificans*) was added to the culture water containing *Artemia*. There was a noticeable improvement in survival contributed by the addition of PHB2 isolate in every form (untreated, pasteurized, frozen and thawed). In this case, the aggregates of PHB cells could easily be taken up by *Artemia*. Additionally, curative effects of the isolate were also noticed. Results from the study led to the conclusion that bacteria accumulating PHB can be used as supplement feed for *Artemia* for protection against vibrio infection.
- Cam et al. (2009) proposed an alternative approach of using a PHB accumulating and homoserine lactone degrading bacterium to protect *Artemia* from pathogenic *V. harveyi*. Organism capable of accumulating PHB and degrading homoserine lactone was isolated using glycerol as the C source and homoserine lactone as nitrogen source. The study validated the importance of PHB on protecting *Artemia* from infection since quorum sensing ability of the enriched culture was less relevant.
- Thai et al. (2014) studied the effect of PHB accumulating *Alcaligenes euthropus* H16 on survival of *Artemia* challenged with *Vibrio harveyi*. The challenged *Artemia* was fed to giant freshwater prawn *Macrobrachium rosenbergii*. Two cultures of PHB accumulating bacteria were used: A10 (10% accumulation) and A80 (80% accumulation). Survival of the larvae fed with *Artemia* enriched medium containing 100–1000 mg/L A80 was significantly higher by 15% in larvae when compared to non-enriched *Artemia*. Even the challenged *Artemia* enriched group showed enhanced survival by 35%. Survival, however, was not significantly different from the unchallenged control group. Larvae fed with

PHB showed a decline in the number of colonies of *Vibrios* in the gut. The study concluded that PHB supplied through live feed improved the survival of crustacean larvae. Strategy of feeding live PHB accumulating cells to *Macrobrachium* feed significantly increased development and disease protection of giant fresh water prawn. Efficiency of amorphous PHB strategy was dependent on combination of PHB content in PHB accumulating bacteria and concentration of the added enrichment feed. Also, through this approach the quantity of *Artemia* feed is lower in comparison to other studies where crystalline PHB was used as direct feed.

- Baruah et al. (2015) investigated the mechanism of protection offered by PHB in gnotobiotic *Artemia franciscana* against challenge *Vibrio* infection. Maximum protection against challenge infection was observed at PHB concentration of 100 mg/L. Reduction in PHB dosage concentration was attributed to the size of the polymer (25–30 μm). Smaller size of the feed enabled higher uptake efficiency by *Artemia*. There are many postulated explanations for shielding effect of PHB, the most popular being the decrease in gut pH. Here, this study proved the triggering of stress protein Hsp70 in *Vibrio* challenged *Artemia* at a concentration sufficient to induce complete protection against the pathogens virulence. Butyrate is a known inducer of Hsp70 and the detection of the same in the gut of *Artemia* proves that the polymer fed was digested to release the monomeric component 3-Hydroxybutyrate. Induction was observed at the protein level and there was no detection at the mRNA level.

12.3.7 PHB as Probiotic

The use of probiotics gains importance given the lack of adaptive immune systems of invertebrates such as shrimp. These organisms need to rely on their innate immune system as primary defence mechanism to resist invading pathogens. Innate immunity is activated when non-self-molecules associated with pathogens are recognized through a process called pathogen-associated molecular patterns (PAMPs) by specific receptors called pathogen/pattern recognition receptors (PRRs). The most important immune mechanism in invertebrates is melanisation (prophenoloxidase (proPO) activation system). In this immune response, microbial PAMPs are recognized by selective PRRs which lead to the activation of a group of serine proteinases leading to cleavage of proPO zymogen resulting into active phenoloxidase (PO) enzyme. Activated PO triggers the release of short lived intermediate molecules like quinones possessing cytotoxic activity towards invading microorganisms, defensive action and wound healing process of damaged tissue (Zhao et al. 2011). In the process, melanin is deposited at the site of infection to engulf the invading microorganisms.

In shrimp, clot formation is an alternative first line of defence which is activated to prevent hemolymph loss and spread of microbial injury. Clotting in crustaceans is immediate suggesting that it is a crucial immune response for survival. A central enzyme involved in the process of clotting is transglutaminase (TGase) which is involved in the final step of stabilization of hemolymph clot. Additionally, heat shock proteins (Hsp) are considered pivotal in conferring tolerance to environmental stress conditions (Kregel 2002).

An effective strategy for protection of shrimp from diseases is through stimulation of immune response genes. Probiotic bacteria such as *Bacillus* sp. have immunostimulatory effects in shrimp which reflected as improved resistance towards pathogenic diseases.

Sugunaa et al. (2014) designed a study to assess the immunostimulatory effect of poly-β hydroxybutyrate–hydroxyvalerate (PHB–HV) extracted *from Bacillus thuringiensis* B.t.A102 on the immune system of *Oreochromis mossambicus*. Four levels of PHB-HV feed were tested: 0, 1, 3 and 5% feed. At regular intervals of every 5 days, the fish were bled and specific immune response was measured in terms of nonspecific immune mechanisms like total peroxidase activity, lysozyme activity and antiprotease activity, and antibody response to sheep red blood cells. Functional immunity was tested via experimental challenge with live virulent *Aeromonas hydrophila*. The study concluded that all levels of feed were effective in stimulating specific and non-specific immune response. However, highest dosage of 5% PHB-HV feed showed better results than 1 and 3%. PHB-HV could potentially be used as an immunostimulant in finfish aquaculture.

Laranja et al. (2017) demonstrated the protective effect of PHB accumulating bacteria on *Paenus monodon* larvae on exposure to pathogenic *V. campbelli.* The authors were successful in signifying that supplementation of PHB accumulating *Bacillus* sp.JL47 enhanced immunity in *P. monodon* significantly before and after *V. campbelli* challenge. A high level of proPO levels was noticed in cultures supplemented with *Bacillus* suggesting a stronger defence mechanism. Levels of TGase were also significantly higher in the test. The authors here suggest an immune priming effect (increase in innate immune activity in the host after first exposure to an immunomodulating agent) by PHB accumulating *Bacillus* JL47. Levels of Hsp70 (a molecular chaperone which repairs denatured proteins during heat shock stress) was upregulated.

12.4 Looking at the Future

Efficiency of PHB as a potential biocontrol agent in protecting commercial marine life and animal husbandry against pathogenic diseases has shown positive results in an experimental scale. Research in the field of radically novel biocontrol avenue is limited to PHB, a scl- polymer. There are few reports outlining the benefits of mcl polymers. Mcl polymer could spearhead a revolutionary containment method given the promising results of medium chain fatty acid use. Since the trend of using PHA is still in its nascent stages, understanding the exact mechanism of curing and protective properties of PHB is crucial. When the mechanism is elucidated, taking the concept of PHB as a biocontrol agent, from lab-to-field can be made commercially available.

12.5 Conclusion

Impetuous use of antibiotics has led to an epidemic of antibiotic resistance. Such a widespread occurrence would have gone by unnoticed if it weren't for the direct and indirect implications on human health. Antibiotic resistance can cause severe illness in humans who consume contaminated poultry and sea food. Although, transmission of antibiotic resistance from contaminated food to humans is unlikely, the infection which results from the pathogen poses severe health risks. Nevertheless, antibiotic use cannot be shunned citing the possibility of developing resistance. But, alternative containment methods can be sought. Short chain length fatty acids, which are constituents of normal gut microbiota, show impressive bactericidal activity. Given their high solubility in water, bacteriocidal effect is shadowed when they are delivered directly. To overcome this problem, SCFAs are administered through different formulations. An alternative approach to deliver SCFA is in the form of polymers containing SCFA monomers. One such polymer is PHB which is broken down to hydroxybutyrate, a bactericidal SCFA. Available literature proves the efficiency of PHB in protecting host organism from specific infections proving that PHB has the potential for wide scale use as a biocontrol agent.

References

Axe DD, Bailey JE (1995) Transport of lactate and acetate through the energized cytoplasmic membrane of *Escherichia coli*. Biotechnol Bioeng 47:8–19. https://doi.org/10.1002/bit.260470103

Baruah K, Tran T, Huy NP, Niu Y, Gupta SK, De Schryver P, Bossier P (2015) Probing the protective mechanism of poly-ß-hydroxybutyrate against vibriosis by using gnotobiotic *Artemia franciscana* and *Vibrio campbellii* as host-pathogen model. Sci Rep 5:9427. https://doi.org/10.1038/srep09427

Berndt BE, Zhang M, Owyang SY, Cole TS, Wang TW, Luther J, Veniaminova NA, Merchant JL, Chen CC, Huffnagle GB, Kao JY (2012) Butyrate increases IL-23 production by stimulated dendritic cells. Am J Physiol Gastrointest Liver Physiol 303(12):G1348–G1392. https://doi.org/10.1152/ajpgi.00540.2011

Cabello FC (2006) Heavy use of prophylactic antibiotics in aquaculture: a growing problem for human and animal health and for the environment. Environ Microbiol 8:1137–1144. https://doi.org/10.1111/j.1462-2920.2006.01054.x

Cam DTV, Hao NV, Dierckens K, Defoirdt T, Boon N, Sorgeloos P, Bossier P (2009) Novel approach of using homoserine lactone degrading and poly-β-hydroxybutyrate-accumulating bacteria to protect *Artemia* from the pathogenic effects of *Vibrioharveyi*. Aquaculture 291:23–30. https://doi.org/10.1016/j.aquaculture.2009.03.009

Chaijamrus S, Udpuay N (2008) Production and characterization of polyhydroxybutyrate from molasses and corn steep liquor produced by *Bacillus megaterium* ATCC 6748. Agric Eng Int X:1–12

Cherrington CA, Hinton M, Pearson GR, Chopra I (1991) Short-chain organic acids at pH 5.0 kill *Escherichiacoli* and *Salmonella* sp. without causing membrane perturbation. J Appl Bacteriol 70:161–165. https://doi.org/10.1111/j.1365-2672.1991.tb04442.x

Davidson PM (2001) Chap. 29. Chemical preservatives and natural antimicrobial compounds. In: Doyle MP, Beuchat LR, Montville TJ (eds) Food microbiology—fundamentals and Frontiers, 2nd edn. American Society for Microbiology, Washington, DC, pp 593–627

Davidson P, Taylor T (2007) Chemical preservatives and natural antimicrobial compounds. In: Doyle M, Beuchat L (eds) Food microbiology: fundamentals and frontiers, 3rd edn. ASM Press, Washington, DC, pp 713–745. https://doi.org/10.1128/9781555815912.ch33

De Schryver P, Sinha AK, Kunwar PS, Baruah K, Verstraete W, Boon N, De Boeck G, Bossier P (2010) Poly-ß-hydroxybutyrate (PHB) increases growth performance and intestinal bacterial range-weighted richness in juvenile European sea bass, *Dicentrarchus labrax*. Appl Microbiol Biotechnol 86:1535–1541. https://doi.org/10.1007/s00253-0092414-9

Defoirdt T, Halet D, Vervaeren H, Boon N, de Wiele TV, Sorgeloos P, Bossier P, Verstraete W (2007) The bacterial storage compound poly-b-hydroxybutyrate protects *Artemia franciscana* from pathogenic *Vibrio campbellii*. Environ Microbiol 9:445–452. https://doi.org/10.1111/j.1462-2920.2006.01161.

Diez-Gonzalez F, Russell JB (1997) The ability of *Escherichia coli* O157:H7 to decrease its intracellular pH and resist the toxicity of acetic acid. Microbiology 143:1175–1180. https://doi.org/10.1099/00221287-143-4-1175

Dobson A, Cotter PD, Ross RP, Hill C (2012) Bacteriocin production: a probiotic trait? Appl Environ Microbiol 78:1–6. https://doi.org/10.1128/AEM.05576-11.

Donohoe DR, Garge N, Zhang X, Sun W, O'Connell TM, Bunger MK (2011) The microbiome and butyrate regulate energy metabolism and autophagy in the mammalian colon. Cell Metab 13:517–526. https://doi.org/10.1016/j.cmet.2011.02.018

Duncan SH, Barcenilla A, Stewart CS, Pryde SE, Flint HJ (2002) Acetate utilization and Butyryl coenzyme a (CoA):acetate-CoA Transferase in butyrate-producing Bacteria from the human large intestine. Appl Environ Microbiol 68(10):5186–5190. https://doi.org/10.1128/AEM.68.10.5186–5190.2002

Duncan SH, Holtrop G, Lobley GE, Calder AG, Stewart CS, Flint HJ (2004) Contribution of acetate to butyrate formation by human faecal bacteria. Br J Nutr 91:915923. https://doi.org/10.1079/BJN20041150

Ferber D (2003) Antibiotic resistance—WHO advises kicking the livestock antibiotic habit. Science 301:1027–1027. https://doi.org/10.1126/science.301.5636.1027

Fischbach MA, Sonnenburg JL (2011) Eating for two: how metabolism establishes interspecies interactions in the gut. Cell Host Microbe 10:336–347. https://doi.org/10.1016/j.chom.2011.10.002

Forni D, Wenk C, Bee G (1999) Digestive utilization of novel biodegradable plastic in growing pigs. Ann Zootech 48:163–171. https://doi.org/10.1051/animres:19990302

Franke A, Roth O, De Schryver P, Bayer T, Garcia-Gonzalez L, Künzel S, Bossier P, Joanna J, Miest, Clemmesen C (2017) Poly-β-hydroxybutyrate administration during early life: effects on performance, immunity and microbial community of European sea bass yolk-sac larvae. Sci Rep 7:15022. https://doi.org/10.1038/s41598-017-14785-z

Freier T, Kunze C, Nischan C, Kramer S, Sternberg K, Sab M, Hopt UT, Schmitz KP (2002) In vitro and in vivo degradation studies for development of a biodegradable patch based on poly (3-hydroxybutyrate). Biomaterials 23:2649–2657. https://doi.org/10.1016/s0142-9612(01)00405-7

Gallo RL, Hooper LV (2012) Epithelial antimicrobial defence of the skin and intestine. Nat Rev Immunol 12:503–516. https://doi.org/10.1038/nri3228

Gantois I, Ducatelle R, Pasmans F, Haesebrouck F, Hautefort I, Thompson A (2006) Butyrate specifically down-regulates salmonella pathogenicity island 1 gene expression. Appl Environ Microbiol 72:946–949. https://doi.org/10.1128/AEM.72.1.946-949.2006

Gowda V, Shivakumar S (2013) Poly(3)hydroxybutyrate (phb) production in *Bacillus thuringiensis* IAM 12077 under varied nutrient limiting conditions and molecular detection of class IV *pha synthase* gene by PCR. Int J Pharm Bio Sci 4:794–802

Gowda V, Shivakumar S (2015) Poly(-β-hydroxybutyrate) (PHB) depolymerase $PHAZ_{Pen}$ from *Penicillium expansum*: purification, characterization and kinetic studies. 3 Biotech 5:901–909. https://doi.org/10.1007/s13205-015-0287-4

Halet D, Defoirdt T, Van Damme P, Vervaeren H, Forrez I, de Wiele TV, Boon N, Sorgeloos P, Bossier P, Verstraete W (2007) Poly-b-hydroxybutyrate-accumulating bacteria protect gnoto-

biotic *Artemia franciscana* from pathogenic *Vibrio campbellii*. FEMS Microbiol Ecol 60:363–369. https://doi.org/10.1111/j.1574-6941.2007.00305.x
Hong YH, Nishimura Y, Hishikawa D, Tsuzuki H, Miyahara H, Gotoh C (2005) Acetate and propionate short chain fatty acids stimulate adipogenesis via GPCR43. Endocrinology 146:5092–5099. https://doi.org/10.1210/en.2005-0545
Hosseini E, Grootaert C, Verstraete W, Van de Wiele T (2011) Propionate as a health-promoting microbial metabolite in the human gut. Nutr Rev 69:245–258. https://doi.org/10.1111/j.1753-4887.2011.00388.x
Humblot C, Bruneau A, Sutren M, Lhoste EF, Dore J, Andrieux C, Rabot S (2005) Brussels sprouts, inulin and fermented milk alter the faecal microbiota of human microbiota-associated rats as shown by PCR-temporal temperature gradient gel electrophoresis using universal, *Lactobacillus* and *Bifidobacterium* 16S rRNA gene primers. Br J Nutr 93:677684. https://doi.org/10.1079/BJN20051372
Israni N, Shivakumar S (2013) Combinatorial screening of hydrolytic enzyme/s and PHA producing bacillus spp., for cost effective production of PHAs. Int J Pharm Bio Sci 4:934–945
Jendrossek D (1998) Microbial degradation of polyesters: a review on extracellular poly (hydroxyalkanoic acid) depolymerases. Polym Degrad Stab 59:317–325. https://doi.org/10.1016/s0141-3910(97)00190-0
Jendrossek D, Handrick R (2002) Microbial degradation of polyhydroxyalkanoates. Annu Rev Microbiol 56:403–432. https://doi.org/10.1146/annurev.micro.56.012302.160838
Kiran GS, Priyadharshini S, Dobson ADW, Gnanamani E, Selvin J (2016) Degradation intermediates of polyhydroxybutyrate inhibit phenotypic expression of virulence factors and biofilm formation in luminescent *Vibrio* sp. PUGSK8. NPJ Biofilms Microbiomes 2:16002. https://doi.org/10.1038/npjbiofilms.2016.2
Kirkpatrick C, Maurer LM, Oyelakin NE, Yoncheva YN, Maurer R, Slonczewski JL (2001) Acetate and formate stress: opposite responses in the proteome of *Escherichia coli*. J Bacteriol 183:6466–6477. https://doi.org/10.1128/jb.183.21.6466-6477.2001
Kregel KC (2002) Heat shock proteins: modifying factors in physiological stress responses and acquired thermotolerance. J Appl Physiol 92:2177–2186. https://doi.org/10.1152/japplphysiol.01267.2001
Laranja JLQ, Amar EC, Ludevese-Pascual GL, Nui Y, Geaga MJ, De Schryver P, Bossier P (2017) A probiotic bacillus strain containing amorphous poly-beta-hydroxybutyrate (PHB) stimulates the innate immune response of *Penaeus monodon* postlarvae. Fish Shellfish Immunol 68:202–210. https://doi.org/10.1016/j.fsi.2017.07.023.
Lawley TD, Walker AW (2013) Intestinal colonization resistance. Immunology 138:1–11. https://doi.org/10.1111/j.1365-2567.2012.03616.x.
Layden BT, Angueira AR, Brodsky M, Durai V, Lowe WL Jr (2013) Short chain fatty acids and their receptors: new metabolic targets. Transl Res 161:131–140. https://doi.org/10.1016/j.trsl.2012.10.007
Leatham MP, Banerjee S, Autieri SM, Mercado-Lubo R, Conway T, Cohen PS (2009) Precolonized human commensal *Escherichia coli* strains serve as a barrier to *E. coli* O157:H7 growth in the Streptomycin-treated mouse intestine. Infect Immun 77:2876–2886. https://doi.org/10.1128/IAI.00059-09
Lemos PC, Serafim LS, Reis MAM (2006) Synthesis of polyhydroxyalkanoates from different short-chain fatty acids by mixed cultures submitted to aerobic dynamic feeding. J Biotechnol 122:226–238. https://doi.org/10.1016/j.jbiotec.2005.09.006
Lin HV, Frassetto A, Kowalik EJ Jr, Nawrocki AR, Lu MM, Kosinski JR (2012) Butyrate and propionate protect against diet-induced obesity and regulate gut hormones via free fatty acid receptor 3-independent mechanisms. PLoS One 7:e35240. https://doi.org/10.1371/journal.pone.0035240
Louis P, Duncan SH, McCrae SI, Millar J, Jackson MS, Flint HJ (2004) Restricted distribution of the butyrate kinase pathway among butyrate-producing bacteria from the human colon. J Bacteriol 186:2099–2106. https://doi.org/10.1128/JB.186.7.2099-2106.2004

Millette M, Cornut G, Dupont C, Shareck F, Archambault D, Lacroix M (2008) Capacity of human nisin- and pediocin-producing lactic acid bacteria to reduce intestinal colonization by vancomycin-resistant *Enterococci*. Appl Environ Microbiol 74:1997–2003. https://doi.org/10.1128/AEM.02150-07

Najdegerami EH, Tran TN, Defoirdt T, Marzorati M, Sorgeloos P, Boon N, Bossier P (2012) Effects of poly-b-hydroxybutyrate (PHB) on Siberian sturgeon (*Acipenser baerii*) fingerlings performance and its gastrointestinal tract microbial community. FEMS Microbiol Ecol 79:25–33. https://doi.org/10.1111/j.1574-6941.2011.01194.x

Nava GM, Friedrichsen HJ, Stappenbeck TS (2011) Spatial organization of intestinal microbiota in the mouse ascending colon. ISME J 5:627–638. https://doi.org/10.1038/ismej.2010.161

Nhan DT, Wille M, De Schryver P, Defoirdt T, Bossier P, Sorgeloos P (2010) The effect of poly-β-hydroxybutyrate on larviculture of the giant freshwater prawn *Macrobrachium rosenbergii*. Aquaculture 302:76–81. https://doi.org/10.1016/j.aquaculture.2010.02.011

Pal A, Prabhu A, Kumar AA, Rajagopal B, Dadhe K, Ponnamma V, Shivakumar S (2009) Optimization of process parameters for maximum poly(-beta-)hydroxybutyrate (PHB) production by *Bacillus thuringiensis* IAM 12077. Pol J Microbiol 58:149–154

Pedron T, Mulet C, Dauga C, Frangeul L, Chervaux C, Grompone G (2012) A crypt-specific core microbiota resides in the mouse colon. MBio 3:116–112. https://doi.org/10.1128/mBio.00116-12

Phillips I, Casewell M, Cox T, De Groot B, Friis C, Jones R, Nightingale C, Preston R, Waddell J (2004) Does the use of antibiotics in food animals pose a risk to human health? A critical review of published data. J Antimicrob Chemother 53:28–52. https://doi.org/10.1093/jac/dkg483

Repaske DR, Adler J (1981) Change in intracellular pH of Escherichia coli mediates the chemotactic response to certain attractants and repellents. J Bacteriol 145:1196–1208. doi: 0021-9193/81/031196-13$02.00/0

Ricke SC (2003) Perspectives on the use of organic acids and short chain fatty acids as antimicrobials. Poult Sci 82:632–639. https://doi.org/10.1093/ps/82.4.632

Roy CC, Kien CL, Bouthillier L, Levy E (2006) Short-chain fatty acids: ready for prime time? Nutr Clin Pract 21:351–366. https://doi.org/10.1177/0115426506021004351

Salehizadeh H, Van Loosdrecht MC (2004) Production of polyhydroxyalkanoates by mixed culture: recent trends and biotechnological importance. Biotechnol Adv 22:261–279. https://doi.org/10.1016/j.biotechadv.2003.09.003

Shivakumar S (2013) Poly-β-hydroxybutyrate (PHB) Depolymerase from *Fusarium solani* Thom. J Chemother 2013:406386. https://doi.org/10.1155/2013/406386

Situmoranga ML, De Schryver P, Dierckens K, Bossiera P (2016) Effect of poly-β-hydroxybutyrate on growth and disease resistance of *Nile tilapiaOreochromis niloticus* juveniles. Vet Microbiol 182:44–49. https://doi.org/10.1016/j.vetmic.2015.10.024

Srividya S (2011) Production of PHB from lactose and whey by *Bacillus thuringiensis* IAM 12077. Res J Biotech 6:12–18

Sugunaa P, Binuramesh C, Abirami P, Saranya V, Poornima K, Rajeswari V, Shenbagarathai R (2014) Immunostimulation by poly-β hydroxybutyrate–hydroxyvalerate (PHB–HV) from *Bacillus thuringiensis* in *Oreochromis mossambicus*. Fish Shellfish Immunol 36:90–97. https://doi.org/10.1016/j.fsi.2013.10.012

Sui L, Liu Y, Sun H, Wille M, Bossier P, De Schryver P (2012) The effect of poly-β-hydroxybutyrate on the performance of Chinese mitten crab (*Eriocheir sinensis Milne-Edwards*) zoealarvae. Aquac Res:1–8. https://doi.org/10.1111/are.12077

Sunkara LT, Achanta M, Schreiber NB, Bommineni YR, Dai G, Jiang W (2011) Butyrate enhances disease resistance of chickens by inducing antimicrobial host defense peptide gene expression. PLoS One 6:e27225. https://doi.org/10.1371/journal.pone.0027225

Sunkara LT, Jiang W, Zhang G (2012) Modulation of antimicrobial host defence peptide gene expression by free fatty acids. PLoS One 7:e49558. https://doi.org/10.1371/journal.pone.0049558

Termén S, Tollin M, Rodriguez E, Sveinsdóttir SH, Jóhannesson B, Cederlund A (2008) PU.1 and bacterial metabolites regulate the human gene CAMP encoding antimicrobial peptide LL-37 in colon epithelial cells. Mol Immunol 45:3947–3955. https://doi.org/10.1016/j.molimm.2008.06.020

Thai TQ, Wille M, Garcia-Gonzalez L, Sorgeloos P, Bossier P, De Schryver P (2014) Poly-ß-hydroxybutyrate content and dose of the bacterial carrier for *Artemia* enrichment determine the performance of giant freshwater prawn larvae. Appl Microbiol Biotechnol 98:5205–5215. https://doi.org/10.1007/s00253-014-5536-7

Tolhurst G, Heffron H, Lam YS, Parker HE, Habib AM, Diakogiannaki (2012) Short-chain fatty acids stimulate glucagon-like peptide-1 secretion via the G-protein-coupled receptor FFAR2. Diabetes 61:364–371. https://doi.org/10.2337/db11-1019

Van Immerseel F, De Buck J, Pasmans F, Velge P, Bottreau E, Fievez V (2003) Invasion of *Salmonella enteritidis* in avian intestinal epithelial cells in vitro is influenced by short-chain fatty acids. Int J Food Microbiol 85:237–248. https://doi.org/10.1016/S0168-1605(02)00542-1

Van Immerseel F, Boyen F, Gantois I, Timbermont L, Bohez L, Pasmans F, Haesebrouck F, Ducatelle R (2005) Supplementation of coated butyric acid in the feed reduces colonization and shedding of *Salmonella* in poultry. Poult Sci 84:1851–1856. https://doi.org/10.1093/ps/84.12.1851

Van Immerseel F, Russell JB, Flythe MD, Gantois I, Timbermont L, Pasmans F (2006) The use of organic acids to combat Salmonella in poultry: a mechanistic explanation of the efficacy. Avian Pathol 35:182–188. https://doi.org/10.1080/03079450600711045

Walter J Ley R (2011) The human gut microbiome: ecology and recent evolutionary changes. In: Gottesman S Harwood CS (ed) Annual review of microbiology. Vol. 65, pp 411–429

Weiner N, Draskoczy P (1961) The effects of organic acids on the oxidative metabolism of intact and disrupted *E. coli*. Pharmacolo Exp Ther 132:299–305. PubMed: 13783909

Wong JM, de Souza R, Kendall CW, Emam A, Jenkins DJ (2006) Colonic health: fermentation and short chain fatty acids. J Clin Gastroenterol 40:235–243. PMID: 16633129

Xiong Y, Miyamoto N, Shibata K, Valasek MA, Motoike T, Kedzierski RM (2004) Short-chain fatty acids stimulate leptin production in adipocytes through the G protein-coupled receptor GPR41. Proc Natl Acad Sci U S A 101:1045–1050. https://doi.org/10.1073/pnas.2637002100

Yu J, Chen L (2005) Kinetics and mechanism of the monomeric products from abiotic hydrolysis of poly[(R)-3- hydroxybuyrate] under acidic and alkaline conditions. Polym Degrad Stab 89:289–299. https://doi.org/10.1016/j.polymdegradstab.2004.12.026

Yu J, Plackett D, Chen LXL (2005) Kinetics and mechanism of the monomeric products from abiotic hydrolysis of poly[(R)-3-hydroxybutyrate] under acidic and alkaline conditions. Polym Degrad Stab 89:289–299. https://doi.org/10.1016/j.polymdegradstab.2004.12.026

Zhao P, Lu Z, Strand MR, Jianf H (2011) Antiviral, anti-parasitic, and cytotoxic effects of 5,6-dihydroxyidole (DHI), a reactive compound generated by phenoloxidase during insect immune response. Insect Biochem Mol Biol 41:645–652. https://doi.org/10.1016/j.ibmb.2011.04.006

Zimmerman MA, Singh N, Martin PM, Thangaraju M, Ganapathy V, Waller JL (2012) Butyrate suppresses colonic inflammation through HDAC1-dependent Fas upregulation and Fas-mediated apoptosis of T cells. Am J Physiol Gastrointest Liver Physiol 302:1405–1415. https://doi.org/10.1152/ajpgi.00543.2011

Applications of PHA in Agriculture 13

Tan Suet May Amelia, Sharumathiy Govindasamy, Arularasu Muthaliar Tamothran, Sevakumaran Vigneswari, and Kesaven Bhubalan

Keywords
Polyhydoxyalkanoates · Agriculture · Bioplastics · Biowastes · By-products · Films

13.1 Introduction

Polyhydroxyalkanoate (PHA) is a well-known biodegradable bacterial polymer. The polymer is produced by some bacteria under stressed growth conditions. In nature, poly (3-hydroxybutyrate) [P(3HB)] is the most commonly found. Nonetheless, research in PHA has resulted in the production of various copolymers with improved properties and modifications to suit a variety of different applications. Identification of new bacteria strains with the ability to produce novel PHA monomers are still on going. Various cheap and renewable carbon feedstock and growth media have been identified. The production of in PHA in industrial scale fermenters have been fine-tuned using statistical approach. The production

T. S. M. Amelia · S. Govindasamy · A. M. Tamothran
School of Marine and Environmental Sciences, Universiti Malaysia Terengganu, Kuala Nerus, Malaysia

S. Vigneswari
School of Fundamental Sciences, Universiti Malaysia Terengganu, Kuala Nerus, Malaysia

K. Bhubalan (✉)
School of Marine and Environmental Sciences, Universiti Malaysia Terengganu, Kuala Nerus, Malaysia

Malaysian Institute of Pharmaceuticals and Nutraceuticals, NIBM, MOSTI, Gelugor, Penang, Malaysia

Institute of Marine Biotechnology, Universiti Malaysia Terengganu, Kuala Nerus, Malaysia
e-mail: kesaven@umt.edu.my

V. C. Kalia (ed.), *Biotechnological Applications of Polyhydroxyalkanoates*,
https://doi.org/10.1007/978-981-13-3759-8_13

efficiency of PHA is still being experimented in order to achieve maximum yield with minimal cost. Among the different applications of PHA, much attention was gained in medical and pharmaceutical fields. This is mainly attributed to the biocompatibility of PHA. However, studies in the application of PHA in agriculture is rather limited. This chapter will survey the efforts of PHA application in agriculture and highlight the successful usage of PHA.

13.2 Mulch Films

Agricultural usage of mulch is integral for crop protection and to increase crop yield. Mulching is used to maintain good soil structure, prevent contamination, moisture retention and weed control (Kasirajan and Ngouajio 2012; Rydz et al. 2014). There are two types of mulch namely natural and synthetic mulch. The inadequate amount of material needed for natural mulch production and lack of weed control made synthetic mulch more favourable for large scale agricultural application. Among synthetic mulch such as paper and plastic, plastic is being used extensively.

The global use of agricultural plastic film is expected to increase to 7.4 million tons in 2019 from 4.4 million tons in 2012 where agricultural mulch accounted for 40% of plastic usage in 2012 (Sintim and Flury 2017). Disposal of used plastic mulch has triggered environmental concerns. Plastic mulch is commonly made of high-density polyethylene (HDPE), low-density polyethylene (LDPE), and linear LDPE that does not readily degrade in the environment (Kasirajan and Ngouajio 2012; Sintim and Flury 2017). Thus, the used plastic mulch often ends up in landfill site or being burned which leads to pollution (Barnes et al. 2009; Sintim and Flury 2017). Studies on developing readily degradable mulch gained momentum in the recent years. The focus of research was on application of materials such as PHA, polybutylene succinate, polylactic acid (PLA), ethylene vinyl acetate and polymer made of corn starch (Niaounakis 2015). There is lack of commercially available PHA-based mulch products. This might be due to high production cost, inferior mechanical properties and inconsistent degradation rate (Niaounakis 2015).

Nodax™ is a patent held by Danimer Scientific and a biodegradable PHA copolymer of poly(3-hydroxybutyrate-*co*-3-hydroxyhexanoate) [P(3HB-*co*-3HHx)] (Hassan et al. 2006). This corporation produced custom-made agricultural mulch using Nodax™. The mulch produced by Nodax™ is not photodegradable and certified compostable to ASTM D6400-99 or ASTM D6868-03 standards (Danimer Scientific 2018a). This product is among the commercially available agricultural mulch. Another commercially available mulch product is made from Mirel™ resin produced by Metabolix, Inc. (Andrews 2014). PHA base polymer of Mirel™ resin is synthesized using natural sugars as carbon source and made into pellets which have been used to produce the mulch (Mirel Bioplastics 2018). Mirel™ films are certified to be compostable to ASTM D6400 by Biodegradable Products Institute (Mirel Bioplastics 2018). However, there are few patent filings regarding the application of PHA in producing agricultural mulch. The patent filings are listed in the Table 13.1.

Table 13.1 List of patents of mulch film utilizing PHA and its derivatives

Patent number	Assignee	Inventor(s)	Title	Publication date
US20130029124A1	BASF SE	Robert Loos, Xin Yang, Jörg Auffermann, Franziska Freese	Biodegradable polyester foil	31 January 2013
CN102140185A	Ruide Yu	Ruide Yu	Biodegradable mulching film and preparation method thereof	3 August 2011
KR20040071992A	SK Chemicals Co. Ltd.	Lee Min Hyeok, Shin Jeong Ju	Degradable mulching film coated water-absorbent material	16 August 2004
WO2001093678 A2	The Procter & Gamble Company	Isao Noda, Michael Matthew Satkowski	Agricultural items and methods comprising biodegradable copolymers	13 December 2001

13.3 Agricultural Nets

The applications of plastic materials in various agriculture sectors ranges from greenhouse films to protection nets (Niaounakis 2015; Guerrini et al. 2017). Nets are essential agricultural products as they increase crop yield and quality, protect crops from meteorological hazards, hailstone, wind, birds and insects, shade crops from sunlight, reduce chemical input, prevent overheating of crops, broaden the selection and planting period of crops, protect low-light plants, and replace the lack of natural shade (Brown 2004; Alvarez et al. 2006; Castellano et al. 2008; Guerrini et al. 2017). Furthermore, artificial shading such as a nursery shading net, is a relatively portable and transitory option compared to natural shading, such as the shading of papaya trees over pineapples via intercropping. Examples of crops that require shade are edible fiddlehead ferns and young cocoa trees, which grow in the shaded rainforest understorey (Brown 2004). Moreover, shading nets also reduce evaporative loss of agricultural water reservoirs. The evaporation rate of a small water body shaded by polyethylene (PE) shade nets were studied, whereby the use of single and double-layer shades conserved 14% and 21% of the daily evaporation losses respectively (Alvarez et al. 2006). They concluded that black PE shade efficiently and economically reduce evaporative loss when water supply is scarce (Alvarez et al. 2006). Accordingly, PE material has been the standard material for agricultural nets due to its many significant uses (Brown 2004).

13.3.1 Application of PHA for Agricultural Nets

Specifically, the commonly used raw material for commercial agricultural nets is HDPE (Castellano et al. 2008). However, petroleum-based plastic is non-biodegradable, harmful to animals, environmentally hazardous and banned in a number of countries (Kasirajan and Ngouajio 2012; Ojanji 2017). In agriculture, traditional plastic products with short-lived application, such as clips, wires, nets and geotextiles, generate large quantities of waste and have a high risk of polluting the agricultural system and its surrounding environment (Guerrini et al. 2017). Thus, biodegradable PHA is currently being considered for application as agricultural net. The major advantage of biodegradable netting is its compostability, which allows direct disposal of the bioplastic in the soil to be composted with organic materials, such as manure, crop remnant and food residue (Castellano et al. 2008). Furthermore, bioplastic netting decomposes into the soil, unlike the highly durable polymeric netting that does not decompose over time and remains above or beneath the soil, which causes problems during subsequent operations by becoming tangled in agricultural equipment (Shelton and Pocher 2005). Subsequently, post-use biopolymeric netting that are highly contaminated with soil do not need to be separated from plant residue, and hence economically save labour and recycling costs

(Guerrini et al. 2017). In view of the advantages of bioplastic nets in agriculture, the United States Patent and Trademark Office (USPTO) have licensed several patents that employed PHA material in agricultural nets (Shelton and Pocher 2005; Havens et al. 2014).

The most important mechanical characteristics of nets are the elongation at break and the tensile strength (Castellano et al. 2008). The commonly used raw material for commercial agricultural nets is HDPE, whereby typical HDPE has the Young's modulus, tensile strength and elongation at break of 30–40 MPa, 16–37 MPa, and 55–600% (Castellano et al. 2008; Kusuktham and Teeranachaideekul 2014; Khalaf 2015). Subsequently, the types of PHA or PHA-blends that harbour similar elongation at break and tensile strength as HDPE are potential material for nets. In general, the potential types of PHA for agricultural nets are poly(4-hydroxybutyrate) [P(4HB)] and some PLA/PHA-blends (Williams et al. 2013; Andrews 2014). The properties of P(4HB) and some PLA/PHA-blends were similar to HDPE, whereby the elongation at break and tensile strength of P(4HB) were 25–1000% and 50–800 MPa (Saito and Doi 1994; Williams et al. 2013; Andrews 2014). Interestingly, a study reported enhanced mechanical properties of ultrahigh molecular weight P(3HB) with a wide range of Young's modulus, elongation at break, and tensile strength of up to 18,100 MPa, 35% and 1320 MPa respectively (Iwata et al. 2004).

There are a few patents licensed by USPTO that utilised PHA in agriculture. A patent by Havens and co-inventors (2014) with publication number US20140245655 A1 disclosed a modified fishing gear to reduce ghost fishing by integrating a PHA component, which degrades faster when continuously soaked in an aquatic environment than when it is periodically removed from an aquatic environment and exposed to light and air. The modified gear has reduced functionality after becoming derelict, and loses the ability to catch and retain fish, which is an effective and economical solution to enforce reckless discarding of ghost nets. Derelict fishing gear could cause negative economic and ecological impact, since it still catches and retains fish continuously. Degradable cull panels made from PHA (Mirel™ P1004 by Metabolix, Inc.) panels that were actively fished reached the 20% weight loss threshold at about 330 days, while PHA panels that were continuously soaked and not regularly fished reached the 20% weight loss threshold at about 90 days, which is 3.5 times faster than actively fished (Havens et al. 2014). Additionally, another patent by Shelton and Pocher (2005) with publication number WO 2005023955 A2 disclosed a modified biodegradable polymer net, comprising a PLA resin with use of additives and other biodegradable resins, include PHA and polypropylene glycol (PPG), as a plasticizer to minimise brittleness of the net (Shelton and Pocher 2005). Other commercial PHA products, namely Mirel™ by Metabolix, Inc., Sogreen™ by Tianjin GreenBio Materials and Nodax™ by Danimer Scientific, offer custom-formulated PHA blends that exhibit the required characteristics, as the physical properties and biodegradability of the PHA are regulatable by blending with synthetic or natural polymers (Lu et al. 2014).

13.4 Agricultural Grow Bags

Bags are major plastic products used in agriculture. The common types of agricultural bags are fertiliser bags and grow bags. One of the most use of polymer grow bags in agriculture is for raising seedlings in nurseries prior to transferring young crop to open fields (Rajdeep and Naithani 2013). Grow bags, also known as seedling bags or planter bags, stabilise the temperature in the soil, retain the moisture in the soil, require less frequent watering of crop, and allow specific use of growing media. Plastic grow bags also minimise the impact of external factors, such as flood, drought, pest, weed and pH, on seedling soil. They are used to isolate plants individually to reduce competition and root disturbance, thus improving the survival rate of crop. Moreover, polymer grow bags save space and are light, hence efficient in the transportation and repositioning of young crop (Donald 1968).

Accordingly, PE bags became the standard planting system for healthy and rapid plant growth for half a century (Donald 1968). Angaji and Hagheeghatpadjooh (2004) also reported that LDPE is the most common PE material applied in agricultural, horticulture and packaging purposes. In a study by Bilck et al. (2014), the traditional use of PE bags in seedling production has justified the use of LDPE bags from Agro-Plast (Brazil) as controls in their study. However, LDPE has been proven to be resistant against degradation and microorganism attacks (Angaji and Hagheeghatpadjooh 2004). The high durability of LDPE grow bags poses detrimental impact on the environment, seedlings and agricultural sector. In consequence, there is a need to produce agricultural bags made of alternative biodegradable material.

13.4.1 Application of PHA for Grow Bags

There are many advantages of using PHA grow bags in agriculture. Firstly, as opposed to non-biodegradable plastic, which releases harmful organic matter and toxic emission when recycled or incinerated (Kolybaba 2003), PHA does not leave any visually distinguishable or toxic residue. PHA undergoes biological degradation during composting at a relatively shorter rate than conventional plastic (ASTM 1998). Therefore, PHA grow bags do not contaminate agricultural fields with harmful toxin.

Next, PHA is a source of reducing power and microbial growth matrix for water denitrification, due to its property as an insoluble and solid substrate in water. According to Hiraishi and Khan (2003), there are several reports where PHA is proven to positively remove nitrogen from water. Hence, the use of PHA grow bags do not contaminate the surrounding water bodies; they denitrify nearby streams, rivers and lakes of agriculture fields. Contrastingly, non-biodegradable PE bags are dumped and indirectly introduced into waterways, thus blocking sewerage systems, causing floods, and providing breeding areas for unwanted vectors, such as mosquitoes (Sanghi 2008; Adane and Muleta 2011). The unmonitored disposal of PE bags causes death of domestic and wild animals after ingestion of PE, as well as deterioration of environment aesthetics.

Furthermore, compostable PHA containers breakdown after direct transplanting of crop into the soil without removing the container. Transplanting and burying discarded PHA products supplement the nutrient cycle in the soil (Huang et al. 1990). On the other hand, PE bags are non-reusable when they are punctured by plant roots and heavily tangled with soil matter, despite its high durability and resistance to UV radiation. In order to prevent the unwanted PE bags from being blown around by wind, the PE bags are usually buried in soil, which leads to reduced water percolation and lack of proper aeration in the soil of agricultural fields (Adane and Muleta 2011).

Moreover, biodegradable grow bags are root-friendly compared to non-biodegradable bags. Bilck and coworkers (2014) reported that no root deformity was observed on plants grown in biodegradable grow bags due to the biodegradability of the bags, whereas root deformation was observed on plants grown in traditional PE bags due to lack of space. Root deformation refers to the spiralling, ensnarling, twisting, kinking and egression of plant roots (Jones 1993; Aldrete et al. 2002; Cedamon et al. 2005). These subsequently affect the crops' growth, vigour, stress resistance, pathogenic immunity, and ability of the plant to anchor quickly into the ground after transplanting (Muriuki et al. 2007). Also, the affected crops are prone to suffer from wind-throw and early dieback (Cedamon et al. 2005). Besides that, while crops that are transplanted in compostable bags are gradually exposed to field soil, crops that are grown in non-biodegradable grow bags risk incidental root damage (Bilck et al. 2014) and transplanting shock, which is due to the sudden drastic change in surrounding soil parameters after bag removal (Muriuki et al. 2014).

Additionally, compostable biodegradable grow bags eliminate double handling and recycling of bags after use (Rajdeep and Naithani 2013). In contrast, the post-use management of PE bags require high labour cost since the heavy contamination of organic material on the bags makes recycling difficult. Due to the lack of an economical approach to properly remove unwanted PE bags, they are usually buried in the soil or burned (Bilck et al. 2014).

The drawbacks of non-biodegradable bags prompted governments to ban or impose taxes on plastic use (UNEP 2005; Adane and Muleta 2011). After plastic used for packaging was banned, the Kenya Forest Service (KFS) mentioned the need to find environmentally-friendly alternatives for tree and crop propagation (Muchangi 2017). The KFS annually produces 150 million tree seedlings, with 5000 private tree nurseries annually producing 115 million seedlings (Muchangi 2017). Other countries that banned the use of plastic or plastic bags are Britain, France (Hodges 2018), India (Johnston 2017), China (Telegraph Reporters 2018) and Taiwan (Liuxin 2017). These measures will make the materials either costlier, if taxed, or unavailable if banned. Application of biopolymer growbags are crucial alternative to agricultural plantations that are actively using traditional PE growbags.

The most common material used for agricultural grow bags is LDPE. The typical tensile strength of LDPE was 7–10 MPa, with a range of elongation at break from 120% to 400%, and Young's modulus or elastic modulus from 0.1 to 0.25 GPa (Crompton 2012; Kormin et al. 2017). Furthermore, although having relatively higher elongation at break and lower Young's modulus than the conventional LDPE grow bags, polybutylene adipate terephthalate (PBAT) grow bags had produced biodegradable, functional and root-friendly grow bags, which were featuring a range of tensile strength from 12 to 13 MPa, elongation at break from 600% to 650%, and a Young's modulus of about 0.06 GPa (Bilck et al. 2014). On the whole, the types of PHA comprising the tensile strength, elongation at break and Young's modulus of 7–13 MPa, 120–600%, and 0.06–0.25 GPa are potential PHA types for agricultural grow bags, whereby the attributes of mcl-PHA is the most similar to the stated properties.

The properties of similar polymeric material observably differed among studies, since each polymer type is known to exhibit properties as ranged values instead of singular and specific values. A publication of the standard ranges of polymer properties is needed. Subsequently, approximate comparison shows that the properties of short-chain-length PHA (scl-PHA) are similar to polypropylene (PP), whereas those of medium-chain-length PHA (mcl-PHA) are similar to LDPE. The scl-PHA consists of 3–5 carbons, meanwhile, mcl-PHA consists of 6–14 carbons (Koller et al. 2013). Theoretically, the production of PHA grow bags is possible because the properties of PHA and its polymeric peers correspond to each other (Table 13.2). In other words, PHA is a suitable additive or building material for biopolymeric products.

A few commercially available PHA-based agricultural grow bags are marketed by Danimer Scientific, Tianjin GreenBio Materials and Greenpoly Co Ltd., along with potentially marketable material that is PHA-distiller's dried grains with solubles (PHA-DDGS) (Greenpoly Co Ltd 2005; Lu et al. 2014; Danimer Scientific 2018a; Tianjin GreenBio Materials Co Ltd 2018). The PHA-DDGS uses resins of PHA combined with fillers of soy protein or lignin to be evaluated for their functionality as biodegradable containers of plants and crops (Lu et al. 2014). Based on the mechanical testing and application trials, they concluded that all materials in their study fulfilled the functional requirements as seedling pots. The material was sufficiently durable for at least 4 months, but biodegrades readily when broken to smaller pieces and installed in the soil. Furthermore, the material PHA-DDGS (80/20) processed and functioned smoothly after initial processing issues were overcome (Lu et al. 2014; Schrader et al. 2016). Due to the difficulty in injection

Table 13.2 General mechanical properties of scl-PHA, mcl-PHA, PP, PE, LDPE and HDPE

Properties	scl-PHA	mcl-PHA	PP	PE	LDPE	HDPE
Tensile strength (MPa)	40	9	26	18	10	32
Elongation at break (%)	6	276	80	350	400	150
Young's modulus (GPa)	3.5	1.4	2	0.5	0.25	1.25

Ojumu et al. (2004), Rai et al. (2011), Crompton (2012), and Panchal et al. (2013 moulding caused by the highly viscous PHA, they made adjustments and processed the PHA blends by setting higher temperatures at the front barrel zones with decreasing temperatures along the injection barrel. This solution improved the processing of the PHA-based composites into seedling pots. However, the viscosity of bioplastic blends is varied, hence each material need to be adjusted or modified individually. Lu et al. (2014) concluded that with management of moisture content, thermal processing profile, temperature, force and rate, all biopolymeric blends are suitable for standard plastic processing and automated robotic handling.

On the other hand, Danimer Scientific is a biopolymer manufacturer in Georgia that was formerly known as Meredian Holdings Group. Danimer Scientific's film resins are non-photodegradable, environmentally friendly alternatives to PE and PP (Danimer Scientific 2018a). With excellent printability, they are also time and cost saving in agriculture and manufacturing as the material is compostable. Their PHA film resins are suitable for manufacturing take-away bags, waste bags and compost bags, agricultural mulch film and water barriers requiring breathable film. Moreover, their custom-formulated hybridised PHA creates biopolymers that meet specific strength, durability and biodegradability requirements (Danimer Scientific 2018a). Danimer Scientific PHA film resins are certified compostable to ASTM D6400-99 or ASTM D6868-03 standards, and have been awarded the 'OK Marine Biodegradable' certification from Vinçotte International, as well as the United States of America Food and Drug Administration Federal Agency's approval of Nodax™ PHA for food contact and as non-hazardous waste after disposal (Danimer Scientific 2018b). Another commercial PHA product is Mirel™ by Metabolix, Inc. As a good substitute for a broad range of petroplastics, Mirel™ is a white granulated P(3HB) plastic that can also be processed into agricultural mulch films, compostable bags, cups, food and cosmetic packaging (Andrews 2014). Accordingly, the major PHA manufacturers notably produce PHA resins for mulch films and packaging, and thus likely require custom-formulated blends to produce PHA grow bags.

The Tianjin GreenBio Material's PHA-based Sogreen™ series was manufactured with different properties for various applications. Sogreen-00X fully degrades to carbon dioxide and water in soil, water bodies and sewage within 3–6 months (Tianjin GreenBio Materials Co Ltd 2018). Alternatively, Sogreen 2013 has high mechanical property, excellent mould-ability and broad applications. Sogreen 2013 is designed for blowing and casting to produce boards, wrapping, packaging, films and other products (Tianjin GreenBio Materials Co Ltd 2018). Sogreen 2013 could be a potential candidate for agricultural bags.

Besides that, PHA is also used in the production of natural vegetable fibre pots by Greenpoly Co Ltd, which manufactures and exports the fibre pots. According to Greenpoly Co Ltd, the fibre pots are made of natural vegetable fibre and biodegradable resins include PHA, P(3HB-*co*-3HV), PLA and polycaprolactam (Greenpoly Co Ltd 2005).

In conclusion, PHA agricultural grow bags can be customised using Danimer Scientific and Metabolix, Inc.'s PHA resins or blends. However, there is a lack in commercial and readily available biodegradable grow bags. To date, the study by

Bilck and coworkers (2014) is the only work related to the functionality and advantages of biodegradable polymer grow bags over conventional LDPE grow bags. However, the biopolymer material studied by Bilck et al. (2014) was PBAT. Consequently, the current biodegradable grow bags and pots commercially available are commonly made of plant fibre fabric (NYP Corp 2018) as the awareness or demand for biodegradable biopolymer grow bags are still lacking.

13.5 Agro-industrial By-Products or Waste as Potential Feedstock for PHA Production

The microorganism selection to produce PHA for industrial purposes varies according to several factors, such as the cell's capability to use inexpensive carbon sources. Attention was shifted to industrial by-product and agricultural waste recently. The cost of carbon sources influences the price of PHA (Chanprateep 2010). The selection of carbon sources should also focus on the availability and global price consistency instead of only on market prices.

The most basic carbon source, glucose, was used to study PHA production in many bacteria, such as *Pseudomonas sp., Cupriavidus necator* (previously known as *Alcaligenes eutrophus*) (Doi et al. 1995) and also *Bacillus sp.* (Valappil et al., 2007). The use of this carbon source in culturing *Cupriavidus necator* produced up to 76% wt of PHA in bacterial cells (Tan et al. 2014). Starches are also an abundant carbon source. Starch has also been used as a bio-based polymer for various types of bioplastics (Chanprateep 2010). It was found that the isolated *Bacillus cereus* could secrete the enzyme amylase and simultaneously produce P(3HB) (Halami 2008). *Bacillus subtilis* and *Bacillus thuringiensis* were also able to produce approximately 60–65 wt% of P(3HB) from starch and industrial food waste (Gowda and Shivakumar 2014).

Inexpensive carbon sources, namely industrial by-product and agricultural or industrial waste, could be used for PHA production by some bacteria (Koller et al. 2010). Example of waste product used for PHA production are waste cooking oil, sweetwater (Wadekar et al. 2012) and glycerol (Palmeri et al. 2012). Meanwhile molasses an industrial by-product of sugar production is among potentially inexpensive carbon source for PHA production. *Bacillus megaterium* was able to produce P(3HB) from cane molasses in shaken flask cultivation. A total of 46.2% of P(3HB) was produced when 3% (w/v) of sugar cane molasses was supplied (Gouda et al. 2001). An attempt to produce P(3HB) from sugar cane molasses by *Bacillus sp.* was reported by Kulpreecha et al. (2009). A higher total sugar concentration of 400 g/L and a C/N molar ratio of 10 were required for the optimal feeding medium in this system. According to Yatim and coworkers (2017), *Bacillus megaterium* UMTKB-1 was proven to produce 15 wt% of P(3HB) using a sole carbon source that is cane molasses. This strain was also able to accumulate 9 wt% of P(3HB) production using sweetwater and this is the first report of the usage of sweetwater as a carbon source for PHA production (Yatim et al. 2017). Since sweetwater is considered to be a potential carbon feedstock, it could be used for biomaterial production (Wadekar et al. 2012; Azemi et al. 2016). Apart from that, studies have been done on potential application of fatty acid and vegetable oil as substrates for PHA production where a theoretical yield of 0.65 wt% of PHA production from fatty acid was reported (Yamane 1993).

Lignocellulosic biomass is produced from agricultural practices, it has to be pre-treated and undergo hydrolysis to convert into sugar monomer (Aslan et al. 2016). Cellulose and hemicellulose from lignocellulosic materials are both excellent carbon sources after hydrolysis to monomeric sugars which can be used in PHA production as carbon sources. A variety of lignocellulosic materials have been investigated for PHA production. According to a study by Nielsen and coworkers (2017), glucose, xylose and arabinose was produced by enzymatic hydrolysis of the cellulose and hemicellulose fractions of ground wheat straw using an ammonia fibre expansion (AFEX) process as the pre-treatment. *Burkholderia sacchari* DSM 17165 undergoes fed-batch fermentation using the hydrolysate as the carbon source obtaining PHA of concentration 72%. Even though lignocellulosic materials require extensive pre-treatment but they have potential as carbon substrate.

Conversion of low valued industrial waste and by-products into value-added material products such as PHA is an important step in reduction of production cost. Furthermore, PHA production using waste stream can reduce the environmental impacts. Although the current price of PHA is an ongoing impediment to its widespread use, the additional cost provides a completely biodegradable products that leaves zero hazardous waste in the environment.

13.6 Future Outlook

This review would have clearly highlighted the potential use of PHA or PHA based polymers in selected agriculture applications. The idea of using biodegradable material to ensure eco-friendly approach in agriculture is well supported by facts shown in previous research. With issues pertaining sustainability and food security, PHA could be developed to address these issues and to ensure global awareness towards environmental issues. The ongoing studies in PHA may result in new findings which could generate new ideas for PHA application in agriculture, a key driver of global economy.

References

Adane L, Muleta D (2011) Survey on the usage of plastic bags, their disposal and adverse effects on environment: a case study in Jimma City, Southwestern Ethiopia. J Toxicol Environ Health Sci 3:234–248 Retrieved from http://www.academicjournals.org/article/article1379595494_Adane%20and%20Muleta%20(1).pdf

Aldrete A, Mexal JG, Phillips R, Nallotton AD (2002) Copper coated polybags improves seedling morphology for two nursery-grown Mexican pine species. Forest Ecol Manag 163:197–204. https://doi.org/10.1016/S0378-1127(01)00579-5

Alvarez VM, Baille A, Martinez JMM, Gonzalez-Real MM (2006) Effect of black polyethylene shade covers on the evaporation rate of agricultural reservoirs. Span J Agric Res 4:280–288. https://doi.org/10.5424/sjar/2006044-205

Andrews M (2014) Mirel™ PHA polymeric modifiers & additives. In: Proceeding from AddCom 2014 conference, Novotel Barcelona, Spain, October 21–22. Retrieved from https://www.slideshare.net/MetabolixInc/metabolix-mirel-pha-polymeric-modifiers-and-additives

Angaji MT, Hagheeghatpadjooh HR (2004) Preparation of biodegradable low density polyethylene by starch urea composition for agricultural applications. Iran J Chem Chem Eng 23:7–11 http://ijcce.ac.ir/article_8148_37e362c7730cd1da4d714c9ffc495a1f.pdf

Aslan A, Ali M, Morad N, Tamunaidu P (2016) Polyhydroxyalkanoates production from waste biomass. IOP Conf Ser Earth Environ Sci 36:012040. https://doi.org/10.1088/1755-1315/36/1/012040

ASTM (The American Society for Testing & Materials) (1998) D883-96: standard terminology relating to plastics. ASTM, New York Retrieved from https://www.astm.org/Standards/D883.htm

Azemi M, Rashid N, Saidin J, Effendy A, Bhubalan K (2016) Application of sweetwater as potential carbon source for rhamnolipid production by marine *Pseudomonas aeruginosa* UMTKB-5. Int J Biosci Biochem Bioinforma 6:50–58. https://doi.org/10.17706/ijbbb.2016.6.2.50-58

Barnes DK, Galgani F, Thompson RC, Barlaz M (2009) Accumulation and fragmentation of plastic debris in global environments. Philos Trans R Soc Lond Ser B Biol Sci 364:1985–1998. https://doi.org/10.1098/rstb.2008.0205

BASF SE (2013) U.S. Patent No. US2013029124 A1. U.S. Patent and Trademark Office, Washington, DC

Bilck AP, Olivato JB, Yamashita F, de Souza JRP (2014) Biodegradable bags for the production of plant seedlings. Polimeros 24:547–553. https://doi.org/10.1590/0104-1428.1589

Brown RP (2004) Polymers in agriculture and horticulture. Rapra Technology, Shawbury Retrieved from https://www.smithersrapra.com/publications/books/browse-by-category/review-reports/polymers-in-agriculture-and-horticulture

Castellano S, Mugnozza GS, Russo G, Briassoulis D, Mistriotis A, Hemming S, Waaijenberg D (2008) J Agric Eng 39:31–42 Retrieved from http://www.j.agroengineering.org/index.php/jae/article/download/jae.2008.3.31/pdf

Cedamon ED, Mangaoang EO, Gregorio NO, Pasa AE, Herbohn JL (2005) Nursery management in relation to root deformation, sowing and shading. Ann Trop Res 27:1–10 Retrieved from https://espace.library.uq.edu.au/view/uq:8612/n28CEDAMO_Nurser.pdf

Chanprateep S (2010) Current trends in biodegradable polyhydroxyalkanoates. J Biosci Bioeng 110:621–632. https://doi.org/10.1016/j.jbiosc.2010.07.014

Crompton TR (2012) Mechanical properties of polymers. Physical testing of plastics. Smithers Rapra, Shropshire, pp 1–148

Danimer Scientific (2018a) Danimer scientific film resins. Retrieved from https://danimerscientific.com/compostable-solutions/a-family-of-biopolymers/film-resins/

Danimer Scientific (2018b) Danimer scientific history. Retrieved from https://danimerscientific.com/about-us/history/

Doi Y, Kitamaru S, Abe H (1995) Microbial synthesis and characterization of poly(3-hydroxybutrate-co-3-hydroxyhexanoates). Macromolecules 28:4822–4828. https://doi.org/10.1021/ma00118a007

Donald DGM (1968) Planting of trees in polythene bags. S Afr For J 67:28–29. https://doi.org/10.1080/00382167.1968.9629198

Gouda M, Swellam A, Omar S (2001) Production of PHB by a *Bacillus megaterium* strain using sugarcane molasses and corn steep liquor as sole carbon and nitrogen sources. Microbiol Res 156:201–207. https://doi.org/10.1078/0944-5013-00104

Gowda V, Shivakumar S (2014) Agrowaste-based polyhydroxyalkanoate (PHA) production using hydrolytic potential of *Bacillus thuringiensis* IAM 12077. Braz Arch Biol Technol 57:55–61. https://doi.org/10.1590/s1516-89132014000100009

Greenpoly Co Ltd (2005) Biodegradable resin. Retrieved from http://www.greenpoly.com/biodegradable_resin.htm

Guerrini S, Borreani G, Voojis H (2017) Biodegradable materials in agriculture: case histories and perspectives. In: Malinconico M (ed) Soil degradable bioplastics for a sustainable modern agriculture. Springer, Germany. https://doi.org/10.1007/978-3-662-54130-2

Halami P (2008) Production of polyhydroxyalkanoate from starch by the native isolate *Bacillus cereus* CFR06. World J Microbiol Biotechnol 24:805–812. https://doi.org/10.1007/s11274-007-9543-z

Hassan MK, Abou-Hussein R, Zhang X, Mark JE, Noda I (2006) Biodegradable copolymers of 3-hydroxybutyrate-co-3-hydroxyhexanoate (NodaxTM), including recent improvements in their mechanical properties. Mol Cryst Liq Cryst 447:23–341. https://doi.org/10.1080/15421400500380028

Havens KJ, Bilkovic DM, Stanhope DM, Angstadt KT (2014) U.S. Patent No. US 20140245655 A1. U.S. Patent and Trademark Office, Washington, DC

Hiraishi A, Khan ST (2003) Application of polyhydroxyalkanoates for denitrification in water and wastewater treatment. Appl Microbiol Biotechnol 61:103–109. https://doi.org/10.1007/s00253-002-1198-y

Hodges H (2018) How rest of world still puts us to shame: France was the first country to ban plastic plates, cutlery and cups. Daily Mail. Retrieved from http://www.dailymail.co.uk/news/article-5261141/How-rest-world-puts-shame-plastic.html

Huang JC, Shetty AS, Wang MS (1990) Biodegradable plastics: a review. Adv Polym Technol 10:23–30. https://doi.org/10.1002/adv.1990.060100103

Iwata T, Aoyagi Y, Fujita M, Yamane H, Doi Y, Suzuki Y, Takeuchi A, Uesugi K (2004) Processing of a strong biodegradable poly[(R)-3-hydroxybutyrate] fiber and a new fiber structure revealed by micro-beam X-ray diffraction with synchrotron radiation. Macromol Rapid Commun 25:1100–1104. https://doi.org/10.1002/marc.200400110

Johnston I (2017) India just banned all forms of disposable plastic in its capital. The Independent. Retrieved from http://www.independent.co.uk/news/world/asia/india-delhi-bans-disposable-plastic-single-use-a7545541.html

Jones N (1993) Essentials of good planting stock, Forests and forestry technical bulletin. World Bank/AGRNR, Washington, DC Retrieved from http://agroforestry.org/2014-03-04-10-06-24/essentials-of-good-planting-stock

Kasirajan S, Ngouajio M (2012) Polyethylene and biodegradable mulches for agricultural applications: a review. Agron Sustain Dev 32:501–529. https://doi.org/10.1007/s13593-011-0068-3

Khalaf MN (2015) Mechanical properties of filled high density polyethylene. J Saudi Chem Soc 19:88–91. https://doi.org/10.1016/j.jscs.2011.12.024

Koller M, Atlić A, Dias M, Reiterer A, Braunegg G (2010) Microbial PHA production from waste raw materials. In: Chen GQ (ed) Plastics from bacteria: natural functions and applications, vol 14. Springer, Berlin/Heidelberg, pp 85–119. https://doi.org/10.1007/978-3-642-03287-5_5

Koller M, Salerno A, Braunegg G (2013) Polyhydroxyalkanoates: basics, production and applications of microbial biopolyesters. In: Kabasci S (ed) BioBased plastics: materials and applications. Wiley, West Sussex, pp 137–170. https://doi.org/10.1002/9781118676646.ch7

Kolybaba M, Tabil LG, Panigrahi S, Crerar WJ, Powell T, Wang B (2003) Biodegradable polymers: past, present, and future, Proceedings from ASAE meeting (Oct 3–4) 12:15–24. Elsevier, Atlanta Retrieved from http://www.biodeg.net/fichiers/Biodegradable%20Polymers%20Past,%20Present,%20and%20Future%20(Eng).pdf

Kormin S, Kormin F, Beg MDH, Piah MBM (2017) Physical and mechanical properties of LDPE incorporated with different starch sources. IOP Conf Ser Mater 226:012157. https://doi.org/10.1088/1757-899X/226/1/012157

Kulpreecha S, Boonruangthavorn A, Meksiriporn B, Thongchul N (2009) Inexpensive fed-batch cultivation for high poly(3-hydroxybutyrate) production by a new isolate of *Bacillus megaterium*. J Biosci Bioeng 107:240–245. https://doi.org/10.1016/j.jbiosc.2008.10.006

Kusuktham B, Teeranachaideekul P (2014) Mechanical properties of high density polyethylene/modified calcium silicate composites. Silicon 6:179–189. https://doi.org/10.1007/s12633-014-9204-4

Liuxin (2017) Taiwan to expand ban on free use of plastic bags in 2018. Xin Hua News Agency. Retrieved from http://www.xinhuanet.com/english/2017-12/27/c_136855626.htm

Lu H, Madbouly SA, Schrader JA, Kessler MR, Grewell D, Graves WR (2014) Novel bio-based composites of polyhydroxyalkanoate (PHA)/distillers dried grains with solubles (DDGS). RSC Adv. https://doi.org/10.1039/C4RA04455J

Mirel Bioplastics (2018) Mirelplastics.com. Retrieved from http://www.mirelplastics.com

Muchangi J (2017) 4 countries seek to emulate Kenya's ban on plastic bags. The Star. Retrieved from https://www.the-star.co.ke/news/2017/12/11/4-countries-seek-to-emulate-kenyas-ban-on-plastic-bags_c1682570

Muriuki JK, Muia B, Munyi A (2007) New methods improve quality of tree seedlings. APA News Asia-Pacific Agrofor Newsl 31:3–5 Retrieved from http://www.worldagroforestry.org/downloads/publications/PDFs/NL07440.PDF

Muriuki JK, Kuria AW, Muthuri CW, Mukuralinda A, Simons AJ, Jamnadass RH (2014) Testing biodegradable seedling containers as an alternative for polythene tubes in tropical small-scale tree nurseries. Small Scale For 13:127–142. https://doi.org/10.1007/s11842-013-

Niaounakis M (2015) Biopolymers: applications and trends. Elsevier, Waltham. https://doi.org/10.1016/B978-0-323-35399-1.00004-1

Nielsen C, Rahman A, Rehman A, Walsh M, Miller C (2017) Food waste conversion to microbial polyhydroxyalkanoates. Microbiol Biotechnol 10:1338–1352. https://doi.org/10.1111/1751-7915.12776

NYP Corp (2018) NYP corporation – national burlap manufacturer. Retrieved from https://nyp-corp.com/our-company/index.html

Ojanji W (2017) Plastic ban to hit seedlings sector as firms adopt new technology. The Daily Nation. Retrieved from https://www.nation.co.ke/business/seedsofgold/Plastic-ban-to-hit-seedlings-sector/2301238-3973850-qvm0ipz/index.html

Ojumu T, Yu J, Solomon B (2004) Production of polyhydroxyalkanoates, a bacterial biodegradable polymer. Afr J Biotechnol 43:18–24 Retrieved from https://tspace.library.utoronto.ca/bitstream/1807/3487/1/jb04003.pdf

Palmeri R, Pappalardo F, Fragala M, Tomasello M, Damigelle A, Catara AF (2012) Polyhydroxyalkanoates (PHAs) production through conversion of glycerol by selected strains of *Pseudomonas mediterranea* and *Pseudomonas corrugate*. Chem Eng Trans 27:121–126 Retrieved from https://pdfs.semanticscholar.org/d6e9/d460ea3faab74d3a131736a6dd00a506ed89.pdf

Panchal B, Bagdadi A, Roy I (2013) Polyhydroxyalkanoates: the natural polymers produced by bacterial fermentation. In: Thomas S, Visakh PM, Mathew AP (eds) Advances in natural polymers: composites and nanocomposites. Springer, Heidelberg/Berlin, pp 397–421. https://doi.org/10.1007/978-3-642-20940-6_12

Rai R, Yunos D, Boccaccini A, Knowles J, Barker I, Howdle S, Tredwell G, Keshavarz T, Roy I (2011) Poly-3-hydroxyoctanoate P(3HO), a medium chain length polyhydroxyalkanoate homopolymer from *Pseudomonas mendocina*. Biomacromolecules 12:2126–2136. https://doi.org/10.1021/bm2001999

Rajdeep SP, Naithani S (2013) Scope of natural and synthetic biodegradable materials in development of eco-friendly substitutes for nursery polybags – a review. eJAFE 1:35–44 Retrieved from https://www.researchgate.net/publication/263751785_Scope_of_natural_and_synthetic_biodegradable_materials_in_development_of_substitutes_for_nursery_polybags-A_review

Rydz J, Sikorska W, Kyulavska M, Christova D (2014) Polyester-based (bio) degradable polymers as environmentally friendly materials for sustainable development. Int J Mol Sci 16:564–596. https://doi.org/10.3390/ijms16010564

Saito Y, Doi Y (1994) Microbial synthesis and properties of poly(3-hydroxybutyrate-co-4-hydroxybutyrate) in *Comamonas acidovorans*. Int J Biol Macromol 16:99–104. https://doi.org/10.1016/0141-8130(94)90022-1

Sanghi S (2008) Use of plastic bags: factors affecting ecologically oriented behaviour in consumers. Found Organ Res Educ 26:1–45 Retrieved from http://www.highbeam.com/doc/1G1-192438179.html

Schrader JA, Behrens JJ, Michel M, Grewell D (2016) Bioplastics and bio composites for horticulture containers: processing, properties, and manufacturing potential. In: Schrader JA, Kratsch HA, Graves WR (eds) Bioplastic container cropping systems: green technology for the green industry. Sustainable Hort Res Consortium, Ames Retrieved https://www.researchgate.net/profile/James_Schrader/publication/311983321_Degradability_of_Bioplastic_Containers_in_Soil_and_Compost/links/5866a3b508ae8fce490f1ed6/Degradability-of-Bioplastic-Containers-in-Soil-and-Compost.pdf

Shelton WS, Pocher JP (2005) WIPO Patent No. WO 2005023955 A2. World Intellectual Property Organization Patent Office, Geneva

Sintim HY, Flury M (2017) Is biodegradable plastic mulch the solution to agriculture's plastic problem? Environ Sci Technol 51:1068–1069. https://doi.org/10.1021/acs.est.6b06042

SK Chemicals Co Ltd (2004) Korean Patent No. KR20040071992A. Korean Intellectual Property Office, Daejeon

Tan GYA, Chen CL, Li L, Ge L, Wang L, Razaas IMN (2014) Start a research on biopolymer polyhydroxyalkanoate (PHA): a review. Polymers 6:706–754. https://doi.org/10.3390/polym6030706

Telegraph Reporters (2018) Plastic waste 'already building up in UK' following China's ban. The Telegraph. Retrieved from http://www.telegraph.co.uk/news/2018/01/02/plastic-waste-already-building-uk-following-chinas-ban9245-3

The Procter & Gamble Company (2001) WIPO Patent No. WO2001093678A2. World Intellectual Property Organization, Geneva

Tianjin GreenBio Materials Co Ltd (2018) Green Bio. Performance index. Retrieved from http://www.tjgreenbio.com/en/Product.aspx?cid=38&title=Products

UNEP (United Nations Environmental Programme) (2005) Plastic bag ban in Kenya proposed as part of new waste strategy. UNEP Press release. Retrieved from http://www.unep.org/documents.multilingual/default.asp?documentid=424&articleid=4734&l=en

Valappil SP, Peiris D, Langley GJ, Herniman JM, Boccaccini AR, Bucke C, Roy I (2007) Polyhydroxylkanoate (PHA) biosynthesis from structurally unrelates carbon sources by a newly characterized *Bacillus* spp. J Biotechnol 127:475–487. https://doi.org/10.1016/j.jbiotec.2006.07.015

Wadekar SD, Kale SB, Lali AM, Bhowmick DN, Pratap AP (2012) Ultilization of sweetwater as a cost-effective carbon source for sophorolipids production by *Sramerella bombicola* (ATCC 22214). Prep Biochem Biotechnol 42:125–142. https://doi.org/10.1080/10826068.2011.577883

Williams SF, Rizk S, Martin DP (2013) Poly-4-hydroxybutyrate (P4HB): a new generation of resorbable medical devices for tissue repair and regeneration. Biomed Tech (Berl) 58:439–452. https://doi.org/10.1515/bmt-2013-0009

Yamane T (1993) Yield of poly-D(-)-3-hydroxybutyrate from various carbon sources: a theoretical study. Biotechnol Bioeng 41:165–170. https://doi.org/10.1002/bit.260410122

Yatim A, Syafiq I, Huong K, Amirul AA, Effendy A, Bhubalan K (2017) Bioconversion of novel and renewable agro-industry by-products into a biodegradable poly(3-hydroxybutyrate) by marine *Bacillus megaterium* UMTKB-1 strain. Biotechnologia 2:141–151. https://doi.org/10.5114/bta.2017.68313

Yu R (2011) China Patent No. CN102140185A. China Trademark Office, Beijing

Polyhydroxyalkanoates in Packaging 14

Neetu Israni and Srividya Shivakumar

Abstract

Synthetic plastics have been very well recognized for the packaging industry. An evident inclination towards development of biodegradable Bioplastics based on biological material and, particularly for bulk packaging applications has been perceived as a strategy to overcome the dependency of packaging sector on fossil fuels and diminish the biospheric plastic waste burden. The packaging industry demands the raw material in quantity and quality both. Quality includes properties such as ample tensile strength, less brittleness, adequate gas, aroma, ultraviolet (UV) and water barrier properties as per the product to be packaged. The inherent hydrophobicity, biodegradability and enormous property range has branded the microbial poly (hydroxyalkanoates) (PHAs) as promising competitors of petro-plastics in the packaging market. This chapter addresses the basic characteristics and alluring advantages of PHAs as packaging materials. The major focus has been on discussing recent developments of PHAs in different material forms of specific relevance to packaging sector.

Keywords

Polyhydroxyalkanoates · Packaging · Permeability · Nanocomposites · Multilayer films · Paper coating · Antimicrobial PHAs

N. Israni · S. Shivakumar (✉)
Department of Microbiology, School of Sciences (SoS) Centre for PG Studies, Jain University, Bangalore, Karnataka, India
e-mail: sk.srividya@jainuniversity.ac.in

V. C. Kalia (ed.), *Biotechnological Applications of Polyhydroxyalkanoates*,
https://doi.org/10.1007/978-981-13-3759-8_14

14.1 Introduction

14.1.1 Plastics: "The Brighter Side"

The term "plastic" existed as early as the 1600s. With the actual debut of synthetic plastics in the 1930s, global dependence on these materials has increased considerably over the years. Global production of plastics has gone up to 380 Mt (Plastics Europe 2016; Geyer et al. 2017). The synthetic or petroleum based plastic industry has rapidly developed catering to the myriad applications covering almost every sphere of human life, making them an indispensable part of our day to day life. Their widespread applications can be attributed to favorable mechanical, thermal and chemical properties, such as high strength, lightness, rigidity or flexibility, stability, durability, being molded into any shape, transparency or opacity, water resistance, resistance to majority of water-borne microorganisms and most importantly cost effectiveness (usually lower than 1€ kg^{-1}) (Valentino et al. 2014; Colombo et al. 2016). It has been over a decade since the conventional materials *viz.* paper, jute, wood, metals, glass and ceramics have been replaced by the synthetic plastics in commercial sectors such as infrastructure, packaging, textile fibers, automotive, agriculture, electronics and medical appliances (North and Halden 2013). Their cost-effectiveness makes them an ideal material for small duration and single-use purposes *viz.* disposable medical devices and packaging. Packaging industry, the major player of the current era draws more than 40% of the global plastic production, of which almost 50% is employed as a food packaging material (Siddiqui and Pandey 2013; Rhim et al. 2013; Khosravi-Darani and Bucci 2015).

14.1.2 Plastics: "The Darker Side"

Of all the plastic products, packaging and carry bags, polyethylene terephthalate (PET) bottles and disposables are the main ubiquitous consumer item and sources of plastic waste on our planet. Due to their rapid increase in production and consumption, improper disposal, intrinsic non-biodegradability or very sluggish degradability, these commercial polymers have long been under environmental scrutiny (Pavani and Rajeswari 2014). As per the study of American Association for Advancement of Science (AAAS), by 2050 approximately 12,000 MMT of plastic waste will be discarded in natural environment. According to the Indian scenario, as estimated by Central Pollution Control Board (CPCB) approximately 5.6 MTPA of plastic waste is generated in our country amounting to 15,342 TPD (tonnes per day), of which approximately over 6137 tonnes of plastic waste remains uncollected and littered, either being land-filled or ending up in streams leading to land, ground water and marine water pollution (Central Pollution Control Board 2013).

Apart from the environmental pollution and waste management problems, an additional growing concern is utilization of limiting fossil feed stocks (petroleum and its allied components) for the production of these recalcitrant synthetic plastics. As a more promising approach, "the biobased – BIOPLASTICS" have emerged as apt sustainable alternatives worldwide (Anjum et al. 2016). The term Bioplastics is a broader term used for a variety of plastics. All biobased plastics are not biodegradable and all biodegradable plastics are not biobased (made from renewable resources). Amongst the diverse bioplastics, owing to their superior performance and consumer liking for sustainability, currently the market is visibly dominated by biobased and non-biodegradable plastics like biobased poly (ethylene), biobased poly (trimethylene terephthalate), biobased poly (ethylene) terephthalate, etc. (Gironi and Piemonte 2011; Chen 2014). However, development of biobased and biodegradable plastics such as polylactic acid (PLA), thermoplastic starch (TPS), cellulose based plastics and plastics from microbial fermentations, "The Polyhydroxyalkanoates (PHAs)" sounds a more environmentally ethical and sustainable solution to the conventional plastics (Saini 2017). The global production capacity of biobased polymers is likely to grow to nearly 7.85 million tonnes by 2020 (Aeschelmann and Carus 2015). The present day materials (PLAs, starch and cellulose based plastics) are short of the diverse material properties embraced by the PHAs (Chen 2014; Wang et al. 2014).

As the materials of choice, the presented chapter is focused on the primary properties of PHAs and a detail outline of their properties appropriate for the packaging applications. The major section covers recent research on the development of polyhydroxyalkanoates (PHAs) for packaging and lastly some case studies wherein PHAs have been used for packaging of varied food products.

14.2 The Microbial Polyhydroxyalkanoates (PHAs)

Polyhydroxyalkanoate(s) (PHA) are a family of natural thermoplastic and elastomeric biopolyesters synthesized and amassed by an extensive range of microbial genera as intracellular carbon/energy storage materials. The polymer formed is accumulated as distinct intracellular granules, number and size varying with producer strain (Keshavarz and Roy 2010; Tan et al. 2014). The current outlook has made known the microbial polyesters PHAs as one of the most scientifically innovative and favorite bioplastics for both mankind and the environment due to the complete green life cycle with 1) bio-based production (using renewable resources as carbon feedstocks); biosynthesis (involving microbes as producers); biospheric biocompatibility (eco-friendly disposal); inherent biodegradability (mineralization by natural microbial degraders to carbon dioxide and water) (Koller et al. 2013; Albuquerque and Malafaia 2018).

Figure 14.1 represents the common structure of polyhydroxyalkanoates. Structurally, PHAs have been classified into three classes namely, short chain (scl) PHAs having C4-C5, medium chain (mcl) PHAs (C6–C14), and long chain length (lcl) PHAs (>C14). This classification is derived from the substrate specificity of the PHA monomer polymerizing enzyme: PHA synthases, which accepts 3HA monomers of a specific carbon length. Additionally, based on the monomeric content of the polymer, PHAs can further be grouped as homo-polymers (either scl or mcl monomers) or copolymers (combination of different scl and/or mcl monomers) (Ali

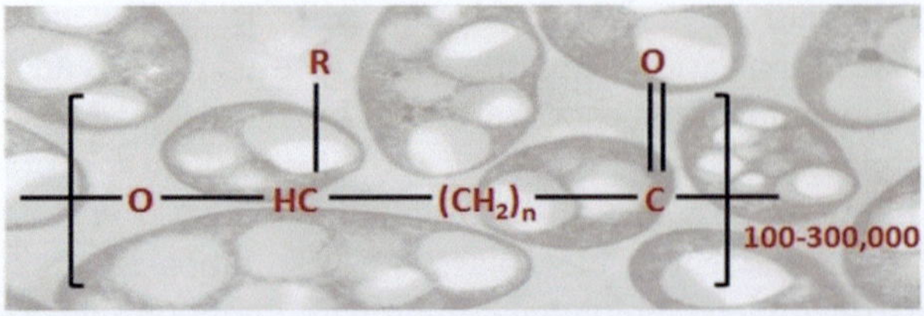

n	R group	Number of carbon atoms in monomer	PHA	Abbreviation
n = 1	Methyl	4	Poly(3-hydroxybutyrate)	P3HB
	Ethyl	5	Poly(3-hydroxyvalerate)	P3HV
	Propyl	6	Poly(3-hydroxyhexanoate)	P3HHx
	Butyl	7	Poly(3-hydroxyheptanoate)	P3HHp
	Pentyl	8	Poly(3-hydroxyoctanoate)	P3HO
	Hexyl	9	Poly(3-hydroxynonanoate)	P3HN
	Heptyl	10	Poly(3-hydroxydecanoate)	P3HD
	Nonyl	12	Poly(3-hydroxydodecanoate)	P3HDD
n = 2	Hydrogen	4	Poly(4-hydroxybutyrate)	P4HB
n = 3	Hydrogen	5	Poly(5-hydroxyvalerate)	P5HV

Fig. 14.1 Structures of Polyhydroxyalkanoates

and Jamil 2016). Till date, a vast number of diverse PHAs, comprising over 150 distinct HA monomer units as constituents, has been isolated, making them the biggest group of natural biopolyesters. The most common monomer is 3-hydroxybutyrate (3HB) (Hazer and Steinbüchel 2007). A detailed compilation of various PHAs with varying length, functional groups, and straight or branched, saturated or unsaturated monomer side chains has been reviewed by Witholt and Kessler (1999). The structural diversity of PHAs is influenced by carbon substrate, medium composition, producer strain, PHA synthase specificity and the biosynthetic pathway (Steinbüchel and Lütke-Eversloh 2003; Mohapatra et al. 2017). Mechanical, physical, thermal properties and molecular weights are usually determined for assessing the PHA materials. PHA material properties such as mechanical, chemical, and thermal depend on their monomeric composition.

As represented in Table 14.1, the family is known to exhibit a variety of physicochemical properties, which for some of the copolymers are quite comparable to the conventional plastics.

The properties range from stiff, brittle, high crystallinity [60–80%]; high melting (Tm) and glass transition temperatures (Tg) (for scl polymers) to flexibility and elasticity with lesser crystallinity of around 25%; reduced tensile strength and high elongation to break; low Tm (40–60 °C) and Tg below room

Table 14.1 Material properties of PHAs and synthetic plastics

Polymer	Tensile strength [MPa]	Elongation at break [%]	Melting temperature [°C]	Glass transition temperature [°C]
P(3HB)	40	3–8	173–180	5–9
P(4HB)	104	1000	60	−50
P(3HB-co- 9% 3HV)	37	–	162	6
P(3HB-co-16% 4HB)	26	444	130	−7
P(3HB-co-3HHx)	20	850	52	−4
P(3HO-co-12% 3HHx)	9	380	61	−35
P(3HB) Biomer P240	18–20	10–17	–	–
P(3HB-co-3HV) Mirel P4001	20	5	–	–
Mirel P5004	25–30	400–500		
P(3HB-co-3HHx) Nodax, Kaneka	10–20	10–100	–	–
Polypropylene (PP)	29.3–38.6	400–900	170–186	−10
HDPE	17.9–33.1	12–700	112–135	−80
LDPE	15.2–78.6	150–600	88–130	−36

temperature (−50 to −25 °C) (for mcl polymers). The properties of some typical PHAs are listed in Table 14.1 (Rai et al. 2011; Rudnik 2012; Bugnicourt et al. 2014; Koller 2014; Anjum et al. 2016; Możejko-Ciesielska and Kiewisz 2016; Wang and Chen 2017).

14.3 Industrial Advancements of PHAs

Efforts for commercial scale production of PHAs can be dated back to the late 1950s, Grace and Company. Further the 1970s petroleum crises had marked the dawn of pilot and industrial scale PHA production by various companies. The first industrialized and commercial PHA: P3HB was called as "The First Generation PHA". Since then diverse polymers are being produced *viz.* poly (3-hydroxybutyrate-co-3-hydroxyvalerate): PHBV; PHBHHx; poly(3-hydroxybutyrate-co-4-hydroxybutyrate): P3HB4HB etc. Later in late 1990, ICI developed a PHA co-polymer Biopol, based shampoo bottles. Its other applications being packaging, woven medical patches, disposable knives and forks, disposable cups, nappy linings, surgical stitches, surgical pins, and disposable razors. As shown in Table 14.2, the present major players of the PHA commercial market are Metabolix Inc. (USA); Meridian Inc. (USA); Biomers (Germany); ADM (with Metabolix) (USA); Kaneka (with Procter and Gambles)(Japan); Tepha (USA), Biocycle (Brazil); Yikeman, Shandong and Tianjin Green Bio-Science (China) etc. Many Chinese companies also focused on large scale PHA production since the late 1990s.

Table 14.2 Current main PHA companies

Company	Country	Trade name	PHA type	Application sectors
Biomers	Germany	Biomer	PHB	Packaging and drug delivery
Metabolix	USA	Mirel	Several PHAs	Packaging
ADM (with Metabolix)		Mirel	Several PHAs	Raw materials, used in paper coating, sheet and foam products etc.
Tepha		TephaFLEX,TephELAST	P4HB	
Meredian		Nodax	Several PHAs	Raw materials
Bio-on	Italy	Minerv-PHA	PHA	N/A
Kaneka (with P&G)	Japan	Nodax	Several PHAs	Packaging
Biocycle	Brazil	Biocycle	PHB	N/A
Yikeman, Shandong	China		PHAs	Raw materials
Ningbo Tian An, Zhejiang			PHBHV	Raw materials, biopol resin for films and coatings
Tianjin Green Bio-Science		Greenbio	P3HB4HB	N/A
Tianan Biological		Enmat	PHBHV	N/A
Material Polyone				

Metabolix Inc. PHA trade name Mirel, can be used to form injection molding, film, thermoforming, sheet products, paper coating, foam and nonwovens. Mirel can be used as an alternative to various forms of PEs, PPs and polystyrene (PS) (Chen 2009; Wang and Chen 2017; Masood 2017). Industrial PHAs have been employed for diverse applications *viz.* packaging and disposable products; biomaterials and drug carriers in medicine; chiral precursors for optically active compounds e.g., drugs; food additives; agricultural purposes (as mulch films, herbicides, insecticides, bacterial inoculants) and the most recent and upcoming biofuels (Fig. 14.2) (Anjum et al. 2016; Wang and Chen 2017).

However, their industrial manufacturing and applicability is not competitive owing to their elevated production costs. Research efforts for developing economic PHA production technologies *via* application of inexpensive carbon substrates, simplification of PHA recovery process, metabolic engineering, synthetic biology, and bioinformatics are underway. Over and above, use of these green and biodegradable PHAs for the production of wide variety of single use materials such as packaging and disposables would be biosphere favorable.

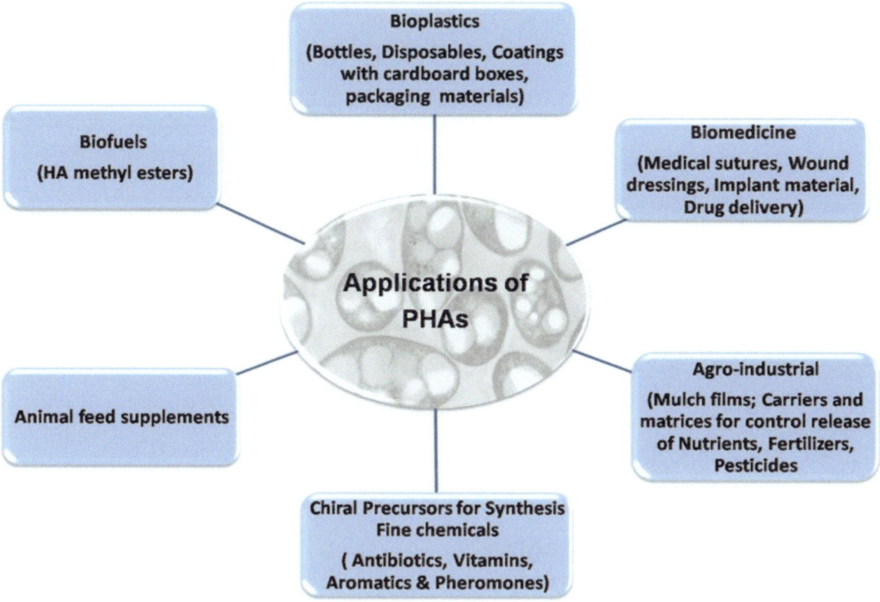

Fig. 14.2 Applications of polyhydroxyalkanoates

14.4 PHAs as Packaging Material

14.4.1 Criteria of Concern

To be applied as packaging's for foods or other materials certain common requirements (Prasad and Kochhar 2014) need to be fulfilled such as:

1. Proper wrapping and shielding from dust, moisture, dehydration, microbial and chemical contamination, UV-radiation and mechanical impairment.
2. Preservation of sensory food characteristics.
3. Use of food grade quality packaging material assisting in preservation of food purity.
4. Maintenance of food stability under extreme storage conditions such as low temperature.
5. Modified atmosphere (such as nitrogen atmosphere) maintenance in the product's headspace particularly for storing ready to eat snack items e.g., Potato chips.
6. During the material storage period, conditions favoring the biodegradation of packaging should be avoided whereas the same should be promoted or arise once the packaging is discarded.

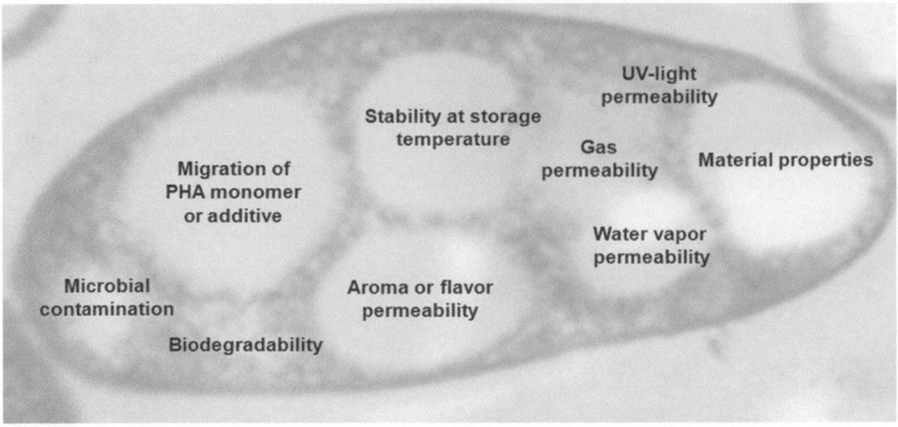

Fig. 14.3 Polyhydroxyalkanoates as packaging material: vital properties considered

14.4.2 Relevance in Packaging

PHAs should be alluring as packaging material due to certain key material and permeability properties discussed below (Fig. 14.3):

14.4.2.1 Flexible Material Properties

The PHAs, crystallinity, mechanical and thermal properties significantly influence the polymer processability and end application. The crystallinity of PHAs, which usually determines the brittleness of polymers, varies from 0% to 60% (highly crystalline to flexible polymers). The thermal properties such as Tg vary from −52 to 4 °C, a Tm above 177 °C is barely observed and thermodegradation temperature (Td) varies from 227 to 256 °C for PHAs. The occurrence of thermal degradation of PHAs nearer to their melting point is a critical parameter to be considered while opting for polymer processing methodology. The mechanical properties usually determined are Young's modulus (defining the polymer stiffness); tensile strength (establishing the total amount of force required to pull a material before it breaks) and elongation at break (defining the stretchability of material before it breaks), respectively. The Young's modulus ranges from 3.5×10^3 MPa (for stiffer scl-PHAs) to 0.008 MPa (for very ductile mcl-PHAs). For PHAs, the tensile strength generally ranges from 8.8 to 10^4 MPa. Elongation at break varies from 2% to 1000% (Chen 2010; Chanprateep 2010; Rai et al. 2011).

This diversity gives an advantage to user to choose an appropriate PHA suiting the target application. For instance, the most commonly biosynthesized and studied polymer, P3HB is water insoluble, moisture resistant, with low oxygen permeability; a high Tm more than 170 °C and a Tg of 4 °C. Its crystallinity, brittleness and low elasticity hinders with its injection molding making the wide applicability difficult (Wang and Chen 2017). Formation of P3HB copolyesters improvises its properties (Table 14.1). Such copolyesters are considered as potential candidates for food packaging application due to their high plasticity and accessibility towards melt extrusion, injection molding or thermoforming (Laycock et al. 2013). A higher HV content contributes to reduced PHB Tm (176–158 °C), brittleness and Young's modulus making the resultant copolyester material more flexible (Thellen et al. 2008; Da Silva et al. 2005; Fu et al. 2014). The introduction of mcl monomers also greatly reduces PHB stiffness thereby improving its flexibility (Albuquerque and Malafaia 2018).

Additionally, formation of varied block and random copolymers, graft polymers and polymer blends leads to novel properties expanding PHA diversity and applicability in various fields. Thus, pliable degree of crystallinity and elasticity, accompanied with appropriate processing/molding methods, the PHAs can be processed to form products ranging from flexible wraps (*via* thermo-forming) to rigid storage boxes and containers (*via* injection molding) (Koller 2014).

14.4.2.2 Permeability

14.4.2.2.1 Water and Gas (Oxygen and Carbon-di-Oxide)

Comparable water vapor permeability values of PHB and its copolymer PHBV films to that of the petrochemical opponents such as (PET) and poly(vinyl chloride) (PVC) is another positive feature for considering these microbial polyesters for packaging applications (Miguel et al. 1997; Miguel and Iruin 1999a, b). Water insoluble PHAs also have the advantage of being non swelling and highly hydrophobic compared to starch, cellulose, chitosan and gluten. Evaluation of water vapor barrier properties of the packaging material is of major importance as physical or chemical deterioration of the packed food is related to their equilibrium moisture content, which plays an important role in maintaining or extending the shelf-life of packaged food. The water vapor barrier is quantified by the water vapor permeability coefficient (WVPC) ($kg*m/(m^2*s*Pa)$) or the water vapor transfer rate (WVTR) ($cc/m^2 \cdot s$ or $g/m^2 \cdot day$), signifying the water vapor amount permeating per unit area and time inside or out of a packaging (Koller 2014). The nature of the packaged food is considered while opting for packaging material (high or low water vapor barrier). Usually, avoiding desiccation is important while packing fresh food products, whereas for the bakery products least or no water permeability is preferred avoiding the packaged food from getting contaminated with fungal infections by molds.

The oxygen permeability value of a packaging material plays a decisive role in quality preservation of fresh products (e.g. meat, fruits, vegetables, salad, various ready to eat foods *etc.*). Oxidative degradation of packaged food may affect its color, flavor and microbial stability. The oxygen barrier is determined in terms of quantified by the oxygen permeability coefficients (OPCs) and/or oxygen transmission rate (OTR), defined as the steady state rate at which oxygen can permeate through a test polymeric material. It indicates the amount of oxygen permeating a packaging material *via* diffusion per unit of area and time ($kg*m/(m^2*s*Pa)$ or nmol $ms^{-1}GPa^{-1}$ etc.), at standardized conditions for temperature (23 °C) and humidity (often 50% relative humidity RH) (Siracusa 2012). An oxygen transmission rate below 2 nmol $ms^{-1}GPa^{-1}$; is desirable and such materials are often labeled as "Barrier polymers". A polymer packaging with low oxygen permeability coefficient, maintains a very low oxygen pressure inside the packaging thereby retarding the oxidation and extending the shelf-life of the product.

The CO_2 transport properties also play an important role in fresh vegetables packaging and the related bacteriostatic properties. The CO_2 barrier is quantified as either CO_2 permeability coefficients (CO_2PC), representing the amount of CO_2 permeating in packaging per unit area and time, denoted as kg m m^{-2} s^{-1} Pa^{-1} or as rate of CO_2 transmission (CO_2TR), expressed in cc m^{-2} s^{-1} (or g m^{-2} day^{-1}) (Siracusa 2012).

Various literature reports have shown the water and oxygen permeability values covering a wide range for the different PHAs. For diverse PHB test samples, oxygen permeability and water vapor permeability has been in the range of 2–10 ml mm m^{-2} day^{-1} atm^{-1} and 1–5 g mm m^{-2} day^{-1}, respectively. For copolyesters such as PHBV, water and oxygen permeabilities have varied from 1–3 g mm m^{-2} day^{-1} and 5–14 ml mm m^{-2} day^{-1} atm^{-1}. The permeability values are reported to increase with increasing molar share of 3 HV in PHBV copolyesters. The water and oxygen permeabilities for other polymers observed were: polylactic acid (PLAs) (5–7 g mm m^{-2} day^{-1} and 15–25 ml mm m^{-2} day^{-1} atm^{-1}), PCL (300 g mm m^{-2} day^{-1} and 20–200 ml mm m^{-2} day^{-1} atm^{-1}), PVC (1–2 g mm m^{-2} day^{-1} and 2–8 ml mm m^{-2} day^{-1} atm^{-1}), LDPE (0.5–2 g mm m^{-2} day^{-1} and 50–200 ml mm m^{-2} day^{-1} atm^{-1}). In general, water and oxygen barrier properties of PHB and PHBV appear to be slightly better than those of PLAs and potentially competitive with that of various synthetic plastics (Miguel et al. 1997; Thellen et al. 2008; Sanchez-Garcia et al. 2008; Plackett and Siró 2011).

A low and comparable CO_2 permeability was reported for PHB with that of polyvinylidene chloride (PVDC). Poley et al. (2005) reported CO_2 diffusion coefficient value of 1×10^{-9} cm^2 s^{-1} (at 25 °C) for PHB which was slightly higher than the determined value of 4.4–4.7×10^{-10} cm^2 s^{-1} (at 30 °C) by Miguel et al. (1997). Miguel and Iruin (1999a) reported CO_2 permeability values of PHB being similar to commodity thermoplastics unplasticized PVC and PET which are usually considered having relatively low CO_2 permeability. Interestingly, at 4 *atm* (the CO_2 pressure in carbonated beverages), PHB outperformed PET in the capture of CO_2.

Recently, Follain et al. (2014), investigated the water and gas transport properties of commercial PHAs: PHB (Biomer P209 and P226) and P (3HB-co-3% 3HV) (TiaNan Enmat Y1000P) films. Higher water and gas permeability values were reported for the solvent cast films compared to the respective compressed counterparts. Enhanced polymer chain motions, microcavity formation and free volume between polymer chains favored permeability of gas molecules with solvent casting. Irrespective of the processing, PP209 film had maximum permeability and the least was of PY1000P film. Similar rankings were reported for diffusivity and solubility also. The CO_2 and O_2 permeability values for PP209 film (0.66, 12.3 and 1.95 Barrer), PP226 film (0.16, 3.04 and 0.49 Barrer) and PY1000P (0.25, 1.2 and 0.35 Barrer) was lower than LDPE (1.0, 28 and 6.9 Barrer) and comparable to PLA with

L:D ratio of 96:4 (1.3, 10.2 and 3.3 Barrer) film, showing promising future of PHAs as ecological packaging.

14.4.2.2.2 Chemical

Weak or strong acidic nature of packaged food may cause hydrolytic reactions to the packaging material. PHAs are known to undergo acid-catalyzed hydrolytic degradation. Interaction between the chemical compounds and the polymers may affect the final mechanical properties of a polymer. Hence, assessment of biopolymer/s performance, suitability and stability with common food simulating solution as a function of time becomes imperative. Miguel et al. (1997) showed PHB films having high barrier properties for completely apolar (n-hexane, isopropyl ether and carbon tetrachloride) or very highly polar (methanol and water) solvents. Whereas, highest permeabilities were exhibited for solvents with moderately polar structures or polarities similar to PHB such as acetone, butyl acetate and toluene.

14.4.2.2.3 Flavoring Substances or Aroma

In many food packaging applications, enhanced barrier to loss of flavor during storage period is highly desirable, especially for beverages. 1-methyl-4-(1-methylethenyl)-cyclohexene [limonene] is usually used as a model aroma compound for testing the suitability of a material for packaging of flavorful foods. In this context, PHB was reported to show high barrier or lower permeability (0.088 * 10^{-}kg m s^{-1} m^{-2} Pa-1) for limonene compared to PET (1.17 * 10^{-13} kg m s-1 m-2 Pa-1); whereas this value was significantly increased with PHBV copolyesters (Sanchez-Garcia et al. 2007).

14.4.3 Migration

Monomers and/or additives migration is of crucial concern in terms of food safety. Bucci et al. (2007) studied production of 500 mL (jar–cap set) PHB food packaging (Commercial grade: BIOCYCLE) through the injection process and also evaluated and compared its performance with polypropylene (PP) containers of the same shape. The authors evaluated the PHB component migration into a range of food simulants *viz.* distilled water, ethanol (15%), acetic acid (3%) and n-heptane at varied conditions. The overall migration rate was lesser than 8.0 mg/dm^2 or 50 mg/kg (suggested limits), thus confirming the suitability of the polymer as a food packaging material. Authors based on migration evaluation indicated the applicability of such PHB containers for product storage under different conditions. Additionally, the light transmission analysis showed natural (zero % TE in 250–350 nm: UV range) and pigmented (Zero % TE in 250–750 nm: full barrier) PHB as better light barriers. Thus, recommending PHB as a packaging for foods such as soy oil and milk. Also, pigmentation and usage of a UV absorber additive could also be avoided with PHB. They also attempted for developing a better quality PHB material with fewer shortcomings and no odor. The biodegradation of the PHB packaging for up to 60 days indicated its immense applicability in packaging sector, circumventing the pollution and waste management issues of conventional plastics.

14.4.4 Biodegradation

The major challenge faced by the "Bioplastic based food packaging industry" is combining the sensorial maintenance along with an acceptable "shelf-life" of the product. Of utmost importance is, environmental conditions leading to packaging biodegradation must be evaded during the food product storage and only be present on subsequent discarding (Bucci et al. 2005).

The biodegradation of PHAs has been extensively studied under both aerobic and anaerobic conditions in varied environments such as soil and compost; fresh, marine and lake water; activated and anaerobic sludge; and laboratories (Goncalves et al. 2009; Correa et al. 2008; Shah et al. 2008). Microorganisms of various bacterial and fungal genera have been identified as potent PHA degraders, dominant being *Pseudomonas*, *Bacillus*, *Streptomyces*, *Azotobacter*, *Penicillium*, *Cephalosporum*, *Paecilomyces*, *Trichoderma* and *Aspergillus* (Shah et al. 2008; Correa et al. 2008; Lodhi et al. 2011; Plackett and Siró 2011). The biodegradation process involves biotic and abiotic hydrolysis followed by bioassimilation (Correa et al. 2008). Various degraders produce extracellular PHA depolymerases hydrolyzing the high molecular weight PHAs to CO_2 and water aerobically (Khanna and Srivastava 2005). Factors influencing the PHA biodegradation rate are polymer characteristics (molar mass, monomeric composition, crystallinity, stereochemistry, amphiphilicity and chain mobility); environmental conditions; microorganism type and load (Khanna and Srivastava 2005). Usually, higher the polymer crystallinity and melting temperature, lower is the degradation rate. Degradation mechanisms are different under aerobic and anaerobic conditions. PHBV was observed to degrade more rapidly than PHB under aerobic conditions (Santos Rosa et al. 2004; Li et al. 2007); however, reverse was reported by Abou-Zeid et al. (2004). For, PHBV, faster degradation has been observed with increasing HV content (Renard et al. 2004). Degradation rate of PHA copolymers has also been represented as a function of monomer side chain i.e. maximum for PHB and least for PHBHHx (Li et al. 2007).

The variations in the permeability or barrier properties of PHAs could be attributed to polymer sample (varying structure, physical and thermal properties), film forming process, varied estimation techniques and equipments used and conditions at the time of measurements. Hence, evaluation of individual PHAs barrier properties for packaging of particular food types would be suggested. Combination of their barrier properties, biodegradability and microbial origin, conforms PHAs as promising substitutes of the long established petroleum plastics. Several research groups have been and are working on additional improvisation of PHAs properties for diverse packaging application. The forthcoming sections will be focused on recent research accomplished in this direction, highlighting the PHA based composites, nanocomposites, paper coatings, multilayer films, antimicrobial active packaging's.

14.5 PHA Nanocomposites

PHA-composites are hybrid materials with fillers such as fiber, clay or particle, for modification of material properties. Usage of minimum one component in nanometer scales (nm, 10^{-9}) defines the hybrids as Nanocomposites. Maiti et al. (2003) reported the fabrication of PHB/OMMT (organomodified montmorillonite) nanocomposites for the first time. The resultant nanocomposite had a improved storage modulus (>40%) and an intercalated morphology in contrast to pure PHB, with still maintaining its biodegradability.

The nanoscaled OMMT layers influence on PHBV crystallization was investigated by Wang et al. (2005). The small quantity addition of OMMT stimulated PHBV crystallization but reduced the chain mobility. Thus, though crystallization rate was increased, the relative crystallinity degree had decreased for the resulting PHBV/OMMT nanocomposites. The increasing OMMT content contributed to reduced Tm and enthalpy of fusion (ΔH_{m}); expansion of the processing temperature range and lastly reduced biodegradability of nanocomposites in soil suspension. The stronger interaction of antimicrobial OMMT with PHBV had adversely affected the nanocomposites biodegradability. Therefore, small quantity addition of OMMT was recommended. Carli et al. (2011) investigated the influence of two types of nanoparticles: OMMT Cloisite® 30B (C-30B) and halloysite (HNT), on PHBV. The PHBV/C-30B nanocomposites had a partially exfoliated structure incontrast HNT dispersed uniformly in the PHBV matrix. The Tm had increased for PHBV/HNT nanocomposites. The Young's modulus was also observed to increase. However, the enhancements for PHBV/C-30B nanocomposites were at the cost of the elongation at break and impact strength. Mohamed El-Hadi (2014) used octadecylamine and trimethyl stearyl ammonium (nanoclay A and B); ATBC and polyvinyl acetate (PVAc) plasticizers for reducing the PHB brittleness. Reduction in Tg from 5 to −13 °C was observed with plasticizer addition. With the nanoclay A addition the crystallization temperature and rate remained unaltered. In contrast nanoclay B addition led to enhanced crystallization rate and thermal stability. The nanoclay served as a nucleating agent resulting in increased nuclear density of the nanocomposites. Martínez-Sanz et al. (2014) developed PHBHV nanocomposites by incorporating bacterial cellulose nanowhiskers (BCNW). An improved water, oxygen and moisture permeability; and decline in Tm, enthalpy, rigidity and stiffness was observed with valerate incorporation in pure polymer. BCNW increased the thermal stability. Furthermore, a significant improvement in crystallinity, thermal stability and processability of PLA-PHB composites was reported with CNCs incorporation (Arrieta et al. 2014a, b, c). Incorporation of either pure or surfactant-modified CNCs was investigated. The ternary nanocomposite had improved oxygen barrier and stretch ability whereas reduced surface wettability UV-light transmission. The degradation of the nanocomposite was slowed due to PHB during composting experiments. Arrieta et al. (2015) worked on making an optically transparent PLA-PHB-ATBC-CNC flexible film, for food packaging with enhanced stretch ability, crystallization, and oxygen barrier; and slightly reduced ultra violet transmission. The nanocomposite film's disintegration was assisted by plasticizer as well as and CNC in compost.

Yu et al. (2014) prepared solvent casted PHBV bionanocomposites reinforcing with PHBV-grafted multi-walled carbon nanotubes (MWCNTs). The obtained nanocomposite films obtained (having 1–10 wt % PHBV-g-MWCNTs) were transparent in the visible wavelength range. The uniform distribution of the MWCNTs had improved the thermal stability, mechanical, barrier, and migration properties of PHBV. The nanocomposite film containing 7 wt % PHBV-g-MWCNTs had the tensile strength and Young's modulus enhanced by 88% and 172% compared to the neat PHBV film. For the same nanocomposite film, the maximum decomposition temperature was improved by 22.3 °C. Thus overall, these nanocomposites had much broader melt-processing phase with diminished water uptake. Also, all nanocomposites migrations in polar as well as non-polar simulants, respectively were within the permissible limits.

A novel biodegradable and renewable PHBHV-Keratin composite material was successfully developed *via* melt compounding by Pardo-Ibáñez et al. (2014). The keratin additive was prepared from poultry feathers. The suitable and good dispersion for low additive loadings was accompanied by barely unmodified optical properties. Also, the composite with keratin (1 wt %) had significantly reduced water, limonene, and oxygen permeabilities. Also an increase of ca. 30% in the elastic modulus for the composite was achieved. This specific composition composite was proposed as a suitable candidate for the development of fully biodegradable renewable food packaging material wherein enhanced barrier properties are necessary. In contrast, composites with higher additive loading were suggested for specific packaging applications wherein transparency is not being considered and exchange of gases and/or water vapor is desirable.

Kiran et al. (2017) reported the development of a PHB: nanomelanin: glycerol polymer film, which was flexible, odorless, nontoxic having antimicrobial and antioxidant activities. Melanin, dark brown ubiquitous photosynthetic pigment has applications in medicine, fabrication of radio-protective materials, cosmetics and food packaging. Spherical nucleated nanomelanin (Nm) particles formed *via* sonication, proved as an effective antimicrobial. The Nm-PHB nanocomposite film was homogeneous, thermostable up to 281.87 °C, and strongly inhibiting *Staphyloccoccus aureus*. Thus, it could be utilized in food packaging sector, for protection against oxidation and bacterial contamination.

Incorporation of natural vermiculite and organoclay to PHBHV, for improving its properties was studied by Reis et al. (2016, 2017). The Scanning Electron Microscopy (SEM) and Optical Microscopy (OM) results showed organoclay vermiculite bionanocomposites to be more homogenous and suitable ones for food packaging. Also, the migration levels being within the permissible limits such materials could be used without compromising with any food security.

Different doses of gamma irradiation influence the physical and biodegradation properties of poly-3-hydroxybutyrate/sepiolite (PHB/SP) nanocomposites (Masood et al. 2018). Modification of sepiolite with vinyltriethoxy silane (VTES) enhanced its compatibility with PHB. The nanocomposite films showed a good network formation and enhanced thermal stability compared to the PHB. It also degraded faster under in-vitro and soil burial biodegradation studies. It suggests the application of these films as biodegradable food packaging material.

It can be concluded that a good amount of progress has already been achieved in terms of PHA-nanocomposites, and literature data shows the potential of such materials to compete with the properties of petro-based plastics, making them usable in the packaging industry. Still more in depth evaluation of every nanocomposites material and development of novel ones is needed to meet up the existing packaging industry requirements in all perspectives.

14.6 PHAs in Multilayer Films

Multilayer films comprise of hydrophilic and hydrocolloidal synthetic polymers sandwiched between the layers of better barrier properties possessing hydrophobic and biodegradable polymers. Limited literature is available on fabrication of PHA-based multilayer films. Fabra et al. (2013) developed multilayer films with electrospun zein nanofibers sandwiched between PHBV using compression molding and casting methodology. A multilayer film prepared by the compression molded multilayer film had better mechanical properties and water vapor barrier than the film prepared by the casting technique. However, the influence on the water vapour and limonene permeability was depended on zein content of the interlayers. Also, irrespective of the processing technique used zein nanostructure addition was reported to enhance the oxygen barrier properties of the multilayer films. Thellen et al. (2013) developed outer-PHAs and inner- polyvinyl alcohol (PVOH) multilayer films with coextrusion process. The peel strength of the multilayer improved by twofolds on grafting maleic anhydride to PHAs (reaction initiator was dicumyl peroxide). Significant degradation of unmodified and maleated PHAs was observed in contrast to the non biodegradable PVOH. Fabra et al. (2014a) reported significant improvement of the flexibility and oxygen barrier; and decreased transparency of PHB or PHBV mutlilayers with nanostructured zein interlayers addition. For the control multilayer PLA/PET structures, no influence of zein interlayer incorporation on the water and oxygen barrier was observed. Fabra et al. (2014b) also reported unaltered transparency, crystallinity, water and oxygen permeability of PHBV/zein multilayer films post storage. At 100% RH and over the 3 month storage period the *Listeria monocytogenes* growth was promoted.

14.7 PHAs as Paper and Cardboard Coating

Biodegradable packaging comprising of paper has also been well recognized, the only limitation being its hygroscopic nature. Paper based packaging's are often coated with hydrophobic polymers such as PVC, poly(ethylene-co-vinyl alcohol), polyolefin and oriented PET, PLA and PHB (Rastogi and Samyn 2015) Such composites not only minimize disposal costs but also maintain the dimensional stability in wet environments. Very few research groups have evaluated PHAs as paperboard coatings. Both PHB and PHBV have been tested as paper and cardboard coating effectively resulting in significant lowering of the moisture absorption and water permeability. Bilayer films comprising of PHB and cellulose paper (solvent casted); PHB and cellulose cardboard (compression molded), respectively were developed by Cyras et al. (2007, 2009). Both studies reported improvement of the water barrier. The elastic modulus, tensile strength, and strain at break also improved with PHB layering in a concentration dependent manner. The study presented the PHB coatings as a better alternative for Tetra Paks. In another study, natural as well as artificially synthesized PHB and PHBV were applied for paper sizing and coatings. The pressing and heating of impregnated papers for a short duration resulted in thin PHB coating on their surface giving a better paper size. Also, PHB was found to be more suitable than PHBHV for paper sizing (Bourbonnais and Marchessault 2010).

Dagnon et al. (2010) described the first attempts producing *scl*-PHA-Kraft paper–composites. Post 8 week soil burial period uncoated Kraft paper had higher weight loss rate compared to the P(3HB-co-4HB) coated Kraft paper. Also P(3HB-co-4HB) biopolymer coating had significantly improved the thermo mechanical properties of Kraft paper. Such materials could be employed as liners for fabrication of fiberboard boxes intended for varied packaging applications wherein mechanical properties as well as environmental disposal is of prime concern. Rastogi and Samyn (2017) investigated the development of structured PHB microparticles (PHB-MP) and unstructured sub-micron particles (PHB-SP), respectively as hydrophobic coatings for packaging papers. Additionally, the carnauba plant wax was also used as a hydrophobizing additive. Phase-separation methodology was employed for making structured PHB-MP. Filter papers dipped in the particle suspension were additionally sized with the wax. Increase in the static contact angles from 105–122° to 129–144° corresponded to inherent hydrophobicity improvement of the PHB-MP-coated papers with wax addition. The maximum PHB and non-solvent concentrations favored enhancement of contact angles. For, unstructured PHB-SP, the particles prepared was mixed with aqueous NFC suspensions of varying 0–7 wt %, respectively. Static contact angles varied from 112 to 152° for filter papers dip-coated in PHB-SP/NFC suspensions and sized with wax solution. The progressive increase in hydrophobicity observed was governed more by NFC, which was responsible for entrapment of polymer PHB particles, thus assisting its anchorage to the paper fibers. An efficient improvement of the paper surface hydrophobicity was achieved coating with macro to nano-scale PHA particles, without compromising with the biodegradability and biocompatibility of the ecofriendly packaging material "paper".

14.8 Antibacterial Active PHAs for Packaging

An innovative way to protect minimally processed and/or "ready to eat" products are the use of packaging with antimicrobial properties. Recently, the inclusion of antimicrobial additives in food packaging materials has drawn significant attention

(Appendini and Hotchkiss 2002). Research on functionalization and rendering of PHAs with antimicrobial activity supports and encourages their development for food packaging applications.

Effective antimicrobial efficacy against diverse foodborne pathogens, food spoilage bacteria, and fungi was observed using eugenol incorporated PHB films with and without combination of crude bacteriocin (pediocin) (Narayanan et al. 2013). The additive eugenol had reduced the melting point, crystallinity and tensile strength; but enhanced the flexibility of the resultant polymer film. The study demonstrated PHB as a suitable bio-based packaging material. Ramachandran et al. (2013) conferred antimicrobial activity to PHB films by incorporation of natural antimicrobial additive, *Clitoria ternatea* seed ethanolic extract. The resultant films showed appreciable antimicrobial action against various multi drug resistance (MDR) human bacterial pathogens. Xavier et al. (2015) studied the antimicrobial efficacy of vanillin (4-hydroxy-3-methoxybenzaldehyde) incorporated PHB films against various food-contaminating bacterial and fungal pathogens. Enhanced vanillin migration into the hydrophobic milieu demonstrated their relevance for fat containing foods. Because of reduced maximum thermal decomposition temperature and tensile strength of the antimicrobial films, the authors suggested such vanillin containing PHB films as secondary films over the primary food wrappings. Fan et al. (2015) reported preparation of antimicrobial PHB fibrous membrane by incorporation of the antimicrobial agent N-halamine, poly [5,5-dimethyl-3-(3′-triethoxysilylpropyl) hydantoin] (PSPH) in PHB material using electrospinning technique. The PHB and PHB/PSPH fibrous membranes had a fairly uniform morphology. Also, additive N-halamine had a small influence on the electrospun PHB film's thermal properties. Chlorine bleach exposure conferred the membranes to be biocidal. The chlorinated PHB/PSPH samples displayed excellent antimicrobial activity against *S.aureus* and *E. coli* O157:H7 within a very short contact time of 30 min, respectively. Such PHB-based antimicrobial fibrous membranes could have great potential for food packaging applications.

Nanocomposite systems employing nanoparticles (metals like Cu, Ag, and Au, respectively); nanomaterials (metal oxides such as ZnO, TiO_2 and MgO, respectively); and carbon nanotubes have also been drawing focus for the development effective antimicrobial packaging materials.

Due to higher surface to volume ratio, the nanosized agents possess better antimicrobial activities compared to their micro- or macroscale counterparts. ZnO nanoparticles are being foreseen as future inexpensive and reliable food packaging solution. Díez-Pascual and Díez-Vicente (2014a) reported the evaluation of ZnO-reinforced PHBV biodegradable nanocomposites as suitable food packaging material by assessing their morphological, material and antibacterial properties. The nanocomposites displayed uniform distribution of nanoparticles without addition of any coupling agents. Raised crystallization temperature and crystallinity index of the matrix with decreasing crystallite size were observed. The nanocomposites also displayed rise in thermal stability; superior mechanical properties such as improved stiffness, strength, toughness, and glass transition temperature; and improved barrier properties such as reduced water uptake, oxygen and water vapor permeability compared to the neat biopolymer. The PHBV/ZnO films also demonstrated significant antibacterial action against human pathogenic bacteria *E.coli* and *S.aureus*. Additionally, the overall migration levels of the nanocomposites were reported to be well below the limits as per current food packaging material norms. Antibacterial efficiency and barrier properties of the nanocomposite system were being influenced by factors *viz.* size, distribution and interaction of nanoparticles with the polymer. The same research group (Díez-Pascual and Díez-Vicente 2014b) also described PHB-ZnO bionanocomposites. The nanoparticle addition led to increase in the crystallization temperature, crystallinity degree, thermal stability and also improved varied mechanical properties. Decreased water uptake and superior barrier properties (gas and vapour) were also exhibited by nanocomposites compared to pure PHB. The antibacterial activity against gram-positive as well as gram-negative microorganisms enhanced with increasing concentration of ZnO. The migration was reported to decrease in food simulants at higher nanoparticle concentration. The authors proposed such sustainable nanomaterials for use as beverage and food containers and also as disposable articles or wrap films.

Kwiecień et al. (2014) reported oligo (3-hydroxybutyrate) (OHB) conjugates with preservatives sorbic acid and benzoic acid. Synthesis of such conjugates was achieved by anionic ring-opening oligomerization of racemic β-butyrolactone initiated by the sodium salt of selected preservatives. Electrospray ionization multistage mass spectrometry (ESI-MSn) structural characterization confirmed the covalent bonding of unaltered preservative molecular structures to OHB. As the preservative molecules in the conjugates were covalently linked to the oligomer chains; limited migration of the antimicrobial additives was assumed. Such preservative-OHB conjugates could potentially be applied as coating of the food packaging surface.

Castro-Mayorga et al. (2016) developed an antimicrobial PHA-Ag nanoparticle multilayer system. Significant reduction of *Salmonella enterica* population was attained even at very low (0.002–0.0005 wt %) silver concentration. The study demonstrated an innovative antimicrobial material for efficient prevention of microbial contamination in food packaging's and contact surfaces.

Salama et al. (2017) investigated the antimicrobial activity of chitosan biguanidine hydrochloride ChG and PHB based graft copolymers (ChG-grafted PHB). The grafts were prepared via condensation reaction between the carboxylic groups of PHB and the amino groups of ChG. The chitosan derivatives showed excellent antimicrobial activity against the test bacterial and fungal strains used, implying their usage as efficient biomedical material. Such derivates could also be evaluated for food spoilage and food borne pathogens thus finding applications of such graft polymeric materials in the packaging sector.

14.9 Application of PHAs in Food Packaging: Demonstration

Very limited studies have done realistic demonstration of using PHAs for packaging of food products, to quote a few:

Kantola and Helén (2001) were the first to evaluate quality changes in organic tomatoes stored in Biopol@ [copolymer poly(HB-co-HV)] -coated paperboard trays and other biodegradable and LDPE packages, respectively at 11 ± 1 °C and 75–85% RH for 3 weeks. The biodegradable packaging materials tested were: Mater-Bi@ type ZF03U (corn starch based); Biopol@ [PHBHV]; cellophane (regenerated cellulose). The varied packagings tested were (1) perforated Mater bag; (2) a Biopol-coated paperboard tray additionally wrapped with a perforated Mater bag, (3) PLA coated paperboard tray additionally wrapped with a perforated Mater bag, and lastly (4) perforated cellophane bag. Post 3 weeks storage period 1.7% weight loss was observed for tomatoes in LDPE-bags whereas >2.5% was for the biodegradable materials' bags. Tomatoes sensory qualities were almost independent of the package type but were significantly influenced by the storage duration.

The study of Haugaard et al. (2003) showed the potential of PHB and PLA packaging cups to be as effective as the conventional HDPE for an orange juice (acidic foods) and dressing (fatty food) simulant storage under fluorescent light or darkness at 4 °C for 10 weeks. The quality changes measured for the orange juice simulant were colour and ascorbic acid degradation; for dressing simulant were colour, peroxide value (POV), development of secondary lipid oxidation products development and α–tocopherols degradation. Due to analogous quality variations observed for both the simulants packed in PHB and PLA and HDPE, respectively; PHB and PLA both could be expected to be used for other commercial juices, acidic beverage/s and fatty food/s packaging. Additionally, the study also revealed that the quality changes were primarily light induced and were being governed by light permeability differences of the packages used.

Bucci et al. (2005) studied and compared the use of injection molded P3HB food packaging with PP (polypropylene) using varied dimensional and mechanical tests. The P3HB was reported as a rigid material than PP with 50% lesser deformation value. Though the PHB performance was inferior to PP at normal freezing and refrigeration conditions, it performed better at higher temperatures. The study stressed on the necessity of designing special molds and/or optimization of injection conditions and processing temperature for PHB. The sensorial investigations with the foods (margarine, mayonnaise and cream cheese) tested indicated no notable difference ($p < 0.05$), displaying PHB performance being equally good to conventional plastic PP for storage of lipophilic products.

Muizniece-Brasava and Dukalska (2006), reported plasticized PHB (plasticizers: dioctylsebacate, bisoflex) material suitable for packaging of sour cream (dairy product) with inconsiderable quality changes in comparison to the conventional dairy packaging material such as lean pouches, PE (with light protective graphite layer) and PS cups, respectively.

Levkane et al. (2008) proposed the use of plasticized PHB and PLA packaging films for *Sous vide* (under vaccum) thermal food treatment of meat and mayonnaise salad samples. At temperatures not more than +63 ± 0.5 °C, consistency of salads was reported to be maintained with reduced total microbial colony count. Influence of pasteurization regime and packaging material for maintaining the meat and mayonnaise salad samples packed in vacuum using conventional, plasticized PHB and PLA packaging films, respectively was focus of their study. Hermida et al. (2008) reported the aptness of PHB for gamma radiation (used for sterilization of food or packaging materials), with no significant decline in PHB properties.

14.10 Conclusion

The presented chapter has demonstrated that PHA biopolyesters very well meet most criteria for developing them as efficient packaging materials for both short-shelf life as well as long-shelf life products. The diversity of barrier properties provides an opportunity to choose from a range of PHA materials based on the type of food or product to be packaged. Ranging from high-quality copolymers to novel composites and multilayered materials with functionalized PHA moieties opens the door for innovative and high-tech, packaging devices with tailored target application in the food sector. Their inherent biodegradability offers a clear advantage making it a superior choice over the existing array of bioplastics. Based on the PHA material type proposed, there is a need for periodic critical evaluation accessing the functionality of these bio-based packaging materials in order to certify them for market launch as sole substitutes for conventional packaging materials. Upcoming research is focused on identification of new PHA compatible antimicrobial compounds with broad spectrum activity and low toxicity for the development of PHAs as active antimicrobial packaging materials. A thoughtful consideration towards migration evaluation of various nano-technology based additives from the PHA films and development of multilayer films is needed for appreciable penetration of these polymer in the packaging sector.

14.11 Opinion

The wonderful family of Polyhydroxyalkanoates (PHAs) may have bright opportunities in the packaging sector if their physical performance and prices both become competitive to synthetic polymers. Global research and development efforts making use of inexpensive substrates for PHA production and simple PHA recovery methods has assisted and is still constantly being worked on worldwide for achieving their competitive production prices. There is a constant need of robust and super PHA producing strains selection, feedstock evaluation, and development of competent fermentation regimes and processing technologies for addressing product quality, aptness, and sustainability and cost issues of PHAs. A close collaboration amongst academia, industries, and government could support these excellent candidates in reaching from packaging production units to the common man.

Acknowledgements The authors wish to thank Council of Scientific and Industrial Research (CSIR), New Delhi, India, for the Senior Research Fellowship. School of Sciences, Jain University is also thankfully acknowledged for providing the necessary facilities.

References

Abou-Zeid DM, Müller RJ, Deckwer WD (2004) Biodegradation of aliphatic homopolyesters and aliphatic± aromatic copolyesters by anaerobic microorganisms. Biomacromolecules 5:1687–1697. https://doi.org/10.1021/bm0499334

Aeschelmann F, Carus M (2015) Polymers in the world: capacities, production, and applications – status quo and trends toward 2020. Industrial Biotechnology (Short report by Nova Institute, Germany on full market study)

Albuquerque PBS, Malafaia CB (2018) Perspectives on the production, structural characteristics and potential applications of bioplastics derived from polyhydroxyalkanoates. Int J Biol Macromol 107:615–625. https://doi.org/10.1016/j.ijbiomac.2017.09.026

Ali I, Jamil N (2016) Polyhydroxyalkanoates: current applications in the medical field. Front Biol 11:19–27. https://doi.org/10.1007/s11515-016-1389-z

Anjum A, Zuber M, Zia KM, Noreen A, Anjum MN, Tabasum S (2016) Microbial production of polyhydroxyalkanoates (PHAs) and its copolymers: a review of recent advancements. Int J Biol Macromol 89:161–174. https://doi.org/10.1016/j.ijbiomac.2016.04.069

Appendini P, Hotchkiss JH (2002) Review of antimicrobial food packaging. Innov Food Sci Emerg Technol 3:113–126. https://doi.org/10.1016/S1466-8564(02)00012-7

Arrieta MP, Fortunati E, Dominici F, Rayón E, López J, Kenn JM (2014a) Multifunctional PLA-PHB/cellulose nanocrystal films: processing, structural and thermal properties. Carbohydr Polym 107:16–24. https://doi.org/10.1016/j.carbpol.2014.02.044

Arrieta MP, López J, Hernández A, Rayón E (2014b) Ternary PLA-PHB-Limonene blends intended for biodegradable food packaging applications. Eur Polym J 50:255–270. https://doi.org/10.1016/j.eurpolymj.2013.11.009

Arrieta MP, López J, Rayón E, Jiménez A (2014c) Disintegrability under composting conditions of plasticized PLA-PHB blends. Polym Degrad Stab 108:307–318. https://doi.org/10.1016/j.polymdegradstab.2014.01.034

Arrieta MP, Fortunati E, Dominici F, López J, Kenn JM (2015) Bionanocomposite films based on plasticized PLA-PHB/cellulose nanocrystal blends. Carbohydr Polym 121:265–275. https://doi.org/10.1016/j.carbpol.2014.12.056

Bourbonnais R, Marchessault RH (2010) Application of polyhydroxyalkanoate granules for sizing of paper. Biomacromolecules 11:989–993. https://doi.org/10.1021/bm9014667

Bucci TZ, Tavares LBB, Sell I (2005) PHB packaging for the storage of food products. Polym Test 24:564–571. https://doi.org/10.1016/j.polymertesting.2005.02.008

Bucci DZ, Tavares LBB, Sell I (2007) Biodegradation and physical evaluation of PHB packaging. Polym Test 26:908–915. https://doi.org/10.1016/j.polymertesting.2007.06.013

Bugnicourt E, Cinelli P, Lazzeri A, Alvarez V (2014) Polyhydroxyalkanoate (PHA): review of synthesis, characteristics, processing and potential applications in packaging. Polym Lett 8:791–808. https://doi.org/10.3144/expresspolymlett.2014.82

Carli LN, Crespo JS, Mauler RS (2011) PHBV nanocomposites based on organomodified montmorillonite and halloysite: the effect of clay type on the morphology and thermal and mechanical properties. Compos Part A 42:1601–1608. https://doi.org/10.1016/j.compositesa.2011.07.007

Castro-Mayorga JL, Fabra MJ, Cabedo L, Lagaron JM (2016) On the use of the electrospinning coating technique to produce antimicrobial polyhydroxyalkanoate materials containing in situ-stabilized silver nanoparticles. Nanomaterials 7. https://doi.org/10.3390/nano7010004

Central Pollution Control Board (2013) Overview of plastic waste management. Annual report (2011–12) on implementation of PWM

Chanprateep S (2010) Current trends in biodegradable polyhydroxyalkanoates. J Biosci Bioeng 110:621–632. https://doi.org/10.1016/j.jbiosc.2010.07.014

Chen GQ (2009) A microbial polyhydroxyalkanoates (PHA) based bio-and materials industry. Chem Soc Rev 38:2434–2446. https://doi.org/10.1039/b812677c

Chen GQ (2010) Plastics completely synthesized by bacteria: polyhydroxyalkanoates. In: Chen GQ (ed) Plastics from bacteria: natural functions and applications, microbiology monographs. Springer, Berlin, pp 17–37. https://doi.org/10.1007/978-3-642-03287_5_2

Chen YJ (2014) Bioplastics and their role in achieving global sustainability. J Chem Pharm Res 6:226–231
Colombo B, Pepè Sciarria T, Reis M, Scaglia B, Adani F (2016) Polyhydroxyalkanoates (PHAs) production from fermented cheese whey by using a mixed microbial culture. Bioresour Technol 218:692–699. https://doi.org/10.1016/j.biortech.2016.07.024
Correa MCS, Rezende ML, Rosa DS, Agnelli JAM, Nascente PAP (2008) Surface composition and morphology of poly(3-hydroxybutyrate) exposed to biodegradation. Polym Test 27:447–452. https://doi.org/10.1016/j.polymertesting.2008.01.007
Cyras VP, Commisso SM, Mauri AN, Vázquez A (2007) Biodegradable double-layer films based on biological resources: polyhydroxybutyrate and cellulose. J Appl Polym Sci 106:749–756. https://doi.org/10.1002/app.26663
Cyras VP, Soledad CM, Analía V (2009) Biocomposites based on renewable resource: acetylated and nonacetylated cellulose cardboard coated with polyhydroxybutyrate. Polymer 50:6274–6280. https://doi.org/10.1016/j.polymer.2009.10.065
Da Silva MG, Vargas H, Poley LH, Rodriguez RS, Baptista GB (2005) Structural impact of hydroxyvalerate in polyhydroxyalkanoates (PHAscl) dense film monitored by XPS and photothermal methods. J Braz Chem Soc 16:790–795. https://doi.org/10.1590/S0103-50532005000500017
Dagnon KL, Thellen C, Ratto JA, D'Souza NA (2010) Physical and thermal analysis of the degradation of poly (3-hydroxybutyrate-co-4-hydroxybutyrate) coated paper in a constructed soil medium. J Polym Environ 18:510–522. https://doi.org/10.1007/s10924-010-0231-y
Díez-Pascual AM, Díez-Vicente AL (2014a) Poly(3-hydroxybutyrate)/ZnO bionanocomposites with improved mechanical, barrier and antibacterial properties. Int J Mol Sci 15:10950–10973. https://doi.org/10.3390/ijms150610950
Díez-Pascual AM, Díez-Vicente AL (2014b) ZnO-reinforced poly(3-hydroxybutyrate-co-3-hydroxyvalerate) bionanocomposites with antimicrobial function for food packaging. ACS Appl Mater Interfaces 6:9822–9834. https://doi.org/10.1021/am502261e
Fabra MJ, Lopez-Rubio A, Lagaron JM (2013) High barrier polyhydroxyalcanoate food packaging film by means of nanostructured electrospun interlayers of zein. Food Hydrocoll 32:106–114. https://doi.org/10.1016/j.foodhyd.2012.12.007
Fabra MJ, Lopez-Rubio A, Lagaron JM (2014a) Nanostructured interlayers of zein to improve the barrier properties of high barrier polyhydroxyalkanoates and other polyesters. J Food Eng 127:1–9. https://doi.org/10.1016/j.jfoodeng.2013.11.022
Fabra MJ, Sanchez G, Lopez-Rubio A, Lagaron JM (2014b) Microbiological and ageing performance of polyhydroxyalkanoate-based multilayer structures of interest in food packaging. LWT-Food Sci Technol 59:760–767. https://doi.org/10.1016/j.lwt.2014.07.021
Fan X, Jiang Q, Sun Z, Li G, Ren X, Liang J, Huang TS (2015) Preparation and characterization of electrospun antimicrobial fibrous membranes based on polyhydroxybutyrate (PHB). Fibers Polym 16:1751–1758. https://doi.org/10.1007/s12221-015-5108-1
Follain N, Chappey C, Dargent E, Chivrac F, Crétois R, Marais S (2014) Structure and barrier properties of biodegradable polyhydroxyalkanoate films. J Phys Chem C 118:6165–6177. https://doi.org/10.1021/jp408150k
Fu XZ, Tan D, Aibaidula G, Wu Q, Chen JC, Chen GQ (2014) Development of *Halomonas* TD01 as a host for open production of chemicals. Metab Eng 23:78–91. https://doi.org/10.1016/j.ymben.2014.02.006
Geyer R, Jambeck JR, Law KL (2017) Production, use, and fate of all plastics ever made. Sci Adv 3:e1700782. https://doi.org/10.1126/sciadv.1700782
Gironi F, Piemonte V (2011) Bioplastics and petroleum-based plastics: strengths and weaknesses. Energy Sources A: Recover Utilization Environ Eff 33:1949–1959. https://doi.org/10.1080/15567030903436830
Goncalves SPC, Martins-Franchetti SM, Chinaglia DL (2009) Biodegradation of the films of PP, PHBV and its blend in soil. J Polym Environ 17:280–285. https://doi.org/10.1007/s10924-009-0150-y

Haugaard V, Danielsen B, Bertelsen G (2003) Impact of polylactate and poly(hydroxybutyrate) on food quality. Eur Food Res Technol 216:233–240. https://doi.org/10.1007/s00217-002-0651-6
Hazer B, Steinbüchel A (2007) Increased diversification of polyhydroxyalkanoates by modification reactions for industrial and medical applications. Appl Microbiol Biotechnol 74:1–12. https://doi.org/10.1007/s00253-006-0732-8
Hermida EB, Mega VI, Yashchuk O, Fernández V, Eisenberg P, Miyazaki SS (2008) Gamma irradiation effects on mechanical and thermal properties and biodegradation of poly(3-hydroxybutyrate) based films. Macromol Symp 263:102–113. https://doi.org/10.1002/masy.200850313
Kantola M, Helén H (2001) Quality changes in organic tomatoes packaged in biodegradable plastic films. J Food Qual 24:167–176. https://doi.org/10.1111/j.1745-4557.2001.tb00599.x
Keshavarz T, Roy I (2010) Polyhydroxyalkanoates: bioplastics with a green agenda. Curr Opin Microbiol 13:321–326. https://doi.org/10.1016/j.mib.2010.02.006
Khanna S, Srivastava A (2005) Recent advances in microbial polyhydroxyalkanoates. Process Biochem 40:607–619. https://doi.org/10.1016/j.procbio.2004.01.053
Khosravi-Darani K, Bucci DZ (2015) Application of poly (hydroxyalkanoate) in food packaging: improvements by nanotechnology. Chem Biochem Eng Q 29(2):275–285. https://doi.org/10.15255/CABEQ.2014.2260
Kiran GS, Jackson SA, Priyadharsini S, Dobson ADW, Selvin J (2017) Synthesis of Nm-PHB (nanomelanin-polyhydroxy butyrate) nanocomposite film and its protective effect against biofilmforming multi drug resistant *Staphylococcus aureus*. Sci Rep 7:9167. https://doi.org/10.1038/s41598-017-08816-y
Koller M (2014) Poly(hydroxyalkanoates) for food packaging: application and attempts towards implementation. Appl Food Biotechnol 1:3–15
Koller M, Salerno A, Muhr A, Reiterer A, Braunegg G (2013) Polyhydroxyalkanoates: biodegradable polymers and plastics from renewable resources. Mater Technol 47:5–12
Kwiecień I, Adamus G, Bartkowiak A, Kowalczuk M (2014) Synthesis and structural characterization at the molecular level of oligo(3-hydroxybutyrate) conjugates with antimicrobial agents designed for food packaging materials. Des Monomers Polym 17:311–321. https://doi.org/10.1080/15685551.2013.840505
Laycock B, Halley P, Pratt S, Werker A, Lanta P (2013) The chemomechanical properties of microbial polyhydroxyalkanoates. Prog Polym Sci 38:536–583. https://doi.org/10.1016/j.progpolymsci.2012.06.003
Levkane V, Muizniece-Brasava S, Dukalska L (2008) Pasteurization effect to quality of salad with meat and mayonnaise. Foodbalt:69–73
Li Z, Lin H, Ishii N, Chen GQ, Inoue Y (2007) Study of enzymatic degradation of microbial copolyesters consisting of 3-hydroxybutyrate and medium-chain-length 3-hydroxyalkanotes. Polym Degrad Stab 92:1708–1714. https://doi.org/10.1016/j.polymdegradstab.2007.06.001.
Lodhi AF, Hasan F, Shah Z, Hameed A, Faisal S, Shah AA (2011) Optimization of culture conditions for the production of poly(3-hydroxybutyrate) depolymerase from newly isolated Aspergillus fumigatus from soil. Pak J Bot 43:1361–1372
Maiti P, Batt CA, Emmanuel G (2003) Renewable plastics: synthesis and properties of PHB nanocomposites. Polym Mater Sci Eng 88:58–59
Martínez-Sanz M, Villano M, Oliveira C, Albuquerque MG, Majone M, Reis M, Lopez-Rubio A, Lagaron JM (2014) Characterization of polyhydroxyalkanoates synthesized from microbial mixed cultures and of their nanobiocomposites with bacterial cellulose nanowhiskers. New Biotechnol 31:364–376. https://doi.org/10.1016/j.nbt.2013.06.003
Masood F (2017) Polyhydroxyalkanoates in the food packaging industry. In Oprea AE, Grumezescu AM (eds) Nanotechnology applications in food. Academic/Elsevier, London, pp 153–177. ISBN:978-0-12-811942-6
Masood F, Aziz M, Haider H, Shakil O, Yasin T, Hameed A (2018) Biodegradation of gamma irradiated poly-3-hydroxybutyrate/sepiolite nanocomposites. Int Biodeterior Biodegrad 126:1–9. https://doi.org/10.1016/j.ibiod.2017.09.012

Miguel O, Iruin JJ (1999a) Evaluation of the transport properties of poly(3-hydroxybutyrate) and its 3-hydroxyvalerate copolymers for packaging applications. Macromol Symp 144:427–438. https://doi.org/10.1002/masy.19991440140

Miguel O, Iruin JJ (1999b) Water transport properties in poly(3-hydroxybutyrate) and poly(3-hydroxybutyrate-co-3-hydroxyvalerate) biopolymers. J Appl Polym Sci 73:455–468. https://doi.org/10.1002/(SICI)1097-4628(19990725)73:4<455::AID-APP1>3.0.CO;2-Y

Miguel O, Fernandez-Berridi MJ, Iruin JJ (1997) Survey on transport properties of liquids, vapors, and gases in biodegradable poly(3-hydroxybutyrate) (PHB). J Appl Polym Sci 64:1849–1859. https://doi.org/10.1002/(SICI)1097-4628(19970531)64:9<1849::AID-APP22>3.0.CO;2-R

Mohamed El-Hadi A (2014) Investigation of the effect of nano-clay type on the non-isothermal crystallization kinetics and morphology of poly(3(R)-hydroxybutyrate) PHB/clay nanocomposites. Polym Bull 71:1449–1470. https://doi.org/10.1007/s00289-014-1135-0

Mohapatra S, Maity S, Dash HR, Das S, Pattnaik S, Rath CC, Samantaray D (2017) *Bacillus* and biopolymer: prospects and challenges. Biochem Biophys Rep 12:206–213. https://doi.org/10.1016/j.bbrep.2017.10.001

Możejko-Ciesielska J, Kiewisz R (2016) Bacterial polyhydroxyalkanoates: still fabulous? Microbiol Res. https://doi.org/10.1016/j.micres.2016.07.010

Muizniece-Brasava S, Dukalska L (2006) Impact of biodegradable PHB packaging composite materials on dairy product quality. In: Proceedings of the Latvia University of Agriculture LLU Raksti 16:79–87

Narayanan A, Neera M, Ramana KV (2013) Synergized antimicrobial activity of eugenol incorporated polyhydroxybutyrate films against food spoilage microorganisms in conjunction with pediocin. Appl Biochem Biotechnol 170:1379–1388. https://doi.org/10.1007/s12010-013-0267-2

North EJ, Halden RU (2013) Plastics and environmental health: the road ahead. Rev Environ Health 28:1–8. https://doi.org/10.1515/reveh-2012-0030.

Pardo-Ibáñez P, Lopez-Rubio A, Martínez-Sanz M, Cabedo L, Lagaron JM (2014) Keratin–polyhydroxyalkanoate melt-compounded composites with improved barrier properties of interest in food packaging applications. J Appl Polym Sci 131:39947/1–39947/10. https://doi.org/10.1002/app.39947

Pavani P, Rajeswari TR (2014) Impact of plastics on environmental pollution. National Seminar on impact of toxic metals, minerals and solvents leading to environmental pollution. J Chem Pharm Sci 3:87–93

Plackett D, Siró I (2011) Polyhydroxyalkanoates (PHAs) for food packaging. In: Lagarón JM (ed) Multifunctional and nanoreinforced polymers for food packaging. Woodhead Publishing, Cambridge, pp 498–526. https://doi.org/10.1533/9780857092786.4.498

Plastics Europe (2016) Plastics-the facts 2016: an analysis of European plastics production, demand and waste data

Poley LH, Siqueira APL, Da Silva MG, Sanchez R, Prioli R, Mansanares AM, Vargas H (2005) Photothermal methods and atomic force microscopy images applied to the study of poly(3-hydroxybutyrate) and poly(3-hydroxybutyrate-co-3- hydroxyvalerate) dense membranes. J Appl Polym Sci 97:1491–1497. https://doi.org/10.1002/app.21891

Prasad P, Kochhar A (2014) Active packaging in food industry: a review. IOSR J Environ Sci Toxicol Food Technol (IOSR-JESTFT) 8:01–07

Rai R, Keshavarz T, Roether J, Boccaccini AR, Roy I (2011) Medium chain length polyhydroxyalkanoates, promising new biomedical materials for the future. Mater Sci Eng R 72:29–47. https://doi.org/10.1016/j.mser.2010.11.002

Ramachandran D, Latha RR, Gowri GR (2013) Incorporation of *clitoria ternatea* seed extract into bioplastic sheets having induced plasticity and their antimicrobial activity against multi-drug resistant clinical pathogens. Int J Pharm Pharm Sci 5:543–550

Rastogi VK, Samyn P (2015) Bio-based coatings for paper applications. Coatings 5:887–930

Rastogi VK, Samyn P (2017) Synthesis of polyhydroxybutyrate particles with micro-to-nanosized structures and application as protective coating for packaging papers. Nanomaterials 7:5. https://doi.org/10.3390/nano7010005

Reis DCC, Lemos-Morais AC, Carvalho LH, Alves TS, Barbosa R (2016) Assessment of the morphology and interaction of PHBV/clay bionanocomposites: uses as food packaging. Macromol Symp 367:113–118. https://doi.org/10.1002/masy.201500143
Reis DCC, Oliveira TA, Carvalho LH, Alves TS, Barbosa (2017) The influence of natural clay and organoclay vermiculite on the formation process of bionanocomposites with poly (3-hydroxybutyrate-co-3-hydroxyvalerate). Matéria (Rio J) 22:1186. https://doi.org/10.1590/S1517-707620170004.0220
Renard E, Walls M, Guérin P, Langlois V (2004) Hydrolytic degradation of blends of polyhydroxyalkanoates and functionalized polyhydroxyalkanoates. Polym Degrad Stab 85:779–787. https://doi.org/10.1016/j.polymdegradstab.2003.11.019
Rhim JW, Park HM, Ha CS (2013) Progress in polymer science bio-nanocomposites for food packaging applications. Prog Polym Sci 38:1629–1652. https://doi.org/10.1016/j.progpolymsci.2013.05.008
Rudnik E (2012) Compostable polymer properties and packaging applications In: Ebnesajjad S (ed) Plastic films in food packaging. William Andrew. 2013, pp 217–248. ISBN:978-1-4557-3112-1
Saini RD (2017) Biodegradable polymers. Int J Appl Chem 13:179–196
Salama HE, Saad GR, Sabaa MW (2017) Synthesis, characterization and antimicrobial activity of biguanidinylated chitosan-g-poly[(R)-3-hydroxybutyrate]. Int J Biol Macromol 101:438–447. https://doi.org/10.1016/j.ijbiomac.2017.03.075
Sanchez-Garcia MD, Gimenez E, Lagaron JM (2007) Novel PET nanocomposites of interest in food packaging applications and comparative barrier performance with biopolyester nanocomposites. J Plas Film Sheet 23:133–148. https://doi.org/10.1177/8756087907083590
Sanchez-Garcia MD, Gimenez E, Lagaroân JM (2008) Morphology and barrier properties of nanobiocomposites of poly(3-hydroxybutyrate) and layered silicates. J Appl Polym Sci 108:2787–2801. https://doi.org/10.1002/app.27622
Santos Rosa D, Calil MR, Guedes CGF, Rodrigues TC (2004) Biodegradability of thermally aged PHB, PHB-V, and PCL in soil compostage. J Environ Polym Degr 12:239–245. https://doi.org/10.1007/s10924-004-8151-3
Shah AA, Hasan F, Hameed A, Ahmed S (2008) Biological degradation of plastics: a comprehensive review. Biotechnol Adv 26:246–265. https://doi.org/10.1016/j.biotechadv.2007.12.005
Siddiqui J, Pandey G (2013) A review of plastic waste management strategies. Int Res J Environ Sci 2:84–88
Siracusa V (2012) Food packaging permeability behavior: a report. Int J Polym Sci 2012:302029., 11 pages. https://doi.org/10.1155/2012/302029
Steinbüchel A, Lütke-Eversloh T (2003) Metabolic engineering and pathway construction for biotechnological production of relevant polyhydroxyalkanoates in microorganisms. Biochem Eng J 16:81–96. https://doi.org/10.1016/S1369-703X(03)00036-6
Tan GYA, Chen CL, Li L, Ge L, Wang L, Razaad IMN, Li Y, Zhao L, Mo Y, Wang JY (2014) Start a research on biopolymer polyhydroxyalkanoate (PHA): a review. Polymer 6:706–754. https://doi.org/10.3390/polym6030706
Thellen C, Coyne M, Froio D, Auerbach M, Wirsen C, Ratto JA (2008) A processing, characterization and marine biodegradation study of melt-extruded polyhydroxyalkanoate (PHA) films. J Polym Environ 16:1–11. https://doi.org/10.1007/s10924-008-0079-6
Thellen C, Cheney S, Ratto JA (2013) Melt processing and characterization of polyvinyl alcohol and polyhydroxyalkanoate multilayer films. J Appl Polym Sci 127:2314–2324. https://doi.org/10.1002/app.37850
Valentino F, Beccari M, Fraraccio S, Zanaroli G, Majone M (2014) Feed frequency in a sequencing batch reactor strongly affects the production of polyhydroxyalkanoates (PHAs) from volatile fatty acids. New Biotechnol 31:264–275. https://doi.org/10.1016/j.nbt.2013.10.006
Wang Y, Chen GQ (2017) Polyhydroxyalkanoates: sustainability, production, and industrialization. In Tang C, Ryu CY (eds) Sustainable polymers from biomass. Wiley-VCH Verlag GmbH & Co. KGaA, pp 11–33

Wang S, Song C, Chen G, Guo T, Liu J, Zhang B, Takeuchi S (2005) Characteristics and biodegradation properties of poly(3-hydroxybutyrate-co-3-hydroxyvalerate)/organophilic montmorillonite (PHBV/OMMT) nanocomposites. Polym Degrad Stab 87:69–76. https://doi.org/10.1016/j.polymdegradstab.2004.07.008
Wang Y, Yin J, Chen GQ (2014) Polyhydroxyalkanoates, challenges and opportunities. Curr Opin Biotechnol 30:59–65. https://doi.org/10.1016/j.copbio.2014.06.001
Witholt B, Kessler B (1999) Perspectives of medium chain length poly(hydroxyalkanoates), a versatile set of bacterial bioplastics. Curr Opin Biotechnol 10:279–285. https://doi.org/10.1016/s0958-1669(99)80049-4
Xavier JR, Babusha ST, George J, Ramana KV (2015) Material properties and antimicrobial activity of Polyhydroxybutyrate (PHB) films incorporated with vanillin. Appl Biochem Biotechnol 176:1498–1510. https://doi.org/10.1007/s12010-015-1660-9
Yu HY, Qin ZY, Binsun YXG, Yao JM (2014) Reinforcement of transparent poly(3-hydroxybutyrate-co-3-hydroxyvalerate) by incorporation of functionalized carbon nanotubes as a novel bionanocomposite for food packaging. Compos Sci Technol 94:96–104. https://doi.org/10.1016/j.compscitech.2014.01.018

Approaches for Enhancing Extraction of Bacterial Polyhydroxyalkanoates for Industrial Applications

15

Karine Laste Macagnan, Mariane Igansi Alves, and Angelita da Silveira Moreira

Abstract

Poly(3-hydroxybutyrate) [P(3HB)] and others polyhydroxyalkanoates (PHAs) are recovered from the cells after their accumulation, in the production phase. The impact of PHA extraction may represent more than 50% of total production process. To ensure proper removal of the bioplastic from the bacterial cells several unitary operations must be performed in the extraction process. In addition, one also need to take in to consideration: reduction of product losses during processing, preservation of the physicochemical properties of the biopolymer, obtaining a highly pure product, use of low toxicity solvents/chemicals. In this case, PHA extraction methods include: chemical methods by using solvents, chemical digestion and aqueous system of two phases; mechanical methods with high pressure homogenization, ball mill, French press and ultrasound; biological methods, with enzymatic digestion, digestion by bacteria and predatory larvae or single cell protein; or a combination of these methods. This chapter discusses both classical and innovative methods for recovering/purifying PHAs, emphasizing yield, purity and the effect on the properties of the obtained biopolymers.

K. L. Macagnan
Biotechnology, Technological Development Center, Federal University of Pelotas, Pelotas, Rio Grande do Sul, Brazil

M. I. Alves
Department of Food Science and Technology, Faculty of Agronomy Eliseu Maciel, Federal University of Pelotas, Pelotas, Rio Grande do Sul, Brazil

A. da. S. Moreira (✉)
Biotechnology, Technological Development Center, Federal University of Pelotas, Pelotas, Rio Grande do Sul, Brazil

Department of Food Science and Technology, Faculty of Agronomy Eliseu Maciel, Federal University of Pelotas, Pelotas, Rio Grande do Sul, Brazil

V. C. Kalia (ed.), *Biotechnological Applications of Polyhydroxyalkanoates*,
https://doi.org/10.1007/978-981-13-3759-8_15

Keywords
Bioplastic · Poly(3-hydroxybutyrate) · Recovery methods · Chemical methods · Mechanical methods · Biological methods

15.1 Introduction

Recovery is the last stage of the polyhydroxyalkanoate (PHA) production process. After culturing, the cell suspension containing the bioplastics is extracted from the broth by normal methods, such as filtration, centrifugation or flocculation, for further rupture and the polymer extracted.

The development of a methodology that permits the recovery of PHAs by a simple, efficient and less polluting process is a necessary and attractive proposal. For the recovery of PHAs several methods have been developed, such as: cell disruption and polymer extraction by chemical methods with the use of solvents, chemical digestion and aqueous system of two phases; mechanical methods with high pressure homogenization, ball mill, French press and ultrasound; biological methods with enzymatic digestion, digestion with bacteria and predatory larvae, single-cell protein; or a combination of these methods (Jacquel et al. 2008; Kourmentza et al. 2017).

It is essential that the extraction process provides high yield and is low cost, since this step represents 50% or more of the total cost of the product (Khanna and Srivastava 2005). The high cost of the process, as well as the use of large amounts of solvents, often with high toxicity and danger, hampers the large-scale production of PHAs. However, yield and cost are not the only factors to consider. The extraction and/or purification method may positively or negatively influence the characteristics of the obtained biopolymer.

Poly(3-hydroxybutirate) (P(3HB)) is the PHA that has been the most frequently studied and characterized to date. The high price of P(3HB) and other PHAs, when compared to traditional plastics, discourages their use in conventional products with short life and low added value. However, nobler uses, such as high value-added medical or pharmaceutical products, have redirected the attention of researchers and producers of biodegradable plastics to PHAs (Kourmentza et al. 2017). Properties such as molar mass, crystallinity and speed of biodegradation direct their applicability and may even influence market value (Jacquel et al. 2008). The methods of recovery and purification can alter, to a greater or lesser degree, these properties.

This chapter discusses both classical and innovative methods for recovering/purifying PHAs, highlighting yield, purity and effect on the properties of the obtained biopolymers (Table 15.1).

Table 15.1 Summary of polyhydroxyalkanoates extraction methods

Extraction method	Agent	Microorganism	Results (%, Da)	Coments	References
Solvents	Chloroform	*Bacillus cereus*	Rec. 30, Pur. 92, Mm 1.1×10^6, PDI 1.75	High purity, low recovery, High Mm and low PDI	Valappil et al. (2007)
		Fresh *Ralstonia solanacearum*	Rec. 98, Mm $(2.9–3.3) \times 10^5$, Xc 52	High recovery, lower Mm, highest crystallinity	Macagnan et al. (2017)
		Dry *Ralstonia solanacearum*	Rec. 41.8, Mm $(5.1–6.3) \times 10^5$, Xc 37	Low recovery, highest Mm, lower crystallinity	
	1,2-propylene carbonate	*Cupriavidus necator*	Rec. 95, Pur. 84, Mm 7.4×10^5	High recovery, good purity, high Mm	Fiorese et al. (2009)
	Ethyl acetate	Dry *Ralstonia eutropha*	Pur. 99	High recovery levels and purity	Riedel et al. (2013)
	Methyl isobutyl ketone	Fresh *Ralstonia eutropha*	Rec. 84, Pur. 99	Good recovery, high purity	
	Butyl acetate	*Cupriavidus necator*	Rec. 96, Pur. 98, Mm 1.4×10^6	High recovery, purity and Mm	Aramvash et al. (2015)
Solvents	Ethyl acetate	*Cupriavidus necator*	Rec. 82, Pur. 98, Mm 1.4×10^6	Good recovery, high purity and Mm	Aramvash et al. (2015)
	Acetone		Rec. 82.6, Pur. 98, Mm 6×10^5	Good recovery, high purity, lower Mm	Aramvash et al. (2016)
	Ethanol		Rec. 92, Pur. 99, Mm 1.2×10^6	High recovery and purity, highest Mm	
	Propylene carbonate	*Cupriavidus necator* DSM 545	Rec. 98, Pur. 99	High recovery and purity	Quines et al. (2015)
Chemical digestion	NaCl + NaOH	*Cupriavidus necator*	Rec. 95, Pur, 97, Mm 2.8×10^5	High recovery and purity, low Mm	Anis et al. (2013)

(continued)

Table 15.1 (continued)

Extraction method	Agent	Microorganism	Results (%, Da)	Coments	References
	NaOH + SDS	*P. acidivorans*	Rec. 90, Pur, 99	Good recovery, high purity	Jiang et al. (2015)
	Emal S PASTE	recombinant *Escherichia coli*	Rec. 99, Pur. 98	High recovery and purity	Peng et al. (2013)
Aqueous two-phase systems	Polyethylene glycol/ sodium sulfate/NaCl	*Ralstonia eutropha* H16	Rec. 65	Low recovery	Leong et al. (2015)
	EOPO/NaCl	*Cupriavidus necator* H16	Rec. 94.8	High recovery	Leong et al. (2016)
High- pressure homogenization	–	Recombinant *E.coli*	Rec. 72, Pur. 90.2	Good recovery, high purity	Ling et al. (1998)
	–	*Pseudomonas putida*	Rec. 46, Pur. 80	Low recovery, good purity	Van Hee et al. (2004)
	Acetone	–	Pur. 37.2, Mm 2.1×10^5, PDI 25.5	Low purity and Mm, high PDI	Koller et al. (2013)
	SDS	*Methylobacterium* sp. V49	Rec. 98, Pur. 95	High recovery and purity	Ghatnekar et al. (2002)
	Acetone/ethanol/ propylene carbonate	*Cupriavidus necator*	Rec. 85, Pur. 92	High recovery and purity	Fei et al. (2016)
Ultrasound- assisted	Propylene carbonate	*Cupriavidus necator*	Rec. 99, Pur. 90, Mm 1.38×10^5	High recovery and purity, low Mm	Quines (2015)
	Triton X-100	*Cupriavidus necator*	Rec. 26, Pur. 88.2	Low recovery, high purity	Murugesan and Iyyaswami (2017)
	Acetone/heptane	*Pseudomonas putida* Bet001	Rec. 37.1, Mm 3.9×10^4	Low recovery and Mm	Ishak et al. (2016)
Bead milling	–	*Cupriavidus necator*	Rec. 64, Pur. 89, Mm 7.5×10^5, PDI 1.12	Low recovery and PDI, high purity and Mm	Garcia (2006)
Bead milling	SDS	*Cupriavidus necator*	Rec. 74, Pur. 74	Low recovery and purity	Gutt et al. (2016)

French- press	–	*Ralstonia eutropha*	Rec. 40	Low recovery	Gözde et al. (2012)
	Glycerol	*Pseudomonas extremaustralis*	Rec. 28	Low recovery	Catone et al. (2014)
	Tris–HCl	*Dinoroseobacter* sp. JL 1447	1–2 granules	–	Xiao et al. (2015)
	–	*Aquitalea* sp. USM4	Rec. 27–55	Low recovery	Ng and Sudesh (2016)
	Tris-HCl	*Cupriavidus necator*	Pur. 100	High purity	Murugan et al. (2016)
	Trypsin/chloroform	*Arxula adeninivorans*	Rec. 71.9, Pur. 99.2, Mm 1.7×10^5	Good recovery, high purity, low Mm	Biernacki et al. (2017)
Enzyme igestion	Crude enzyme *Aspergillus oryzae*	*Cupriavidus necator*	Rec. 98, Pur. 96.7, Xc 66	High recovery, purity and crystallinity	Kachrimanidou et al. (2016)
	Lysozyme/methanol	*Cupriavidus necator*	Rec. 75, Pur. 41	Good recovery, low purity	Gutt et al. (2016)
Enzyme digestion	SDS/EDTA/alkaline enzyme	*Cupriavidus necator*	Pur. 90	High purity	Martino et al. (2014)
Cellular lysis with bacterial predator	*Bdellovibrio bacteriovorus* HD100	*Pseudomonas putida* KT2440	Rec. 80	Good recovery	Martínez et al. (2016)
Single cell protein	*Sprague Dawley* rats	*Cupriavidus necator* H16	Rec. 74, Pur. 82	Good recovery and purity	Kunasundari et al. (2013)
	Sprague Dawley rats/ SDS		Pur. 97	High purity	Kunasundari et al. (2017)
	Tenebrio molitor/SDS	*Cupriavidus necator* Re2058/pCB113	Pur. 100	High purity	Murugan et al. (2016)

Rec. Recovery, *Pur.* Purity, *Mm* molar mass, *PDI* polydispersity index, *Xc* degree of crystallinity, *Emal S PASTE* sodium lauryl sulfate, *EOPO* Ethylene oxide-propylene oxide, *SDS* sodium dodecyl sulfate, *EDTA* ethylenediaminetetraacetic acid

15.2 Chemical Methods

The chemical methods of extracting PHAs from bacteria involve the isolated or combined use of solvents and chemical digesters.

15.2.1 Solvents

One of the oldest method to recover PHAs is by using solvent. It is the most widely used technique due to its simplicity, obtaining polymers with high purity and reduced reduction of the molar mass of the product (Gumel et al. 2013).

The solvent is the substance capable of dissolving the solute in order to form a solution of the solute in the solvent (Mantellato and Durão 2008). Solubilization of the polymer by a given solvent is possible when the interactions between the solvent molecules and the polymer chains have a magnitude greater than the magnitude of the interactions between the polymer chains (Jacquel et al. 2008). The solvent acts in two steps: (i) the cell membrane permeability modification, and (ii) the solubilization of the lipids, including the PHAs, forming a suspension of the cellular debris in the polymer and solvent solution (Jacquel et al. 2008). After solubilization of the polymer in the solvent, the polymer is separated from the cell waste by filtration or centrifugation and the final product is usually obtained after evaporation of the solvent, or by precipitation of the polymer by addition of a "non-solvent" such as an alcohol or water (Peng et al. 2013). A non-solvent is defined as a substance that is unable to dissolve the biopolymer, causing the precipitation thereof to occur by induction and/or by weakening the solvent dilution power (Peng et al. 2013).

Halogenated solvents such as chloroform, dichloroethane, chloropropane and methyl chloride have been widely used for the extraction of PHAs from bacteria, with chloroform being the most studied. However, other solvents have also been studied for this purpose, such as acetone, methyl isobutyl ketone, propylene carbonate, ethyl acetate and isoamyl alcohol (Valappilet al. 2007; Fiorese et al. 2009; Ibrahim and Steinbüchel 2009; Budde et al. 2011; Riedel et al. 2013; Rodríguez-Contreras et al. 2013; Quines et al. 2015; Macagnan et al. 2017).

Valappil et al. (2007) evaluated methods of PHA recovery using a dispersion of hypochlorite in chloroform with *Bacillus cereus*. Variations of these methodologies were the use of the solvent chloroform (100 mL/g) with or without Soxhlet reflux, or a mixture of 1:1 sodium hypochlorite and chloroform (100 mL/g). High purity rates were obtained, above 92%, but very low yields, around 30% in all methodologies. Comparing the weight average molar mass (Mm) and polydispersity index (PDI) of the polymer isolated using the Soxhlet extraction technique (Mm 1.1×10^6, PDI 1.75) to that of the chloroform–hypochlorite dispersion method (Mm 0.885×10^6, PDI 3.1) and the chloroform extraction method (Mm 0.882×10^6, PDI 2.6) confirms that the Soxhlet extraction results in higher Mm and lower PDI (Valappil et al. 2007).

Macagnan et al. (2017) optimized the extraction methodology of P(3HB) using chloroform as solvent from *Ralstonia solanacearum* cells. In this study, the time of heating, the separation method of the cell residue, solvent (filtration or separation of liquid-liquid phase), state of biomass (fresh or dried cell concentrate) and solvent/biomass proportion were determined. The process was improved using dry cells, and the ratio of solvent/biomass was 40:1, the heating time was reduced to 30 min and the liquid–liquid phase separation method was more efficient, recovering 98% of the accumulated polymer. The molar mass (Mm) of the polymers which were extracted from the fresh cell concentrate (FCC) were higher than those recovered from the dry cell concentrate (DCC). The values ranged from (2.9–3.3) × 10^5 Da for the DCC polymers and (5.1–6.3) × 10^5 Da for the FCC polymers. The extraction conditions with the highest yield utilizing DCC produced P(3HB) with 52% crystallinity, higher compared to the P(3HB) extracted from FCC (37%) (Macagnan et al. 2017). The degree of crystallinity directly influence the PHA biodegradability. Polymers with low crystalline structure are preferred due to their lower biodegradation times and they also possess desirable mechanical properties for medical applications such as sutures and pharmaceuticals such as controlled-release of drugs (Lee 1996; Serafim et al. 2002).

The main problem with the use of solvents is its hazardous nature causing trouble for the environment and operators. This is partly because solvent extraction is widely used in laboratories, however, their usage in pilot plant and large-scale processing is restricted. An excellent recovery method should have low toxicity, not be expensive nor generate high volumes of liquid waste and also not cause unwanted reduction of the molar mass by product fragmentation. The heating time should be the shortest and use the lowest temperature possible.

Efforts to reduce these problems, especially in relation to toxicity, extraction methods using non-halogenated solvents have been developed. Fiorese et al. (2009) evaluated a procedure on extraction with 1,2-propylene carbonate for the recovery of P(3HB) produced by *Cupriavidus necator*, especially, temperature (100–145 °C) and contact period (15–45 min) on the extraction efficiency of P(3HB) and biopolymer characteristics was evaluated. Higher yield (95%) and purity (84%) were obtained by a combination of a temperature of 130 °C and a contact period of 30 min in a precipitation period of 48 h. Here, the P(3HB) had a molar mass of 7.4 × 10^5; higher temperatures yielded polymer with lower molar weight and incomplete extraction yields. For higher temperatures (145 °C) were obtained lower values (2.2 × 10^5). This process uses large quantities of chemicals, which adds to recovery costs and makes the process environmentally unfriendly, however polymer molar mass and purity were improved by pretreating biomass with pH/heat combination (Fiorese et al. 2009).

Riedel et al. (2013) evaluated the solvents methyl isobutyl ketone, butanone, butyl acetate, ethyl acetate and isoamyl alcohol for the extraction of poly(3-hydroxybutyrate-co-3-hydroxyhexanoate) copolymer from dry and fresh biomass of *Ralstonia eutropha*. With the use of ethyl acetate solvent and dry cells, high recovery levels and high product purity (99%) were obtained. However, to reduce the recovery time and cost of the process, the extraction with fresh cells is shown as a possible alternative. For the process using fresh cells, the authors recommended the solvent methyl isobutyl ketone, achieving purity of up to 99% and total recovery yields of up to 84%. Butyl acetate may alternatively be used because of its low miscibility with water, low flammability and higher boiling points. A real problem while using butyl acetate is it's hydrolysis in the presence of water, which reduces the level of sustainability since solvent is continuously recycled (Riedel et al. 2013).

Aramvash et al. (2015) proposed a new efficient and environmentally friendly method of extracting P(3HB) from the fresh biomass of *Cupriavidus necator* using non-toxic solvents. The time and temperature of the extraction with butyl and ethyl acetate (100 mL/g) were evaluated, by which they achieved high polymer yields of 96% and 82%, respectively, and a high purity rate of 98% in both solvents. Here, P(3HB) recorded the highest molecular weight of 1.4×10^6 in comparison to recovery using chloroform) (Aramvash et al. 2015). Subsequently, this same research group tested the use of acetone and ethanol at different times and temperatures for the extraction of the polymer. They achieved better results using ethanol for 32 min at 76 °C: 92% yield and 99% purity (Aramvash et al. 2016). Moreover, the molar mass of P(3HB) samples obtained using acetone (6×10^5 at 76.3 °C within 32.5 min) was lower than that of the ethanol-extracted (1.2×10^6 at 50 °C within 60 min). The P(3HB) molar mass was influenced by temperature and duration of heat treatment during the course of extraction (Chan et al. 2004). However, is possible set a limit to the temperature, as the solvent begins to degrade the biopolymer at temperatures above the boiling point of the solvent (Liu et al. 2005). Those authors, by setting the temperature to near the boiling point, observed a 20–30% decrease in molecular weight compared to the standard procedure.

Quines et al. (2015) evaluated the effect of extraction time, heating temperature and amount of solvent during the extraction process of P(3HB) using propylene carbonate as solvent from *Cupriavidus necator* DSM 545. Excellent recovery (98%) and purity (99%) results were obtained, and concluded that the most important factor in these responses was the temperature, which should be between 110 and 150 °C.

15.2.2 Chemical Digestion

The use of chemical digestion for recovery of PHAs produced by bacteria involves the solubilization of the cell wall materials, leaving the PHAs granules in suspension. The non-polymeric materials digested in this process are lipids, phospholipids, peptidoglycan, proteinaceous materials (debris) and nucleic acids.

Several studies report the use of chemical digesters for PHA recovery, but the efficacy of the method depends on the microbial species from which the extraction will be performed (Suzuki et al. 2018). The most commonly used digesters are sodium dodecyl sulfate (SDS), sodium hypochlorite (NaClO), sodium hydroxide (NaOH) and sulfuric acid (H_2SO_4) (Anis et al. 2013; Peng et al. 2013; Rathi et al. 2013; Jiang et al. 2015; Villano et al. 2013; López-Abelairas et al. 2015).

Anis et al. (2013) aimed to develop a economical, simple and environmentally friendly procedure for extraction bioplastic using NaOH digestion from recombinant *Cupriavidus necator* cells. The release of the non-polymeric protein material occurs under alkaline conditions and NaOH increasing the permeability of the cell membrane through the saponification of the lipid layer. For a high-volume recovery, different strategies have been studied, where centrifugation can be avoided to reduce energy consumption. The best strategy, with high recovery of PHA (95%) and purity (90%), was the use of NaOH for digestion and subsequent addition of ethanol. In one of the strategies, a pre-treatment with NaCl was used; although this combination generated high purity (97%), greater degradation of bioplastic was obtained. The molar mass (Mm) was reduced by about 50% (2.8×10^5 Da) in comparison to the PHA recovery with chloroform (4.6×10^5 Da). In addition, significant changes in polymer extracted and purity were not recorded when fresh or lyophilized cells were used (Anis et al. 2013).

Jiang et al. (2015) investigated the feasibility of purifying poly(3-hydroxybutyrate) from chemical treatments based on alkaline biomass of *P. acidivorans*. The alkaline treatment (0.2 M NaOH) along with SDS (0.2% w/v) allowed almost 99% purity, and 90% recovery. The loss of P(3HB) can be attributed mainly to the hydrolysis of the polymer during alkaline treatment. The biomass pretreatment (freeze-drying or lyophilization) or effective cell lysis (adjustment of alkali concentration) are alternatives to increase the crystallinity of the granules within the cells and decrease the hydrolysis of P(3HB). However, this recovery and purification method greatly influenced the quality of the product especially for thermoplastic applications. P(3HB) purified using the solvent showed thermal stability which was as good as that recorded in commercial P(3HB). However, alkali treatment deteriorated the thermal stability of the P(3HB), in spite of higher recovery and purity of the yield. This procedure needs to be studied further for other applications (Jiang et al. 2015).

Peng et al. (2013) evaluated the isolation of poly(3-HB) from recombinant *Escherichia coli* using six types of commercial surfactants: Emal 10P, Emal S PASTE (sodium lauryl sulfate), Emal AD-25 (ammonium lauryl sulfate), Neopelex SS (dodecyl benzene sulfonic acid), Triton X-100 (polyoxyethylene-*p*-tert-octylphenol) and CTAB (cetyltrimethylammonium bromide). Factors such as temperature and reaction time, proportion of non-polymeric cellular material (NPCM)/surfactant and pH were investigated to reach optimal conditions. Emal 10P, Emal AD-25, Emal S PASTE and Triton X-100 are suitable surfactants for this purpose. Using the proportion 0.67 g/g NPCM/Emal S PASTE reached 98% purity and recovery of 99% P (3HB) at 70 °C for 30 min (Peng et al. 2013).

Although the PHA recovery and purity results obtained from extraction with the combination of chemical digesters and solvents of microbial cells are high, they do not compensate for problems related to excessive volumes of chemicals used (often with high toxicity), making the cost process high and generating large volumes of waste water (Chen et al. 2001; Quines et al. 2015).

15.2.3 Aqueous Two-Phase Systems (ATPS)

Aqueous two-phase systems (ATPS) are new methods of extraction and purification of various materials which have many advantages especially for the separation of sensitive biomaterials, since the ATPS has high water content (70–90% w/w); the materials forming the different phases/layers of ATPS are safe and environmentally friendly. In addition, it has high purification capacity with reduced numbers of steps; and large-scale purification can also be achieved. ATPS is a viable solution to the industrial scale production with high economy and efficiency (Leong et al. 2015, 2016, Iqbal et al. 2016; Kourmentza et al. 2017).

Among PHA recovery methods, ATPS promotes the accumulation of biopolymers using environmentally friendly phase-forming components. Leong et al. (2015) have developed a method with an ability to handle high operations and to reduce downstream processing volume. In this study, the effect of pH and addition of salts on *Ralstonia eutropha* H16 cultures was analyzed, with a recovery yield of 65%, yielding 0.139 g/L of PHA, using an ATPS with polyethylene glycol 8000/sodium sulfate at pH 6 and addition of 0.5 M NaCl (Leong et al. 2015).

In order to improve the recovery performance of the biopolymer by the ATPS method, Leong et al. (2016) carried out a new study evaluating the effect of the thermal separation of the phases during the recovery process. The vapor-point extraction (CPE) technique allows the re-use of the phase-forming component, and takes advantage of the thermo-responsive polymer. The extraction of PHA from *Cupriavidus necator* H16 via CPE was investigated. A recovery yield of 94.8% and purification factor of 1.42 were achieved using 20.0 wt % of ethylene oxide–propylene oxide (EOPO), addition of 10 mM NaCl, temperature of 60 °C and gross raw material limit of 37.5 wt%. In addition, this process has the ability to recycle and reuse EOPO 3900 (Leong et al. 2016).

15.3 Physical Methods

Mechanical rupture methods arise as interesting alternatives because they are economical for large-scale recovery techniques (Middelberg 1995).

15.3.1 High- Pressure Homogenization

The high- pressure homogenization processes are based on cell disruption through the passage of the biomass through a valve or narrow orifice, followed by depressurizing and large increase of flow velocity, with high shear stress and consequent cavitation causing the deformation or the rupture of cells that are in suspension (Masson 2010; Maresca et al. 2011).

Cellular disruption with high-pressure homogenization presents advantages such as the operation pressure, the number of passes of the cell suspension by the equipment and the cellular concentration used in the process. However, as a disadvantage, the process can degrade the product and form very fine cellular debris causing impurity in the recovered biopolymer (Ghatnekar et al. 2002; Van et al. 2004).

A method the recovery of P(3HB) from recombinant *E.coli* with a high-pressure homogenizer was proposed by Ling et al. (1998), with an operating pressure of 550 bar, four passes of the biomass by the equipment and subsequent centrifugation for the recovery of the sedimented biopolymer. The results obtained in this work were 72% recovery and 90.2% purity. By using high pressure (1600 bar), followed by centrifugation of the broth at 30,000 × g for 3 h for separation of the polymer material, Van Hee et al. (2004) report the use of a high-pressure homogenizer for extraction of PHA produced by *Pseudomonas putida*, with recovery of 46% and purity 80%; a lower recovery than that obtained by Ling et al. (1998), probably due to losses during the process. Koller et al. (2013), extracted short chain polyhydroxyalkanoates (scl-PHAs) from acetone under high temperature and high pressure using a steel-containing equipment in its aluminum, connection and acetone-proof seals. In the process 21 g of biomass were used in 700 mL of acetone at a temperature of 120 °C for 20 min under continuous stirring at a pressure of 7 bar. The molar mass and the thermal characteristics of the produced biopolymer were determined. The molar mass presented by the method proposed by the authors was 2.0×10^5 Da and polydispersity was 25.5, too much higher than the other methods tested that were reported in the work. For the thermal characteristics, the tested method obtained results similar to those found by classical recovery methods, but after the PHA separation process, the result was lower than the others tested, with 37.2% purity, a value far below that found by the authors quoted above.

When evaluating production and recovery of P(3HB) by *Methylobacterium* sp. V49, Ghatnekar et al. (2002) used a two-cavity high pressure homogenizer equipped with a ceramic rupture valve with a pressure ranging from 100 to 500 Kg/cm^2 and from one to three passes. After the process, the sample was centrifuged for further separation of P(3HB). Maximum purity and yield (95% and 98%) were obtained at an operating pressure of 400 Kg/cm^2 after two passes and using an SDS solution of 5% (w/v). The authors report that the recovery process with the high-pressure homogenizer associated with SDS may become promising in the biopolymer recovery process as it is a short and efficient method.

Fei et al. (2016) created a system for recovering P(3HB) from the *Cupriavidus necator* with a capacity of 600 g of cells. An extraction was carried out in a self-assembled stainless-steel reactor, which had a welded stainless-steel mesh plate without a tube reactor to serve as carrier or filter paper. The solvents acetone, ethanol, propylene carbonate, isopropanol, butanol, as well as mixtures thereof were tested. The thermal characteristics were analyzed in the work, obtaining in all the recovery processes values for the enthalpy around 84.75 J/g and melting point approximately 174.77 °C. The best results of polymer yield (85%) and purity (92%) were obtained with a mixture of acetone/ethanol/propylene carbonate for 48 h extraction at a temperature of 120 °C.

15.3.2 Ultrasound-Assisted

The ultrasound-assisted process has the purpose of breaking the cell by mechanical disturbance, in which ultrasonic waves are transferred by a metallic tip and dissipated in a liquid system (cell suspension), through cavitation bubbles, forming a kind of field, resulting in increased mass and heat transfer into the liquid medium, thereby producing a velocity gradient and a force capable of destabilizing the cell wall (Michelon et al. 2012).

In order to reduce the cost of recovering PHAs, it is necessary to obtain a large amount of polymer and to extract it quickly and with the minimum possible volume of solvent. For this objective to be achieved, it is necessary to transfer to an extraction system that can be incorporate the use of ultrasound-assistance (Ishak et al. 2016; Kosseva and Rusbandi 2017).

However, the ultrasound-assisted process for the recovery of bioproducts is used only at laboratory scale; at industrial scale its use is limited by the cost, due to the need to use several probes during the process and to install an efficient refrigeration system (Harrison et al. 1991).

When studying the ultrasound-assisted extraction of P(3HB) produced by *Cupriavidus necator* with propylene carbonate, Quines (2015) used a pre-treatment volume of 35 mL with three cycles of ultrasound application (59 s with pulse interval of 30 s) and 200 W of potency at 50% amplitude. After the process, the sample was centrifuged and dried in an oven. Determination of molar mass of the polymer together with purity analysis was carried out by gas chromatography (CG). The authors report that the extraction temperature influenced the degradation of the molar mass of the recovered polymer, since, in the extraction at 130 °C for 30 min, the polymer obtained had molar mass of 1.38×10^5 Da, while the extraction temperature of 150 °C caused a reduction in the molar mass of the polymer extracted to 1.03×10^5 Da. The best result was obtained with a purity close to 99% and 92% recovery. Nevertheless, Murugesan and Iyyaswami et al. (2017) subjected the fermented broth produced by *Cupriavidus necator* with the studied surfactants to a sonication process at a frequency of 2–10 kHz for 2–10 min with a pulse interval of 2 s. After the process, the phases were centrifuged. The frequency that presented the best result was 6 kHz for 6 min, recovering 88.23% of PHA, a lower value than that obtained by Quines (2015), probably because the high pulse interval.

Ishak et al. (2016) studied rapid ultrasonic extraction with medium solvent/non-marginal solvent (acetone and heptane) in the production of poly(3-hydroxyalkanoates) by *Pseudomonas putida* Bet001, where they used 35 g of dry-cell mass. Frequencies of 37–80 kHz with variable power output from 30 to 90 min (%) and a temperature of 25 ± 0.2 °C were established in the sonicator. The results showed that the frequency of 37 kHz was more efficient than 80 kHz, at a time of 5 min, with a molecular weight of 3.9×10^5 Da, a value below the control used in the study, and a recovery of 37.1%.

15.3.3 Bead Milling

Used with glass beads in a container containing the cell suspension, bead milling causes cell disruption by the spheres colliding with the biomass, promoting a transfer of energy and consequently cellular disruption. Some interferences must be taken into account, such as the size of the spheres, speed of agitation and cellular concentration (Tamer and Moo-Young 1998). The efficiency of the mill depends on the residence time, shear forces, type of microorganism, concentration of the cell suspension and speed of agitation. The disadvantage is that there are a large number of interferents, such as break chamber geometry, agitator type and speed, bead size, number of beads, cell concentration and temperature, and the possibility of breaking the polymer chains (Garcia 2006).

Garcia (2006) proposed an alternative mechanical method for extraction of P(3HB) from *Cupriavidus necator* using bead milling for cell breaking. Glass beads and chloroform were added to the biomass and the suspension was subjected to constant heating and stirring. They were centrifuged, and three distinct phases were formed, the lower phase consisting of the glass beads, the intermediate phase containing a solution of polymer and chloroform, and the upper phase formed by the cellular debris. This mechanical and chemical method developed for extraction of P(3HB) from *C. necator* showed a recovery of 64% of polymer with a purity of 89%, with the best result for average molar mass for use of the purified mechanical method and unpurified product yielding around 7.5×10^5 for both, and with polydispersity at 1.12. Results larger were found by Gutt et al. (2016) using the ANOVA models to compare and optimize P(3HBHV) extraction protocols by *C. necator*, used a mill containing 300 metal beads at 500 rpm for 2 min, where 1 g of the biomass was dispersed in 10 ml of buffer solution. After the procedure, the samples were centrifuged for 30 min at 10,000 × g, obtaining a 74% accumulation result at 46% purity. The authors also performed ball milling in combination with other methods, where 1–4 g biomass was used in buffer solutions ranging from pH 2 to 7.5 under stirring at 500 rpm for 2–5 min. After the SDS addition at 1.6–18.4% concentration and centrifugation for 60 min at 10,000 × g, the best purity was 74% at pH 7.5.

15.3.4 French Press

Gözde et al. (2012) used electrofiltration as a strategy for the purification of poly(3-hydroxybutyrate) produced *by Ralstonia eutropha* where the cells were passed through a French press for two cycles at 2000 psi. The best recovery in this study was about 40%.

In studying whether high P(3HB) production in *Pseudomonas extremaustralis* is linked with the expression of PHA synthase genes, Catone et al. (2014) used the French press loaded onto a glycerol gradient (87% and 50% v/v) resulting in a better yield of 28%.

By observing in vivo and in vitro the granules of P(3HB) formed by *Dinoroseobacter* sp. JL 1447, Xiao et al. (2015) used the French press process to verify microscopically the presence or absence of the biopolymer in the cellular biomass. The authors report that the granules of P(3HB) were not formed until 3 days of cultivation; their best result was on the 4th day, when there were 1–2 granules of the biopolymer present in the cell.

Ng and Sudesh (2016) identified, *Aquitalea* sp. USM4 (JCM 19919), as a new PHA producer and characterized its PHA synthase. In the process, the authors used the French press as a method to break the cells, obtaining at the end of the study a yield of 27–55%.

Murugan et al. (2016) performed the passage of the washed and resuspended cells with Tris-HCl (pH 7.2) with three cycles in the French press at a pressure of 1000 psi, resulting in polymers almost 100% pure.

Reinforcing poly(3-HB-HV) accumulation in *Arxula adeninivorans* by stabilizing the production, Biernacki et al. (2017) used the press in the extraction process, resulting in 71.9% recovery and 99.2% purity, having a number average mass of 8.6×10^5 Da and weight average of 1.7×10^5 Da molar mass (values considered by the authors to be very low) and a polydispersity of 2.

15.4 Biological Methods

Biological methods of extracting bacterial PHAs are an alternative to methods that use chemical reagents, and are based on the application of enzymes, such as lysozymes, nucleases and proteases, and the use of living organisms for biopolymer recovery (Kapritchkoff et al. 2006; Suzuki et al. 2008).

15.4.1 Enzyme Digestion

Enzymatic digestion is an advantageous method because it is a selective and energy efficient process, besides offering a lower risk of damage to the product when compared to other methodologies, but it is not an economically attractive process (Jacquel et al. 2008).

Kachrimanidou et al. (2016) produced crude enzyme by solid-state fermentation using the fungal strain *Aspergillus oryzae*, which was used to lyse *C. necator* cells, allowing the recovery of poly(3HB-3-HV) [P(3HB-co-12 mol% 3HV). A central composite design was used to optimize the temperature (50 °C) and pH (6), achieving the greatest lysis of the cells (88.9%). Enzymatic lysis of bacterial cells was also performed at the ideal temperature and uncontrolled pH, leading to recovery yield and purity of co-polymer of 98% and 96.7%, respectively. These authors compared the thermal and chemical characteristics of the PHAs obtained from the enzymatic methodology developed and the conventional chemical methodology using chloroform. From this methodology a PHA with a higher degree of crystallinity (Xc) (66%) was obtained compared to the standard methodology with chloroform (40%) (Kachrimanidou et al. 2016).

Gutt et al. (2016) evaluated different PHAs recovery protocols; among them the enzymatic methodology, where they used 1 g of dried cell biomass (dBM) dispersed in lysozyme solution (2 mg/mL). The dBM-lysozyme suspension was incubated for 1 h at 37 °C. Subsequently, the samples were centrifuged for 30 min at 10,000 × g. The recovery percentage was 75% and purification of P3HBHV was performed by washing the cell pellet three times with methanol, reaching a low purity of 41% (Gutt et al. 2016).

Martino et al. (2014) recovered the PHB granules from *Cupriavidus necator* biomass with a solvent-free solution using sodium dodecylsulfate (SDS) (0.3 g/g), ethylenediaminetetraacetic acid (EDTA) (0.01 g/g) and the alkaline enzyme (0.3 AU/g) in an aqueous medium. The extracted P(3HB) granules showed higher purity of 90% and no crystallization (Martino et al. 2014).

15.4.2 Cellular Lysis by Bacterial Predator

Martínez et al. (2013) began studies with the microorganism *Bdellovibrio bacteriovorus* HD100, a predator that grows within the Gram-negative bacteria, including mcl-polyhydroxyalkanoate (mclPHA) producing *Pseudomonas putida* (Martínez et al., 2013). Later, Martínez et al. (2016) evaluated the potential of this predatory bacterium as external cell-lipid agent for the recovery of valuable intracellular bioproducts such as PHAs produced by *P. putida* and *C. necator*, or by genetically modified *E. coli* strains. This innovative process allowed the recovery of up to 80% of PHAs, indicating that B. bacteriovorus can act as a bio-agent for the economic extraction, on an industrial scale, of intracellular bioplastics of Gram-negative cultures (Martínez et al. 2016). Also in this work, the molar mass of recovered PHAs was compared with the co-culture of *B. bacteriovorus* HD100 and *P. putida* KT2440 and from *P. putida* KT2440 cells submitted to cell lysis using the French press. It was observed that the physical method recovered PHAs with higher molar mass (Mm) and lower polydispersity index (PDI) than the extracted with the biological method. This indicates that, with co-cultivation, the bioplastic was partially degraded, and the heterogeneity of the PHA chains was increased.

15.4.3 Single Cell Protein

Kunasundari et al. (2013) in one study demonstrated for the first time that of P(3HB) can be readily extracted using only the biomass produced as Single Cell Protein (SCP) without the use of solvents and chemicals. Besides that, they reported the safety and tolerance limits of the animal model – Sprague Dawley rats, with the freeze-dried cells of *Cupriavidus necator* H16. Fecal granules of whitish color were observed in test animals. The granules were determined to contain about 82–97% by weight of P(3HB) in 74% recovery yield, and had a molar mass (Mm) of about 930 kg/mol, which was similar to the polymer extracted with chloroform (950 kg/mol) (Kunasundari et al. 2013). Subsequently, this research group optimized this recovery methodology using 2% SDS detergent for polymer purification. They concluded that biological recovery of P(3HB) developed in combination with a low detergent concentration purification procedure could efficiently isolate high purity P(3HB) without solvents or strong chemicals. The Mm and the thermal characteristics of the P(3HB) were comparable with P(3HB) extracted with chloroform (Kunasundari et al. 2017).

Murugan et al. (2016) reported the use of flour beetle larva (*Tenebrio molitor*) to extract PHA granules from *C. necator* Re2058/pCB113. The larvae readily consumed lyophilized *C. necator* cells and excreted PHAs granules as whitish stools. One hundred percent pure PHA granules were obtained using purification with water, 1% SDS detergent and heat (50 °C). The methodology developed did not decrease the molar mass and the polydispersity index of the recovered PHA molecules compared to the traditional method using chloroform (Murugan et al. 2016).

15.5 Conclusion

In this chapter, several PHAs extraction methods have been reviewed, such as the chemical method, that using chemical solvents and digesters and that are extensively used to provide significant recovery percentages. However, most of these studies using toxic and volatile highly chemicals and are contradictory to the fact that these bioplastics are widely studied to minimize the environmental aggressions caused by petrochemical polymers. In this context, the physical and biological methods are a positive alternative for the recovery of PHAs, since there is no need to insert chemicals into the process, but present lower polymer recovery results than those found in the other methods. On the other hand, these methods have their efficacy increased when combined with chemical. Besides, they are costly and difficult to apply on an industrial scale. The strategies for PHAs extraction under development discussed in this chapter clearly indicate that most methods are primarily aimed at obtaining high percentages of recovery and purity of biopolymer. Thus, it is essential to performed out studies that relate the efficiency of the process and its influence on the characteristics of the final product. For these reasons, the extraction step requires: reducing product losses, obtaining a product of purity high and preserved physicochemical characteristics, obtaining a product that fits the term sustainable, reducing process costs and developing a method applied on an industrial scale.

References

Anis SNS, Iqbal NM, Kumar S, Amirul AA (2013) Effect of different recovery strategies of P(3HB-co-3HHx) copolymer from *Cupriavidus necator* recombinant harboring the PHA synthase of *Chromobacterium* sp. USM2. Sep Purif Technol 102:111–117. https://doi.org/10.1016/j.seppur.2012.09.036

Aramvash A, Gholami-Banadkuki N, Moazzeni-Zavareh F, Hajizadeh-Turchi S (2015) An environmentally friendly and efficient method for extraction of PHB biopolymer with non-halogenated solvents. J Microbiol Biotechnol 25(11):1936–1943. https://doi.org/10.4014/jmb.1505.05053

Aramvash A, Gholami-Banadkuki N, Seyedkarimi MS (2016) An efficient method for the application of PHA-poor solvents to extract polyhydroxybutyrate from *Cupriavidus necator*. Biotechnol Prog 32(6):1480–1487. https://doi.org/10.1002/btpr.2346

Biernacki M, Marzec M, Roick T, Pätz R, Baronian K, Bode R (2017) Enhancement of poly(3-hydroxybutyrate-co-3-hydroxyvalerate) accumulation in *Arxula adeninivorans* by stabilization of production. Microb Cell Factories 16:144. https://doi.org/10.1186/s12934-017-0751-4

Budde CF, Riedel SL, Willis LB, Rha C, Sinskey AJ (2011) Production of poly(3-hydroxybutyrate-co-3-hydroxyhexanoate) from plant oil by engineered *Ralstonia eutropha* strains. Appl Environ Microbiol 77(9):2847–2854. https://doi.org/10.1128/AEM.02429-10

Catone MV, Ruiz JA, Castellanos M, Segura D, Espin G, López N (2014) High Polyhydroxybutyrate production in *Pseudomonas extremaustralis* is associated with differential expression of horizontally acquired and core genome polyhydroxyalkanoate synthase genes. PLoS One 9:e98873. https://doi.org/10.1371/journal.pone.0098873

Chan CH, Kummerlöwe C, Kammer HW (2004) Crystallization and melting behavior of poly(3-hydroxybutyrate)-based blends. Macromol Chem Phys 205:664–675. https://doi.org/10.1002/macp.200300062

Chen Y, Yang H, Zhou Q, Chen J, Gu G (2001) Cleaner recovery of poly(3-hydroxybutyric acid) synthesized in *Alcaligenes eutrophus*. Process Biochem 36(6):501–506. https://doi.org/10.1016/S0032-9592(00)00225-9

Fei T, Cazeneuve S, Zhiyou W, Wu L, Wang T (2016) Effective recovery of Poly-b-Hydroxybutyrate (PHB) biopolymer from *Cupriavidus necator* using a novel and environmentally friendly solvent system. AIChE J 32:678–685. https://doi.org/10.1002/btpr.2247

Fiorese ML, Freitas F, Pais J, Ramos AM, Aragão GMF, Reis MAM (2009) Recovery of polyhydroxybutyrate (PHB) from *Cupriavidus necator* biomass by solvent extraction with 1,2-propylene carbonate. Eng Life Sci 9(6):454–461. https://doi.org/10.1002/elsc.200900034

Garcia MCF (2006) Proposta de um processo de extração de poli(3-hidroxibutirato) produzido por Cupriavidus necator e seu efeito sobre as características do polímero Dissertação (Mestre em Engenharia de Alimentos) – Departamento de Engenharia Química e Engenharia de Alimentos, Universidade Federal de Santa Catarina, Santa Catarina, 158f

Ghatnekar MS, Pai JS, Ganesh M (2002) Production and recovery of poly-3-hydroxybutyrate from *Methylobacterium* sp V49. J Chem Technol Biotechnol 77:444–448. https://doi.org/10.1002/jctb.570

Gözde G, Prechtl C, Krischhöfer F, Mothes G, Ondruschka J, Breener- Weiss G, Obst U, Posten C (2012) Electrofiltration as a purification strategy for microbial poly-(3-hydroxybutyrate). Bioresour Technol 123:272–278. https://doi.org/10.1016/j.biortech.2012.07.039

Gumel AM, Annuar MSM, Chisti Y (2013) Recent advances in the production, recovery and applications of polyhydroxyalkanoates. J Polym Environ 21(2):580–605. https://doi.org/10.1007/s10924-012-0527-1

Gutt B, Kehl K, Ren Q, Boesel LF (2016) Using ANOVA models to compare and optimize extraction protocols of P3HBHV from *Cupriavidus necator*. Ind Eng Chem Res 55(39):10355–10365. https://doi.org/10.1021/acs.iecr.6b02694

Harrison ST (1991) Bacterial cell disruption: a key unit operation in the recovery of intracellular products. Biotechnol Adv 9:217–240. https://doi.org/10.1016/0734-9750(91)90005-G

Ibrahim MHA, Steinbüchel A (2009) Poly(3-Hydroxybutyrate) production from glycerol by *Zobellella denitrificans* MW1 via high-cell-density fed-batch fermentation and simplified solvent extraction. Appl Environ Microbiol 75(19):6222–6231. https://doi.org/10.1128/AEM.01162-09

Iqbal M, Tao Y, Xie S, Zhu S, Chen D, Wang X, Huang L, Peng D, Sattar A, Shabbir MAB, Hussain HI, Ahmed S, Yuan Z (2016) Aqueous two-phase system (ATPS): an overview and advances in its applications. Biol Proced Online 18:1–18. https://doi.org/10.1186/s12575-016-0048-8

Ishak KA, Annuar MSM, Heidelberg T, Gumel AM (2016) Ultrasound-assisted rapid extraction of bacterial intracellular medium-chain-length poly(3-hydroxyalkanoates) (mcl-PHAs) in medium mixture of solvent/marginal non-solvent. Arab J Sci Eng 41(1):33–44. https://doi.org/10.1007/s13369-015-1833-4

Jacquel N, Lo CW, Wei YH, Wu HS, Wang SS (2008) Isolation and purification of bacterial poly(3-hydroxyalkanoates). Biochem Eng J 39:15–27. https://doi.org/10.1016/j.bej.2007.11.029

Jiang Y, Mikova G, Kleerebezem R, Wielen LAMVD, Cuellar MC (2015) Feasibility study of an alkaline-based chemical treatment for the purification of polyhydroxybutyrate produced by a mixed enriched culture. AMB Express 5:5–18. https://doi.org/10.1186/s13568-015-0096-5

Kachrimanidou V, Kopsahelis N, Vlysidis A, Papanikolaou S, Kookos JK, Martínez BM, Rondán MCE, Koutinas AA (2016) Downstream separation of poly(hydroxyalkanoates) using crude enzyme consortia produced via solid state fermentation integrated in a biorefinery concept. Food Bioprod Process 100:323–334. https://doi.org/10.1016/j.fbp.2016.08.002

Kapritchkoff FM, Viotti AP, Alli RCP, Zuccolo M, Pradella JGC, Maiorano AE, Miranda EA, Bonomi A (2006) Enzymatic recovery and purification of polyhydroxybutyrate produced by *Ralstonia eutropha*. J Biotechnol 122:453–462. https://doi.org/10.1016/j.jbiotec.2005.09.009

Khanna S, Srivastava AK (2005) Recent advances in microbial polyhydroxyalkanoates. Process Biochem 40(2):607–619. https://doi.org/10.1016/j.procbio.2004.01.053

Koller M, Bona R, Chiellini E (2013) Extraction of short-chain-length poly-[(R)-hydroxyalkanoates] (scl-PHA) by the "anti-solvent" acetone under elevated temperature and pressure. Biotechnol Lett 35:1023–1028. https://doi.org/10.1007/s10529-013-1185-7

Kosseva MR, Rusbandi E (2017) Trends in the biomanufacture of polyhydroxyalkanoates with focus on downstream processing. Int J Biol Macromol 107(Pt A):762–778. https://doi.org/10.1016/j.ijbiomac.2017.09.054

Kourmentza C, Plácido J, Venetsaneas N, Burniol-Figols A, Varrone C, Gavala HN, Reis MAM (2017) Recent advances and challenges towards sustainable polyhydroxyalkanoate (PHA) production. Bioengineering 4:55–98. https://doi.org/10.3390/bioengineering4020055

Kunasundari B, Murugaiyah V, Kaur G, Maurer FHJ, Sudesh K (2013) Revisiting the single cell protein application of cupriavidus necator h16 and recovering bioplastic granules simultaneously. PLoS One 8(10):e78528. https://doi.org/10.1371/journal.pone.0078528

Kunasundari B, Arza CR, Maurer FHJ, Murugaiyah V, Kaur G, Sudesh K (2017) Biological recovery and properties of poly(3-hydroxybutyrate) from *Cupriavidus necator* H16. Sep Purif Technol 172:1–6. https://doi.org/10.1016/j.seppur.2016.07.043

Lee SY (1996) Bacterial Polyhydroxyalkanoates. Biotechnol Bioeng 49:1–14

Leong YK, Koroh FE, Show PL, Lan JCW, Loh HS (2015) Optimisation of extractive bioconversion for green polymer via aqueous two-phase system. Chem Eng Trans 45:1495–1500. https://doi.org/10.3303/CET1545250

Leong YK, Lan JCW, Loh HS, Ling TC, Ooi CW, Show PL (2016) Cloud-point extraction of green-polymers from *Cupriavidus necator* lysate using thermoseparating-based aqueous two-phase extraction. J Biosci Bioeng 123(3):3270–3375. https://doi.org/10.1016/j.jbiosc.2016.09.007

Ling Y, Wong HH, Thomas CJ, Williams DRG, Middelberg APJ (1998) Pilot-scale extraction of PHB from recombinant *E. coli* by homogenization and centrifugation. Bioseparation 7:9–15

Liu JG, Xing JM, Chang TS, Ma ZY, Liu HZ (2005) Optimization of nutritional conditions for nattokinase production by *Bacillusnatto* NLSSE using statistical experimental methods. Process Biochem 40(8):2757–2762. https://doi.org/10.1016/j.procbio.2004.12.025

López-Abelairas M, García-Torreiro M, Lú-Chau T, Lema JM, Steinbüchel A (2015) Comparison of several methods for the separation of poly(3-hydroxybutyrate) from *Cupriavidus necator* H16 cultures. Biochem Eng J 93:250–259. https://doi.org/10.1016/j.bej.2014.10.018

Macagnan KL, Rodrigues AA, Alves MI, Furlan L, Kesserlingh SM, Moura AB, Oliveira PD, McBride AJA, Moreira AS, Vendruscolo CT (2017) Simplified recovery process of *Ralstonia solanacearum*-synthesized polyhydroxyalkanoates via chemical extraction complemented by liquid-liquid phase separation. Quim Nova 40(2):125–130. https://doi.org/10.21577/0100-4042.20160162

Mantellato PE, Durão NAS (2008) Process for extracting and recovering polyhydroxyalkanoates (phas) from cellular biomass. U.S Pat. 20080193987 A1

Maresca P, Donsı F, Ferrari G (2011) Application of a multi-pass high pressure homogenization treatment for the pasteurization of fruit juices. J Food Eng 104:364–372. https://doi.org/10.1016/j.jfoodeng.2010.12.030

Martínez V, Jurkevitch E, García JL, Prieto MA (2013) Reward for *Bdellovibrio bacteriovorus* for preying on a polyhydroxyalkanoate producer. Environ Microbiol 15(4):1204–1215. https://doi.org/10.1111/1462-2920.12047

Martínez V, Herencias C, Jurkevitch E, Prieto MA (2016) Engineering a predatory bacterium as a proficient killer agent for intracellular bioproducts recovery: the case of the polyhydroxyalkanoates. Sci Rep 6:24381. https://doi.org/10.1038/srep24381

Martino L, Cruz MV, Scoma A, Freitas F, Bertin L, Scandola M, Reis MAM (2014) Recovery of amorphous polyhydroxybutyrate granules from *Cupriavidus necator* cells grown on used cooking oil. Int J Biol Macromol 71:117–123. https://doi.org/10.1016/j.ijbiomac.2014.04.016

Masson LMP (2010) Desenvolvimento de bebida láctea fermentada submetida ao processamento térmico e/ou à homogeneização à ultra-alta pressão. 109p. Tese (Doutorado em Engenharia Química) – Escola de Química, Universidade Federal do Rio de Janeiro, Rio de Janeiro

Michelon M, Borba TM, Rafael RS, Burket CAV, Burket JFM (2012) Extraction of carotenoids from *Phaffia rhodozyma*: a comparison between different techniques of cell disruption. Food Sci Biotechnol 21:1–8. https://doi.org/10.1007/s10068-012-0001-9

Middelberg APJ (1995) Process-scale disruption of microorganisms. Biotechnol Adv 13:491–551. https://doi.org/10.1016/0734-9750(95)02007-P

Murugan P, Han L, Gan CY, Maurer FHJ, Sudesh K (2016) A new biological recovery approach for PHA using mealworm, *Tenebrio molitor*. J Biotechnol 239:98–105. https://doi.org/10.1016/j.jbiotec.2016.10.012

Murugesan S, Iyyaswami R (2017) Low frequency sonic waves assisted cloud point extraction of polyhydroxyalkanoate from *Cupriavidus necator*. J Chromatogr B Analyt Technol Biomed Life Sci 1060:207–214. https://doi.org/10.1016/j.jchromb.2017.06.009

Ng LM, Sudesh K (2016) Identification of a new polyhydroxyalkanoate (PHA) producer *Aquitalea* sp. USM4 (JCM 19919) and characterization of its PHA synthase. J Biosci Bioeng 122:550–557. https://doi.org/10.1016/j.jbiosc.2016.03.024

Peng YU, Lo CCW, Wu HS (2013) The isolation of poly(3-hydroxybutyrate) from recombinant *Escherichia coli* xl1-blue using the digestion method. Can J Chem Eng 91(1):77–83. https://doi.org/10.1002/cjce.20685

Quines LKM (2015) Extração de poli(3-hidroxibutirato), produzido por *Cupriavidus necator*, utilizando carbonato de propileno combinado com métodos mecânicos. Tese de doutorado - Programa de Pós-Graduação em Engenharia de Alimentos, Universidade Federal de Santa Catarina, Florianópolis

Quines LKM, Ienczak JL, Schmidt M, Zanfonato K, Rodrigues MI, Schmidell W, Aragão GMF (2015) Extração de poli(3-hidroxibutirato), produzido por *Cupriavidus necator*, com carbonato de propileno. Quim Nova 38(2):214–220. https://doi.org/10.5935/0100-4042.20140297

Rathi DN, Amir HG, Abed RMM, Kosugi A, Arai T, Sulaiman O, Hashim R, Sudesh K (2013) Polyhydroxyalkanoate biosynthesis and simplified polymer recovery by a novel moderately halophilic bacterium isolated from hypersaline microbial mats. J Appl Microbiol 114:384–395. https://doi.org/10.1111/jam.12083

Riedel LS, Brigham CJ, Budde CF, Bader J, Rha C, Stahl U, Sinskey AJ (2013) Recovery of poly(3-hydroxybutyrate-co-3-hydroxyhexanoate) from *Ralstonia eutropha* cultures with non-halogenated solvents. Biotechnol Bioeng 110(2):461–470. https://doi.org/10.1002/bit.24713

Rodríguez-Contreras A, Koller M, Dias MMS, Calafell M, Braunegg G, Marqués-Calvo MS (2013) Novel poly[(r)-3-hydroxybutyrate]-producing bacterium isolated from a bolivian hypersaline lake. Food Technol Biotechnol 51(1):123–130

Serafim LS, Lemos PC, Levantesi C, Tandoi V, Santos H, Reis MAM (2002) Methods for detection and visualization of intracellular polymers stored by polyphosphate-accumulating microorganisms. J Microbiol Methods 51(1):1–18. https://doi.org/10.1016/S0167-7012(02)00056-8

Suzuki DV, Carter JM, Rodrigues MFA, Silva ES, Maiorano AE (2018) Purification of polyhydroxybutyrate produced by *Burkholderia cepacia* IPT64 through a chemical and enzymatic route. World J Microbiol Biotechnol 24:771–775. https://doi.org/10.1007/s11274-007-9537-x

Tamer IM, Moo-Young M (1998) Disruption of *Alcaligenes latus* for recovery of poly(β-hydroxybutyric acid): comparison of high-pressure homogenization, bead milling, and chemically induced lysis. Ind Eng Chem Res 37:1807–1814

Valappil SP, Misra SK, Boccaccini AR, Keshavarz T, Bucke C, Roya I (2007) Large-scale production and efficient recovery of P(3HB) with desirable material properties, from the newly characterized *Bacillus cereus* SPV. J Biotechnol 132:251–258. https://doi.org/10.1016/j.jbiotec.2007.03.013

Van HP, Middelberg AP, Van Der Lans RG, Van Der Wielen LA (2004) Relation between cell disruption conditions, cell debris particle size, and inclusion body release. Biotechnol Bioeng 88:100–110. https://doi.org/10.1002/bit.20343

Villano M, Valentino F, Barbetta A, Martino L, Scandola M, Majone M (2013) Polyhydroxyalkanoates production with mixed microbial cultures: from culture selection to polymer recovery in a high-rate continuous process. New Biotechnol 31(4):289–296. https://doi.org/10.1016/j.nbt.2013.08.0

Xiao N, Jiao N, Liu Y (2015) In vivo and in vitro observations of polyhydroxybutyrate granules-formed by *Dinoroseobacter* sp. JL 1447. Biol Macromol 74:467–475. https://doi.org/10.1016/j.ijbiomac.2014.11.045

Nanofibers from Polyhydroxyalkanoates and Their Applications in Tissue Engineering

16

Sumitra Datta and Gopalakrishnan Menon

Abstract

Nanofibers are one-dimensional nanomaterials with a host of research and commercial applications. They possess exceptional physicochemical characteristics with fiber diameter ranging from tens to hundreds of nanometers. They have large surface area-to-volume ratio and can form highly porous interconnecting network. Natural, synthetic, carbon-based, semiconducting, and composite/blend polymers have been used to synthesize them. By virtue of their special properties like lower acidity and bioactivity, non-toxic degradation, biocompatibility, tunable surface modifications, wide ranging mechano-physico-chemical characteristics, and non-carcinogenicity, polyhydroxyalkanoates (PHAs) have emerged as a potential candidate in the field of tissue engineering and regenerative medicine.

Tissue engineering and regenerative medicine combines biological systems with engineering expertise, to restore the healthy functions at cellular levels. Their system development is based on cells, biomolecules, and biomaterials. Fabrication of nanofibrous 2D and 3D scaffolds have revolutionised the field of tissue engineering with the development of skin grafts, vascular grafts, bone implants and corneal tissue replacements. Research on fabrication of scaffolds and biodegradable polymers/blends required for fabrication is on a rise with their ever-increasing applications.

Keywords

Nanofibers · Electrospinning · Bioprinting · Scaffold · Fabrication · Tissue engineering

S. Datta (✉)
Amity Institute of Biotechnology, Amity University, Kolkata, West Bengal, India
e-mail: sumitra.datta@gmail.com

G. Menon
Vipragen Biosciences Private Limited, Mysore, Karnataka, India

V. C. Kalia (ed.), *Biotechnological Applications of Polyhydroxyalkanoates*,
https://doi.org/10.1007/978-981-13-3759-8_16

16.1 Introduction

Nanobiotechnology has essentially become the next-generation technology preferred globally by both science and economy. It has successfully created solutions to a wide range of problems posed by health, environment, energy generation and storage and textile sectors (Haider et al. 2015). The fascinating world of nanomaterials and nanocomposites include the zero-dimensional nanoparticles also known as quantum dots; one-dimensional nanowires, nanorods, nanofibers, and nanotubes; in addition to two-dimensional nanosheets which exhibit extraordinary physico-chemical features (Kenry and Lim 2017).

Nanofibers are polymeric fiber whose diameters range from 0.01 to 0.1 μm. At this dimension they exhibit intriguing properties such as high surface to volume ratio, flexibility in surface functionalities, greater porosity and better mechanical performance like stiffness and tensile strength when compared to any other known form of the same polymeric material (Huang et al. 2003). Unlike the conventional rigid porous structures, nanofiberous porous structures form dynamic arrangements where the pore sizes and shapes can be altered by optimizing the start materials and synthesis parameters. When required they can also be converted into rigid structures by linking them across (Ramakrishna et al. 2006).

Polymeric nanofibers have hence become the best suited candidates for multifarious applications. Especially when they are assembled or fabricated into membranes nanofibers they extend wide ranging applications in biomedical (scaffolding used in tissue engineering, wound dressing, drug delivery, artificial organs, vascular grafts), environmental protection (water treatment and ultrafiltration, photocatalysis, chemical and gas sensing), nanosensor, electronic (composites and structures for nano-electronic machines), protective clothing (protective shields and masks in speciality fabrics), separation industry (filter media for submicron particles) and energy generation and storage (batteries, fuel cells, supercapacitors, solar cells, hydrogen storage and generation, piezoelectricity) fields (Frenot and Chronakis 2003; Haider et al. 2015; Kenry and Lim 2017; Thenmozhi et al. 2017). Ligand molecules, biomacromolecules, and cells can either be attached or hybridized with membranous nanofiber systems to be used as affinity membranes for protein purification and waste water treatment; as membrane bioreactors for industrial enzymatic catalysis or synthesis and as biosensors for chemical analysis and diagnostics (Ramakrishna et al. 2006). Furthermore, nanofibers can be aligned to construct unique functional nanostructures such as nanotubes and nanowires whilst their non-woven composites are used in filters (Frenot and Chronakis 2003).

16.2 Use of Biopolymers for Nanofiber Fabrication

The multifaceted applications of using nanofibers in various fields can be clearly understood from the never-ending list of materials which are used for their synthesis. They can be broadly classified under natural and synthetic polymers, carbon compounds and semiconducting and composite materials (Kenry and Lim 2017).

Biopolymers or green polymers like polysaccharides (cellulose, chitin, chitosan, dextrose, alginate, dextran, hyaluronic acid (HA)), proteins (casein, collagen, zein, egg albumen, human and bovine fibrinogen, gelatin, silk, wool etc.), DNA, PHAs, poly (3-hydroxybutyrate-co-3-hydroxyvalerate) (PHBV) and derivatives/composites of two or more biopolymers are also increasingly used as fabrication materials (Schiffman and Schauer 2008; Xuezhu et al. 2012; Haider et al. 2015; Farokhi et al. 2018;) . Biopolymers are preferred owing to their sustainability, cost-effectiveness, eco-efficiency, ease of extraction, antibacterial activity, biodegradability, biocompatibility and renewable nature. However, using biopolymers can be challenging as a number of properties like molecular weight, degree of deacetylation, purity, distribution of charged groups and crystallinity of the same polymer will vary when derived from different sources (Schiffman and Schauer 2008). Lately, attempts have been made to utilize wide ranging blends of natural and synthetic polymers for synthesizing nanofibers (Haider et al. 2015).

16.3 PHA as a Biomaterial

PHAs are linear polyesters synthesized naturally by both Gram positive and negative bacteria under physical or nutritional stress conditions. They are biodegradable, biocompatible with desired mechanical strength and can be easily procured in copious amounts by growing bacterial isolates in optimum physio-chemical parameters. Additionally, a host of different PHAs and their co-polymers can be obtained by either altering the bacterial species used for production or by manipulation of growth conditions. Depending on the number of carbons in monomer, PHAs can be classified as short chain length (scl) PHAs (like Poly(3-hydroxybutyrate) (PHB), poly(3-hydroxyvalerate) (PHV), poly(3-hydroxybutyrate-co-valerate) (PHBV)), medium chain length (mcl) PHAs (like, polyhydroxyoctanoate (PHO), polyhydroxynonanoate (PHN), polyhydroxyhexanoate (PHHx), polyhydroxyheptanoate (PHHp)) and long chain length (lcl) PHAs (Ishii et al. 2009; Gadgil et al. 2017). Amongst all PHAs, PHB and PHBV have been extensively employed for various tissue regeneration applications like sutures, barriers, valves, patches, grafts, scaffolds, pins and guides (Ali and Jamil 2016; Gadgil et al. 2017). Figure 16.1 depicts the various fields of tissue engineering and regeneration that has been revolutionized by the use of PHA nanofibers. Diverse modification methods like epoxidation, carboxylation, chlorination, hydroxylation, and pyrolysis have been used on PHAs to increase their bioactivity, biodegradability, biocompatibility and flexibility while reducing crystallinity and hydrophobicity (Gadgil et al. 2017). It has been found that in comparison to poly lactic acid (PLA) and poly-glycolic acid (PGA), PHAs have lower acidity and bioactivity, thus posing minimal risk when used in vivo (Tang et al. 2010; Ali and Jamil 2016). The monomers of PHA which are either 3-hydroxybutyric acid (3HB) or 4-hydroxybutyric acid (4HB) are less acidic than those of PGA, PLA, and poly(lactic acid-coglycolic acid) (PLGA) (i,e. α-hydroxy acids) and can be effortless expelled from our body in within an hour. Degradation of PHAs in vivo is carried out by hydrolytic enzymes like lipases. Cytotoxicity

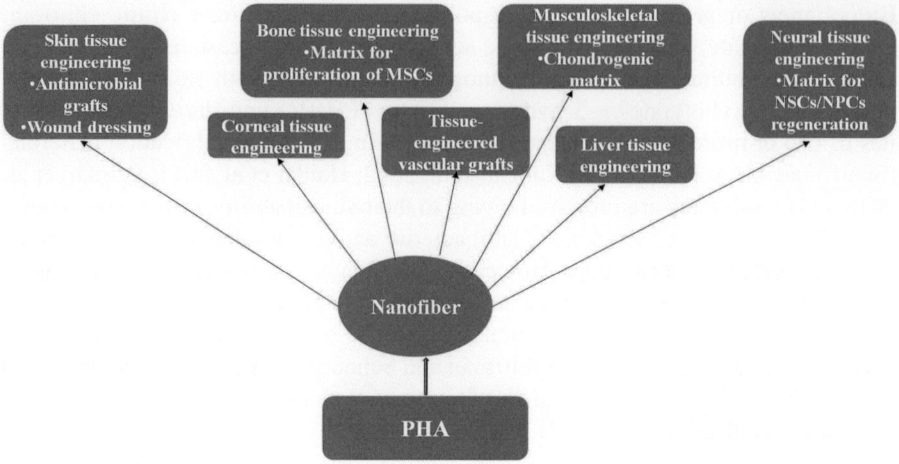

Fig. 16.1 Multifarious applications of PHA in tissue engineering and regeneration

studies of PHAs have shown negligible effects, however, if any exists, they seem to decrease inversely with the length of the oligomeric side chain (Ali and Jamil 2016).

16.4 Nanofiber Fabrication Techniques

In recent years, fabrication of nanofibers has been done by using a few techniques like; drawing, template synthesis, phase separation, self-assembly and electrospinning. While drawing produces single long nanofiber, template synthesis employs nanoporous membrane templates to make solid or hollow tubular nanofibers. Phase separation is a relatively lengthy process that transforms solid polymer into nanoporous foam. In self-assembly, however individual, pre-existing entities organize themselves into patterns. Like phase separation, this technique too is time consuming (Huang et al. 2003). Electrospinning is a simple, but versatile and cost-effective technique based on the principle of "electrostatic attraction" of charges and has been widely used in fabricating non-woven fibers with significant and desirable porosity and surface area. There are certain parameters which critically govern the electrospun nanofiber morphology. They are physiochemical characteristics of polymer (like concentration, viscosity and molecular weight), applied voltage, tip-to-collector distance, and solvent. Electrospinning is preferred over rest of the fabrication techniques as it has been found that electrospun nanofibers possess higher surface area to volume ratio and inter/intra fibrous porosity. This technique was first patented in the US in 1902 by J.F. Cooley (Ramakrishna et al. 2006**;** Haider et al. 2015).

The basic electrospinning setup comprises of one glass syringe containing the polymer solution attached to a metallic needle, power supply and the metallic collector onto which nanofibers deposit or aligns. The process begins as electric charges move into the polymer solution and build up at the tip of the metallic needle causing induction of charges on the polymer droplet. The reciprocal repulsion of charges opposes the surface tension, making the polymer solution to flow along the electric field. This further result in formation of Taylor cone from which nanofibers emerge and are collected on a oppositely charged metallic collector kept at an optimized distance. Finally, the internal and external charge forces cause the whipping of the liquid jet in the direction of the collector which results in the creation of fibers with nanoscale diameters. Currently, attempts are being made to electrospin nanofiber with greater cell infiltration property for production of 3D scaffolds (Haider et al. 2015; Kenry and Lim 2017; Thenmozhi et al. 2017).

Lately, some emerging nanofiber synthesis strategies have also been reported like CO_2 laser supersonic drawing, solution blow spinning, plasma-induced synthesis, centrifugal jet spinning, and electrohydrodynamic direct writing which are capable of fabricating highly porous mesh like network of nanofibers with significant interconnective pores, making them an ideal candidates for a wide ranging applications (Kenry and Lim 2017).

16.5 Nanofibers in Tissue Engineering

Although electrospun nanofibers have broad ranging applications as drug carriers, tissue scaffolds, wound dressings, reinforcement materials, filters, and protective clothing, however, this chapter will principally focus on the multifarious use of nanofibers for fabrication of an array of scaffolds in tissue engineering. As nanofibers mimic structural and functional characteristics of native extracellular matrix (ECM), hence they have been widely used for tissue engineering scaffolds. Apparently, natural and synthetic polymers, biomimetic composites, ceramics, and metals electrospun into nanofibers find hosts of applications in tissue engineering and regenerative medicine (Ingavle and Leach 2014).The success rate of artificial extracellular matrix scaffolds depends on its biocompatibility, osteoconductivity, degradability, high surface to volume ratios and mechanical properties. The porosity of scaffold is a crucial factor as it has direct effects on the number of cell anchoring points and proliferation, wettability, and rate of degradation (Schiffman and Schauer 2008). In addition, with the introduction of the 3D scaffolds profound impacts have been found in the process of mechanical signalling in terms of positive cell receptor ligation, intercellular signalling and cellular migration and differentiation. 3D scaffolds have also been found to control diffusion and adhesion of proteins, growth factors, and enzymes, largely influencing cell viability and function (Nisbet et al. 2009; Pilehvar-Soltanahmadi et al. 2016).

16.6 Current Scenario of PHA-Nanofibrous Scaffolds in Tissue Engineering and Regenerative Medicine

16.6.1 Skin Tissue Engineering

Apart from being the largest organ of our body skin also acts as first line of defence. ECM comprising of collagen, elastin, and glycosaminoglycans, and fibroblasts form the dermis which provides skin with strength and elasticity, besides supporting the vascularisations, neural network and the lymphatic system. Skin is predominantly damaged by burn injuries, chronic wounds, acute trauma, and other dermatological conditions like pigmentation and scar. Skin epidermis has basal layer of epidermal stem cells which helps in damage recovery. However, deep wounds spanning both epidermis and dermis fail to heal by themselves. Wound healing process progresses with hemostasis, inflammation, proliferation, and finally maturation. Non-woven polymeric nanofibrous scaffolds mimic the morphological, fibrillar, and topographical aspects of the native ECM of the skin. Nanofiber scaffolds owing to their small pore size can prevent entry to microbial pathogens while restoring structural integrity and inhibiting wound contraction (Kalia and Purohit 2011; Kalia 2014; Koul et al. 2016; Pilehvar-Soltanahmadi et al. 2016; Wang et al. 2018). Modified PHAs attached with a thioester group in the side chains and PHB co-electrospun with antibiotics have exhibited excellent antibacterial properties against potential pathogen like *Staphylococcus aureus* (Agarwala et al. 2014; Gui et al. 2014; Kalia 2015; Kalia and Kumar 2015; Luef et al. 2015; Ali and Jamil 2016; Jeyanthi and Velusamy 2016; Azman et al. 2017; Kalia et al. 2017). PHB scaffold blends of chitosan have been found to promote cell attachment and proliferation (Pilehvar-Soltanahmadi et al. 2016; Wang et al. 2018). PHB and its copolymers like poly(3-hydroxybutyrate-*co*-3-hydroxyvalerate) (PHBHV) and poly(3-hydroxybutyrate-*co*-3-hydroxyhexanoate) (PHBHHx) have been used for skin regeneration (Ahiwale et al. 2017; Grande et al. 2017). In recent times, negative pressure wound therapy (NPWT) or vacuum assisted closure, topical negative pressure therapy or microdeformational wound therapy is employed in healing burns (Wang et al. 2018).

16.7 Bone Tissue Engineering

Bone is complex connective tissue composed of calcified bone matrix, cells and bioactive factors. HAp which is osteoconductive and osteointegrative is the main component of bone matrix formed of collageneous fibers (Jang et al. 2009; Wang and Yeung 2017; Zhang et al. 2018a). Osteoblast, osteoclast and progenitor cells are present in the bone matrix. Osteoblasts synthesize organic components of the bone ECM, like type I collagen (90–95% of organic material and serves as a template), vitamin-K dependent proteins, osteocalcin and matrix Gla protein, osteopontin (for cell adhesion) and alkaline phosphatise. Osteoclasts aided with lysosomal secretions help in mineralization and degradation of the organic phase.

Thus, bone ECM should be an organic–inorganic nanocomposite that would perform biomechanical functions. Mesenchymal stem cells (MSCs), contribute to the regeneration and osteogenic differentiation into mesenchymal tissues, like bone, cartilage, muscle, ligament, tendon by acquiring osteoblastic markers and secreting extracellular matrix and calcium crystals (Jang et al. 2009; Holzwarth and Ma 2011; Wang and Yeung 2017). Electrospun PHA, PHB and PHBV nanofibers have exhibited enhanced cell growth behavior (SaOS-2 cell line) than on their flat film counterparts (Jang et al. 2009). Electrospun PHB scaffolds, with electrosprayed HA nanoparticles (nHA) showed enhanced mechanical properties like porosity and direct biceramic exposure to the human MSCs. This assisted them to proliferate till reaching confluence (Ramier et al. 2014). Elevated biomineralization rate is also reported in a similar study where nHA-sprayed PHA scaffold was used facilitating osteoblastic differentiation. Moreover, it was found that epoxydation of PHA scaffolds immobilizes vital enzymes and proteins promoting cell adhesion and proliferation (Grande et al. 2017).

16.8 Musculoskeletal Tissue Engineering

As bone tissue engineering has been previously discussed, hence this section focuses on skeletal muscle, tendon, cartilage, and ligament tissue engineering. When compared to fracture, large segmental defects are difficult to heal. This is because the articular cartilage lacks self-reparative potential (i.e., unavailability of chondrocytes and progenitor cells) and although tendons and ligaments heal slowly but is impaired to a certain degree. Thus, erosion of articular cartilage and meniscus tissue might lead to degenerative osteoarthritis (Vasita and Katti 2006; Lin et al. 2018). Most of the tendon injury reports are associated with tearing of the interdigitating tendons of the rotator cuff. In general, MSCs derived from bone marrow, umbilical cord, muscle satellite cells and induced pluripotent stem cells embedded in self-assembled or 3D nanofibrous scaffolds help in regeneration of musculoskeletal tissue by mimicking the native ECM with growth factor sequestering nanofibers (Vasita and Katti 2006; Lin et al. 2018; Loebel and Burdick 2018). It has been reported that planar P(3HB-co-3HHx) films exhibit noteworthy cellular response interms of chondrogenic initiation of bone marrow MSCs. PHA binding protein combined with an mcl-PHA might also prove an excellent scaffold for these cells (Ali and Jamil 2016). Poly(3-hydroxybutyrate-co-3-hydroxyhexanoate) (PHBHHx) is yet another member of PHA family that can be tailor made to possess excellent mechanical and thermal properties required for both bone and soft tissue regeneration (Chang et al. 2014).

16.9 Neural Tissue Engineering

Electrospun and self assembled nanofibrous scaffolds have been used both in vitro and in vivo for nerve repair and regeneration. Even though polymers like collagen, gelatin, laminin, chitosan, PGA, poly (L-lactic acid) (PLLA), polycaprolactone

(PCL) and their copolymers/blends have been electrospun into scaffolds for growing neural stem cells/neural progenitor cells (NSCs/NPCs), PHB is mostly preferred owing to its biocompatibility and low cytotoxic nature (Cao et al. 2009; Lu et al. 2013). However, reduction in cell adhesion and proliferation has been found due to poor hydrophilicity of PHA. Hence, various approaches like grafting, ultraviolet exposure, plasma treatment and surface hydrolysis have been applied to PHA for improving its hydrophilicity resulting in stronger cell attachment (Lu et al. 2013). In another study, it was found that three different PHAs like PHB, copolymer of 3-hydroxybutyrate and 4-hydroxybutyrate (P3HB4HB), and PHBHHx when fabricated into 2D and 3D nanofibrous scaffolds using a modified phase separation method supported NSC growth and differentiation mimicking the natural ECM (Xu et al. 2010).

16.10 Tissue-Engineered Vascular Grafts (TEVGs)

TEVGs have successfully substituted damaged vessels. They have the required mechanical properties, blood compatibility, endothelium friendliness and biodegradability. Unlike conventional vascular grafts made of polytetrafluoroethylene (PTFE) and polyurethane (PU), cells seeded on the TEVGs can give rise to a newly regenerated blood vessel as the newly formed ECM with slowly replace the biodegradable scaffold. Fabrication techniques involve 3D printing, solvent casting, phase separation and electrospinning (Sell et al. 2009; Wu et al. 2018). PHA treated by salt leaching had been used for heart valve scaffold fabrication which had three valve leaflets (Sodian et al. 2000). The valve is a complex structure made out of single spongy layer sandwiched between laminar anisotropic fibrous layers on either side. PHA-polymers, like poly-3-hydroxybutyrate (poly(3HB)), poly-3-hydroxybutyrate-co3-hydroxyvalerate (poly(3HB-co-3 HV), poly-R-3-hydrooxyoctanoate-co-R-3-hydroxyhexanoate (poly(3-HO-co-3HH)), poly-4- hydroxybutyrate (poly(4HB)) and poly-R-3-hydroxybutyrate-co4-hydroxybutyrate (poly(3-HO-co-4HB)) are being used by companies including Metabolix and Tapha for fabrication of TEVGs (Jana et al. 2014).

16.11 Corneal Tissue Engineering

Stroma comprises 90% of the cornea and is made up of collagen types I, V and VI. Of late stromal defects are treated by either corneal transplantation or artificial corneas. Hence, tissue engineering approaches are being sought after (Matthyssen et al. 2018). (PHBV)/gelatine nanofirbrous scaffolds have facilitated attachment, proliferation and formation of corneal epithelium with improved transparency (Kong and Mi 2016).

16.12 Liver Tissue Engineering

Primary hepatocytes cultured on 2D collagen coated matrix have been found to produce honeycomb like morphology with increased gene expression. Enhanced hepatocyte aggregation, spreading, and metabolism were exhibited when primary hepatocytes were seeded in 3D scaffold. Mesenchymal, embryonic, and induced pluripotent stem cells have been widely used (Zhang et al. 2018b). A recent study reported the recovery of injured mice liver when a Poly (3-Hydroxybutyrate-Co-3-Hydroxyvalerate-Co3-Hydroxyhexanoate) (PHBVHHx) scaffold loaded with human umbilical cord Wharton's jelly (WJ) MSCs was transplanted (Li et al. 2015).

16.13 Future Prospects

The inherent property of the natural polymers to be tailor made for consumer specific usage has created a unique niche for them as compared to their counterparts (Schiffman and Schauer 2008). Currently most of the nanofibrous scaffolds are dependent on electrospinning for their fabrication. It should be noted that this technique also has some bottlenecks, like less throughput, cumbersome equipment setup, high voltage, and conducting collectors, and complex in situ nanofiber deposition. Sometimes, electrospun nanofibrous systems exhibit low mechanical strength owing to low crystallinity and random alignment. Therefore, it is quintessential to develop high novel throughput nanofiber fabrication methods integrating the current and the emerging ones. New variations such as multi-needle and needleless electrospinning maybe employed. For acquiring usage specific functionalities in nanofibers appropriate start materials coupled with novel fabrication techniques might help in manipulating nanofiber in terms of smaller diameter, interfiber adhesion and surface functionalization (Kenry and Lim 2017). Approaches like 'click chemistry' and co-axial electrospinning show promising future for introducing surface functionality modulation and variable configurations like core-shell, multilayer, and multicomponent to nanofibers respectively (Konwarh et al. 2013; Kenry and Lim 2017). In addition, it is imperative to study the degradation rates, tissue regeneration and mechano-spatio-geometric requirements of the new class of PHA like PHBHHx and interactions of other copolymers with different bioceramics (Misra et al. 2006; Yang et al. 2014). In conclusion, research has to be focussed on improving PHA yield in the cells, so that large scale production of the polymer could be done from cheap substrates (Luef et al. 2015). As a matter of fact, researchers are strongly focusing on development of innovative fabrication technologies using biomaterials to reduce reliance on fossil fuels (Schiffman and Schauer 2008).

Despite the relentless progress in engineered constructs, they fail to offer the growing cells a holistic cell-cell and cell-ECM interaction due to loads of exogenous signals. This issue has been addressed by bioprinting which constructs reproducible and flexible models with accurately arranged live cells. However, there still exit several drawbacks like printing resolution, advanced cell/printing permissive bioink, spatiotemporal control with proper supply of nutrients and oxygen that are to be addressed (Varsha et al. 2016; Zhuang et al. 2018).

References

Agarwala M, Choudhury B, Yadav RN (2014) Comparative study of antibiofilm activity of copper oxide and iron oxide nanoparticles against multidrug resistant biofilm forming uropathogens. Indian J Microbiol 54:365–368. https://doi.org/10.1007/s12088-014-0462-z

Ahiwale SS, Bankar AV, Tagunde S, Kapadnis BP (2017) A bacteriophage mediated gold nanoparticle synthesis and their anti-biofilm activity. Indian J Microbiol 57:188–194. https://doi.org/10.1007/s12088-017-0640-x

Ali I, Jamil N (2016) Polyhydroxyalkanoates: current applications in the medical field. Front Biol 11:19–27. https://doi.org/10.1007/s11515-016-1389-z

Azman AS, Othman I, Fang C-M, Chan K-G, Goh B-H, Lee L-H (2017) Antibacterial, anticancer and neuroprotective activities of rare actinobacteria from mangrove forest soils. Indian J Microbiol 57:177–187. https://doi.org/10.1007/s12088-016-0627-z

Cao H, Liu T, Chew SY (2009) The application of nanofibrous scaffolds in neural tissue engineering. Adv Drug Deliv Rev 61:1055–1064. https://doi.org/10.1016/j.addr.2009.07.009

Chang HM, Wang ZH, Luo HN, Xu M, Ren XY, Zheng GX, Wu BJ, Zhang XH, Lu XY, Chen F, Jing XH, Wang L (2014) Poly(3-hydroxybutyrate-co-3-hydroxyhexanoate)-based scaffolds for tissue engineering. Braz J Med Biol Res 47:533–539. https://doi.org/10.1590/1414-431X20143930

Farokhi M, Mottaghitalab F, Samani S, Shokrgozar MA, Kundu SC, Reis RL, Fatahi Y, Kaplan DL (2018) Silk fibroin/hydroxyapatite composites for bone tissue engineering. Biotechnol Adv 36:68–91. https://doi.org/10.1016/j.biotechadv.2017.10.001

Frenot A, Chronakis IS (2003) Polymer nanofibers assembled by electrospinning. Curr Opin Colloid Interface Sci 8:64–75. https://doi.org/10.1016/S1359-0294(03)00004-9

Gadgil BST, Killi N, Rathna GVN (2017) Polyhydroxyalkanoates as biomaterials. Med Chem Commun 8:1774–1787. https://doi.org/10.1039/C7MD00252A

Grande D, Ramier J, Versace DL, Renard E, Langlois V (2017) Design of functionalized biodegradable PHA-based electrospun scaffolds meant for tissue engineering applications. New Biotechnol 37:129–137. https://doi.org/10.1016/j.nbt.2016.05.006

Gui Z, Wang H, Ding T, Zhu W, Zhuang X, Chu W (2014) Azithromycin reduces the production of α-hemolysin and biofilm formation in *Staphylococcus aureus*. Indian J Microbiol 54:114–117. https://doi.org/10.1007/s12088-013-0438-4

Haider A, Haider S, Kang IK (2015) A comprehensive review summarizing the effect of electrospinning parameters and potential applications of nanofibers in biomedical and biotechnology. Arab J Chem. https://doi.org/10.1016/j.arabjc.2015.11.015

Holzwarth JM, Ma PX (2011) Biomimetic nanofibrous scaffolds for bone tissue engineering. Biomaterials 32:9622–9629. https://doi.org/10.1016/j.biomaterials.2011.09.009

Huang Z-M, Zhang Y-Z, Kotaki M, Ramakrishna S (2003) A review on polymer nanofibers by electrospinning and their applications in nanocomposites. Compos Sci Technol 63:2223–2253. https://doi.org/10.1016/S0266-3538(03)00178-7

Ingavle GC, Leach JK (2014) Advancements in electrospinning of polymeric nanofibrous scaffolds for tissue engineering. Tissue Eng Part B Rev 20:277–293. https://doi.org/10.1089/ten.TEB.2013.0276

Ishii D, Yingh TH, Yamaoka T, Iwata T (2009) Characterization and biocompatibility of biopolyester nanofibers. Materials 2:1520–1546. https://doi.org/10.3390/ma2041520

Jana S, Tefft BJ, Spoon DB, Simari RD (2014) Scaffolds for tissue engineering of cardiac valves. Acta Biomater 10:2877–2893. https://doi.org/10.1016/j.actbio.2014.03.014

Jang J-H, Castano O, Kim H-W (2009) Electrospun materials as potential platforms for bone tissue engineering. Adv Drug Deliv Rev 61:1065–1083. https://doi.org/10.1016/j.addr.2009.07.008
Jeyanthi V, Velusamy P (2016) Anti-methicillin resistant *Staphylococcus aureus* compound isolation from halophilic *Bacillus amyloliquefaciens* MHB1 and determination of its mode of action using electron microscope and flow cytometry analysis. Indian J Microbiol 56:148–157. https://doi.org/10.1007/s12088-016-0566-8
Kalia VC (2014) Microbes, antimicrobials and resistance: the battle goes on. Indian J Microbiol 54:1–2. https://doi.org/10.1007/s12088-013-0443-7
Kalia VC (2015) In: Kalia VC (ed) Quorum sensing vs quorum quenching: a battle with no end in sight. Springer, New Delhi. https://doi.org/10.1007/978-81-322-1982-8
Kalia VC, Kumar P (2015) Potential applications of quorum sensing inhibitors in diverse fields. In: Kalia VC (ed) Quorum sensing vs quorum quenching: a battle with no end in sight. Springer, India, pp 359–370. https://doi.org/10.1007/978-81-322-1982-8_29
Kalia VC, Purohit HJ (2011) Quenching the quorum sensing system: potential antibacterial drug targets. Crit Rev Microbiol 37:121–140. https://doi.org/10.3109/1040841X.2010.532479
Kalia VC, Prakash J, Koul S, Ray S (2017) Simple and rapid method for detecting biofilm forming bacteria. Indian J Microbiol 57:109–111. https://doi.org/10.1007/s12088-016-0616-2
Kenry, Lim CT (2017) Nanofiber technology: current status and emerging developments. Prog Polym Sci 70:1–17. https://doi.org/10.1016/j.progpolymsci.2017.03.002
Kong B, Mi S (2016) Electrospun scaffolds for corneal tissue engineering: a review. Materials 9:1–20. https://doi.org/10.3390/ma9080614
Konwarh R, Karak N, Misra M (2013) Electrospun cellulose acetate nanofibers: the present status and gamut of biotechnological applications. Biotechnol Adv 31:421–437. https://doi.org/10.1016/j.biotechadv.2013.01.002
Koul S, Prakash J, Mishra A, Kalia VC (2016) Potential emergence of multi-quorum sensing inhibitor resistant (MQSIR) bacteria. Indian J Microbiol 56:1–18. https://doi.org/10.1007/s12088-015-0558-0
Li P, Zhang J, Liu J, Ma H, Liu J, Lie P, Wang Y, Liu G, Zeng H, Li Z, Wei X (2015) Promoting the recovery of injured liver with poly(3-hydroxybutyrate-co-3-hydroxyvalerate-co-3-hydroxyhexanoate) scaffolds loaded with umbilical cord-derived mesenchymal stem cells. Tissue Eng 21:603–615. https://doi.org/10.1089/ten.tea.2013.0331
Lin J, Zhou W, Han S, Bunpetch V, Zhao K, Liu C, Yin Z, Ouyang H (2018) Cell-material interactions in tendon tissue engineering. Acta Biomater 70:1–11. https://doi.org/10.1016/j.actbio.2018.01.012
Loebel C, Burdick JA (2018) Engineering stem and stromal cell therapies for musculoskeletal tissue repair. Cell Stem Cell 22:325–339. https://doi.org/10.1016/j.stem.2018.01.014
Lu X, Wang L, Yang Z, Lu H (2013) Strategies of polyhydroxyalkanoates modification for the medical application in neural regeneration/nerve tissue engineering. Adv Biosci Biotechnol 4:731–740. https://doi.org/10.4236/abb.2013.46097
Luef KP, Stelzer F, Wiesbrock F (2015) Poly(hydroxy alkanoate)s in medical applications. Chem Biochem Eng Q 29:287–297. https://doi.org/10.15255/CABEQ.2014.2261
Matthyssen S, Bogerd BV, Dhubhghaill SN, Koppen C, Zakaria N (2018) Corneal regeneration: a review of stromal replacements. Acta Biomater 69:31–41. https://doi.org/10.1016/j.actbio.2018.01.023
Misra SK, Valappil SP, Roy I, Boccaccini AR (2006) Polyhydroxyalkanoate (PHA)/inorganic phase composites for tissue engineering applications. Biomacromolecules 7:2249–2258. https://doi.org/10.1021/bm060317c
Nisbet DR, Forsythe JS, Shen W, Finkelstein DI, Horne MK (2009) Review paper: a review of the cellular response on electrospun nanofibers for tissue engineering. J Biomater Appl 24:7–29. https://doi.org/10.1177/0885328208099086
Pilehvar-Soltanahmadi Y, Akbarzadeh A, Moazzez-Lalaklo N, Zarghami N (2016) An update on clinical applications of electrospun nanofibers for skin bioengineering. Artif Cells Nanomed Biotechnol 44:1350–1364. https://doi.org/10.3109/21691401.2015.1036999

Ramakrishna S, Fujihara K, Teo W-E, Yong T, Ma Z, Ramaseshan R (2006) Electrospun nanofibers: solving global issues. Mater Today 9:40–50. https://doi.org/10.1016/S1369-7021(06)71389-X
Ramier J, Bouderlique T, Stoilova O, Manolova N, Rashkov I, Langlois V, Renard E, Albanese P, Grande D (2014) Biocomposite scaffolds based on electrospun poly(3-hydroxybutyrate) nanofibers and electrosprayed hydroxyapatite nanoparticles for bone tissue engineering applications. Mater Sci Eng C 38:161–169. https://doi.org/10.1016/j.msec.2014.01.046
Schiffman JD, Schauer L (2008) A review: electrospinning of biopolymer nanofibers and their applications. Polym Rev 48:317–352. https://doi.org/10.1080/15583720802022182
Sell SA, McClure MJ, Garg K, Wolfe PS, Bowlin GL (2009) Electrospinning of collagen/biopolymers for regenerative medicine and cardiovascular tissue engineering. Adv Drug Deliv Rev 61:1007–1019. https://doi.org/10.1016/j.addr.2009.07.012
Sodian R, Sperling JS, Martin DP, Egozy A, Stock U, Mayer JE, Vacanti JP (2000) Fabrication of a trileaflet heart valve scaffold from a polyhydroxyalkanoate biopolyester for use in tissue engineering. Tissue Eng 6:183–188
Tang HY, Ishii D, Sudesh K, Yamaoka T, Iwata T (2010) Nanofibrous scaffolds of bio-polyesters: in vitro and in vivo characterizations and tissue response. Kumar A Nanofibers, INTECH, Rijeka, pp 189–212. ISBN:978-953-7619e-86-2
Thenmozhi S, Dharmaraj N, Kadirvelu K, Kim HY (2017) Electrospun nanofibers: new generation materials for advanced applications. Mater Sci Eng B 217:36–48. https://doi.org/10.1016/j.mseb.2017.01.001
Varsha KK, Nishant G, Sneha SM, Shilpa G, Devendra L, Priya S, Nampoothiri KM (2016) Antifungal, anticancer and aminopeptidase inhibitory potential of a phenazine compound produced by Lactococcus BSN307. Indian J Microbiol 56:411–416. https://doi.org/10.1007/s12088-016-097-1
Vasita R, Katti DS (2006) Nanofibers and their applications in tissue engineering. Int J Nanomedicine 1:15–30
Wang W, Yeung KWK (2017) Bone grafts and biomaterials substitutes for bone defect repair: a review. Bioact Mater 2:224–247. https://doi.org/10.1016/j.bioactmat.2017.05.007
Wang Y, Beekman J, Hew J, Jackson S, Issler-Fisher AC, Parungao R, Lajevardi SS, Li Z, Maitz PKM (2018) Burn injury: challenges and advances in burn wound healing, infection, pain and scarring. Adv Drug Deliv Rev 123:3–17. https://doi.org/10.1016/j.addr.2017.09.018
Wu J, Hu C, Tang Z, Yu Q, Liu X, Chen H (2018) Tissue-engineered vascular grafts: balance of the four major requirements. Colloid Interface Sci Commun 23:34–44. https://doi.org/10.1016/j.colcom.2018.01.005
Xu XY, Li XT, Peng SW, Xiao JF, Liu C, Fang G, Chen KC, Chen GQ (2010) The behaviour of neural stem cells on polyhydroxyalkanoate nanofiber scaffolds. Biomaterials 31:3967–3975. https://doi.org/10.1016/j.biomaterials.2010.01.132
Xuezhu X, Jiang L, Zhou Z, Wu X, Wang Y (2012) Preparation and properties of electrospun soy protein isolate/polyethylene oxide nanofiber membranes. ACS Appl Mater Interfaces 4:4331–4337. https://doi.org/10.1021/am300991e
Yang Q, Wang J, Zhang S, Tang X, Shang G, Peng Q, Wang R, Cai X (2014) The properties of poly(3-hydroxybutyrate-co-3-hydroxyhexanoate) and its applications in tissue engineering. Curr Stem Cell Res Ther 9:215–222. https://doi.org/10.2174/1574888X09666140213160853
Zhang D, Wu X, Chen J, Lin K (2018a) The development of collagen based composite scaffolds for bone regeneration. Bioact Mater 3:129–138. https://doi.org/10.1016/j.bioactmat.2017.08.004
Zhang J, Zhao X, Liang L, Li J, Demirci U, Wang S (2018b) A decade of progress in liver regenerative medicine. Biomaterials 157:161–176. https://doi.org/10.1016/j.biomaterials.2017.11.027
Zhuang P, Sun AX, An J, Chua CK, Chew SY (2018) 3D neural tissue models: from spheroids to bioprinting. Biomaterials 154:113–133. https://doi.org/10.1016/j.biomaterials.2017.10.002

MIX
Papier aus verantwortungsvollen Quellen
Paper from responsible sources
FSC® C105338

If you have any concerns about our products,
you can contact us on
ProductSafety@springernature.com

In case Publisher is established outside the EU,
the EU authorized representative is:
Springer Nature Customer Service Center GmbH
Europaplatz 3, 69115 Heidelberg, Germany

Printed by Libri Plureos GmbH
in Hamburg, Germany